# Rhizoctonia Solani, Biology and Pathology

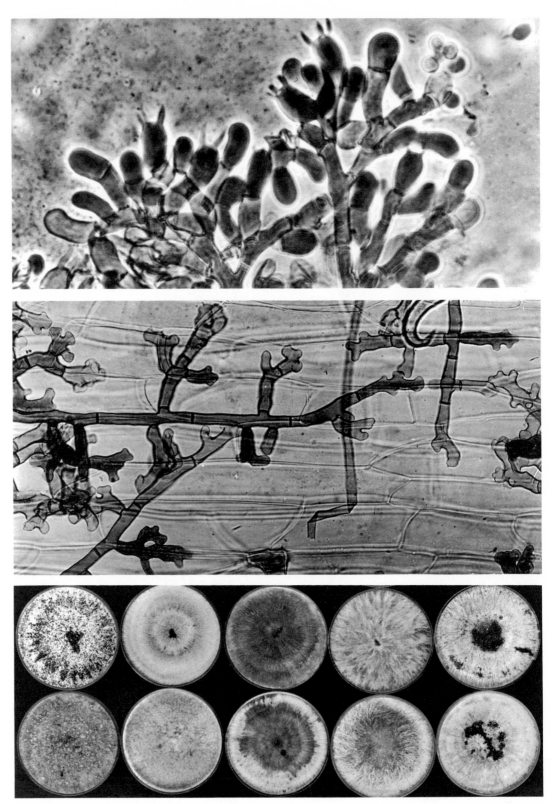

The perfect state of *Rhizoctonia solani* (*Thanatephorus cucumeris*), the vegetative state as it appears on the surface of a host plant, and agar cultures showing the wide range of cultural types obtained from field isolations.

# Rhizoctonia Solani, Biology and Pathology

Based on an
American Phytopathological Society
Symposium on *Rhizoctonia solani*
held at the Miami meeting of the Society, October, 1965.

## Edited by J. R. Parmeter, Jr.

University of California Press
*Berkeley, Los Angeles and London 1970*

Sponsored and financially supported by the
American Phytopathological Society

University of California Press
Berkeley and Los Angeles, California

University of California Press, Ltd.
London, England

# Contents

# Preface

This volume is the outgrowth of a symposium on "*Rhizoctonia solani* and Related Forms" held in conjunction with the 1965 Annual Meeting of the American Phytopathological Society at Miami, Florida. The idea for such a symposium came out of a 1963 meeting of the society's Soil Microbiology Committee, where it was decided that some sort of synthesis needed to be made from the mass of publications concerning major plant pathogens. The *Rhizoctonia solani* complex was chosen as a first attempt.

Organizing the symposium and publishing this volume based on it was a joint effort of the following subject-matter committees of the A.P.S.: Chemical Control, Disease and Pathogen Physiology, Host Resistance and Microbial Genetics, Mycology, and Soil Microbiology. These groups functioned through a special symposium committee consisting of D. F. Bateman, E. E. Butler, N. T. Flentje, J. D. Menzies, R. T. Sherwood, and J. R. Parmeter Jr., chairman.

Following the symposium, North Carolina State University hosted a three-day meeting of symposium participants and others interested in *R. solani*. These informal discussions, ranging over many aspects of taxonomy, biology, and pathology, helped to coordinate terminology and content of the final papers.

This volume, developed from the symposium papers, was prepared with two main goals in mind: (1) to bring together and integrate all of the available information on *R. solani* in all of its various aspects, and (2) to compile a literature list that, directly or indirectly, provides access to all of the important works on *R. solani*. Hopefully, any worker needing information or contemplating research on *R. solani* can start with this volume, knowing that the literature has been thoroughly reviewed and condensed for him. Many aspects that need research, and directions such research might take, are also indicated.

To achieve these goals, more than 1,000 references are included, spanning 11 decades from J. G. Kühn's original, crude description of *R. solani* in 1858 to very detailed studies on biochemistry, physiology, pathology, ecology, genetics, and control published in 1967. Much original and previously unpublished data are also included.

The papers included in this volume were organized around three major aspects: *R. solani*, the organism; *R. solani*, the soil saprophyte; and *R. solani*, the plant pathogen. Each paper was reviewed by all of the participants and all shared jointly in the task of minimizing duplication and insuring complete coverage. During this process of revision, new references were added as they appeared. Verification of literature citations was left to the individual authors. Citations follow *The Style Manual for Biological Journals* and the abbreviation system of *Chemical Abstracts* in so far as possible.

In its final form, this volume goes beyond the symposium in both scope and time.

J. R. PARMETER, JR.
October 15, 1967

# Acknowledgments

The presentation of the symposium on *Rhizoctonia solani* and the preparation of this volume developed from the symposium papers was a major undertaking, requiring help from many sources. The sponsoring subject matter committees of the American Phytopathological Society — Chemical Control, Disease and Pathogen Physiology, Host Resistance and Microbial Genetics, Mycology, and Soil Microbiology —are especially to be thanked for their support in this first attempt to bring together, in book form, all of the available information on a major plant disease fungus.

Participation of foreign contributors was made possible by a grant of funds from the Rockefeller Foundation to assist with travel expenses.

The counsel and cooperation of Mr. Ernest Callenbach and the editorial assistance of Sara Berenson are gratefully acknowledged. The very helpful guidance of Dr. K. F. Baker and the many hours of assistance from Mr. W. D. Platt in handling editorial details deserve special appreciation.

Preparation of this volume required patience, cooperation, and criticism from all symposium participants, and it is through their concerted efforts that the separate contributions were coordinated and unified.

Part I.

*Rhizoctonia solani:* the organism

# Introduction: THE FIRST CENTURY OF RHIZOCTONIA SOLANI

J. D. MENZIES—*Soils Laboratory, Soil and Conservation Research Division, Agricultural Research Service, United States Department of Agriculture, Beltsville, Maryland.*

Slightly more than 100 years ago, Julius Kühn observed a fungus on diseased potato tubers and named it *Rhizoctonia solani* (Kühn, 1858). Since then his fungus has gained the reputation of being a widespread, destructive, and versatile plant pathogen. It has been the subject of hundreds of research papers, but in spite of this (or perhaps because of it), there is much confusion and disagreement on the taxonomy and nomenclature of the fungus as well as on many aspects of its pathology. This group of symposium papers is an attempt to organize our present knowledge on *R. solani* and related forms into a comprehensive, but condensed treatise. It may not clear away all confusion or settle all disagreements, but it should, at least, serve as a useful starting point for future work.

The fungi generally grouped as *R. solani* types occur in all parts of the world and are probably indigenous to uncultivated areas. *Rhizoctonia solani* is capable of attacking a tremendous range of host plants, causing seed decay, damping-off, stem cankers, root rots, fruit decay, and foliage diseases.

How is it that this fungus can be so destructive in so many crops, and yet retain such a vigorous saprophytic capability in soil? The only root pathogen that probably exceeds *R. solani* in vigor of saprophytic growth is *Rosellina*, but the latter grows more in tropical surface litter than down in the soil where *R. solani* competes so well (Garrett, 1956). It is this combination of competitive saprophytic ability with lethal pathogenic potential and almost unlimited host range that makes *R. solani* such an economically important pathogen.

Because of wide variation in morphology, pathogenicity, and physiology, the taxonomy and nomenclature of *R. solani* have been sources of confusion and controversy for many years. The mycelium of this fungus is multinucleate, and much recent study has been devoted to the phenomena of nuclear division, movement of nuclei within the mycelium, and nuclear exchange by anastomosing. These attributes of *R. solani* have caused its selection for many of the more recent investigations on basic physiology, ecology, and genetics of pathogenic fungi. It has been used in studies on the mechanism of antibiosis and other microbial interactions in soil; nuclear behavior in multinucleate fungi; hyphal anastomosis and heterokaryosis; ultrastructure of fungus cells, particularly the structure of the septal pore; the enzyme chemistry of parasitism and pathogenesis; and the biochemical basis for host resistance.

A comprehensive review of *Rhizoctonia* was done once before, just 50 years ago, by B. M. Duggar (1915). In explaining why he felt it was time for a monographic treatment of *Rhizoctonia*, Duggar wrote, "The literature of *Rhizoctonia* diseases has grown enormously in the past 15 years, yet some unnecessary confusion and difference of opinion exists regarding the two main species and their distribution and relation to diseases in plants." Duggar listed this literature exhaustively, as was the custom in those days, and compiled 145 references, only 48 of which referred to *R. solani*. Of those, 12 were dated prior to 1900. By 1940 the reference-card files of the Mycology Division of the United States Department of Agriculture listed more than 1,000 publications dealing with this fungus. A complete bibliography today would have at least 4,000 entries. To Duggar, the appearance of 36 papers in 15 years was an "enormous increase." It is not surprising that 50 years later, some 20 participants were needed in this symposium to assimilate the publications that have appeared since Duggar's time.

A brief search through *Biological Abstracts* and *Review of Applied Mycology* reveals that papers mentioning *Rhizoctonia* are currently appearing at the rate of around 100 per year. Half of these list *Rhizoctonia* in their titles. Considering the annual increase in scientific publications generally, it seems safe to predict a similar increase in Rhizoctonia papers, especially in view of the current interest in physiology of parasitism and the biochemistry of specific metabolic processes. Obviously we are going to have an almost unmanageable literature on this fungus in the near future. This is probably the last time that all aspects of *R. solani* research can be brought together for publication under one cover. This will soon be true for other major pathogenic genera. It may already be too late for *Fusarium*.

A review of the earliest publications on *Rhizoctonia* may be of historic interest. Kühn's original description appeared in 1858 in his book on crop diseases. This was during the golden age of biological discovery, when the first microscopes were in use, fungi were being recognized as plant pathogens, and the practice of binomial Latin naming of organisms was developing. A few years earlier (1815), the great French mycologist, A. P. De Candolle, had described the genus *Rhizoctonia* for the violet root-rot organism. A translation of De Candolle's paper includes the following: "I propose in this memoir to describe a new genus of fungus which I call *Rhizoctonia* (from the Greek 'death of roots') because it rapidly attacks and kills the roots of phanerogamic plants." De Candolle set up the genus to separate the violet root-rot fungus from truffles and puffballs, pointing out that *Rhizoctonia* had much smaller and simpler tubercles, or what we now call sclerotia.

Kühn was using his microscope on scabby spots on potato tubers when he observed black sclerotia adhering to the surface of the tuber and connected with dark, ramifying fungus hyphae. He decided that this was a species of *Rhizoctonia* and provided us with the original illustration that accompanied his naming of the fungus (Fig. 1). Apparently he in-

ing of feeder roots, pits on the tubers, tubers but no tops, and knobby tubers, as well as the better known stem and stolon lesions (Heald, 1933).

The average pathologist working with *R. solani* is not an expert in taxonomy or fungus nomenclature, his interest in these matters being primarily that of knowing what name to use. He is not in a good position to evaluate rival proposals. One of the main objectives of this symposium has been to strive for a consensus among present-day *Rhizoctonia* taxonomists. Insofar as the participants speak for all students of taxonomy, this consensus has been reached. In the following papers, *R. solani* Kühn is accepted as a valid name for the imperfect state and *Thanatephorus cucumeris* (Frank) Donk is favored for the perfect state. Time will tell whether these recommendations are generally accepted, but at least the nonexpert can use these names with confidence that

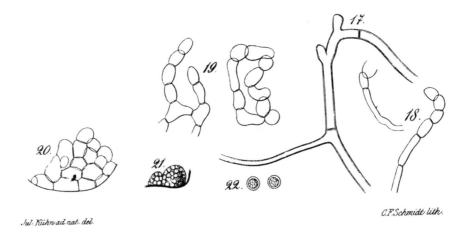

*Fig. 1.* *Rhizoctonia solani* as illustrated by Kühn in 1858. The structures 21 and 22 are evidently contaminants.

cluded a few spores and structures from other fungi, and his depictions of the sclerotia and hyphae would scarcely be considered diagnostic today. The reader will find in these symposium papers electron micrographs of structural details undreamed of by Kühn when he looked through his primitive microscope.

Kühn also described, in great detail, the pathology of his fungus on potato tubers. Unfortunately he seems to have been observing potato scab or some similar lesion on the tuber. In fact, he referred to the disease as "Der Schorf oder Grind" which, I believe, translates into English as "scurf or scab." We now know that *R. solani* does not cause this kind of tuber lesion. Kühn did not describe the stolon and stem lesions that are the real parasitic expressions of this fungus on potato. It was not until 1918 that clear experimental proof was furnished that Kühn's fungus really was pathogenic on potato (Edson and Shapovalov, 1918). This symposium presents an assessment of the present status of the disease complexes attributed to *R. solani*. For example, on potato alone, *R. solani* has been blamed for scabby lesions on the tubers, virus-like symptoms on the tops, kill-

others will know what fungus he means. There is a certain ironic satisfaction in the retention of Kühn's Latin binomial. If he could look at the saprophytic mycelium and sclerotia, confuse them with a few unrelated fungous structures, describe the wrong symptoms, and still reach the correct conclusion that he was dealing with a potato pathogen, then surely he deserves the honor of naming it.

The durability of the generic name for the imperfect state of this fungus, in spite of the sketchy original description, is all the more surprising when we consider the succession of names used for the basidal state. When first discovered in 1891 by Prillieux and Delacroix, it was called *Hypochnus solani*. Later it became *Corticium vagum*, then *Pellicularia filamentosa*, and now *Thanatephorus cucumeris*. Perhaps in the case of *Rhizoctonia*, as with other pathogens whose asexual states are the parasitic ones, the persistence of the early name has been favored by the pathologist's resistance to change and the taxonomist's preoccupation with sexual phenomena.

*Rhizoctonia solani* undoubtedly is responsible for

a major share of root diseases, but it is surely blamed for disease damage in cases where it is only an innocent bystander. Possibly, *Pythium, Fusarium,* and the soft-rot bacteria should also be included in this group of pathogens whose reputation often exceeds their deeds. One can scarcely fail, in examining diseased roots, to observe what appears to be the characteristic mycelium of *R. solani,* but careful research is necessary to prove a causal relationship, or even to be sure of the identification of the fungus. The problem of identifying *R. solani* is more thoroughly discussed in papers that follow. These papers emphasize that the identification is not so simple as many of us have supposed. Fortunately they also provide guidelines to aid in correct identification.

Even when *R. solani* is really present on the diseased root, it may not be pathogenic. Kühn evidently jumped to conclusions when he associated *Rhizoctonia* mycelium with scab lesions, and a number of persons have repeated his mistake. It is easy to isolate *R. solani* because the organism grows so fast in culture that it often dominates isolation plates. In many practical cases, where diagnosis does not go beyond microscopy and isolation in culture, a root-disease situation can too easily be attributed to "*Rhizoctonia* sp." In the *Index of Plant Diseases in the United States* (1953), *R. solani* or *Rhizoctonia* sp. were reported as pathogens on approximately 54% of the genera listed (Weiss and O'Brien, 1953). It would be of more exciting interest to be able to state that one has found a higher plant species that cannot be parasitized by *R. solani* or related forms.

The practical reason for so much past research on *R. solani* has been to find better methods for controlling the diseases it produces in cultivated crops. Because of the peculiar versatility of this fungus, the classical approaches to control have not been very effective. Its predominantly subterranean habit has made chemical control difficult, its wide host range has tended to defeat attempts to breed for resistance, and its pronounced saprophytic ability has weakened the effectiveness of crop rotation. Even so, as described in detail in these proceedings, some success can be achieved with all these approaches. Furthermore, *R. solani,* like other root-disease fungi, must exist, grow, and parasitize in close association with other components of the soil microflora. The many studies, reviewed in the papers that follow, on the ecological relations of this pathogen keep alive the hope of someday being able to manipulate these microbial associations to achieve practical disease control.

The *R. solani* complex is only one component of the soil-borne root-disease situation that remains an unsolved problem in plant pathology. It is hoped that this synthesis of what is known about an important pathogen will not only stimulate more effective research toward its control, but will also contribute to a better understanding of the root-disease fungi in general.

# Taxonomy and Nomenclature of the Imperfect State

J. R. PARMETER JR. and H. S. WHITNEY—*Department of Plant Pathology, University of California, Berkeley, California, and Department of Forestry and Rural Development, Forest Research Laboratory, Calgary, Alberta.*

The genus *Rhizoctonia* was erected by de Candolle (1815) to accommodate the nonsporulating root pathogen *R. crocorum* D.C. ex Fr. The basic characters of the genus, as set forth by de Candolle, were production of sclerotia of uniform texture with hyphal threads emanating from them and the association of the mycelium with roots of living plants. Nearly 100 species have since been designated, many possessing neither of the above characters and having little in common aside from the absence of conidia. As a result, the genus contains a heterogeneous mixture of fungi of diverse relationships (Fig. 1). One of these many species is *R. solani* Kühn.

obtain and errors in communication will be frequent.

Our present state of knowledge does not permit the development of a system of taxonomy and nomenclature that removes all the major sources of error. Too few mycelia have been studied and even fewer perfect states have been induced with them. As a result, there is no assurance that all the mycelia fitting the description of *R. solani* are actually *R. solani*. Recent evidence indicates that at least two distinct fungi have frequently been studied together as *R. solani* (Parmeter, et al., 1967). In addition, several workers have noted that the mycelia of some ascomycetous fungi may closely resemble *R. solani*

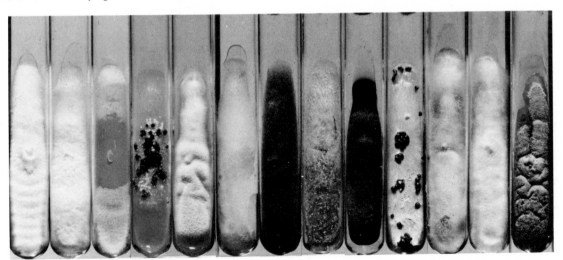

Fig. 1. PDA cultures of *Rhizoctonia* species, showing a range of variation in cultural characteristics. Left to right: *R. lilacina* (Saksena #1142), *R. repens* (Saksena #2583), *R. rubigenosa* (Saksena #2747), *R. carotae* (ATCC), *R. lanuginosa* (ATCC), *R. globularis* (ATCC), *R. hiemalis* (CBS), *R. callae* (CBS), *R. crocorum* (CBS), *R. ferrugena* (ATCC), *R. endophytica* (ATCC), *R. endophytica* (ATCC), and *R. rubi* (ATCC).

*R. solani* has long been recognized as a destructive pathogen on a wide variety of plants throughout the world. Maximum precision in the identification, taxonomy, and nomenclature of *R. solani* is essential to the accumulation and dissemination of needed information on this fungus. Unfortunately a number of species of *Rhizoctonia* are very similar to *R. solani*, and until the relationships among these fungi are clearly understood, precision will be difficult to

(Duggar, 1915; Moreau and Moreau, 1956; Whitney and Parmeter, 1964). Until the mycelia of many fungi have been studied in detail and until additional mycelial characters have been delineated by which mycelia similar to that of *R. solani* can be distinguished, the taxonomy and nomenclature of *R. solani* and similar forms must be considered tentative.

In preparing this paper, we have examined descriptions of nearly 100 species of *Rhizoctonia*. Where possible, we have worked with cultures from the American Type Culture Collection (ATCC), the Commonwealth Mycological Institute (CMI), the Centraalbureau voor Schimmelcultures (CBS), or from the describer. We have also drawn on our own experience with several hundred isolates from various parts of the world. The following treatment is a synthesis of these studies. We recognize the tentative nature of this treatment and we can only hope that it will provide a beginning for the future clarification of the problems discussed.

THE SPECIES CONCEPT IN R. SOLANI. — The taxonomy and nomenclature of *R. solani* was critically reviewed by Duggar (1915), and it is largely from this work that the current species concept was established. Kühn's (1858) original description was too brief to allow certain identification and it apparently involved elements of a second organism, as indicated by his illustration of sporelike bodies. Since no type material was preserved, it is difficult to determine with any assurance that Kühn's fungus was what we now recognize as *R. solani*. Duggar, however, visited Kühn in 1899-1900, and from their discussions, Duggar concluded that Kühn's *R. solani* was the same as the fungus he was studying in connection with damping-off and root diseases of various crops in the United States. So, while there is no way of knowing with certainty that Kühn's description actually referred to what we now call *R. solani*, Duggar's conclusion has generally been accepted by subsequent workers.

In his 1915 paper and one the following year, Duggar reviewed the earlier literature on *R. solani* and provided a detailed discussion of the types of diseases associated with it, its morphology and growth habit, and its probable relationship to other *Rhizoctonia* species. From the work of Duggar and those he cited, a species concept of *R. solani* has evolved without benefit of formal diagnosis beyond that provided by Kühn. Various workers have placed greater stress on some features than on others, but in general most workers have recognized *R. solani* as being characterized by:

1. Pale to dark brown, rapidly growing mycelium of relatively large diameter with branching near the distal septum of hyphal cells, often at nearly right angles in older hyphae
2. Constriction of branch hyphae at the point of origin
3. Formation of a septum in the branch near the point of origin
4. Production of monilioid cells, often called barrel-shaped cells or chlamydospores, in chains or aggregates sometimes referred to as sporodochia
5. Production of sclerotia of nearly uniform texture and varying in size and shape from small, round sclerotia, often less than 1 mm in diameter, to thin crusts several centimeters across
6. Pathogenicity to a wide range of hosts, resulting in a variety of symptoms including damping-off, rotting of roots and other underground plant parts, blighting of hypocotyls, stems, and leaves, and decay of fruits and seeds
7. Possession of a basidiomycetous perfect state variously referred to as *Hypochnus*, *Corticium*, *Pellicularia*, *Ceratobasidium*, or *Thanatephorus* (see Talbot, this vol.).

In addition to these generally recognized characteristics of *R. solani*, we have found two additional features to be valuable diagnostically:

8. Possession of a prominent septal pore apparatus
9. Possession of multinucleate cells in actively growing hyphae

VARIATION IN R. SOLANI. — The taxonomy and nomenclature of *R. solani* cannot be discussed without first considering variations in the characters listed above. Some of these characters vary widely, whereas others appear to be uniform and stable. For this reason, some characteristics are more reliable diagnostic features than others. The literature on variation is voluminous, so only selected references illustrating specific points are included here. We have also drawn on our own experience with several hundred isolates collected from many parts of the world.

Variation in hyphal characteristics, monilioid cells, and sclerotia are discussed in detail by Butler and Bracker (this vol.). Additional information on variation in growth rate and hyphal diameters is also included by Sherwood (this vol.). These papers should be consulted for additional detail.

*Hyphal characteristics*. — Pigmentation of hyphae appears to be restricted to various shades of brown. Young colonies on nutrient media may be white or nearly so, but all of the older colonies we have observed show some shade of brown. Some colonies remain pale brown, whereas others eventually become a very dark brown. Sclerotia of some isolates are so dark as to appear nearly black. While it is possible that complete albinos may exist, we have not seen one, nor have we seen any bright red, violet, orange, blue, green, or bright yellow pigments in *R. solani* (pale brown may appear to be dull yellow). For these reasons, the brown pigment appears to be a stable diagnostic feature. Any mycelia remaining permanently white or showing any pigmentation other than various shades and hues of brown are not likely to be *R. solani*.

Hyphal diameters, while usually relatively large, vary widely both within and among isolates. No standardized methods for comparing hyphal diameters have been developed, and it is therefore difficult to interpret the results of various authors. Measurements by Matsumoto (1921), Thomas (1925), Schultz (1936), Richter and Schneider (1953), and others indicate that most isolates fall within the 5-14$\mu$ range, with a usual average of about 6-10$\mu$. The diameters of hyphae within a colony vary widely with respect to age, rank, and position. In addition, Palo (1926) demonstrated that the composition of the medium may affect hyphal diameters, and Monteith and Dahl (1928) provide evidence that temperature may also affect hyphal diameters. In the absence of standardized measuring procedures that account for age, rank, and position of hyphae, temperature, and composition of media, data on hyphal diameters have only limited usefulness. We can only agree with Chen (in Kernkamp, et al., 1952) that, within broad limits, hyphal diameters have little diagnostic value. Available evidence suggests, however, that while some hyphae in a colony of *R. solani* may be less than 5-6$\mu$, colonies in which hyphae consistently are less than 5-6$\mu$ wide are not likely to be *R. solani*.

The angle of branching is also quite variable among isolates, particularly at the advancing margins of colonies, where hyphae often branch at acute angles. While right-angle branching is sometimes cited as a diagnostic feature, it cannot be considered reliable. Other features of branching are diagnostically more useful.

Branching almost invariably occurs near the distal septum of a cell in young, advancing hyphae. In older hyphae, branching may occur any place along the cell. Constriction of branch hyphae at the point of origin and formation of a septum in the branch near the point of origin appear to be stable, reliable characteristics of *R. solani*. The absence of this branching habit in a mycelium indicates that it is not likely to be *R. solani*. This type of branching is common in many other fungi, however, and the presence of *R. solani*-like branching is not in itself a certain indication that a mycelium is *R. solani*.

The septal pore apparatus (Bracker and Butler, 1963) is a prominent feature in *R. solani*. This apparatus is readily observed with phase microscopy at 800-1,600 ×, and while it is not always evident in every septum, we have observed it in many septa of every verified isolate of *R. solani* we have examined, including a large number of single-spore isolates. This apparatus appears to be a uniform and reliable feature of the species, and mycelia in which a septal pore apparatus cannot be demonstrated by phase microscopy, either because the apparatus is absent or too small to see, can be excluded from *R. solani*. This apparatus is also common in other basidiomycetes, and its presence is not in itself indicative of *R. solani*.

Numbers of nuclei per cell also provide valuable diagnostic data. Many studies, such as those of Müller (1924), Schultz (1936), Chen (in Kernkamp, et al., 1952), Sanford and Skoropad (1955), Flentje, Stretton and Hawn (1963), and Flentje and Stretton (1964), indicate that the cells of young, actively growing hyphae are multinucleate. We have examined numerous isolates representing a wide range of cultural types, all producing the perfect state and therefore reliably identified. In every case, these isolates and their single-spore progeny have been found to be multinucleate. This therefore, appears to be a stable characteristic that can be used to distinguish *R. solani* from similar fungi with regularly binucleate cells (Parmeter, et al., 1967).

*Monilioid cells.* — Some workers, notably Saksena

and Vaartaja (1960, 1961), have placed considerable emphasis on the use of monilioid cells in distinguishing species of *Rhizoctonia*. However, the wide variation in the dimensions of monilioid cells reported by Matsumoto (1921), Gratz (1925), Townsend and Willetts (1954), and Saksena and Vaartaja (1961) suggest that such measurements do not provide suitable criteria for distinguishing species. No comprehensive studies on variations in monilioid cells have been made, but the same range of variability in monilioid cells may be expected as is found in hyphal diameters or in sizes and shapes of sclerotia formed from monilioid cells (see Butler and Bracker, this vol.). In addition, many fungi, including both ascomycetes and basidiomycetes, produce monilioid cells similar to those of *R. solani*. Among *R. solani* isolates growing on water agar, we have observed that monilioid cells may be numerous or rare, loosely grouped or tightly clumped, and may tend to form in or on the substrate, depending on the isolate. Without additional research into variability among monilioid cells, there is no adequate basis on which to attach special taxonomic significance to these variations.

*Sclerotia.*—The nature of the sclerotium of *Rhizoctonia* is a basic character by which the genus is distinguished from *Sclerotium*. Sclerotial cells of *R. solani* are essentially similar and, while the outer cells may be darker and thicker walled, there is no obvious differentiation into a rind and a medulla. Mycelia with differentiated, *Sclerotium*-type sclerotia can be readily excluded from *Rhizoctonia*. Since other members of the genus *Rhizoctonia* produce sclerotia of the *R. solani* type, mycelia with this type of sclerotium are not necessarily *R. solani*. In addition, sclerotia may be absent in some *R. solani* isolates under certain cultural conditions (Meyer, 1965), therefore the absence of sclerotia does not automatically exclude a mycelium from *R. solani*.

Sclerotia of *R. solani* are quite variable. Studies by Exner and Chilton (1943), Exner (1953); Hawn and Vanterpool (1953), Kernkamp, et al. (1953), Whitney and Parmeter (1963), and Papavizas (1965) on variations among single-spore isolates show that sclerotia of sibling isolates vary widely in size, shape, shade of brown, and distribution in agar plates. Whitney and Parmeter (1963) have shown that a parent with numerous small and very dark sclerotia may give rise to progeny with large distinct sclerotia, small distinct sclerotia, massive sclerotial crusts, or few sclerotia at all (Fig. 2). They also have shown that single-spore isolates with diverse sclerotial types may produce heterokaryons with yet other sclerotial types. These observations indicate that the usefulness of the sclerotium as a taxonomic character is limited by its variability. The basic structure of the sclerotium may serve to exclude a mycelium from *R. solani*, but the size or shape of the sclerotium, within broad limits, does not.

In addition to the morphological characters possessed by *R. solani*, the presence of characters unknown in *R. solani* can be considered diagnostic in a negative way. *Rhizoctonia solani* has never been

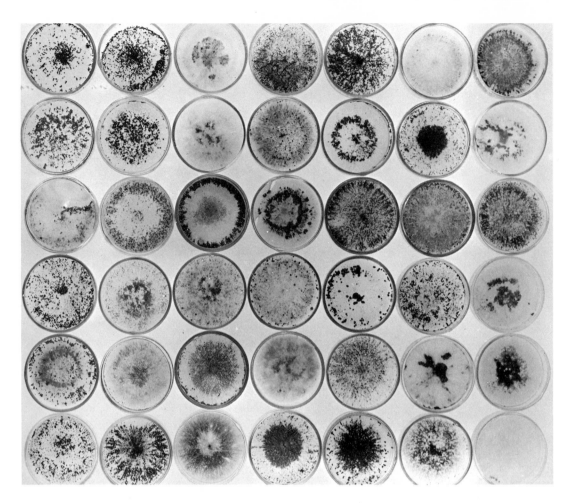

*Fig. 2.* PDA cultures of single-spore isolates from 1 parent (*R. 43, upper-left corner*), showing variations in size, shape, and distribution of sclerotia.

found to produce clamp connections, conidia, spermatia, true chlamydospores, or a perfect state other than *Thanatephorus cucumeris* (Frank) Donk. The production of any of the above structures or of any other perfect state by a mycelium indicates that the fungus is not *R. solani.*

*Growth rate and temperature relationships.* — Numerous studies have been made on the growth rates of *R. solani* isolates and on the relationships of temperature to growth rate. Older literature citations can be found in Togashi (1949), Kernkamp, et al. (1952), and Richter and Schneider (1953). Recent studies include those of Exner (1953), Ui, et al. (1963), Whitney (1963), Meyer (1965), Papavizas (1965), and Sherwood (this vol.). The temperatures and substrates used by different workers are sufficiently different that their results are difficult to compare directly. In general, growth rates of isolates of *R. solani* may vary from a few hundredths of a millimeter per hour to at least 1 millimeter per hour. Slower growth rates have usually been found among single-spore isolates (Kernkamp, et al., 1952; Exner, 1953; Papavizas, 1965), but since very slow

growing strains would be difficult to isolate in nature, there is no satisfactory data to provide good information on the range of variation in natural field populations.

Cardinal temperatures are also quite variable. In general, optimum growth rate usually lies between 20-30°C, though higher and lower optima are occasionally reported. Growth curves vary from fairly sharply defined peaks to broad flat peaks in which growth rate is essentially the same at several temperatures. Comprehensive studies such as those of Richter and Schneider (1953) and Sherwood (this vol.) indicate that minima may range from at least 0-10°C and maxima from at least 29-38°C. Thomas (1925), Schultz (1936), and Richter and Schneider (1953) demonstrated that isolates from warm areas or from greenhouses tend to have higher minima and maxima than do isolates from colder areas. Because of the wide variation in cardinal temperatures and the probable effects of geographic origin and laboratory substrate on these temperatures, they do not appear to provide reliable characteristics for taxonomic distinctions.

*Pathogenicity.*—Pathogenicity and types of diseases caused by *R. solani* are covered in another section of this symposium and need not be discussed here in detail. From a taxonomic standpoint, three characteristics are significant: (1) isolates may cause several types of diseases, including aerial blights, damping-off, seed decay, fruit decay, and root rot; (2) isolates may range from avirulent to aggressively virulent; and (3) host range among isolates may vary from limited to extremely wide. Testing problems pose nearly insurmountable obstacles to precision in pathogenicity characterization of isolates. Since *R. solani* has such a wide host range, it is difficult to assume that a given isolate is avirulent because it fails to attack a number of hosts. The tests may not have included plants on which that isolate is virulent. By the same token, host specificity is difficult to determine with precision. Variations in virulence shown by some isolates when tested repeatedly (LeClerg, 1939*b*; Maier, 1959*b*) or when tested in different soils (Sanford, 1938*a*) further complicate evaluation of host specificity and virulence, as does the fact that prior nutrition may influence virulence (Sims, 1960; Weinhold and Bowman, 1967).

Survey of the literature on variations in virulence and host range, summarized well by Kernkamp, et al (1952), Richter and Schneider (1953), Flentje and Saksena (1957), and Luttrell (1962), indicates that these features are useful to characterize *R. solani* only in general terms. Demonstrated variations among field isolates and single-spore isolates derived from them (Exner, 1953; Papavizas, 1965; Garza-Chapa and Anderson, 1966) show that representatives of the species *R. solani*, while usually virulent on one or more hosts, may be avirulent, weakly virulent on a few hosts, or highly virulent on a wide range of hosts. Thus, while pathogenicity to a wide range of plants is characteristic of *R. solani*, avirulence and restricted host range do not exclude isolates from *R. solani*.

*Anastomosis.*—Anastomosis, while not a diagnostic feature in itself, has been used as a criterion for the recognition of species (Forsteneichner, 1931). Studies by Matsumoto (1921), Matsumoto, et al. (1932), Schultz (1936), Kernkamp, et al. (1952), Richter and Schneider (1953), Ito, et al. (1955), and Flentje and Stretton (1964) have shown that hyphae of the same isolate fuse readily. All found that, among field isolates of *R. solani*, hyphae of some isolates will not fuse with hyphae of other isolates or groups of isolates. Thus, failure to anastomose does not provide an adequate basis for the exclusion of isolates from *R. solani*.

Matsumoto, et al. (1932) early recognized that anastomosis in *R. solani* is a complex process, and that there are varying degrees of anastomosis. They found that some isolates fused readily to produce mixing of the cytoplasm. Other isolates showed cell-wall fusion but no mixing of the cytoplasms of the fused cells. Yet other isolates failed to fuse at all.

Flentje and Stretton (1964) and Stretton, et al. (1967) have made similar observations and have shown in addition that cytoplasmic mixing often leads to the death of fusion cells and sometimes of adjacent cells.

Kernkamp, et al. (1952) and Flentje and Stretton (unpub.) have also shown that hyphae of single-spore isolates from the same parent may fail to anastomose. In light of these observations plus our nearly complete ignorance of the mechanisms underlying the complex processes of anastomosis in *R. solani*, particularly with regard to the genetics of success or failure, it is clear that the failure to anastomose cannot be used as a criterion for species distinction.

Successful anastomosis, however, can be considered as evidence that mycelia represent the same species, particularly in a species such as *R. solani*, where anastomosis apparently involves complex compatibility factors. Anastomosis as it relates to sub-specific grouping in *R. solani* is discussed in a later section of this paper.

VARIABILITY AND THE IDENTIFICATION OF R. SOLANI.—From the foregoing sampling of the literature on variation, it is clear that *R. solani* is a highly variable fungus and that any satisfactory species concept must accommodate this variability (Fig. 3). It is also clear that our present state of knowledge does not permit the formulation of a precise species concept, since we yet do not know the entire range of variability in the species.

No single feature, excepting the perfect state, serves to distinguish *R. solani* from other similar fungi. Recognition of the species depends rather on the presence of a combination of several features. Mycelia possessing the usual rapid growth, branching habit, color, septation, nuclear condition, monilioid cells, sclerotia, and pathogenic habit of *R. solani* can be assigned to that species with confidence. Mycelia in which one or more of these features are wanting or vary from the usual are more difficult to place. Available evidence indicates that the mycelium of *R. solani* invariably possesses multinucleate cells, branching near the distal septum of a cell, constriction of the branch and formation of a septum near the point of branch origin, and a prominent septal pore apparatus in the septum. Mycelia lacking any of these features cannot be assigned to *R. solani*. Monilioid cells, sclerotia, rapid growth, virulence, or similar variable features may be lacking in some isolates of *R. solani* and cannot be considered essential to the placement of a mycelium in *R. solani*. These features are usually present in field isolates, however, and care should be exercised in including in *R. solani* fungi in which one or more of these features are absent.

As has been suggested (Parmeter, 1965), it is difficult if not impossible to describe a mycelium with absolute assurance that another worker can identify that mycelium from the written description. It is possible only to provide a description with sufficient detail as to exclude most other mycelia. The follow-

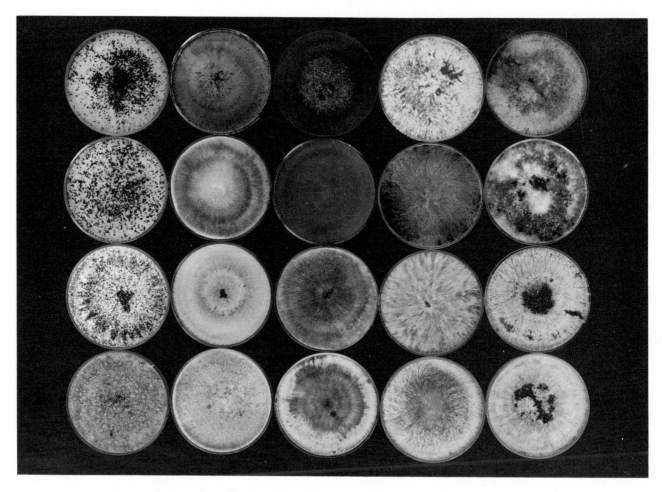

*Fig. 3.* PDA cultures of various field isolates of *R. solani,* showing range of cultural characteristics and types of sclerotia.

ing summary of characteristics for *R. solani* serves to delineate the species with as much accuracy as present information permits.

Characteristics consistently present:

1. Multinucleate cells in young vegetative hyphae
2. Prominent septal pore apparatus
3. Branching near the distal septum of cells in young vegetative hyphae
4. Constriction of the branch and formation of a septum in the branch near the point of origin
5. Some shade of brown

Characteristics usually present, but one or more of which may occasionally be lacking in individual isolates (usually single-spore isolates):

1. Monilioid cells
2. Sclerotia (without differentiated rind and medulla)
3. Hyphae greater than 5μ in diameter
4. Rapid growth rate
5. Pathogenicity

Characteristics never possessed:

1. Clamp connections
2. Conidia
3. Sclerotia differentiated into a rind and medulla
4. Rhizomorphs
5. Red, green, blue, bright yellow, orange, or other pigments except brown
6. Any perfect state other than *T. cucumeris*

One feature, habitat, was not included in the above list despite the fact that most workers associate *R. solani* with soil. Although many strains are soil inhabiting, we know very little about the behavior of aerial strains. Some of these may not be associated with soil. Furthermore, fungi frequently are obtained from soil as casual contaminants. Thus, the origin of an isolate in or outside of the soil is not a definitive taxonomic feature.

Species of Rhizoctonia. — *Species of Rhizoctonia excluded from consideration.* — In considering the taxonomy and nomenclature of *R. solani,* it has been necessary to examine descriptions and, where possible, cultures or herbarium specimens of some 100 species of *Rhizoctonia* to which we found reference. Many of these species possess spore states, clamp connections, sclerotia of the Sclerotium type, and similar features clearly distinguishing them from *R. solani*. The absence of *R. solani*-type branching, septation, mycelial habit, and similar characters excludes many additional species. *R. crocorum* D.C. ex Fr. and its synonyms are excluded, as are some 20 species associated with orchids. Some of the orchid fungi have mycelial characteristics similar to *R. solani* and may be closely related or may be *R. solani* (Thomas, 1925; Downie, 1957, 1959a,b). However, in order to narrow the scope of this work, they have been arbitrarily dismissed. A number of species are so inadequately described as to preclude identification or placement and therefore are not considered here.

After the above species have been excluded, there remains a group of species with many of the characteristics of *R. solani*. The following discussion concerns these species and their relationship to *R. solani*. One species without an official *Rhizoctonia* designation, *Hypochnus sasakii* Shirai (1906), is also discussed, since its mycelium has long been recognized as in or near *R. solani*.

*Taxonomy and nomenclature of fungi with characteristics of R. solani.*—In considering the taxonomy and nomenclature of these fungi, certain limitations imposed by lack of type cultures, deficiencies in written descriptions, and inadequacies in our knowledge of mycelial systems must be recognized.

Where possible, we have examined cultures from the ATCC, CMI, CBS, or from the describer. Such cultures, when deposited by the authority for a species, provide adequate material for study. Many cultures in these collections were isolated, identified, and deposited by workers other than the authority and therefore do not provide suitable material for identification, since they may represent misdeterminations. Many species have no type cultures, particularly the older species.

In the absence of type cultures, it has been necessary to rely on written descriptions, most of which are sufficiently lacking in details as to make identification difficult. As has been pointed out, fungi wholly unrelated to *R. solani* may look very much like it. Some fungi resemble *R. solani* so closely (Fig. 4) that they can be differentiated adequately only after laboratory study (Parmeter, et al., 1967). There may be other fungi resembling *R. solani* even more closely that have not yet been studied. We have been able to differentiate some of the *R. solani*-like fungi on the basis of nuclear condition and perfect state. However, few descriptions contain this information, and it is therefore doubtful that the identity of some species can ever be determined.

The following treatment is conservative to the extent that type cultures or general concensus among workers regarding the identity of a species is considered necessary before that species is reduced to

synonymy. Thus, *R. microsclerotia* is considered identifiable without type cultures. *R. melongena* is not considered positively identifiable because no body of literature or concensus among workers exists. This approach excludes from the list of synonymy a number of fungi that are very probably *R. solani*. Until more is known about the mycelial states of *Ceratobasidium* spp. and their pathogeni-

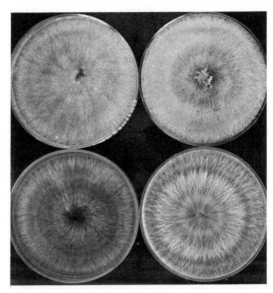

*Fig. 4.* PDA cultures of *R. solani* (*top*) and *Ceratobasidium* sp. (*bottom*), showing similarity in cultural appearance.

city, a conservative approach is unavoidable since several descriptions may apply as well to the mycelia of *Ceratobasidium* spp. as they do to *R. solani* (Parmeter, et al., 1967).

*Rhizoctonia napaeae* West. and Wall. (1846)

Westendorp and Wallays in 1846 collected on decaying turnips in storage a fungus they described and preserved in their *Herbier Cryptogamique* (Fasc. 5, No. 225, 1846) as *R. napaeae*. Duggar (1915) examined preserved material and concluded that it was identical to *R. solani*. He therefore reduced it to synonymy. Since *R. napaeae* was validly described 12 years prior to *R. solani*, the identity of this fungus has a marked bearing on the nomenclature of *R. solani*.

We have examined preserved specimens of *R. napaeae* (Westendorp and Wallays, 1846), and we agree that it may be *R. solani*. *Rhizoctonia solani*-like hyphae are present on the substrate (presumably the cortex of the turnip root) and the sclerotial crusts are like those arising from large masses of monilioid cells. In the absence of living cultures, however, it must be acknowledged that an element of doubt exists since other fungi, *R. tuliparum* (Klebh.) Whetzel and Arthur or *Ceratobasidium* sp. (Parmeter, et al., 1967) for example, have similar characteristics. Joshi (1924) pointed out the similarity between *R. napi* (= *R. napaeae*) and *Sclero-*

*tinia sclerotiorum.* It is unlikely that *R. napaeae* can be established beyond question as *R. solani.* For this reason, *R. napaeae* is considered only as a probable synonym.

### Rhizoctonia rapae West. (1852)

*Rhizoctonia napaeae* was later described by Westendorp (1852) as *R. rapae.* His description was based on the same specimens and was a verbatim copy of the earlier description of *R. napaeae. Rhizoctonia rapae* is therefore an obligate synonym of *R. napaeae. Rhizoctonia napi* (Saccaado, 1899) and *R. napae* (Kickx, 1867) are variant spellings.

### Rhizoctonia betae Eidam (1887)

Duggar recognized *R. betae* as a synonym of *R. solani.* From Eidam's description (we have seen only the 1888 abstract), it is probable that the fungus is *R. solani* but positive identification from this description is not possible.

### Hypochnus sasakii Shirai (1906)

*Hypochnus sasakii* was described on camphor trees in Japan. The similarity between the mycelial state of this fungus and *R. solani* has long been recognized. Matsumoto, et al. (1932) made extensive tests to find a means by which the two species could be readily distinguished. Since only one isolate of *R. solani* was used for comparison, their results are of questionable value. Variations among *H. sasakii* isolates in tolerance to chemicals and ability to anastomose were great and did not offer satisfactory means for separation of *R. solani* from *H. sasakii.* A number of workers subsequently have questioned the separation from *R. solani* of the mycelium of *H. sasakii* (Malaguti, 1951; Ito, et al., 1955; Sato and Shoji, 1957; Akai, et al., 1960; Ogura, et al., 1961; Kontani and Mineo, 1962).

Exner (1953), in comparing *H. sasakii* with *C. microsclerotia* and *C. solani,* was unable to demonstrate any mycelial, sclerotial, or behavioral characteristics by which the three species could be recognized. She concluded that *H. sasakii* was a form of *Pellicularia filamentosa* (= *T. cucumeris*). Ito, et al. (1955) and Sherwood (this vol.) also failed to demonstrate any basis for recognizing *H. sasakii* as a distinct species. We have examined the CMI isolate of Exner's forma *sasakii* (# 61793) and we agree that it is *R. solani.*

In addition, Sherwood and Parmeter (unpub.) studied anastomosis of Exner's CMI deposition and of two isolates (Va. 29 and S240) considered by Sherwood to represent *H. sasakii.* All three isolates fall into an anastomosis group that includes R43 of Whitney and Parmeter (1963), ATCC 13248, *P. filamentosa* f. *timsii* (CMI 61796), *P. filamentosa* f. *microsclerotia* (Sherwood's 220), Müller's CBS deposition of *R. solani* v. *hortensis,* and two isolates representing f. *microsclerotia* from blighted bean leaves in Costa Rica (Echandi, 1965). These data indicate that isolates identified as *H. sasakii* or *P. filamentosa* f. *sasakii* show a close relationship to several diverse isolates of *R. solani* and cannot be distinguished from them on the basis of anastomosis.

The mycelium of *H. sasakii* is therefore considered to be *R. solani.* The perfect state *H. sasakii* is also considered to be synonymous with *T. cucumeris* (Talbot, this vol.).

### Rhizoctonia potomacensis Wollenw. (1913)

Wollenweber (1913) described this fungus from tomatoes in the United States. The species was distinguished from *R. solani* mainly on the occurrence of subepidermal sclerotia on tomato. *Rhizoctonia solani* is known to produce sclerotia in tissues (Christou, 1962), and the distinction is therefore questionable. The species is considered as a probable synonym of *R. solani.*

### Rhizoctonia microsclerotia Matz (1917)

*Rhizoctonia microsclerotia* was described from fig in Florida in 1917 (Matz, 1917, 1921). No type cultures are available, but the work of Exner (1953), Echandi (1965), and Sherwood (this vol.) provide some agreement on identification. Exner (1953) compared single-spore progeny of *R. microsclerotia* with those of *R. solani* and found wide variation and overlap in cultural characteristics and pathogenicity. She concluded that there was no satisfactory basis for recognizing *R. microsclerotia* as a species. She did, however, recognize it as a forma of *P. filamentosa* (= *T. cucumeris*). Sherwood has shown that isolations from aerial infections give a group of cultures corresponding to Matz' *R. microsclerotia* and that these cultures possess sufficient similarities to form a generally recognizable group. He also found that variability within this and other groupings of *R. solani* was sufficiently great as to preclude recognition of *R. microsclerotia* as a distinct species.

We have examined numerous isolates of the microsclerotia type, including isolates of Exner, Echandi, and Sherwood, and compared them to single-spore progeny of various isolates of *R. solani.* The overlap in morphology and cultural characteristics is great. In addition, the Costa Rican isolates from bean have the pathogenic habit and sclerotial type of *R. microsclerotia* on the host, but they do not have the cultural characteristics described for *R. microsclerotia.* It thus appears that *R. microsclerotia* is variable and that isolates show gradations and intermediate characteristics that preclude separation from *R. solani.* Anastomosis data (discussed above under *H. sasakii*) also indicate that *R. microsclerotia* is *R. solani.*

### Rhizoctonia macrosclerotia Matz (1921), R. dimorpha Matz (1921), R. melongena Matz (1921)

Matz (1921) described several species of *Rhizoctonia,* three of which appear to be *R. solani.* No type cultures are available, however, and some of his descriptions lack essential details, particularly regarding pathogenicity, salient mycelial features, and adequate illustrations. In addition, no body of work has developed to provide any agreement as to the identity of these fungi. Their identity is therefore open to some question. All three have characteristics well within the known range of variation of *R. solani. Rhizoctonia melongena* was erected mainly on the basis of substrate, decaying egg plant. *Rhizoc-*

*tonia dimorpha* and *R. macrosclerotia* were erected mainly on the basis of sclerotial variations in culture. As was pointed out by Kernkamp, et al. (1952) with specific reference to Matz' species, and as has been shown by other authors studying variation in *R. solani*, the criteria of sclerotial size, shape, color, and position in or on the substrate are of questionable value, since these characteristics vary greatly. On the basis of present information, these three species cannot be excluded from *R. solani*, but some reservation in placement is necessary.

We have examined the CBS isolate of *R. melongena*, and while it has many of the characteristics of *R. solani*, its cultural appearance is sufficiently different to leave some doubt. Until more work has been done on collection and characterization of *R. solani*-like isolates from Puerto Rico, *R. melongena*, *R. dimorpha*, and *R. macrosclerotia* can only be considered as probable synonyms of *R. solani*.

*Rhizoctonia alba* Matz (1921), *R. ferrugena* Matz (1921), *R. grisea* (Stev.) Matz (1921), *R. palida* Matz (1921)

Matz' remaining four species do not appear to be *R. solani*. *Rhizoctonia ferrugena* produces white zonate colonies with rusty-red sclerotia buried in the medium. The CBS isolate of *R. ferrugena* has clamp connections (not mentioned by Matz) and sclerotia resembling those of *S. rolfsii*. Since this is not a type culture, it may be a misidentification, but it serves to indicate that Matz may have been dealing with a fungus resembling *S. rolfsii*. *Rhizoctonia alba* might be a pale or albino *R. solani*, but the fact that the sclerotia and mycelium remain white or only slightly yellow suggests that *R. alba* is not *R. solani*. *Rhizoctonia palida* has small hyphae and both the hyphae and sclerotia are yellow. *Rhizoctonia grisea* reportedly resembles *R. palida*. It is likely that these last two fungi are the same species and that neither represent variants of *R. solani*.

*Rhizoctonia gossypii* Forst. (1931)

*Rhizoctonia gossypii* was described from cotton in the Near East by Forsteneichner (1931). This fungus has the cultural and pathological characteristics of *R. solani*, and it was distinguished from *R. solani* mainly on the basis of its failure to anastomose with a group of German isolates. Schultz (1936), again on the basis of failure to anastomose with other isolates, accepted *R. gossypii* as a species. It should be noted that Forsteneicher recognized two varieties of *R. gossypii* because his species could be separated into two groups that failed to anastomose, thus he considered failure to anastomose as a specific character; then he recognized groups within his species that would not anastomose. As indicated earlier, failure to anastomose cannot be considered as an adequate basis for specific separation. We have examined Forsteneicher's isolates from the CBS and find them morphologically and pathologically indistinguishable from *R. solani*.

*Rhizoctonia alpina* Cast. (1935), *R. callae* Cast. (1935), *R. fraxini* Cast. (1935), *R. lupini* Cast. (1935), *R. muneratii* Cast. (1936), *R. pini-insignis* Cast. (1935), *R. quercus* Cast. (1935)

Castellani (1935, 1936) described seven new species of *Rhizoctonia* from Italy. We have examined his CBS depositions of *R. callae*, *R. fraxini*, *R. muneratii*, *R. pini-insignis*, and *R. quercus* and they all have binucleate hyphal cells. *Rhizoctonia callae*, *R. muneratii*, and *R. fraxini* resemble *R. solani* closely and could easily be mistaken for it, but the binucleate condition indicates that none of these species is *R. solani*. We have not seen cultures of *R. lupini* or *R. alpina*, but Castellani's description suggests that *R. alpina* is not *R. solani*. It has white to lemon-yellow colonies, grows poorly on organic medium, and is apparently nonpathogenic. *Rhizoctonia lupini* possesses the characteristics of *R. solani* and is reportedly a virulent pathogen. Without examination of cultures, it is not possible to determine the identity of this fungus, but it probably is *R. solani*.

*Rhizoctonia aderholdii* (Ruhl.) March. (1946)

This fungus, described by Ruhland (1908) as *Moniliopsis aderholdii*, was also considered by Duggar (1916) to be synonymous with *R. solani*. The literature on *R. aderholdii* is much confused, and it is doubtful that the many authors who have used the name were all working with the same fungus. Ruhland's description and those of subsequent workers do not permit certain distinction between *R. solani*, *Ceratobasidium* sp., and possibly others. The identity of *R. aderholdii* is therefore in doubt. Thomas (1925) and Rasulev (1959), for instance, considered *R. solani* and *R. aderholdii* to be the same. Baldacci and Cabrini (1937) considered *M. aderholdii* to be identical to their *R. solani* v. *ambigua* (the culture of this fungus obtained from CBS was *Fomes annosus*). Marchionatto (1946) considered the two fungi to be distinct. Kharitinova (1958b) not only accepted the separation of the two species, she suggested that *R. aderholdii* might include two different species.

In the absence of type cultures, the identity of *R. aderholdii* cannot be determined. It may represent a variant of *R. solani* or it may be *Ceratobasidium* sp. or something similar. It is therefore listed only as a probable synonym of *R. solani*.

*Rhizoctonia aderholdii* is sometimes cited in Russian literature as *R. aderholdii* (Ruhl.) Kolosh. (Khartinova, 1958b), *R. aderholdii* Koloschina (Dementyeva, 1962) and *R. aderholdii* (Ruhl.) Naumov (Verderevski, 1959). We have been unable to trace the original sources for these citations.

*Rhizoctonia dichotoma* Saks. and Vaar. (1960), *R. endophytica* Saks. and Vaar. (1960), *R. globularis* Saks. and Vaar. (1960), *R, hiemalis* Saks. and Vaar. (1960)

These four species were described by Saksena and Vaartaja (1960) from forest-tree nurseries in Canada. We have examined the ATCC and CBS depositions of these species, as well as isolates obtained from the Forest Insect and Disease Laboratory, Saskatoon, Canada. *Rhizoctonia hiemalis* has been shown to be the mycelium of a discomycete (Whitney and Parmeter, 1964; Warcup and Talbot, 1966). The mycelial cells of *R. globularis* and *R. endophytica* are binucleate. *Rhizoctonia endophytica* is similar to and will anastomose with the mycelium of a species of *Ceratobasidium*. *Rhizoctonia globularis* includes mycelia of two different basidiomycetes, *C. cornigerum* (Bourd.) Rogers and *Sebacina* sp. (Warcup and Talbot, 1966). *Rhizoctonia dichotoma* is similar in all respects to *R. solani*. Our observations on septation, nuclear condition, cultural characteristics, and pathogenicity indicate *R. globularis* is *R. solani*.

### *Rhizoctonia praticola* Saks. and Vaar. (1961)

Kotila (1929) described as *Corticium praticola* a fungus he referred to as a "new spore-forming *Rhizoctonia*." He did not, however, provide a *Rhizoctonia* designation, nor did he refer at any time to *R. praticola*. Although some workers have referred to *R. praticola* Kotila (Papavizas and Ayers, 1965), *R. praticola* (Kotila) Flentje (Papavizas, 1965), or to *R. praticola* (Kotila) Saksena and Vaartaja (Stretton, et al. 1964), these citations are questionable because *R. praticola* was not erected by Kotila nor can *R. praticola* be considered a new combination for *C. praticola*. The suggestion of Saksena and Vaartaja (1961, p. 637) can be considered as intent to erect the new species *R. praticola* based on the mycelium of *C. praticola*. The correct citation thus becomes *R. praticola* Saks. and Vaar. (1961).

*Corticium praticola* was described by Kotila as having hyaline, *Rhizoctonia*-like hyphae 6-8μ in diameter and brown sclerotia generally smaller and less abundant than those formed by *R. solani*. He found that maximum mycelial growth occurred at about 24°C. Flentje (1956) emphasized that *C. praticola* had smaller hyphae than those of *R. solani*, grew somewhat more rapidly, and produced colonies with a white mealy surface texture. Vanterpool (1953) and Saksena and Vaartaja (1961) also have considered *C. praticola* to possess distinct and reliable differences from *R. solani*.

Kernkamp, et al. (1952), Luttrell (1962), and Papavizas (1965) have challenged the validity of mycelial characters as means of distinguishing the two species. We have examined a wide range of isolates varying from white mealy colonies (when young) to dark-brown colonies with abundant aerial hyphae. These produce a continuous series of intergradations with no suitable point of separation. In addition, examination of numerous single-spore isolates from these cultures has shown that flat mealy parents may give rise to progeny with abundant aerial hyphae. As indicated earlier, some of these single-spore isolates produce sclerotia of highly variable size and shape. In no isolates have the hyphae remained hyaline. Brown pigment has invariably developed in some hyphae as colonies age. Both hyphal diameters and growth rates have been found by several authors to be highly variable. Thus, none of the characteristics supposedly distinguishing the mycelium of *C. praticola* from *R. solani* can be considered reliable.

Studies on variation in the perfect state have led to the conclusion that *C. praticola* is synonymous with the perfect state of *R. solani* (Papavizas, 1965; Talbot, this vol.). Since the perfect states are the same and no reliable method of distinguishing the two fungi in mycelial state exists, recognition of *C. praticola* or of the proposed *R. praticola* (Saksena and Vaartaja, 1961) is presently unacceptable.

### *Rhizoctonia fragariae* Hussain and McKeen (1963)

This fungus, isolated by Hussain and McKeen (1963) from strawberry, is described as differing from *R. solani* mainly in the absence of sclerotia. We have examined one of McKeen's isolates and a number of similar isolates obtained from California strawberries by Wilhelm. All have binucleate mycelial cells and several have produced a *Ceratobasidium* perfect state (Parmeter, et al., 1967). *Rhizoctonia fragariae* is therefore distinct from *R. solani*.

SYNONYMY.—In preparing a list of synonymy, it is necessary to recognize serious problems in identification. Because of the lack of type cultures, and uncertainties involved in identifications based solely on written diagnoses, few species can be assigned with confidence. Those species for which lack of suitable type material or adequate investigation presently precludes confident assignment are listed as probable synonyms.

Synonymy:

RHIZOCTONIA SOLANI Kühn (1858:24)
*Rhizoctonia microsclerotia* Matz (1917:117)
*Rhizoctonia gossypii* Forsteneichner (1931: 385)
*Rhizoctonia dichotoma* Saksena and Vaartaja (1960:934)
*Rhizoctonia praticola* Saksena and Vaartaja (1961:637)

Probable synonymy:
*Rhizoctonia napaeae* Westendorp and Wallays (1846:225)
*Rhizoctonia rapae* Westendorp (1852:402)
*Rhizoctonia betae* Eidam (1887: we have seen only the abstracted description in *Bot. Centraalb.*, 35:303-304, 1888)
*Rhizoctonia potomacensis* Wollenweber (1913:30)
*Rhizoctonia macrosclerotia* Matz (1921:19)
*Rhizoctonia dimorpha* Matz (1921:20)
*Rhizoctonia melongena* Matz (1921:29)
*Rhizoctonia lupini* Castellani (1935:70)
*Rhizoctonia aderholdii* (Ruhl.) Marchionatto (1946:4)

In addition to the above species, it should be noted that the mycelium of *H. sasakii* also is *R. solani*.

SUBSPECIFIC DIVISION OF R. SOLANI. — Several workers have attempted to recognize varieties, formae, or informal subgroupings of *R. solani*. Some of these, such as the varieties of *H. solani* erected by Schultz (1936) or the formae of *P. filamentosa* erected by Exner (1953), were designated as sub-specific taxa of the perfect state and therefore have no *Rhizoctonia* designations. These subspecific taxa have been based mainly on differences in the *Rhizoctonia* state, however, and are therefore discussed here.

*Subspecific divisions based on anastomosis.*—Anastomosis has been used by Forsteneichner (1931) and Schultz (1936) to recognize varieties of *R. solani*. Forsteneichner erected two varieties of *R. gossypii* (= *R. solani*), *R. gossypii* v. *aegyptica* and v. *anatolica*, mainly on the basis of their failure to anastomose. Schultz erected three varieties of *R. solani*: v. *hortensis*, v. *fuchsiae*, and v. *lycopersicae*, and three varieties of *H. solani*: v. *brassicae*, v. *typica*, and v. *lactucae*, also based on anastomosis groupings. Richter and Schneider (1953) described six groups of isolates that could be differentiated on the basis of anastomosis compatibility and to a lesser degree on cultural variations. These groups were given no formal taxonomic designations, however.

There are several obstacles to the acceptance of varieties or other subspecific groupings based on anastomosis. As indicated earlier, we know little about anastomosis or the genetic factors controlling its success or failure. It has been shown that some sibling single-spore isolates do not anastomose (Kernkamp, et al., 1952; Flentje, personal communication). These would become different varieties under a system such as that developed by Schultz or by Richter and Schneider. We have found that some very similar isolates do not anastomose. Richter and Schneider found 23 isolates that could not be placed in any of their anastomosis groupings. Two of Schultz' varieties were represented by only one isolate each. These workers also found that anastomosis groupings did not always correspond to groupings based on cultural appearance or similar physical features.

Because of these inconsistencies, the variability in *R. solani*, and the lack of suitable characteristics by which to identify anastomosis groupings, workers could not confidently place an isolate in the groupings of Schultz or Richter and Schneider without reference cultures. And as collections of *R. solani* from around the world increase, the accumulation of many isolates fitting none of these groupings is probable. For these reasons, formal subspecific groupings based on anastomosis are presently questionable. Until the phenomenon of anastomosis and its relationship to subspecific grouping has received more study, the erection of varieties or formae based on anastomosis appears to be premature.

*Subspecific division based on substrate.* — Ciferri (1938) described *R. solani* f. *paroketea* from starch residue of processed roots of *Manihot esculenta* Cranz. This forma was based mainly on substrate and apparent *Moniliopsis* state. It cannot be determined from the description whether the fungus was *R. solani* or a fungus with mycelium similar to *R. solani*. In either case, designation of the forma *paroketea* does not appear to be justified.

*Subspecific division based on cultural characteristics.*—Cultural differences have been used by several workers to recognize subspecific divisions of *R. solani* or its perfect state. Exner erected four formae of *P. filamentosa*: f. sp. *solani* (Kühn), f. sp. *microsclerotia* (Matz), f. sp. *sasakii* (Shirai), and f. sp. *timsii*, based mainly on cultural characteristics. She recognized, however, that these formae overlapped in both pathological and cultural characteristics. *Rhizoctonia solani* v. *cedri-deodare* was erected by Castellani (1935) on the basis of host and cultural characters. Castellani's description indicates no satisfactory cultural criterion by which this variety can be recognized. Baldacci and Cabrini (1937) erected *R. solani* v. *ambigua*, which they considered to be identical with *Moniliopsis aderholdii*. Again, there are no satisfactory criteria by which this variety can be recognized.

Several other workers, notably Houston (1945), Luttrell (1962), and Sherwood (this vol.) have organized cultures of *R. solani* into various groups according to cultural and pathological characteristics. Again, because of the extreme variability among isolates of *R. solani* (see Fig. 3), these groupings show certain inconsistencies and do not accommodate many isolates. Sherwood indicated that no single character could be used to place an isolate in one of his groups, but he felt that a combination of several characters was sufficient to place an isolate. His approach provides a step toward the possible recognition of subdivisions of *R. solani*, but his system is not yet usable, and future collections may well show that there are as many "groups" as there are possible combinations of characters.

In addition to the varieties discussed above, we have found reference to *R. solani* v. *graminis* Bunschoten, for which we have not been able to locate the original description. We cannot therefore comment on this variety at this time.

CONCLUSIONS. — From the accumulated data on variation in *R. solani*, it is possible to reach a few tentative conclusions. Perhaps the most important conclusion that can be drawn from our present knowledge of variation in *R. solani* is that we yet have far too little information to consider any division of the species into additional species or subspecific groups. The fungus occurs throughout most, if not all, of the world's land mass. Recent evidence (Flentje and Saksena, 1957; Flentje and Stretton, 1964; Baker [K.] et al., 1967) suggest that variants of *R. solani* are not easily introduced into new soils and that basidiospores do not serve as a ready means of disseminating variant strains. Thus, it appears likely that numerous variants have adapted to the numerous climate, soil, and vegetation regimes of the world and that these variants are to some degree isolated by their ecologies, leading to certain geographical and ecological discontinuities in distribution

Local variation in *R. solani* populations also must be considered. Flentje and Saksena (1957) found that, within a given soil, isolates of the same pathogenic strain were uniform and readily recognizable. But isolates of the same strain from a different soil were different and distinguishable from the other group of isolates. Ui, et al. (1963) found that two strains were involved with diseases of flax in the same field; one strain was active in May-June, the other in July-August. Thus, there are variants associated with large geographic regions down to different fields in the same region or even different seasons in the same field.

Until most of the possible variant types have been collected and studied, attempts to delimit specific or subspecific units within *R. solani* can only be considered questionable. Present evidence suggests that within the species there is continuous variation in characteristics and combinations of characteristics, and that no one characteristic or special combination of characteristics serves to distinguish any subgrouping. Because of the overlapping and intergrading of these characteristics and their combinations, the recognition of described varieties and forma or the erection of new ones appears premature. No such taxa have yet been characterized with sufficient clarity to avoid the problem of intermediates. Even where rather distinct and identifiable groupings have been found within a given area, examination of data for the species as a whole has shown that intermediates occur and that these groupings become obscured by the total range of variation within the species.

Attempts to separate *R. solani* into species, varieties, or formae may actually have decreased precision in communication. References to *Rhizoctonia* sp. occurs frequently in the literature. This designation means little more than "mycelium," since the genus contains many diverse elements. The use of this term by some workers undoubtedly stems from uncertainty as to what is meant by *R. microsclerotia, C. praticola, H. sasakii,* etc., and to the frequent occurrence of isolates that fall between these types. The use of "*Rhizoctonia sp*" is, however, totally unacceptable and serves only to encumber the literature with confusing and meaningless references. *Rhizoctonia* sp. might be anything from the mycelium of a discomycete to the mycelium of a basidiomycete.

The problem of precision in communication among workers dealing with *R. solani* and *R. solani*-like fungi has two facets: how to indicate clearly that a fungus is *R. solani,* and how to indicate a particular variant type of *R. solani.* As noted, there are fungi that resemble *R. solani* to a striking degree, and there is strong evidence that at least one other fungus has been frequently confused with *R. solani.* Future study may well show that there are several fungi that have been confused with *R. solani.*

Precision in identification and communication is greatly increased by use of the perfect-state designation, *T. cucumeris.* Among any group of isolates, some can usually be induced to sporulate (for methods of inducing sporulation, see Sinclair, this vol.). Production of the *Thanatephorus* state and use of the perfect-state designation in the literature precludes possible confusion between *R. solani* and the mycelia of other basidiomycetes.

When the perfect state cannot be induced, data on nuclear condition and septal pore characteristics should be included in any report on *R. solani.* We have yet no assurance that there are no other *R. solani*-like basidiomycetes with multinucleate hyphal cells, but present evidence indicates that confusion between *R. solani* and the mycelium of *Ceratobasidium* sp. (Parmeter, et al., 1967) can be eliminated by checking the nuclear condition. Until the hyphal systems of many additional basidiomycetes, particularly those pathogenic forms associated with corticioid or pellicular fruiting habits, have been studied, an element of question must be accepted in the identification of some isolates as *R. solani.* This is especially true of tropical and neotropical isolates since these types, frequently involved with aerial diseases, have not received sufficient study. Again, clarification of the relationship of these mycelia to *R. solani* rests ultimately with study of perfect states.

Precision in the characterization of variant types within *R. solani* poses an even more difficult problem. While there is nearly continuous variation among isolates, certain clusters of variation can be recognized. Thus, many isolates show rapid growth, appressed mealy appearance when young, gray-brown colonies, somewhat small hyphal diameters, wide host ranges, high virulence, and a tendency to fruit on agar. Isolates with this combination of characteristics have frequently been called *C. praticola,* but because of intermediates and integradations, this combination of characters does not define adequately any taxonomic unit. It is rather a stereotype into which some isolates of *R. solani* may be fitted. Such stereotypes are useful if it is clearly understood that they refer informally to certain combinations of characteristics and not to discreet, formally recognizable taxa.

Such designations as "praticola type," "sasakii type," and "solani type" (see Sherwood, this vol.) provide convenient, "short-hand" references to variant types, but they should not be used as substitutes for adequate description. Research reports on any isolates of *R. solani* should be accompanied by a thorough description of the isolates. Such descriptions have a twofold purpose: to increase precision in communication, and to accumulate data that may ultimately contribute to the development of a system whereby subspecific taxa, if they exist, can be recognized. The utility of anastomosis grouping and its relationship to possible sub-specific taxa remains to be determined.

One final conclusion that can be drawn from the foregoing discussion is that erection of species of *Rhizoctonia* is often useless and that there is grave danger of creating an almost impossible confusion if new species continue to be described. The concept of a dual nomenclature for fungi with both perfect and imperfect states is workable only if the imperfect state includes fruiting structures that provide reasonably good characteristics for identification. The extension of this concept of dual nomenclature to include the idea that the mycelia of fungi should be given designations in the *Mycelia Sterilia* is untenable. This can lead only to the erection of a multitude of *Rhizoctonia* species to accommodate mycelial states of the numerous basidiomycetes and ascomycetes possessing mycelia with the general characteristics of the genus.

*Rhizoctonia solani* is a firmly established designation and it is unlikely that it will be abandoned in favor of the perfect-state designation *T. cucumeris.* For the less firmly established designations, every attempt should be made to discover the perfect states and to use the perfect-state designations to the exclusion of the *Rhizoctonia* designations. In this way, the practice of maintaining mycelial designations may be gradually discarded and the extremely confusing suggestion of kinship among the diverse fungi with similar mycelial states can be precluded.

ACKNOWLEDGMENTS. — We wish to thank Miss Brigitta Flick and Mrs. V. Langenberg for valuable assistance in the translation of German and Italian papers, the American Type Culture Collection for a grant of cultures, and W. D. Platt for assistance in research and manuscript preparation.

# Taxonomy and Nomenclature of the Perfect State

P. H. B. TALBOT—*Waite Agricultural Research Institute, Adelaide, South Australia.*

SUMMARY.—The perfect state of *Rhizoctonia solani* Kühn is considered to be *Thanatephorus cucumeris* (Frank) Donk (basionym *Hypochnus cucumeris* Frank). *Hypochnus solani* Prill. and Delacr. and *H. filamentosus* Pat. are regarded as synonyms of *T. cucumeris*. The generic name *Pellicularia* Cooke is rejected as nomenclaturally invalid. *Botryobasidium* Donk and *Ceratobasidium* Rogers are readily differentiated from *Thanatephorus*, while the genera *Corticium* and *Hypochnus* are not acceptable for both taxonomic and nomenclatural reasons.

*Thanatephorus cucumeris* is treated as a collective species that includes *T. praticola* (Kotila) Flentje, *Corticium microsclerotia* Weber, and probably *C. sasakii* (Shirai) Matsumoto, *C. areolatum* Stahel and *Pellicularia filamentosa* (Pat.) Rogers f.sp. *timsii* Exner. As a collective species, it has no fundamental systematic importance and its nomenclature is unlikely to remain stable. Its disposal in *Thanatephorus* is, however, likely to persist.

BASIONYM OF THE PERFECT STATE.—The connection of *R. solani* Kühn (1858) with a Basidiomycete perfect state was first reported by Rolfs (1903). Burt (in Rolfs, 1903) determined the fruiting material as *Corticium vagum* Berk. and Curt. var. *solani* Burt. Later Burt (1918, 1926) treated this variety,

and also *Hypochnus solani* Prillieux and Delacroix (1891) and *C. botryosum* Bresadola (1903), as synonyms of the species *C. vagum* Berkeley and Curtis (1873). As Burt certainly misinterpreted *C. vagum* and *C. botryosum*, their names need be considered no further; both would now be regarded as species of the genus *Botryobasidium* (Rogers, 1935, 1943; Donk, 1958a; Eriksson, 1958b) and have no bearing on the perfect state of *R. solani. Hypochnus solani*, however, remains as a possible basionym for the species under discussion.

The names of two other species of *Hypochnus* have subsequently competed with *H. solani* as basionyms for this species: Rogers (1943) identified this species with *H. filamentosus* Patouillard (in Patouillard and Lagerheim, 1891), and Donk (1956a, 1958a) with *H. cucumeris* Frank (1883).

In the literature, it is very generally agreed that the perfect state of *R. solani* fits the diagnosis of *H. solani*. This has been confirmed in the present study by examination of authentic material (Fig. 1) of *H. solani* (Herb. Mus. Paris, on *Solanum tuberosum*, Paris, Oct., 1891).

Rogers (1943) examined type material of *H. filamentosus* and, finding it "quite indistinguishable" from recent collections of the perfect state of *R.*

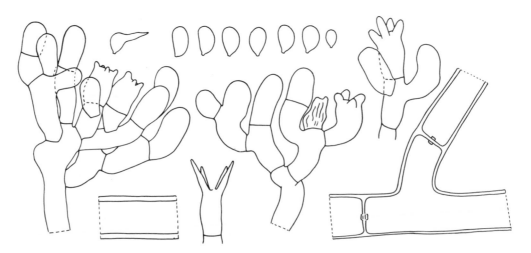

Fig. 1. *H. solani*, from authentic material in Herb. Mus. Paris, on *Solanum tuberosum*, Paris, Oct., 1891.

*solani* on potato, expressed surprise that *H. solani* had not immediately been detected as a synonym of *H. filamentosus* as these species had been published in successive issues of the same journal. Donk (1958*a*), however, regards the species *H. filamentosus* as insufficiently known and its name as a *nomen dubium;* he draws attention to the fact that *H. filamentosus* was described from leaves without mention of its occurring in soil or on the lower parts of plants; secondly to the fact that its fructification was described as *"roseus vel albidus,"* and pink is a color never associated with the perfect state of *R. solani.* The second objection may have some validity but it is also true that color descriptions are often inaccurate and that "roseus" could have covered various off-white tints in a discontinuous fructification overlying a reddish-brown mycelium. Donk states: "One thing about *H. solani* . . . we know . . . is that it is a soil fungus. This knowledge we lack altogether in connection with *H. filamentosus.*" It is true that *H. solani* was described from the lower parts of potato stems but a soil connection was not mentioned. The type specimen of *H. filamentosus* (Fig. 2; Farlow Herbarium, on *Dianthus caryophyllus* leaves, leg. G. de Lagerheim, Quito, March, 1891) has been examined in the present studies. Intermingled with the fructifications are pieces of grit, which might suggest that lower leaves were involved; moreover, the habit of a *Dianthus* plant would readily permit lower leaves to come into contact with soil. The writer agrees with Rogers that the type of *H. filamentosus* is "quite indistinguishable" from the perfect state of *R. solani.*

Donk (1956*a*, 1958*a*) identified the perfect state of *R. solani* with *H. cucumeris* Frank (1883), of which there is regrettably no type material in existence. A translation of Frank's description merits inclusion here:

A fungus, easily observable with the naked eye . . . which covers the lower parts of the plants. From soil level upward a fibrous grey or brownish-grey fungal membrane covers the upper parts of roots, stems and even the petioles of the lower leaves up to several centimeters from the soil surface, creeping upward with a floccose or radiating outer margin. This membrane is the fungal mycelium, the older parts of which are covered with a hymenium in which more or less densely grouped basidia are to be found, each bearing a single, one-celled, ovate, hyaline spore on each of the four sterigmata. The spores give the upper surface a powdery appearance. The mycelial mat loosely covers the abovenamed parts of the cucumber plant and can easily be removed, exposing fresh and healthy-looking tissue although the fungus has invaded it at some place on the lower part. . . . The mycelial layer consists of very thickly interwoven branched hyphae, 0.006-0.009 mm wide, with numerous septa, running in various directions in a more or less tortuous manner. The hyphal walls are quite thick, hyaline or pale brown. The hymenium develops from the surface hyphae which first form numerous septa. The hyphae are thus divided into very short cells which are nearly as wide as they are long. On the outer side of these hyphae numerous short branches are formed, again having one or more septa, and further branching into tuft-like clusters. The ovate outer cells form the basidia. These are mostly grouped in quite dense clusters with their apices pointing outward. Four thin sterigmata are formed on these apices, as in the Hymenomycetes, and each bears a single spore. The spores are ovate with a mean length of 0.009 mm, single-celled and hyaline. In drops of water the spores germinate after as little as 24 hr, producing a germ tube at one or both ends.

The merit of this description is vividly brought out if it is compared with the hardly more informative descriptions of the perfect state of *R. solani* in many modern plant pathology textbooks (e.g. Butler [E. J.] and Jones, 1949).

Since there is no type specimen for reference, it may be held that *H. cucumeris* is a *nomen dubium,*

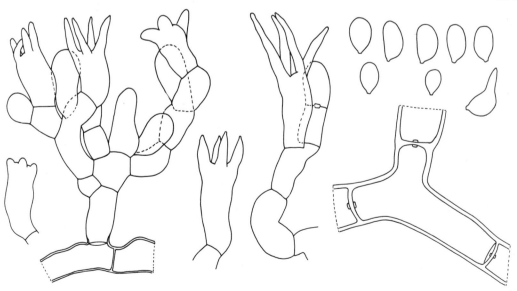

*Fig. 2. H. filamentosus,* from the type in Farlow Herbarium, on *Dianthus caryophyllus,* G. de Lagerheim, Quito, March, 1891.

i.e. a name based on a species that is now difficult or impossible to recognize. Olive (1957) considered that to be the case and took up *H. filamentosus* as the basionym of the species under discussion. Nevertheless, for the times in which it was written, Frank's description of the fungus and its relationship to the host plant is exceptionally fine and, as far as it goes, is wholly in accord with the perfect state of *R. solani* as we know the species today. In this context, the pathological evidence is as important as the mycological and the writer is not aware of any other species that would fit Frank's diagnosis; accordingly he recommends adoption of *H. cucumeris* Frank as the basionym of the perfect state of *R. solani,* with *H. filamentosus* and *H. solani* as later synonyms. This species will be designated as *T. cucumeris* (Frank) Donk henceforth in this paper. Reasons for rejecting other genera and adopting *Thanatephorus* are now presented.

HISTORICAL AND CURRENT GENERIC CONCEPTS. — The synonymy of *T. cucumeris,* given later, reveals that this species has been classed by various authors in six different genera, which will now be considered in turn. Some of these genera are nomenclaturally invalid; the limits of other genera have been narrowed and more closely defined in recent years, resulting in change of classification and nomenclature of the species under discussion. Such a process of generic restriction is by no means confined to this group but is in progress throughout the Hymenomycetes as an inevitable result of increased taxonomic knowledge.

*Hypochnus.* — *Hypochnus* was commonly used in the past (e.g. by Burt, 1916) for a genus of resupinate Aphyllophorales with loose floccose fructifications and usually with roughened colored spores. This concept, "as unworkable as it is unnatural" (Rogers, 1943), would embrace many disrelated forms and, in addition, the name is nomenclaturally unacceptable.

The Nomenclature Committee of the British Mycological Society (Wakefield, 1939) recommended that: "as the name *Hypochnus* has been used in different senses and is a permanent source of confusion it should be placed on a list of *nomina ambigua.*" Rogers (1939, 1949) agreed with this conclusion and recommended rejection of a proposal to conserve the generic name. Donk (1941, 1957a) also pointed out that *Hypochnus* as applied to Hymenomycetes is a later homonym of *Hypochnus* applied to lichens. It is clear that *Hypochnus* is not nomenclaturally acceptable, and the proposal for its conservation was rejected by ballot of the Special Committee for Fungus Nomenclature in 1953.

*Corticium.*—*Corticium* Fr. has long been used for a highly heterogeneous genus of resupinate Aphyllophorales characterized by a "smooth" hymenium and the absence of notable hymenial elements other than basidia and gloeocystidia, a wide circumscription that has resulted in a "taxonomic monstrosity" (Rogers, 1949). The process of segregating groups of structurally similar *"Corticium"* species into well-defined genera is not new, but has been accelerated recently as a result of publications by Rogers (1949), Donk (1956a, 1957a,b), and Eriksson (1958b), in which it was shown that *"Corticium"* in its generally accepted sense is nomenclaturally invalid; residual species that cannot at present be referred to segregated genera are still retained in *"Corticium"* as a temporary measure.

*Botryobasidium.* — In 1911, Bourdot and Galzin divided the genus *"Corticium"* into 16 sections. One of these, *Corticium* sect. *Botryodea* Bourd. and Galz., included such species as *C. solani* (Prill. and Delacr.) Cost. and Dufour and was characterized by its hypochnoid, pellicular or submembranous texture, by wide hyphae branching at a wide angle, and by broad basidia bearing 2-8 sterigmata and grouped in discontinuous candelabrum-like clusters. Donk (1931) regarded sect. *Botryodea* as a natural genus for which he proposed the name *Botryobasidium* Donk. *Hypochnus solani* was originally included as a species of *Botryobasidium,* but when this genus was later emended, Donk (1958a) transferred this species to *Thanatephorus* as a synonym of *T. cucumeris.*

The differences between *Botryobasidium,* as emended, and *Thanatephorus* are readily comprehended. In *Botryobasidium,* the fructification becomes hypochnoid with a relatively thick layer of ascending hyphae; the basidia are urniform or subform with a median constriction that results from their development in two phases; there are (4-)6(-8) small curved sterigmata per basidium; the basidiospores are not repetitive; no rhizoctonia-state or sclerotia are formed; conidial states, when present, belong to the genus *Oidium* (as emended by Linder [1942]). In contrast, the fructification of *Thanatephorus* is relatively thin and more discontinuous; the basidia are not constricted about the middle; the (2-)4(-7) sterigmata per basidium are large, relatively straight, about as long as or longer than the metabasidium; the basidiospores are capable of repetition; conidial states are not known but sclerotia occur commonly and rhizoctonia-states regularly.

In the apparent belief that the hyphal construction and the subcymose clusters of basidia were of primary taxonomic significance, Rogers (1935) merged the genera *Botryobasidium* and *Botryohypochnus* Donk (1931), describing eight species under the former name. The resulting genus contained species with diverse basidial morphology that would, in the opinion of many, have been better kept apart. Rogers also proposed the genus *Ceratobasidium* Rogers for four species, two of which were transferred from *Corticium* sect. *Botryodea.*

*Pellicularia.*—Rogers (1943) placed *Botryobasidium* in synonymy with the genus *Pellicularia* Cooke (1876b) on the grounds that their type species were congeneric and that *P. koleroga* Cooke was based on an acceptable type specimen. Both of these premises are questionable.

Cooke (1876a, b, c, 1881) erected a genus *Pellicularia* with the single species *P. koleroga* associated with a web- or thread-blight of coffee leaves in India. His description laid stress on a gelatinous matrix and vegetative hyphae, and on spores that have subsequently been shown to belong to a mold; but he denied the presence of basidia and basidiospores. These were later found in the type specimen and illustrated for Burt (1918, 1926) by Wakefield. As the type specimen was certainly a mixture of at least two species—a mold and a Basidiomycete—the typification of *P. koleroga* has been controversial.

Rogers (1943, 1951) typified *P. koleroga* by the Basidiomycete element in Cooke's type specimen, contrary to the recommendation of the Code of Nomenclature that in such cases the principal basis for the selection of a satisfactory type is the original *description* of the taxon and not the type specimen. In his several articles on *P. koleroga*, Cooke consistently emphasized the matrix and the mold spores, commenting but little on the vegetative hyphae and denying the presence of basidia; thus, on the basis of the original description, a Basidiomycete should not have been selected from the mixed type specimen to typify *P. koleroga*. In effect, Cooke had given the name *P. koleroga* to the mold element and vegetative hyphae in the type specimen (Venkatarayan, 1949). Based as it is on a mixed type specimen and a confused original description, *P. koleroga* is regarded by Donk (1954a, 1958a), Warcup and Talbot (1962), and Talbot (1965) as a *nomen confusum* and thus inapplicable to any taxon whatever; it follows that the genus *Pellicularia* is also inadmissible. Several authors have tacitly rejected *Pellicularia sensu* Rogers; they include Olive (1957), Pilát (1957), Eriksson (1958a,c), Christiansen (1959, 1960), Saksena (1961a,b), Flentje, Dodman, and Kerr (1963), Flentje, Stretton, and Hawn (1963), Whitney and Parmeter (1963), Flentje and Stretton (1964), and Whitney (1964a).

It is legitimate to use the Basidiomycete element of Cooke's type specimen of *P. koleroga* to typify a new species, as was done by Donk (1958a) in proposing his new genus and species *Koleroga noxia*. *Pellicularia koleroga sensu* Rogers thus became an obligate synonym of *K. noxia*. However, Talbot (1965) has evidence that this species belongs properly in the genus *Ceratobasidium* and therefore that *Pellicularia sensu* Rogers, and *Koleroga*, are synonyms of *Ceratobasidium*. The effect of accepting Rogers' typification of *Pellicularia* would merely be to make *Ceratobasidium* a synonym of *Pellicularia*; it would still not permit *T. cucumeris*, the perfect state of *R. solani*, to be included in either *Pellicularia* or *Ceratobasidium*, for it is generically distinct.

As used by Rogers (1943) in his influential monograph, the genus *Pellicularia* has a wide circumscription and includes species with several types of highly characteristic basidia (Fig. 3). These species can mostly be divided now among more homogeneous genera, which also have the merit of being nomenclaturally unassailable.

*Ceratobasidium.* — Olive (1957) considered that Frank's description of *Hypochnus cucumeris* was "too vague to permit a positive identification of his species" and he therefore regarded *H. filamentosus*, which he recombined as *Ceratobasidium filamentosum* (Pat.) Olive, as the perfect state of *R. solani*. Pilát (1957) made the combination *C. solani* (Prill. and Delacr.) Pilát for the same species. Olive stated that "two important features indicating . . . relationship to *Ceratobasidium* and to the Tremellales have not been sufficiently emphasized; i.e. the Ceratobasidium-like basidia with their characteristic horn-like sterigmata, a type of sterigma common to so many of the Tremellales, and the germination of the basidiospores by repetition, a feature common to the majority of the Tremellales but not to the higher Basidiomycetes." These are in fact the primary diagnostic characters assigned to *Ceratobasidium* by Rogers (1935) when he proposed the genus for "delicate corticioid fungi whose holobasidia show the division into hypo- and epibasidia and whose spores the germination by repetition together characterizing the subclass Heterobasidiomycetes"; but, in addition, the basidia were defined as subglobose or short, stout claviform, and the fructifications with distinct, loose hyphae forming a scanty subiculum.

It is now clear that quite a large number of holobasidiate species occupy a position intermediate between Heterobasidiomycetes and Homobasidiomycetes with respect to basidial morphology and spore repetition. One solution to the problem of their classification might be to place such species in a single genus (e.g. *Ceratobasidium*), but this is certain to lead to a "taxonomic monstrosity." Another solution, adopted by Donk (1956a, 1958a, 1964),

Fig. 3. Some generic segregates from *Pellicularia sensu* Rogers: A. *Ceratobasidium* (*C. cornigerum*). B. *Thanatephorus* (*Hypochnus solani*). C. *Uthatobasidium* (*U. fusisporum*). D. *Botryobasidium* (*B. subcoronatum*). E. *Botryohypochnus* (*B. isabellinus*).

Eriksson (1958c), Christiansen (1959, 1960), and Talbot (1965), is to recognize a number of smaller homogeneous genera at the outset, by taking into account critical basidial morphology, growth habit, hyphal characteristics, host relationships, and any other apparently significant combinations of features shared by groups of species.

The genus *Ceratobasidium* becomes rather homogeneous when attention is focused on the typically sphaero- or pyropedunculate metabasidia that are two or three times the width of their supporting hyphae and that arise singly or in small irregular groups from short and scanty suberect hyphae that branch from the loose reticulum of wide repent hyphae lacking clamp connections. In contrast, the basidia of *Thanatephorus* are barrel-shaped or subcylindrical, about the same width or little wider than their supporting hyphae which arise in asymmetrical cymes or racemes from discontinuous tufts of ascending hyphae derived from wide, Rhizoctonia-like basal hyphae lacking clamp connections. With these generic criteria, there is no difficulty in separating *Ceratobasidium* and *Thanatephorus*.

It is unsound to lay stress on whether the basidia show an affinity with the Heterobasidiomycetes by being divided into hypo- and epibasidia. This demands the ability to distinguish between an "epibasidium" and a voluminous sterigma, a feat that some mycologists claim is impossible (Donk, 1954b; Talbot, 1954). In *Ceratobasidium* and *Thanatephorus*, the sterigmata are essentially similar and to regard them in the one genus as sterigmata, but in the other as epibasidia, has sometimes led to the two genera being placed in different subclasses of the Basidiomycetes; yet superficially the similarity between the two genera is so great that at other times they have been considered synonymous. *Peniophora heterobasidioides* Rogers (1935) is a particularly instructive species in this connection: Rogers stated that "the sterigmata [are] so clearly only sterigmata [i.e. not "epibasidia"] that the fungus seems best placed on the homobasidial side of the line"; but Olive (1957) erected the genus *Heteromyces* (= *Oliveonia* [Donk, 1958a]) for this species and placed it in the Ceratobasidiaceae (Tremellales, Heterobasidiomycetae).

*Thanatephorus.* — This genus was proposed by Donk (1956a), with *H. solani* Prill. and Delacr. (= *T. cucumeris* [Frank] Donk) as the type species. Reasons for adopting the specific epithet *cucumeris* rather than *solani* were given in a fuller taxonomic treatment (Donk, 1958a) that also revealed clearly the generic differences to be found in the tulasnelloid group of fungi. Talbot (1965) has confirmed most of Donk's conclusions in regard to these genera and both authors have provided generic keys.

THE TAXONOMIC POSITION OF THANATEPHORUS. — *Thanatephorus* and *Ceratobasidium*, as well as certain other genera, have basidial features intermediate between those of Heterobasidiomycetes and Homobasidiomycetes, namely, a combination of holobasidia with repetitive spores and voluminous sterig-

mata that sometimes become adventitiously septate. In a sense, it is unimportant on which side of the dividing line these genera should be classed, as Martin (1957) has pointed out; but conventional taxonomy depends on discontinuity for separating taxa and has no satisfactory method of dealing with intermediates. To erect a distinct family, Ceratobasidiaceae (Martin, 1948, 1957), is only a partial solution to the taxonomic problem because the family still has to be accommodated in a higher taxon and this is likely to be somewhat arbitrary despite the best endeavors of authors to look for relationship with taxa whose classification is not controversial.

Rogers (1943) placed *Pellicularia* in the Thelephoraceae, a family that has been greatly restricted in recent years and would not now accommodate *Pellicularia sensu* Rogers. Martin (1945) first placed *Ceratobasidium* in the Tulasnellaceae but later (1948, 1957) he placed it in the family Ceratobasidiaceae as a member of the Tremellales (Heterobasidiomycetes). Jackson (1949) assigned *Pellicularia* and *Ceratobasidium* to the Ceratobasidiaceae at the base of the Homobasidiomycetes. Olive (1957) placed *Ceratobasidium*, inclusive of *Thanatephorus*, in the Ceratobasidiaceae (Tremellales, Heterobasidiomycetes). Christiansen (1959, 1960) included the Ceratobasidiaceae in the Heterobasidiomycetes but regarded *Thanatephorus* as a member of the Corticiaceae (Homobasidiomycetes). Donk (1956a, 1958a, 1964) and Eriksson (1958b, c) do not recognize the Ceratobasidiaceae as a distinct family but include its members in the Corticiaceae. Talbot (1965) rejected the Ceratobasidiaceae, including both *Ceratobasidium* and *Thanatephorus* in the Tulasnellaceae *emend.* (Tulasnellales) and recommended that the subclasses Heterobasidiomycetes and Homobasidiomycetes should no longer be recognized. The writer agrees with Martin (1957) that the existence of intermediate taxa does not imply that heterobasidiate and homobasidiate lines are not valid phyletic concepts; nevertheless, in practical taxonomy, one must be able to recognize and separate taxa and this system breaks down with intermediate categories.

It is clear that the taxonomic position of *Thanatephorus* is highly controversial and likely to remain a matter of personal interpretation for a considerable time.

TAXONOMIC CHARACTERS AND THEIR USEFULNESS. —The genus *Thanatephorus* is distinguished by a combination of characters given in the generic diagnosis below. Some of these are variable, e.g. the soil habitat, the parasitic habit, and the presence of sclerotia; such characters are supplementary, not diagnostic for the genus.

*Hymenium.* — The discontinuous hymenium, consisting of imperfect cymes or racemes of basidia, is distinctive though variable; it is essentially similar in some other genera (notably *Botryobasidium s.str.*, *Uthatobasidium* and *Botryohypochnus*). In the fructification of *Thanatephorus*, the system of erect hyphae is less extensive than in *Botryobasidium* and

*Uthatobasidium,* and more extensive than that of *Ceratobasidium;* these features are, however, relative and rather variable according to the age of the fructification.

*Basidia.*—Basidial morphology taken in combination with the general growth form is diagnostic. In *Thanatephorus,* the holobasidia are relatively short and wide, barrel-shaped to subcylindrical, not uniform nor constricted about the middle and little wider than the supporting hyphae. The sterigmata are stout and usually straight; they vary in number per basidium, in the presence or absence of adventitious septation, and in length, but are usually as long as or longer than the metabasidia when mature.

*Basidiospores.*—The spores in *Thanatephorus* are smooth, thin-walled, prominently apiculate, not amyloid, hyaline; they tend to be oblong to broad-ellipsoid and unilaterally flattened, while often they are widest toward the distal end. Spores of this kind are also found in other genera, notably *Ceratobasidium.*

Basidiospore repetition is commonly but not invariably present in fungi with typical phragmobasidia, but it is unknown in association with all but comparatively few holobasidial fungi, of which *Thanatephorus* is one. By classing holobasidial fungi exhibiting spore repetition with the Heterobasidiomycetes, Patouillard (1900) emphasized spore repetition rather than the septation of the basidia as the basis for separating the subclasses Homobasidiomycetes and Heterobasidiomycetes. With the development of the concept of an "epibasidium" (Neuhoff, 1924), the presence of stout variable sterigmata ("epibasidia") was also taken to indicate a Heterobasidiomycete when in association with repetitive spores, but would not necessarily be regarded as significant in the absence of spore repetition. Later, with the example of the *Tulasnella* basidium in which there is frequently a septum at the base of each sterigma, the concept of a phragmobasidium was broadened to include basidia whose sterigmata occasionally develop adventitious septa, not necessarily basal in position (as in *Thanatephorus* and *Ceratobasidium*). Thus, the Ceratobasidiaceae became widely regarded as Heterobasidiomycetes allied to, if not members of, the Tremellales. Nevertheless, there is much controversy on the systematic position of the Ceratobasidiaceae (see above section on the taxonomic position of *Thanatephorus*). Donk (1958b, 1964) and Talbot (1965) have given evidence that adventitious septation in the sterigmata is not a sign of a phragmobasidium or necessarily of a member of the Heterobasidiomycetes.

Donk (1956a, 1964) has reviewed the usefulness of spore repetition as a taxonomic character in the Basidiomycetes and, while doubting its value at the family level, he (1958a) has employed it for generic delimitation in combination with other characters. Olive (1957) and Pilát (1957) also considered spore repetition to be of generic value.

It is clear that the character of spore repetition has been widely regarded as important in taxonomy although mycologists are aware that it may be inconstant even in species whose spores are known to be capable of repetition. *Thanatephorus cucumeris* is one such example. In this species, Whitney (1964b) has shown that certain ether-soluble substances bring about spore repetition. He concludes that "repetition is probably determined genetically but its usefulness as a taxonomic criterion is valid only if the right environment is provided."

The author has experienced exceedingly few instances in which he has been unable to find direct evidence of spore repetition in species known to exhibit this phenomena and in species where other taxonomic considerations would suggest that the spores might be repetitive. One useful pointer is marked variation in size of morphologically similar, free-lying basidiospores in the preparation, as when some spores are perhaps only one-half the size of the majority. In these instances, direct evidence of repetition can usually be found by diligent searching.

In a genus of wide circumscription, such as *Pellicularia sensu* Rogers, it may be desirable to know the method of spore germination in order to determine the species. Such knowledge is unnecessary for determining species of *Thanatephorus* and allied genera as currently segregated, for each genus is homogeneous in regard to the method of spore germination. Since each of these genera is based on a summation of several taxonomic characters, it is not essential for generic determination to know whether the spores are repetitive or not.

SPECIFIC CHARACTERS AND HISTORICAL SPECIES CONCEPTS. — Species referable to the genus *Thanatephorus* have in the past been differentiated by combining the characters of their perfect and imperfect states. This paper is confined to a discussion of their perfect states.

Specific characters of the perfect states would include: shape and size of basidia and basidiospores, length of sterigmata and their average number per basidium, type of hymenial branching, maximum width of hyphae in the fructification, and color.

Table 1 summarizes the range of variation of measurements recorded for the basidia, sterigmata, basidiospores, and hyphae of some species referred to *Thanatephorus* or informally allied with this genus pending further investigation. Although some authors have indicated the number of measurements made to record the range of variation, and the mean of measurements made for each organ, these details are lacking in most descriptions and thus the data cannot be analyzed statistically. It is possible that statistical analysis might reveal significant differences in the sizes of basidia and spores of different species (as found, for example, by Flentje [1956]), but such are not apparent from inspection of Table 1, which gives the broad impression that the measurements recorded for *T. cucumeris* overlap those of the other species. As regards spore size, the matter is complicated by the fact that spores in these species are capable of repetition and, depending on the proportion of primary and secondary spores present, very

TABLE 1. Range of variation of measurements (in $\mu$) recorded in the literature for organs of *T. cucumeris* and allied species.

| Species | Basidia | Sterigmata | | Spores | Hyphae diameter |
| | | No. per basidium | Length | | |
|---|---|---|---|---|---|
| *T. cucumeris* | 10-25.2×6-19 | 4(2-7) | 5.5-36.5 | 6-14×4-8 | 4.5-17 |
| *T. praticola* | 13.5-22×5.8-8.7 | 3(1-5) | 10.4-43.7 | 5.2-10.8×4-7.5 | 6-12 |
| *C. microsclerotia* | 6.8-18×5.4-9.4 | 4(2) | 5.0-10.2 | 7-12×4-7.2 | 6-8 |
| *C. sasakii* | 5.6-16×5-10 | 2-4 | 2.5-8.4 | 6-11×4-8.4 | 5-11 |
| *C. areolatum* | 10-14×8-10 | 4(2) | 5.5-13 | 7.5-9×4.5-5 | 5-10 |

different values could be obtained for the range and mean of spore size. It would not appear possible to separate these species on the basis of microscopic measurements of their most important taxonomic features. These species will now be reviewed:

    1. THANATEPHORUS PRATICOLA ( Kotila ) Flentje *apud* Flentje, Stretton and Hawn (1963:451), Flentje and Stretton (1964), and Talbot (1965, f.13)
    *Corticium praticola* Kotila (1929:1065, f.5-6), Rogers (1943:115), Flentje (1952:892), Hawn and Vanterpool (1953), Boidin (1958: 100), and Vanterpool (1953:488)
    *Pellicularia praticola* (Kotila) Flentje (1956: 353, f.2), and Saksena (1960:165-167)
    *Ceratobasidium praticola* (Kotila) Olive (1957:431), and Saksena (1961a:717, 1961b: 749)
    *Rhizoctonia praticola* Saksena and Vaartaja (1961:637), *nom. anam.*

This species was originally described (Kotila, 1929) as saprobic in soil and parasitic on alfalfa. Kotila noted that very humid conditions were needed for fruiting, but that the basidia, sterigmata, and spores were identical in fructifications formed on potted alfalfa plants and in agar culture. The species was differentiated from *T. cucumeris* on the following grounds: (1) its hymenium formed in small plaques and hymenial branching was subracemose instead of subcymose; (2) its sterigmata averaged three, instead of four, per basidium and were of great length, very often more than three times the length of the largest basidiospores; (3) its vegetative hyphae were subhyaline and 6-8 $\mu$ in diameter, while those of *T. cucumeris* were brown and wider; (4) its spores tended to be somewhat smaller than those of *T. cucumeris;* and (5) it grew faster in culture than typical isolates of *T. cucumeris.*

Several authors (cited in the above synonymy) have confirmed these differences in some of their isolates and have maintained two species. There are also other indications that isolates may be separable into two species. Boidin (1958) found differences in enzyme activity between isolates referred to *T. praticola* and *T. cucumeris.* Differences in nuclear complement of the vegetative cells have been found (Sanford and Skoropad, 1955; Boidin, 1958; Saksena, 1961a, b; Flentje, Stretton, and Hawn, 1963) and in the nuclear complement of the basidiospores (Saksena, 1961a; Flentje, Stretton, and

Hawn, 1963; Flentje and Stretton, 1964). The ability to fruit on agar has also been investigated: Stretton, et al., (1964) found that all their isolates of *T. praticola* fruited on agar but not on soil; pathogenic isolates of *T. cucumeris* fruited on soil, not on agar, while nonpathogenic ones fruited on both agar and soil.

Flentje (1956) considered that sterigmatal length is of little use in differentiating *T. cucumeris* and *T. praticola* because, in a strain of the former, the sterigmata could be lengthened to a maximum of 46.8 $\mu$, thus exceeding the maximum recorded for *T. praticola*, by changing the substrate and cultural conditions. Flentje analyzed his measurements of basidia, spores, and hyphae statistically and was able to show significant differences in some of the isolates, which were to a certain extent correlated with the taxonomic division into two species on other grounds.

Several authors have doubted that the two species are distinguishable. Rogers (1943) listed *T. praticola* under *"species inquirendae"* and suggested that cultural conditions, particularly humidity, had influenced the size and form of the basidia and spores in the authentic material he studied, and that it was not strictly comparable with field specimens of *T. cucumeris.* Boidin (1958) despite finding differences in growth rate, nuclear number, and enzyme activity, tended to minimize their importance and to regard the two species as very close to one another. Luttrell (1962), Whitney and Parmeter (1963), and Whitney (1964a) have suggested that these species are not separable, or at least are so close that they may be discussed together. Papavizas (1965) has compared single-spore isolates of *T. cucumeris* and *T. praticola* in much detail. He concluded that:

The principal, in fact the only, apparent distinguishing characters between wild clones of *P. praticola* and *P. filamentosa* in the present studies were the number and length of sterigmata on each basidium. In *P. praticola* . . . most basidia developed 3 sterigmata longer than 22 $\mu$, although 1, 2, 4 and 5 sterigmata were also observed. In *P. filamentosa* . . . most basidia developed 4 sterigmata less than 16 $\mu$ in length . . . although several basidia were observed with 2, 3 and 5 sterigmata. A few single-basidiospore isolates of *P. praticola* that formed the perfect state, however, developed basidia with 2, 3, 4 or 5 sterigmata per basidium and the sterigmata were shorter (9.6-19.2 $\mu$) than the sterigmata of the parent isolates. Thus it appears that there was no valid and constant distinction between the perfect states of the 2 species and that the number and length of the sterigmata . . . cannot be considered as valid criteria for separation of the 2 species.

Papavizas found, in addition, that the asexual characters, growth rates, and morphology and size of the thick-walled moniliform cells ("chlamydospores") completely overlapped in the two species.

Numerous authors have recorded that *T. cucumeris* and *T. praticola* can exist in many strains that differ in cultural characters and pathogenicity, or that single-spore isolates from a single parent culture can also differ (Briton-Jones, 1924; Müller, 1924; Sanford, 1938a; Ryker and Exner, 1942; Exner and Chilton, 1943; Houston, 1945; Exner, 1953; Papavizas and Davey, 1962b; Daniels, 1963; Papavizas, 1964, 1965). Anastomosis in opposed strains of *T. cucumeris* was reported by Flentje and Stretton (1964), but the effect was not permanent and as yet no interfertility tests between strains of *T. cucumeris* and *T. praticola* have been reported.

Until very recently, the author (Talbot, 1965) had regarded *T. cucumeris* and *T. praticola* as distinct species separable by Kotila's criteria. However, the mounting evidence of intergradation between isolates assigned to these species, and also personal experience of an isolate (Waite Institute No. 87)

with the growth rate and cultural characters of *T. praticola* but the basidial morphology of *T. cucumeris,* have now caused the author to revise his opinion. It would now appear best to treat *T. praticola* as a synonym of *T. cucumeris,* at least until they can be unequivocally separated, recognizing that the latter species can exist in a large number of strains forming an aggregate species that, at present, cannot be satisfactorily subdivided.

2. CORTICIUM MICROSCLEROTIA Weber (1951:727; 1939:565, *nom. nud.*)
*Pellicularia filamentosa* (Pat.) Rogers f.sp. *microsclerotia* (Matz) Exner (1953:716)
*Rhizoctonia microsclerotia* Matz (1917:117) *nom. anam.*

The imperfect state of this fungus was distinguished from *R. solani* principally by its smaller sclerotia and narrower hyphae (Matz, 1917). Although Weber (1939) cited certain authors who considered *R. microsclerotia* and *R. solani* to be different species, and others who thought that *R. microsclerotia* was only a strain of *R. solani,* he (1939, 1951) made no comparison of their perfect states, and the main justification for erecting a new

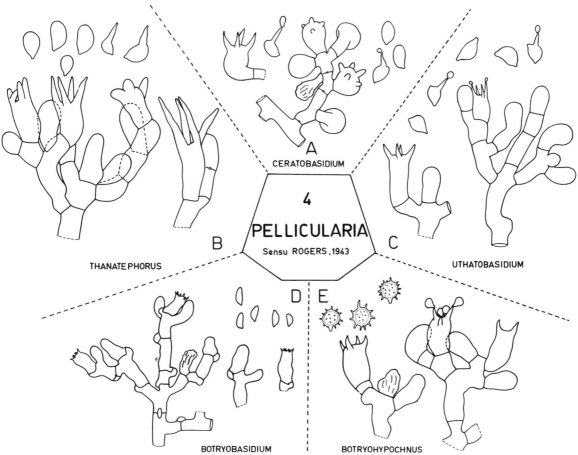

Fig. 4. *C. microsclerotia,* from Herb. Florida Agric. Expt. Stn. No. F6621, on *Crotalaria spectabilis,* Geo. F. Richey, Gainesville, 1944; and on *Bauhinia saigonensis,* Gainesville, 1943 (det. Erdman West).

species for the perfect state of *R. microsclerotia* still remained the vegetative characters of the imperfect state.

Rogers (1943) compared the perfect states and regarded *C. microsclerotia* as a synonym of *P. filamentosa* ( = *T. cucumeris*). Cunningham (1963) also considered these synonymous. In her careful comparison, Exner (1953) could not distinguish these species by characters of the perfect states but, finding that they produced diseases that were "reasonably distinct" and that a "rather clear-cut cultural strain was associated with each disease," she divided *P. filamentosa* into a number of *formae speciales* of which *C. microsclerotia* was one. For taxonomic purposes, Exner separated these specialized forms on the basis of cultural characters, sclerotial size, and diseases produced, not on morphological differences in the perfect states. Although the average spore measurements of *C. microsclerotia* were at the lower end, and those of *P. filamentosa* at the higher end of the range of species studied, there are in fact negligible differences in the figures given by Exner.

Weber's drawings and description (1951) show subcymose basidial clusters typical of *Thanatephorus*, nonrepetitive spores measuring 5-6 × 9-11 μ, and hyphae 4-8 μ wide; basidial measurements were not given. The present author has examined material of *C. microsclerotia* (Fig. 4) from the type area and determined by Erdman West, in which the spores were repetitive, (7-)9 (12) × (4-)5-6(-7.2) μ, the basidia 4-spored and measuring 16-18 × 7.5-9 μ, and the hyphae up to 15μ wide in parts. The author has no hesitation in referring *C. microsclerotia* to synonymy with *T. cucumeris* judged on the basis of their perfect states.

### 3. CORTICIUM SASAKII (Shirai) Matsumoto (1934:119)
*Hypochnus sasakii* Shirai (1906:319)
*Pellicularia filamentosa* (Pat.) Rogers f.sp. *sasakii* (Shirai) Exner (1953:717)

Matsumoto (1934) compared isolates of *C. sasakii* in the perfect and imperfect states with "*Corticium vagum*" *sensu* Burt and *R. solani*, respectively. He concluded: "Although the morphological differences between these two fungi in the perfect state are not sufficient to warrant a classification of our fungus as a distinct species, in judging from the vegetative characters of our fungus this cannot be included in the species *C. vagum*, and should be separated from the latter as a distinct species rather than as a biologic form."

The comparison with *C. vagum sensu* Burt was unfortunate since this comprises more than one species, but Matsumoto's description of the perfect state of *C. sasakii* would permit its inclusion in *T. cucumeris*. Exner (1953) came to the same conclusion in regarding *C. sasakii* as a *forma specialis* of *P. filamentosa* ( = *T. cucumeris*) based on cultural and vegetative characters rather than on any differences in their perfect states.

Donk (1958a) associated *C. sasakii* with the genus *Thanatephorus* and thought it possible that *Pachysterigma griseum* Racib. referred to the same species; if so, *P. griseum* would be the first validly published name for this species in the perfect state.

The author has not seen specimens of *C. sasakii*. It is clear, however, that means of differentiating it from *T. cucumeris* are unlikely to be found in their perfect states.

### 4. PELLICULARIA FILAMENTOSA (Pat.) Rogers f.sp. *timsii* Exner (1953:717)
Tims and Mills (1938) recorded a Rhizoctonia leaf blight of fig, which was studied and induced to fruit in culture by Exner (1953). Exner regarded the perfect state as indistinguishable from *P. filamentosa* but proposed the *forma specialis* on vegetative and cultural differences. It would seem likely therefore that this *forma specialis* may be referred to *T. cucumeris*, but the author has not seen specimens of it.

### 5. CORTICIUM AREOLATUM Stahel (1940:129) *nec Corticium areolatum* Bresadola (1925)
This name is invalid because of the existence of an earlier homonym.

Rogers (1943, f. 11, e-g) and Cunningham (1963) listed this species as a synonym of *P. filamentosa* ( = *T. cucumeris*), and Rogers' figures are typical of a *Thanatephorus*. Donk (1958a) tentatively placed this species under *Thanatephorus*. Talbot (1965) thought it possible that the species might belong to *Ceratobasidium* because Stahel's figure (Stahel, 1940, f.3) suggested this genus; on the other hand, Stahel's figure 4(A, B and C) shows basidia more typical of *Thanatephorus*. The author has so far been unable to obtain fruiting material of this species for study.

GENERIC SYNONYMY AND DIAGNOSIS OF THANATEPHORUS.—

> THANATEPHORUS Donk (1956a:376, 1958a: 28), Eriksson (1958c), Christiansen (1960), and Warcup and Talbot (1962:500)
> *Corticium* sect. *Hypochnoidea* Killermann (1928:136)
> *Pellicularia* Cooke *sensu* Rogers (1943) *pro parte*

Typically parasitic on plant parts in or near soil but often saprobic in soil or rotten wood, forming a rhizoctonia-state and often forming sclerotia. Fructification resupinate, pruinose-pellicular, flaky to somewhat tufted or almost hypochnoid. Hyphae wide (sometimes up to 17 μ), branching at a wide angle and often forming cruciform cells, monomitic; basal hyphae longer celled and often colored and with thickened walls; ascending hyphae shorter celled, thin-walled, barrel-shaped, bearing basidia in discontinuous clusters of small asymmetrical cymes or less typically racemes; clamp connections absent. Basidia short, barrel-shaped to subcylindrical, not uniform nor constricted about the middle, about the same diameter as the supporting hyphae or little wider; sterigmata (1-)4(-7) per basidium, straight, stout, reaching the same length as the metabasidia or longer, rarely becoming adventitiously septate. Basidiospores capable of repetition, not amyloid, smooth, hyaline, prominently apiculate. No known conidial states.

SYNONYMY AND DESCRIPTION OF THANATEPHORUS CUCUMERIS.—

THANATEPHORUS CUCUMERIS (Frank) Donk (1956*a* : 376, 1958*a* : 28-34), Christiansen (1960:68, f.48), Warcup and Talbot (1962: 500, f.3); and Flentje, Stretton and Hawn (1963:450-467)

*Hypochnus cucumeris* Frank (1883:62)

*Hypochnus filamentosus* Patouillard *apud* Patouillard and Lagerheim (1891:163)

*Hypochnus solani* Prillieux and Delacroix (1891:220)

*Corticium vagum* var. *solani* Burt *apud* Rolfs (1903:729)

*Corticium solani* (Prill. and Delacr.) Costantin and Dufour (1895:228), Bourdot and Galzin (1911:248)

*Corticium vagum sensu* Burt (1918:128, f.3*a;* 1926:295, f.3*a*) *pro parte, nec C. vagum* Berkeley and Curtis (1873), and Ullstrup (1939)

*Corticium vagum* subsp. *solani* (Prill. and Delacr.) Bourdot and Galzin (1928:242)

*Botryobasidium solani* (Prill. and Delacr.) Donk (1931:117), and Rogers (1935:18)

*Pellicularia filamentosa* (Pat.) Rogers (1943: 113, f.11) *pro parte,* Kotila (1945*a,b;* (1947), Cunningham (1953:328, f.3, 1963:45, f.8), Sims (1956:472, f.1), Boidin (1958:99), Flentje (1956:343, f.1-3), and Talbot (1958: 136, f.8)

*Pellicularia filamentosa* f.sp. *solani* (Kühn) Exner (1953:716)

*Ceratobasidium filamentosum* (Pat.) Olive (1957:431)

*Ceratobasidium solani* (Prill. and Delacr.) Pilát (1957:81)

*Rhizoctonia solani* Kühn (1858:224) *nom. anam.*

*Thanatephorus praticola* (Kotila) Flentje *apud* Flentje, Stretton and Hawn (1963:451), see also synonymy of this species, above

*Corticium microsclerotia* Weber (1951:727, 1939:565, *nom. nud.*), and see also synonymy of this species, above

Other possible synonyms are: *Corticium sasakii, Pellicularia filamentosa* f.sp. *timsii,* and *Corticium areolatum* (see above).

Fructification resupinate, creamy to grayish-white, loosely attached to the substratum, composed of arachnoid repent hyphae that give rise to thin hypochnoid or submembranous fertile patches in which clusters of basidia are produced terminally in a discontinuous hymenium formed by imperfectly cymose or racmose branching of erect hyphae. Hyphae: (4.5-)9-12(-17) $\mu$ wide, smooth, not encrusted, lacking clamp connections, often anastomosed or cruciately branched in the median part of the fructification; basal hyphae widest, with brown to subhyaline walls up to 2-3 $\mu$ thick, branching at a wide

angle with the septum in the branch hypha near its junction with the main axis; median and sub-hymenial hyphae narrower, thin walled, hyaline. Basidia (9-)14(-25) × (5-)9(-12) $\mu$ (sometimes abnormally shortened and widened), barrel-shaped to subcylindrical, without a median constriction, broadly attached, little wider than the supporting hyphae; sterigmata (5-)13.5(-25-45) $\mu$ long, up to 3(-4) $\mu$ wide at the base, somewhat divergent, straight, attenuated toward the apex, usually 4 per basidium, rarely 2, 3, or up to 7 per basidium. Spores (5-)9 (-12-14.5) × (4-)5.5(-8) $\mu$, hyaline, smooth, thin walled, not amyloid, oblong to ellipsoid and dorsally flattened, or broad ovoid and commonly widest toward the distal end, with a prominently truncate apiculus, capable of forming smaller secondary basidiospores of the same shape by repetition.

*Specimens examined.*—As *Hypochnus filamentosus*: Type, G. de Lagerheim, on *Dianthus caryophyllus,* Quito (Farlow Herbarium).

As *Hypochnus solani*: Authentic material, on *Solanum tuberosum* stem, Paris, Oct., 1891 (Herb. Mus. Paris).

As *Thanatephorus cucumeris*: N. T. Flentje, Waite Institute Isolates 48, 82, 85, 87, on soil, South Australia; Isolate 16, wheat roots, and Isolate 69, cabbage stem, South Australia.

As *Pellicularia filamentosa*: J. C. Hopkins, on *Solanum tuberosum,* Southern Rhodesia (Myc. Herb. Dept. Agric. S. Rhodesia, No. 5151); Martin, on *Amarantus paniculatus,* South Africa (National Herbarium, Pretoria, No. 41434).

As *Corticium microsclerotia* (all from Herb. Florida Agric. Expt. Stn.): Erdman West, on *Bauhinia saigonensis,* Gainesville, Florida, 1943; Geo. E. Richey, F.6621, on *Crotalaria spectabilis,* Gainesville, Florida, 1944; Weber and Kelbert, F.14869, on *Phaseolus lunatus* var. *macrocarpus,* La Crosse, Florida, 1932; other collections showing only sclerotia and vegetative hyphae.

As *Thanatephorus praticola*: N. T. Flentje, Waite Institute Isolate 42, lettuce stems, Slough, England.

STABILITY OF NOMENCLATURE. — Stability of nomenclature depends entirely on stability in taxonomic concepts, for when the classification of an organism is changed as a result of new knowledge or because different emphasis is placed on knowledge already obtained, so must the nomenclature of that organism change. A corollary of this is that nomenclature is most subject to change in those groups that attract a large number of workers.

To recognize and determine morphological species is one of the chief functions of a taxonomist, but his task is gravely complicated by the existence of microevolutionary variation within populations of a species; orthodox herbarium taxonomy has, as yet, no satisfactory way of dealing with such variants. Whatever the type of variation, no system of categories such as species or varieties can adequately express the actual form of microevolutionary change taking place (Heslop-Harrison, 1953; Gilmour,

1961). Field collections, kept in herbaria, are usually not very extensive and it is quite possible to have herbarium specimens of two well-defined "species" that in fact merely represent the extremes of a graded series discovered only when the population as a whole is carefully analyzed. Type specimens chosen more or less fortuitously from living populations are seldom fully representative of the species, and, being dead, are sometimes difficult or impossible to relate conclusively to living populations or cultures. The population is dynamic and subject to continuous change but a type specimen is static. The concept of a particular species must therefore be static too; this concept includes minor variants noted in large herbarium collections, but the cause or extent of their variation cannot be assessed from herbarium material and as soon as the variation becomes apparently discontinuous, a different species is indicated. The orthodox taxonomist recognizes that his species is largely an illusion, although a useful one, for he cannot compare organisms against a continually changing standard.

Reverting now to *Thanatephorus,* it is probable that naturally occurring field collections would show less variation than cultures derived from them in the laboratory. The generation time is greatly speeded up in the laboratory and evolutionary changes are likely to occur more rapidly. Selection pressures that would tend to keep field populations relatively stable may be absent in the laboratory, but of this we have no definite information. A further factor is that laboratory cultures are subjected to many environmental conditions that induce expression of variation (Snyder and Hansen, 1939), some of which are deliberately manipulated in the hope of striking conditions that will cause sterile mycelia to form fructifications for identification purposes. On the other hand, as Biggs (1937) has pointed out, "without a careful cultural analysis it is never permissible to regard wide morphological variations within a species as the expression of chance environmental fluctuations"; they may indicate an aggregate species that we have not the ability to analyze into its components.

Many of the experimental disciplines that give information on the causes, nature, and extent of variation cannot be usefully employed in orthodox taxonomy. It is true that they can point to variable features which would be bad to use in classification and so cause one to look more closely at morphological differences, but otherwise they suffer much the same disabilities as the use of morphological criteria alone. Experimental taxonomy can show, as can morphological taxonomy, that two organisms are likely to be of the same species, but one still has to decide whether two very similar, but not identical, organisms should be placed in the same or a different unit of classification. The final decision is quite as arbitrary as that made when two organisms are found to be morphologically very similar; the use of experimental methods may merely substitute cause for indecision. Moreover, as Heslop-Harrison (1953) and Gilmour (1961) have made quite clear, the ex-

perimental taxonomist tends to use the units of orthodox classification (e.g. a species) for the particular type of unit that he regards as important (e.g. a unit defined on an interfertility basis); often these units are very close, but essentially they cannot be equated.

Variations in *Thanatephorus* have been observed and investigated almost entirely by morphological methods and mostly in culture. Isolates are brought into culture, rarely under comparable conditions, grown for a period and compared, and the net result is a broad spectrum of variants whose morphological characters appear to overlap the characters of several species described from field collections. In our present state of knowledge, we are thus forced to regard these species as synonymous, or at least as comprising an aggregate species. But it is also possible that our laboratory procedures, rather than inducing variation, are separating field types into some of their genetic components that can exist under cultural conditions but that would not survive in nature. The aberrant forms seen in the laboratory may not necessarily be variants of the field types, but simply parts of them.

The success of Snyder and Hansen (1940) in their reclassification of *Fusarium* species was largely due to the methods they employed, namely, a process of analyzing for genetic variation and of regrouping the clones (synthesis) into larger units. This was an intellectual synthesis of the data obtained by experiment and was made possible by the fact that in *Fusarium* the macro- and microconidia are uninucleate and the technique of single-spore culturing therefore provides a ready means of analyzing a thallus for its nuclear mixtures; the resulting clones are then grouped according to the similarities common to them, instead of being separated by their dissimilarities. With *Thanatephorus,* a similar method of analysis is likely to be difficult; Flentje, Stretton, and Hawn (1963) have shown that while most basidiospores are uninucleate, some isolates may have 35% or more of binucleate basidiospores. They conclude: "It would appear that genetic studies on any isolates of *Thanatephorus* should be preceded by cytological studies to determine the type and frequency of aberration in the behaviour of nuclei in the basidia and their consequent effects on the characteristics of single-spore cultures." A different form of synthesis can perhaps be employed with *Thanatephorus,* namely, attempting to reconstitute field types by macerating together what appear to be variant or aberrant cultural types and culturing from the macerate. Such a method has been employed successfully for reconstituting the wild type from biochemical mutants of certain fungi.

Work on the mechanisms of variation in *Thanatephorus* is dealt with in another section of this volume and is without doubt of great importance in planning an attack on some of the outstanding taxonomic problems, with the reservation, however, that experimental and orthodox taxonomic units do not always coincide. At present we do not know just what our collections and isolates represent in terms of species.

With *Thanatephorus*, another important question is whether one is justified in considering several described species or strains to represent a single aggregate species when their perfect states appear to be almost identical but their imperfect states appear to show reasonably constant differences. Exner's (1953) division of *P. filamentosa* ( = *T. cucumeris*) into several *formae speciales* based on differences in the imperfect states, diseases produced, and specificity to particular parts of the hosts, is one way of providing names for some of the more stable variants or strains, provided that the differences noted are reasonably constant. This type of work should have prompted plant pathologists to a detailed investigation of the real host range, and specificity to particular parts of the hosts, of their isolates; it could well be that such isolates are not infinitely variable in this respect but can be separated into a number of pathogenic strains. We have perhaps been looking too much at differences and overlooking similarities. Rogers (1943) adopted the conservative policy of placing all strains or species with similar perfect states in synonymy under one aggregate species; the same procedure has been followed in this paper.

It is quite clear that the situation in *Thanatephorus* is analogous with that found in numerous other fungi and perhaps best expressed by Biggs (1937) in connection with her study of *Corticium coronilla* ( = *Sistotrema brinkmannii*). Biggs found that the growth varieties she described were but four of an indefinite number of similar varieties, and that *C. coronilla* "considered in the widest sense is a very complex entity made up of innumerable different and more or less well-defined strains." The practical difficulties of species identification would have been almost insuperable if such groups had been given recognition as species. Biggs concluded: "When specimens referred to a single species show a wide range of morphological variation, one may suspect that this is a collective entity. These collective species are convenient provisional groups made up of innumerable more or less constant genetic strains. Such collective species can have no fundamental systematic importance."

Davis and Heywood (1963) have stated that: "The practice of taxonomy involves making decisions on the materials and resources available. All a taxonomist's decisions are subject to revision in time . . . Taxonomy is a series of progressive states of knowledge." The genus *Thanatephorus* is likely to persist but it is almost certain that *T. cucumeris*, as comprehended in this paper, is a collective species of the kind discussed by Biggs. It is not to be expected that its nomenclature will remain stable, but rather that species or subspecific units will once again be segregated; this would be advantageous provided that they are soundly based.

ACKNOWLEDGMENTS. — I am deeply grateful to the president, committee, and members of the American Phytopathological Society for entrusting me with the preparation and presentation of this paper, and for the financial support that has enabled me to do so. My sincere thanks are due to my colleagues at the Waite Institute, especially Professor N. T. Flentje and Dr. J. H. Warcup, for providing me with many cultures for study. I am also much indebted to the directors of the following institutions for the loan of valuable specimens, so promptly sent at my request: The Farlow Herbarium; The New York Botanic Garden Herbarium; The Herbarium of the Agricultural Experiment Station, University of Florida; The Landbouwproefstation, Paramaribo, Surinam; The Museum National D'Histoire Naturelle, Paris; and The Herbarium of the Royal Botanic Gardens, Kew.

# Morphology and Cytology of Rhizoctonia Solani

EDWARD E. BUTLER and CHARLES E. BRACKER—*Department of Plant Pathology, University of California, Davis, California, and Department of Botany and Plant Pathology, Purdue University, Lafayette, Indiana.*

The difficulties encountered in the identification of *Rhizoctonia solani* arise from its complexity and the lack of information in the areas touching speciation. Morphology and cytology are no exceptions. Descriptions of color, size, and shape of hyphae and sclerotia were made by pathologists while dealing with diseases caused by *R. solani*, and though many investigators have described the gross appearance of colonies, very few have described in detail the characteristics of the mycelium in its many forms. No paper has dealt specifically with gross morphology and fewer than 10 with the morphology and behavior of organelles. Moreover, it is probable that important segments of knowledge on the morphology and cytology do not present an adequate view of the species as a whole. This is particularly true of nuclear cytology and ultrastructure, where only a few isolates represent the species.

This review is divided into three main sections: gross morphology, cytology with emphasis on the nucleus, and ultrastructure. In evaluating the literature, we have followed the concept of the species given by Duggar (1915). The vegetative state of *Hypochnus sasakii* Shirai and *R. microsclerotia* Matz are considered as synonyms of *R. solani* as suggested by Parmeter and Whitney (this vol.).

GROSS MORPHOLOGY.—*Diagnostic characteristics of vegetative hyphae.*—Hyphae that serve to identify *R. solani* are aerial, on the surface of culture media or present on external surfaces of a host (unless otherwise noted, all of our discussions of hyphae will refer to this type). In our opinion, reliable diagnosis can only be made with mycelium in pure culture. The difficulties in identifying *R. solani* from a host were discussed by Matz (1921). In evaluating the literature, it was difficult to determine the source of hyphae from which descriptions were made. Differences in hosts or growth media from which hyphae were taken for observation might affect morphology, but there are few data to support this.

Duggar (1915), in laying the groundwork for our present concept of the species, stated that young hyphal branches are inclined in the direction of growth (Fig. 4) and are ". . . invariably somewhat constricted at the point of union with the main hyphae. A septum is formed in the branch near the constriction. As hyphae mature they become uniform and rigid and branch at right angles." He did not

mention a constriction in older hyphae but in an earlier paper Duggar and Stewart (1901) stated, in connection with mature hyphae, ". . . the constriction at the place of union may not be so marked."

With respect to the morphology of young hyphae, nearly all workers who have described this vegetative state confirmed Duggar's observations (Peltier, 1916; Bourn and Jenkins, 1928a; Walker, 1928; Matsumoto, et al., 1932; Wei, 1934; Valdez, 1955). Palo (1926), working with isolates from rice, emphasized the diagnostic importance of the young vegetative hyphae. He added to Duggar's description by noting that "In some cases the young branches arise at right angles to the main hyphae but they later bend toward the direction of the growth of the main filaments."

Although Duggar (1915) stated that branches of mature hyphae arise at right angles to the main hyphae, this is not the only pattern characteristic of *R. solani*. Duggar furnished two illustrations of vegetative hyphae, one showing nearly all branches at right angles; the other at acute angles approaching 45°. Probably both of his illustrations were of mature hyphae and not newly formed young hyphae. Other workers described or illustrated mature branches inclined approximately 45° to the main hyphae (Peltier, 1916; Rosenbaum and Shapovalov, 1917; Matsumoto, 1921, 1934; Matz, 1921; Palo, 1926; Flentje, et al., 1961; Flentje, Stretton, and Hawn, 1963). Thus, there is good evidence that in mature hyphae, branches arise at right and at acute angles, i.e. near 45° to the main branch.

Prior to this symposium, we recorded the type of branching of mature hyphae in 20 cultures (potato-dextrose agar [PDA], 25°C, 7 days) from our collection. Of 2,000 branches, about half arose at right angles (Figs. 2, 3), and half at acute angles (Fig. 1).

The evidence for a septum in the branch near the point of origin is unanimous, but the presence of septa in the main hyphae near a branch is variable. Flentje (1956) stated that there are septa in the main hyphae immediately on either side of the branch. Duggar (1915) showed a single septum or none in the main hyphae near a branch. If other septa occurred, they were far removed or not yet formed. A number of workers illustrated a single septum in the mature main hyphae near a branch (Duggar, 1915; Rosenbaum and Shapovalov, 1917; Palo, 1926; Bourn and Jenkins, 1928a; Matsumoto,

1934) and it appears that this is the most common pattern of septum location in connection with branches.

Secondary septa (Fig. 1) were described by Flentje, Stretton, and Hawn (1963). These are formed in older hyphae and ". . . were easily distinguished from the primary septa, the latter being thicker at the junction with the cell wall, whereas the secondary septa joined the cell walls without increase in thickness and were thinner than the original septa." Little is known about the prevalence of secondary septa; it is doubtful that they have taxonomic value.

There is no unanimous agreement on the occurrence of a constriction in hyphal branches at the junction with main hyphae, but it is commonly illustrated and described and probably is characteristic of most isolates of *R. solani* (Figs. 1-3). Duggar (1915) did not emphasize the constriction but in one of his illustrations two of the four branches shown are constricted at the point of origin. His drawings show only a slight constriction, as do illustrations of Rolfs (1904), Peltier (1916), and Palo (1926). Matsumoto, et al. (1932), in discussing the point of branching in *H. sasakii*, stated ". . . but a constriction at this point is not so remarkable as in the case of *Rhizoctonia solani*." On the other hand, Flentje (1956) noted that the point of attachment of a branch is restricted, but the base of this branch is swollen and cut off by a septum. Daniels (1963) agreed with Flentje (1956), as did Bourn and Jenkins (1928a) who showed deep constrictions in hyphal branches near the point of origin.

*Runner hyphae.*—Beyond the characters previously described, there is little to distinguish morphologically the hyphae of *R. solani* from those of other fungi. Duggar (1915) remarked that as hyphae mature they become "uniform and rigid." It seems likely that he was referring to more or less parallel thick walls as viewed with a microscope. Whereas there is agreement that mature hyphae have thickened walls, uniformity in hyphal diameter could not in any sense be considered unique to *R. solani*. We have observed many cultures where the form of aerial vegetative hyphae was mostly irregular, i.e. undulating, swollen, constricted, or with erratic dimensions.

On the other hand, many isolates produce hyphae with relatively thick and parallel walls that may run for several millimeters on the surface of a host (Flentje, et al., 1961; Daniels, 1963), in soil (Warcup, 1957), or in culture. Branches from these hyphae, which arise sparingly, may give rise to sclerotia (Figs. 11, 12) or infection cushions (Flentje, et al., 1961). Such hyphae have been called runner hyphae (Daniels, 1963), running hyphae (Flentje, 1956; Saksena, 1960), long hyphae (Frederiksen, et al., 1938), or distributive hyphae (Samuel and Garrett, 1932).

*Dimensions of vegetative hyphae.*—The diameter of vegetative hyphae and dimensions of cells in vegetative hyphae are so similar to other fungi and so variable that they probably have little diagnostic

value. Even if these dimensions were of taxonomic value, we should have to seek new measurements because, in most cases, neither the age of hyphae, the substrate from which they were taken, nor the location of the hyphae in relation to the substrate, etc., are given. In this regard, Peltier (1916) stated: "Cells varied in size at different ages and on different media." Wei (1934) gave the range in diameter of hyphae as 4-6 $\mu$ on PDA and 6-13 $\mu$ on Hopkins synthetic agar. Palo (1926) measured hyphae on eight different media. His data show that the substrate has a profound influence on cell dimensions.

The diameter of hyphae ranges from approximately 3.0 $\mu$ (Rolfs, 1902; Peltier, 1916) to 17.0 $\mu$ (Warcup and Talbot, 1962), but most are between these extremes: 8-12 $\mu$ (Duggar, 1915), 6-8 $\mu$ (Matz, 1921), 5.29-14.3 $\mu$ (Valdez, 1955), 6.2-9.5 $\mu$ (Gratz, 1925), 4.3-8.0 $\mu$ (Hansen, 1963), 7-9 $\mu$ (Frederiksen, et al., 1938), 6.0-12.5 $\mu$ (Flentje and Saksena, 1957), 6.5-10 $\mu$ (Saksena, 1960), 6-12 $\mu$ (Flentje, 1956), and 5.5-12.0 $\mu$ (Bourn and Jenkins, 1928a).

A few investigators have given cell length. Duggar (1915) reported that septa in young hyphae occurred at intervals of 100-200 $\mu$. Frederiksen, et al. (1938) stated that cells were often 100 $\mu$ or more long. Other reports are as follows: 100-200 $\mu$ (Buller, 1933), 50-225 $\mu$ (Bourn and Jenkins, 1928a), 15.8-250 $\mu$ (Valdez, 1955), 22-182 $\mu$ (Wei, 1934), 65-180 $\mu$ (Peltier, 1916) and 54-216 $\mu$ (Townsend and Willetts, 1954). Flentje, Stretton, and Hawn (1963) reported the length of tip cells and cells other than tip cells less than 1 day old as 225-250 $\mu$, while cells 7 days old were 55-110 $\mu$. The smaller size of older cells was due to the formation of secondary septa.

There seems to be no relationship between source of culture and diameter or length of hyphae (Peltier, 1916).

*Color of vegetative hyphae.*—Young hyphae from near the periphery of a colony or on or in a host are hyaline (Duggar, 1915; Peltier, 1916; Matz, 1921; Palo, 1926; Walker, 1928; Valdez, 1955). Hyphae within the substrate tend to remain hyaline, as do hyphae within a host (Abdel-Salam, 1933; Ramakrishnan, 1960), but aerial or surface hyphae may become brown. Duggar (1915) referred to mature hyphae as deeply colored but there is no doubt that he meant brown. Peltier (1916), Bourn and Jenkins (1928a), and Walker (1928) noted that as hyphae age, they become first yellowish and then brown. Matz (1921) indicated that the mycelium changes from transparent to brown but that *R. solani* produces a yellow pigment. Kotila (1947) described older hyphae as light yellowish brown. Others stated that mature hyphae are brown: (Palo, 1926; Wei, 1934; Frederiksen, et al., 1938; Valdez, 1955; Saksena, 1960; Saksena and Vaartaja, 1961). According to Briton-Jones (1923), hyphae in 7-week-old cultures are a mixture of hyaline and brown. Thus, the color of aerial hyphae in cultures of *R. solani* may be hyaline, yellowish, or some shade of brown. Brown is a diagnostic color for *R. solani* and cultures lacking or having no history of at least some

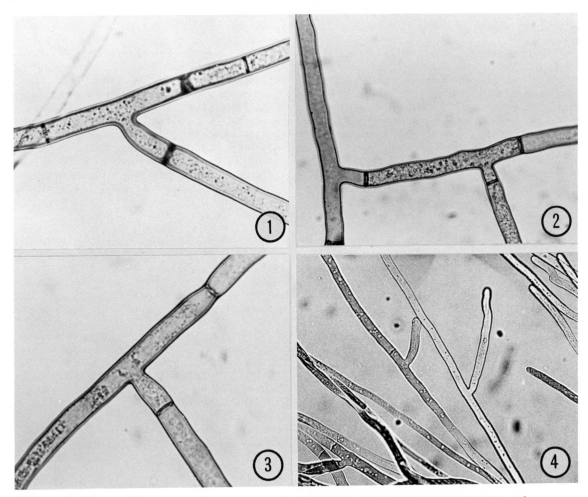

*Figs. 1-3.* Hyphae of *R. solani* from 2-week-old cultures on PDA at 27°C × 750, 620, and 875 respectively.

*Fig. 4.* Hyphae from periphery of young colony on PDA showing characteristic branching. × 275.

brown aerial hyphae are likely to be some other species.

Our observation indicates that the hyphal pigment is in the wall, but no studies have been made to demonstrate either its location or chemical nature.

*Monilioid cells.*—In addition to ordinary vegetative hyphae, which have a length-to-width ratio generally greater than 5:1, *R. solani* produces simple or branched chains of short broad cells with a length-to-width ratio of approximately 1-3:1. These cells may be hyaline or brown, barrel-shaped, pyriform, irregular, or lobate. The morphology of six isolates is shown in Figs. 5-10 and by Duggar (1915), Peltier (1916), Rosenbaum and Shapovalov (1917), Matsumoto (1921), Dodge and Stevens (1924), and Palo (1926).

Monilioid cells have been called doliform cells, barrel-shaped cells, short cells, sclerotial cells, and chlamydospores. The formation of new monilioid cells occurs differently from that of new cells in vegetative hyphae. They arise as buds or blown-out ends of preexisting cells (Figs. 5-10). The diameter of the point of origin of new cells is smaller than the diameter of the cell from which it arises. This provides a constriction between cells. The septum, if present, is formed at the constriction. Monilioid cells may develop relatively thick walls but do not necessarily differ from other hyphal cells in this respect.

Monilioid cells have the appearance of conidia, but there is no evidence that they serve this function. Saksena and Vaartaja (1961) referred to them as chlamydospores and suggested that their morphology is characteristic for each species of *Rhizoctonia*. This is possible, but the range in shapes and sizes of monilioid cells (Figs. 5-10) suggest that they are of limited taxonomic value.

Monilioid cells form on branches that arise from long hyphae, as illustrated by Duggar (1915). Their size is variable: 20-22 × 30-35 $\mu$ (Townsend and Willetts, 1954), 10-15 × 20-25 × 27-40 $\mu$ (Saksena and Vaartaja, 1961), 11 × 19 to 23 × 32 $\mu$ (Gratz, 1925), and 14 × 38 to 27 × 40 $\mu$ (Matsumoto, 1921).

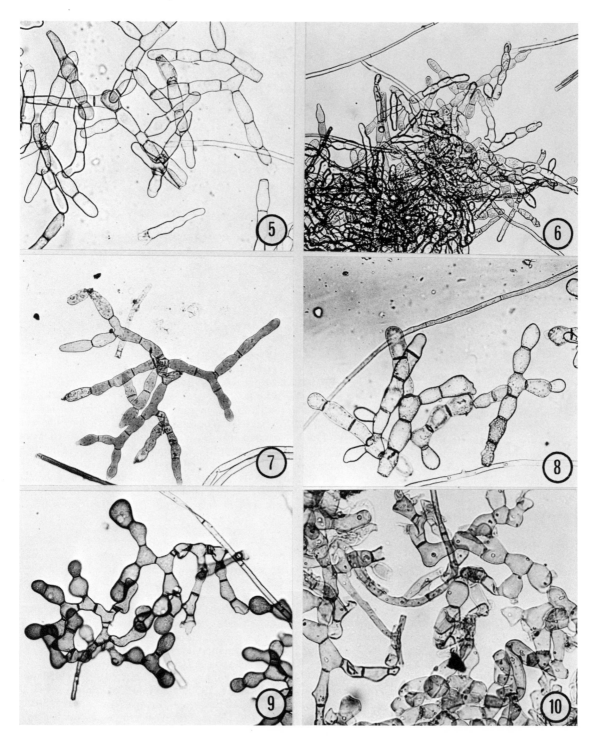

Figs. 5-10. Monilioid cells from 6 different isolates of *R. solani*. × 225.

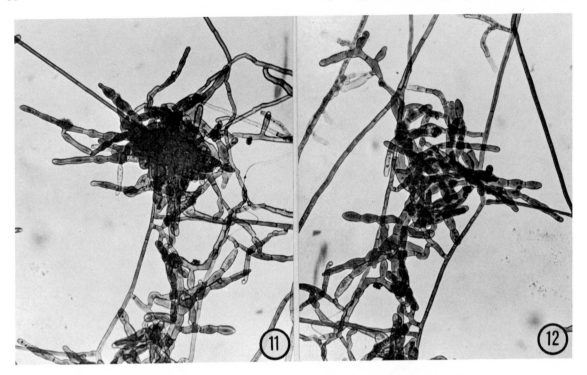

*Figs. 11-12.*   Early stages in sclerotial development of *R. solani.* × 150.

Chains of monilioid cells are formed in various patterns, mainly on or above the surface of a substrate, but also within host tissue. They may be few in number and scattered, or form loose to semicompact masses of varying sizes, or they may be aggregated into sclerotia (Figs. 11-12).

SCLEROTIA. — *Structure.* — Sclerotia are composed principally of compact masses of monilioid cells (Rolfs, 1904; Duggar, 1915; Peltier, 1916; Palo, 1926). Most observations agree with this, but Saksena and Vaartaja (1961) stated: "Moreover, sclerotia are not always made up of chlamydospores . . ." (monilioid cells). Their contention is not documented, but our observations confirm this statement. On sterile soil, we have observed sclerotia composed solely of undifferentiated hyphae, but further study would be required to establish this mode of formation for the species. Many sclerotia contain a few ordinary vegetative hyphae. Townsend and Willets (1954), in describing types of sclerotial development, referred to *R. solani* as the "loose type" and stated that, "In the formation of the sclerotial initials there is no definite pattern of organization of the hyphae, and the resulting sclerotium is very loosely constructed. . . . The mature sclerotium is much less closely interwoven than that of other fungi studied and there are no well-defined zones . . . toward the margin the hyphae are more loosely arranged." Others confirmed the homogeneous structure of the sclerotium: (Duggar, 1915; Peltier, 1916; Matz, 1921; Walker, 1928).

We have sectioned sclerotia of several different isolates of *R. solani.* The structure near the centers of sclerotia from four different isolates is shown in Figs. 13-16. The chains of cells are loose to compact and free from one another. We also confirmed this structure by examining sclerotia formed on PDA, subsequently boiled for 20 minutes in a potassium hydroxide solution, and dissected with fine glass needles. Intact chains of monilioid cells were also recovered in this manner from sclerotia formed on potato tubers.

*Color, shape, size, and topography.* — The basic color of mature sclerotia is brown. Truly black sclerotia are not characteristic of *R. solani.* Numerous shades of brown have been employed to describe them: deep chestnut brown (Duggar, 1915); dark brown (Flentje, 1956); grayish brown to cinnamon-brown (Gratz, 1925); natal brown (Houston, 1945); bone brown to light seal brown (Matsumoto, 1934); Mars brown or chocolate (Matsumoto, 1921), etc. Young sclerotia are at first white, turning dark with maturity (Peltier, 1916; Matz, 1921; Matsumoto, 1934; Weber, 1939; Valdez, 1955). Sclerotia may be darker near the center than on the surface; Briton-Jones (1923) described sclerotia of one isolate as buff-colored outside and brown inside.

Although it is possible to characterize the color of sclerotia, the shape is not definable within narrow limits either on a host or in culture. Duggar (1915) described them from potato as irregular with a smooth surface. They also have been characterized as irregular by Ullstrup (1936), Frederiksen, et al. (1938), and Flentje (1956). Houston (1945) described sclerotia as globose with irregular surface or as slightly raised areas radiating from the center of

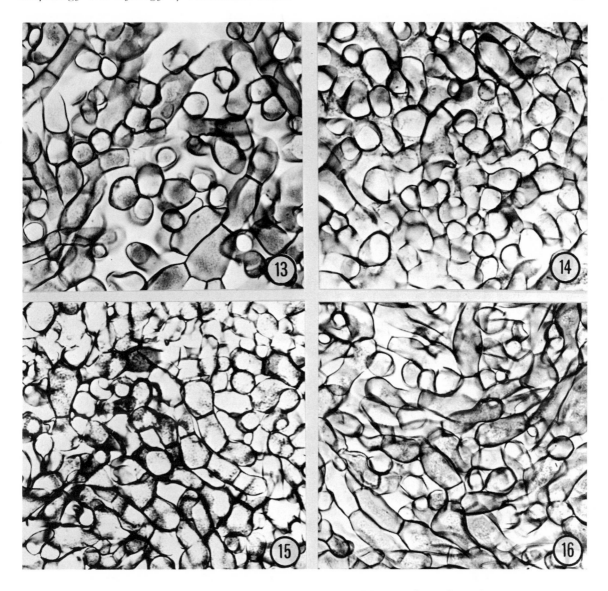

*Figs. 13-16.* Vertical cross sections through sclerotia of 4 different isolates of *R. solani.* × 250.

a culture. According to Matz (1921), sclerotia are irregular and flat, tending to elongation. Valdez (1955) observed them in the form of crusts on PDA, and Weber (1939) described them as subglobose. Peltier (1916) illustrated sclerotia from soil culture as irregularly globose with pitted surfaces. Ryker (1939) described them as flattened on the bottom and rounded on top.

The size of sclerotia also varies. Frederiksen, et al. (1938) stated that sclerotia on potato tubers range from about 1 mm diameter to crusts or scales over the entire surface of the tuber. In culture, they range from the size of a pinhead to 5-6 mm in diameter, but through the confluence of several sclerotia, a crust of several cm may be formed. Duggar (1915) gave the size as "scarcely visible" to 1-2 cm in diameter; Gratz (1925), minute to large compact masses; Peltier (1916), on potato, 1-5 mm, on

carnation, 5-8 mm; Rolfs (1902), from a speck to 0.5 inch or more.

Sclerotia are generally superficial, but they may also be formed inside host tissue (Cristou, 1962). According to Duggar (1915), "On the majority of hosts, however, sclerotia formation is relatively rare." On the other hand, Peltier (1916) stated that sclerotial formation in nature is rather common in many hosts, potato tubers being the best known. Sclerotia are illustrated by Rolfs (1902), Morse and Shapovalov (1914), Peltier (1916), Chamberlain (1931), Pittman (1937), and Frederiksen, et al. (1938).

*Germination of sclerotia.* — The mechanism of sclerotium germination has not been reported, perhaps because it is "common knowledge" that sclerotia germinate. But whether all cells of a sclerotium, or only certain cells, can give rise to new thalli is unknown. There is ample evidence that the monilioid cells, of which sclerotia are comprised, do germinate. Duggar (1915) did not mention germination, but he stated that chains of monilioid cells readily break up into short lengths or single cells. Peltier (1916), Palo (1926), and Walker (1928) described germination of such fragments. According to Peltier (1916), "Germination generally takes place by the protrusion of a tube through the septum of a cell where it has broken away from an adjacent cell. In some cases the hyphae of the germinating cells pass through adjacent cells, which are apparently empty." Bourn and Jenkins (1928a) illustrated a germ tube arising from the apex of a terminal monilioid cell.

*Morphology of the colony.*—The basic color of the colony of *R. solani* is brown. Within the substrate, the mycelium is generally hyaline, irregular in shape, and not unlike the mycelium of many other septate fungi. Substrate mycelium may be sparse or consist of a thick stroma-like layer. The aerial portion, i.e. on or above the substrate, may consist of hyaline or brown runner hyphae, irregular hyaline or brown hyphae, large or small masses of monilioid cells and sclerotia. The occurrence and arrangements of these components are many. Detailed descriptions of isolates from different hosts are given by Gratz (1925), Ullstrup (1936), Exner (1953), and Flentje and Saksena (1957). For excellent photographs of colonies, see Kernkamp, et al. (1952) and Flentje and Stretton (1964).

Houston (1945), Richter and Schneider (1953), and Luttrell (1962) attempted to place large collections of cultures from various hosts into cultural groups. Houston found that isolates obtained from common, commercially grown crops in California could readily be placed in one of three groups. Type A produced, on the surface of the medium, a heavy stroma-like layer which was white at first, and later pale gray. On the surface of this layer were light cinnamon drab strands of hyphae. Sclerotia were natal brown and formed as slightly raised areas radiating from the inoculum. Aerial mycelium was sparse. Type B cultures produced no stroma-like layer, few globose natal-brown sclerotia with irregular surfaces and with little aerial mycelium. Type C cultures were characterized by a slight stroma-like layer which was at first vinaceous buff, but turned wood-brown with age. Sclerotia were natal brown, globose, with a very irregular surface. The three groups also differed in growth rate. Type A was highly pathogenic on several hosts, types B and C weakly pathogenic. According to Flentje (1956), Houston's type A cultures were probably *Pellicularia praticola* (Kotila) Flentje, now recognized as the perfect state of *R. solani* Kühn. Luttrell (1962) placed isolates from forage legumes and grasses into four groups, but he experienced some difficulties. He made the following comments: "Although a classification of the isolates into four cultural types was made, it was possible to do so only by recognizing small groups of extreme forms (types A, B, C) and

leaving a large central group of variable forms (type D) that integrated with all other types." Luttrell's type A was probably similar to Houston's type A.

Richter and Schneider (1953) classified cultures from different hosts into six groups based on colony morphology and the capacity of the hyphae of isolates to anastomose. There was a positive correlation between anastomosis groups and colony morphology. Their groups were based on essentially the same characteristics employed by Houston (1945), but it is doubtful that many of their cultures were *R. solani*, and the validity of their groupings is questionable. In their group IV, the color of sclerotia was yellowish to cream, and their group III lacked sclerotia. In addition, they stated that the mean diameter of the hyphae differed for each group.

Although attempts to classify cultures of *R. solani* into groups may serve as a record of culture types, the great variability in this species makes the task difficult and not practical.

CYTOLOGY AS VIEWED WITH THE LIGHT MICROSCOPE.—Although there are a few scattered observations on the internal gross appearance of the hyphae of *R. solani*, there is little critical work at the level of the light microscope on their extranuclear contents. The contents of young hyphae have been described as "vacuolate" (Duggar, 1915; Palo, 1926), granular or vacuolate (Matsumoto, et al., 1932), "filled with granular cytoplasm" (Matz, 1921), and filled with protoplasm (Frederiksen, et al., 1938). Duggar and Stewart (1901) noted that young hyphae are often strongly vacuolate, but later they usually become uniformly granular. Buller (1933) stated that the cytoplasm ". . . is very faintly clouded with fine almost imperceptible particles."

According to Matz (1921) and Palo (1926), older hyphae become vacuolate and finally empty. Frederiksen, et al. (1938) concurred with the observation that vacuoles are formed in older hyphae. Buller (1933) described the enlargement of vacuoles in hyphae that had ceased to grow and observed the flow of protoplasm in hyphae. Matsumoto (1934) showed the monilioid cells of *R. solani* to be densely protoplasmic while Bourn and Jenkins (1928a) illustrated them to be highly vacuolate. Valdez (1955) and Walker (1928) described some cells of a sclerotium to be without contents.

We have observed the young and old hyphae of several isolates from different hosts. Young tip cells in rapidly growing cultures are densely protoplasmic without conspicuous vacuoles, but large vacuoles may appear in young cells other than the tip. Mature aerial hyphae may appear to lack cell contents, to possess many large vacuoles, or to be densely protoplasmic without vacuoles (Fig. 2).

From the foregoing descriptions, the occurrence of vacuoles appears to be without a pattern, but systematic observations have not been recorded.

According to Dodge and Stevens (1924) and Frederiksen, et al. (1938), monilioid cells of sclerotia contain oil globules.

*The nucleus.*—Fewer than 12 research papers deal with the nuclear cytology of *R. solani*. They agree that the cells of the vegetative hyphae are multinucleate, but there is great variation in the number of nuclei per cell. Sanford and Skoropad (1955) counted the numbers of nuclei in tip cells, "Y"-shaped cells, and unbranched hyphae other than tip cells. Tip cells generally contained 4-8 nuclei, with a range of 2-15; "Y" cells mostly 7-11, with a range of 2-25; and straight hyphae mostly 6-11, with a range of 3-19. They observed fewer nuclei in older cells and no nuclei in older yellowish hyphae. Ito and Kontani (1952), on the other hand, illustrated three of what appear to be mature runner-type hyphae with nine, eight, and ten nuclei, respectively. Hawn and Vanterpool (1953) show two tip cells each with eight nuclei, and two older cells each with two nuclei. Saksena (1961 *b*) gives the number of nuclei per vegetative cell as 4-25, but he does not specify the type of cell.

Flentje, Stretton, and Hawn (1963) recorded the numbers of nuclei in cells of different ages and types. Tip cells mostly contained 5-7, with a range of 3-9; cells other than tip cells and less than a day old contained mostly 5-8, with a range of 2-12; cells 7-10 days old, mostly 2-4, with a range of 2-8. The reduced number of nuclei in older cells was due to the formation of secondary septa reducing the length of the cell. The number of nuclei in monilioid cells was similar to that of tip cells. Hawn and Vanterpool (1953) and Flentje, Stretton, and Hawn (1963) observed the nuclear condition of cells connected with the sexual stage. According to Flentje, Stretton, and Hawn, (1963), cells destined to become basidia (prebasidial cells) become binucleate by pairing of nuclei and formation of septa. "In the terminal binucleate cells the two haploid nuclei fuse to form a diploid, which then divides by meiosis to give four haploid nuclei." This is not always the pattern and the exceptions are described by them in some detail. When the sterigma reached half or more of its final length, the elongated nucleus moved into it and eventually migrated into the developing spore. Similar nuclear behavior is described by Hawn and Vanterpool (1953). In the early stages of basidiospore germination, they found that the spore is uninucleate, although the binucleate condition has also been observed.

Nuclei in living hyphae have been observed with the aid of the phase-contrast microscope by Saksena (1961*b*), Flentje, Stretton, and Hawn (1963), and Shatla and Sinclair (1966). Flentje, Stretton, and Hawn reported that ". . . nuclei measured 3.1 by 2.0 $\mu$ and were oval, but readily changed shape as they moved in the cell. A circular dark patch 1.2 by 1.2 $\mu$, possibly the nucleolus, was obvious within the white cloudy mass of each nucleus." Saksena (1961*b*) observed the nuclei in hyphae, germ tubes, and basidiospores as spherical or oval bodies. The interdivisional nucleus contained an optically dense central body, the nucleolus, around which was a less dense "halo." He found that the nucleolus rotates slowly within the halo and the entire nucleus shows

slow movements. He also observed nuclear division with the aid of the phase-contrast microscope and described it in some detail. Shatla and Sinclair (1966) described resting nuclei of living vegetative cells as spherical, delimited by a nuclear membrane, and with an optically dense central body, a nucleolus. They also observed the nucleolar halo described by Saksena (1961*b*) and nuclear movement within the cell.

The morphology of the stained nucleus has been recorded by Fukano (1932), Hawn and Vanterpool (1953), Sanford and Skoropad (1955), Saksena (1961*b*), Flentje, Stretton, and Hawn (1963), and Shatla and Sinclair (1966), but there is disagreement on nuclear structure and behavior. Nevertheless, chromosomes have been described and illustrated for *R. solani*. Hawn and Vanterpool (1953) stated: "There appear to be 12 chromosomes in the diploid nucleus in the basidium." According to Flentje, Stretton, and Hawn (1963), ". . . The chromosomes were clearest during meiosis and there appeared to be six pairs." Shatla and Sinclair (1964), on the other hand, propose four for the haploid chromosome number at metaphase. Saksena (1961*b*) would not agree that chromosomes, in the classical sense, exist in *R. solani*, and in place of typical mitosis, he described nuclear division as an elongation, constriction, and separation of the two parts without aid of a metaphase plate or spindle fibers. Shatla and Sinclair (1964) disagree with Saksena but their evidence does not support their contention.

Overall, the evidence for the occurrence of discrete chromosomes in vegetative cells of *R. solani* is inconclusive. We hope that with improved techniques, this problem will be clarified in the near future.

ULTRASTRUCTURE.—Structures recognizable in *R. solani* at the subcellular level include nuclei, mitochondria, vacuoles, endoplasmic reticulum (ER), ribosomes, lipid inclusions, probable glycogen granules, microtubules, lomasomes, cytoplasmic vesicles, hyphal wall, and the components of the septal pore apparatus. Protoplasmic organization appears different depending on the age of hyphae (Figs. 17-19). Previous studies by Bracker and Butler (1963, 1964) and Shatla (1965), as well as some new information presented here, characterize the internal organization of five isolates of *R. solani*. Electron micrographs in this paper are of specimens cultured on PDA, killed with potassium permanganate ($KMnO_4$) and embedded in Araldite epoxy resin (Bracker and Butler, 1963).

*Membranes.*—As in all eukaryotic cells, membranes provide the major element of protoplasmic compartmentalization and structural organization. Although there may be differences between membranes of different cell components, the pattern of membrane structure is basically the same. In low magnification electron micrographs, membranes appear as a single dark line (Fig. 30). At higher magnification, membranes are resolved into two electron-dense zones separated by an electron-transparent region (Fig.

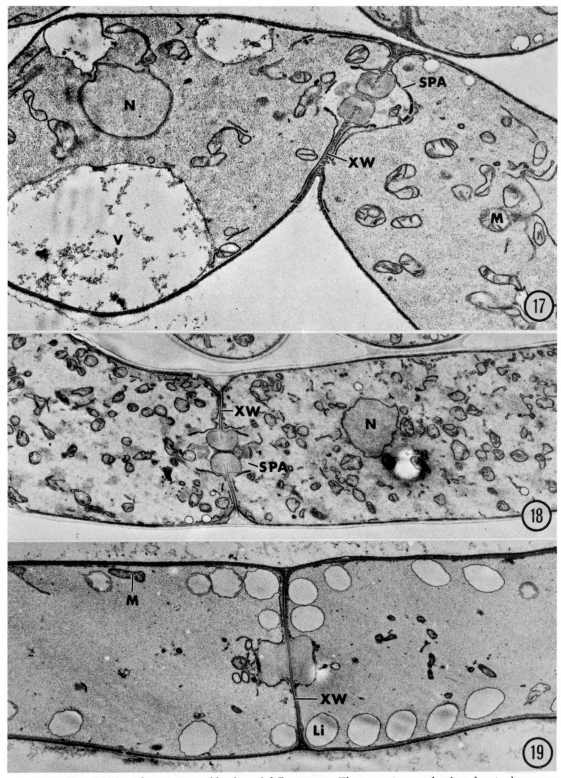

*Figs. 17-19.* Ultrastructure of hyphae of different ages. Electron micrographs show longitudinal sections including septa. *Fig. 17.* Vacuolate hypha back from hyphal tips in region 1-12 hr old. Septal pore apparatus is well developed. × 13,000. *Fig. 18.* Mature hypha from region of hypha approximately 24 hr old. × 6,500. *Fig. 19.* Aged hypha from 20-day-old colony. Section is nonmedian and does not include septal pore. × 8,200. Cross wall, *XW*; septal pore apparatus, *SPA*; lateral hyphal wall, *LW*; mitochondrion, *M*; nucleus, *N*; vacuole, *V*; lipid inclusion, *Li.*

31). The total thickness of membranes may vary from 60-100 A with the plasma membrane and tonoplast being thicker than membranes of the nuclear envelope, ER, or mitochondria.

*Mitochondria.* — Mitochondria of *R. solani* are typical in their basic organization, but vary somewhat in their form. The ultrastructural organization of the mitochondria is very much like that shown in other Basidiomycetes (Girbardt, 1961; Moore and McAlear, 1963; Wells, 1964; Giesy and Day, 1965; Wilsenach and Kessel, 1965). Structurally limited by an outer membrane, the mitochondria possess an inner membrane with infoldings called cristae. The cristae are platelike (Fig. 20) and sparsely distributed around the inner periphery of the mitochondria.

Mitochondria are typically polymorphic in *R. solani* as in other fungi and higher plants. In hyphal tips and young mycelium of living cultures of *R. solani*, many mitochondria appear in the form of long rods (up to 7 $\mu$) when viewed with phase contrast. In older hyphae, the mitochondria are shorter, often approaching a spheroid form. In living hyphae, mitochondria exhibit a dynamic pleomorphism, changing form from moment to moment. Because of the thinness (400-900 A) of sections, the image is essentially two-dimensional. Thus, a rod-shaped mitochondrion is likely to appear oval or round depending on the plane of the section. Likewise, portions of a long curved mitochondrion may be intercepted more than once by the same section giving the initial impression of two or more mitochondria. The doughnut-shaped mitochondrion ($M_1$) and the dumbbell-shaped mitochondrion ($M_2$) in Fig. 20 are interpreted as perpendicular sections through cup-shaped mitochondria. The narrow bridge between the swollen ends ($M_1$, Fig. 20) is without cristae. In *R. solani*, these cup-shaped mitochondria occur in young hyphae.

*Nuclei.*—Although nuclei are difficult to detect in living hyphae, they are readily apparent in thin sections. After $KMnO_4$ fixation, the nucleoplasm may appear variegated but it is most commonly seen as a homogeneous matrix (see Figs. 17, 27, 28). The nuclear envelope has infrequent pores and connections to the endoplasmic reticulum. Dividing nuclei have not been observed in *R. solani* with the electron microscope, but we have seen nuclei joined by continuous nuclear envelopes (Fig. 27).

*Endoplasmic reticulum (ER).* — This is the most variable component in *R. solani* hyphae. Its basic construction is that of two membranes, approximately parallel, forming a closed cisternal system. It is commonly oriented beside the cross walls (Figs. 17, 34, 37) where it is continuous with the septal pore cap (Figs. 17, 18, 37, 38, 47). Other strands of ER ramify throughout the cytoplasm. Various configurations have been observed, from continuous sheets to porous cisternae and, less commonly, as a tubular system. The ER may form coiled whorls (Fig. 22) or lamellar stacks (Fig. 23) as well as single unas-sociated cisternae. Elaborate configurations that may be continuous with simpler forms (Figs. 21, 24) also occur. Thus, no single ER form is typical of *R. solani*.

In transverse section, porous ER may appear to be discontinuous or vesicular but in tangential section (Fig. 22), it is more evident that the ER consists of a fenestrated sheet. Vesiculation of the ER may occur, but requires careful analysis of serial sections for verification.

Endoplasmic reticulum is plentiful in young hyphae but less so in the older mycelium or in sclerotia. In the aged cells, fenestrated cisternae or elaborate configurations are infrequent and the ER appears primarily as small continuous sheets. In young hyphae, the ER is often adjacent to the cross walls. In older mycelium and sclerotia, it is fragmented or absent, particularly near the septa. In old hyphae, the clear continuity between the cap and the ER is also absent (Figs. 43, 46).

*Golgi apparatus.*—We have not observed a Golgi apparatus in *R. solani*. Golgi apparatus have been reported in some fungi (Moore, 1965a; Grove,, et al., 1966), most of these consisting of perinuclear assemblages of cisternae. The Golgi apparatus is a cell component consisting of interassociated dictyosomes (Mollenhauer and Morré, 1966). The identification of dictyosomes rests on a combination of morphological criteria that can be judged by electron microscopy (Grove, et al., 1966; Mollenhauer and Morré, 1966). Using these criteria, the lamellar stacks of double membranes in hyphae of *R. solani* (Fig. 24) are not considered to be dictyosomes. Similar structures were observed by Wells (1964) in *Exidia nucleata* (Schw.) Burt. Evidence suggests that these lamellar membrane complexes represent one of the many ER configurations found in fungous protoplasts. However, this does not necessarily preclude the existence of a functional equivalent for the Golgi apparatus (Mollenhauer and Morré, 1966) in *R. solani*.

*Lomasomes.* — Lomasomes (Moore and McAlear, 1961) occur in the thallus of *R. solani*. We have observed them adjacent to lateral walls and cross walls as well as on the septal swelling (Fig. 36), but they are not regularly occurring structures.

*Vacuoles.* — These are found against the wall in cells back from the hyphal tips (Fig. 17) as reported by Buller (1933). During preparation for the electron microscope these vacuoles may be lost through shrinkage. Occasional stellate vacuoles are interpreted as fixation artifacts rather than representing the true form of hyphal vacuoles. Vacuoles in living hyphae appear fully distended, without corners or irregular projections. When vacuoles are well preserved, they retain a smooth outline and may contain small amounts of flocculent electron-dense material (Fig. 17).

*Cytoplasm.*—The ground cytoplasm appears differently depending on the preparative methods employed. In living hyphae, viewed with phase contrast, it is a clear liquid with organelles suspended in it. After osmic acid ($OsO_4$) or aldehyde plus $OsO_4$ fixation, ribosomes are a prominent structural feature. Ribosomes may be attached to the surface of ER and nuclear envelope, but most are suspended in the cytoplasmic matrix. Dark-staining rosette particles, resembling glycogen granules (Fawcett, 1966), are conspicuous. In $KMnO_4$-fixed hyphae, ribosomes are not preserved and the ground cytoplasm appears a homogeneous granular matrix (Figs. 17, 19, 20, 37). In older hyphae, some regions of the cytoplasm are more electron-dense than others. These electron-dense regions surround organelles and intracytoplasmic membranes (Fig. 26). They are interpreted as a condensed hyaloplasmic phase and may differ from the rest of the cytoplasm by their physical state.

*Lipid inclusions.* — In old hyphae and monilioid cells of sclerotia, lipids are stored as spherical inclusions. After $KMnO_4$ fixation, these lipid bodies are electron-transparent and seem to have a limiting membrane (Figs. 19, 45, 52). High-magnification views substantiate that this is not a true membrane, as is found around organelles, but is a densely staining boundary layer. After $OsO_4$ fixation, the inclusions are homogeneous electron-dense bodies lacking a limiting membrane. They thus conform to published descriptions of lipid inclusions in other cells (Frey-Wyssling, et al., 1963; Wells, 1964; Fawcett, 1966). In *R. solani*, they are located predominantly around the periphery of the protoplast (Figs. 19, 52).

*Hyphal wall.*—Hyphae are coated by an amorphous layer lying on the outer surface of the wall (Figs. 17, 22, 26). This layer may be partially sloughed during specimen preparation (Bracker and Butler, 1963), but it remains intact as an intercellular coating between sclerotial cells (Fig. 49).

The wall near the hyphal tip consists of a single layer about 80 A thick (Fig. 30). With increasing age, the wall thickens and additional layering is detected (Fig. 22). Wall thickening is thought to occur by centripetal deposition. This has been demonstrated for the cross walls of *R. solani* (Bracker and Butler, 1963). The walls of some hyphae undergo extensive secondary thickening (Figs. 29, 32), becoming more than $1 \mu$ thick. These are the brown hyphae observed by light microscopy. The cross wall is composed of two electron-dense plates. The plates are continuous with the lateral wall (Figs. 33, 34) and may contain from one to several lamellae, depending on their age and stage of development.

*Septal pore apparatus.*—The electron microscope has been particularly effective in resolving the structure of septa in *R. solani*. The septal pore apparatus of *R. solani* confirms its basidiomycetous affinities even when the perfect state is lacking (Bracker and Butler, 1963). Moore and McAlear's (1962) demonstration of an elaborate pore apparatus, which they termed the dolipore septum, has been confirmed re-

peatedly for Basidiomycetes (Girbardt, 1961; Beneke, 1963; Bracker and Butler, 1963, 1964; Wells, 1964; Berliner and Duff, 1965; Geisy and Day, 1965; Wilsenbach and Kessel, 1965). Exceptions to this general rule are the Ustilaginales and the Uredinales where the dolipore septum has not been demonstrated (Moore, 1965*b*). The more simple perforate septum is characteristic of Ascomycetes.

First recognized by Buller (1933) as hemispherical pads on either side of the pore, the septal pore apparatus of *R. solani* has been characterized in greater detail by electron microscopy (Bracker and Butler, 1963, 1964). An amorphous annular swelling occurs near the center of the septum and forms the boundary of the septal pore (Figs. 17-19, 36-42, 46, 47). The septal pore is about $0.1$-$0.2 \mu$ diam. A perforate membranous septal pore cap occurs on each side of the septum as a dome-shaped structure. The plasma membrane is continuous from one cell to the next, passing around the septal swelling and through the pore (Figs. 36-39).

The formation of a cross wall by invagination from the lateral hyphal walls takes approximately 10 minutes in young hyphae of *R. solani* (Buller, 1933; Bracker and Butler, 1963). During the invagination, the septal swelling is absent, as determined by phase-contrast (Bracker and Butler, 1963) and electron microscopy (Fig. 35). The septal swelling and pore cap become evident shortly after the cross wall has formed (Bracker and Butler, 1963). When the young swelling is observed by electron microscopy, it has a diameter of about $0.6$-$0.7 \mu$, is bounded by the plasma membrane, and is situated on the innermost edge of the cross wall (Fig. 36). At this stage, the layers of the cross wall do not protrude into the septal swelling. Further maturation results in enlargement of the swelling until it reaches a diameter greater than $2 \mu$. Growth of the swelling is largely centrifugal. Thus, the layers of the cross wall appear to protrude into the swelling in mature septa. Throughout the development of the septum, a distance of about $0.1$-$0.2 \mu$ is maintained between the end of the cross wall and the plasma membrane lining the septal pore. The pore is not closed by growth of the swelling, it remains $0.1$-$0.2 \mu$ diam. Protoplasmic continuity is retained between hyphal compartments, and blockage, when it occurs, results from plugging or walling off of the septal pore.

Increases in cross-wall thickness and diameter of the swelling occur concomitantly. When lamellae of the cross wall are evident, fine lines emanate from the terminus of the cross wall into the swelling (Fig. 40). Presumably, these are fibers from the cross-wall lamellae. In hyphal cross sections, where the septal swelling appears doughnut-shaped, the fine lines from the cross wall are concentrically arranged (Fig. 39). The concentric rings extend into the swelling only as far as the cross wall protrudes. A clear zone between the pore and the concentric rings corresponds to the distance between the edge of the cross wall and the plasma membrane, as seen in longitudinal sections (Fig. 38). In young septa, the concentric rings are lacking. Several lines of evidence indicate that the septal swelling of *R. solani* is both chemically and physically different from the wall proper (Bracker and Butler, 1963).

The septal pore cap is consistently associated with the septal pore apparatus, regardless of age (Figs. 17, 18, 36-38, 41-43, 46-48). A membranous cap is present on each side of the septum. The cap is continuous with endoplasmic reticulum and appears to be modified ER, but details of its genesis and chemical nature are not available. Typically, the septal pore cap is thicker than ER (about 750 A) and the matrix between the limiting membranes contains electron-dense staining material. Internal structures (lamellae) have been reported for the cap in *R. solani* (Bracker and Butler, 1963).

The cap of *R. solani* is consistently perforate (Figs. 37, 38, 41, 42, 46, 47). The apertures are up to 1 μ, giving the cap the appearance of a discontinuous membrane in section.

Within the bounds of the septal pore caps, the ground cytoplasm frequently stains different from the cytoplasm in the rest of the hypha. This cytoplasmic region has been termed the subcap matrix (Bracker and Butler, 1963). It usually stains lighter (Fig. 38) than the surrounding cytoplasm, but in aged hyphae or sclerotia, it may be darker (Fig. 46).

Occasionally, incomplete septa occur in old hyphae (Fig. 51). These are simple (i.e. without the complex septal pore apparatus) and appear as thickened ingrowths from the lateral walls. No evidence of septal degeneration, as occurs in *Coprinus lagopus* (Geisy and Day, 1965), has been observed in *R. solani*. These simple septa in *R. solani* must be a type of secondary septa, but they do not agree with the description of secondary septa given by Flentje, Stretton, and Hawn (1963).

Plasmodesmata are not evident in the cross walls of *R. solani* hyphae or sclerotia. The occurrence of plasmodesmata, or plasmodesma-like strands, between adjacent cells of fungous thalli is rare (Hawker, et al., 1966; Kirk and Sinclair, 1966).

Protoplasmic streaming has been observed often in living hyphae of *R. solani* (Buller, 1933; Bracker and Butler, 1964; and personal observations). Unidirectional streaming is regarded as an important means of food transport and translocation of protoplasm in filamentous fungi (Buller, 1933; Schutte, 1956; Kamiya, 1959), and it may also play a role in the migration of nuclei through fungous thalli (Buller, 1933; Snider, 1963). Mitochondria and other cell particulates pass from one cell to the next through the septal pore of *R. solani*. Thus, the complex nature of the septal pore apparatus does not appear to deter mass flow through the hyphae. Although examinations of electron micrographs show the pore to be mostly 0.1-0.2 μ in diameter, in some it is as large as 0.5 μ (Figs. 41, 42) (Bracker and Butler, 1964). In nearly all instances of the latter, the pore contains membranes and constricted organelles that appear to have been moving through the pore at the time of fixation. The potential diameter of the pore is limited by the inner edge of the cross wall—not by the septal swelling. Thus, the

flexibility of the septal swelling permits expansion of the pore from 0.1 μ to about 0.5 μ. The discontinuities in the septal pore cap (Figs. 37, 38, 42) remove the cap as a potential barrier to protoplasmic flow. These features of the septal pore apparatus, coupled with the inherent plasticity of organelles, expedite protoplasmic movement through the pore and make feasible Buller's statement (1933) that "as the protoplasm came up to the septum it seemed to pass through it with the greatest of ease." In other fungi, where the cap is continuous or has small apertures or when the swelling is inflexible, the pore apparatus may be a physical barrier to protoplasmic movement.

Most of the monilioid cells of sclerotia observed by electron microscopy contain protoplasm (Fig. 52). Although the cells differ in shape from older hyphal cells, their internal organization is similar. Mitochondria are small, with few cristae, and most do not exhibit the elongate or branching form seen in hyphal tips. Membrane components are generally sparse, as would be expected in quiescent protoplasts. The septal pore apparatus is common and is assumed to occur at each septum.

In aged cells or in cells under stress, the septal pore may become plugged (Buller, 1933; Bracker and Butler, 1963). Plugging seems to involve two types of blockage. The septal pore becomes filled with material that is electron-dense after $KMnO_4$ fixation (Figs. 47, 48). Also, at each end of the pore, an amorphous electron-dense plug occupies a position of blockage beneath the septal pore cap. The plug is not a simple sphere, it has a small rim, protruding like the brim of a hat. It contacts the plasma membrane around the septal swelling. Crescent-shaped amorphous structures occur over each end of the pore in younger unplugged hyphae (Figs. 38, 47). Because of their position and shape, these structures appear to be precursors for plug formation. In some old hyphae and sclerotia, the pore is plugged and the subcap matrix may stain darker than the surrounding cytoplasm. The caps are not continuous with ER, but the discontinuities in them remain. Additionally, the caps are drawn close to the septal swelling. The entire apparatus is covered with a sheet of ER on each side of the septum (Fig. 46). These septal characteristics are in contrast to the pore apparatus in active hyphae (Figs. 17, 37-42).

When a hyphal compartment becomes evacuated, either because of age or injury (Buller, 1933), the septal pore apparatus in the evacuated compartment is labile and disappears (Figs. 43-45). The cap and the swelling are absent from such compartments, but the cross wall and its layers in the pore region are intact and the pore is plugged. In older hyphae, walling off of the intact compartment occurs (Fig. 45). New wall material is deposited inside the existing lateral walls and across the septum, in effect sealing off the compartment adjacent to the evacuated portion of the hypha. This is similar to the walling off reported by Wells (1964) for *Exidia nucleata*. Regrowth from these walled-off portions may occur in the form of intrahyphal hyphae (Fig. 29). On occasion even intra-intrahyphal hyphae occur (Fig. 50).

When discussing cell ultrastructure, the question of artifact invariably arises. To be viewed in the electron microscope, a specimen must pass through treatments that render it dead, immobile, dehydrated, largely substituted, and stained. Thus, what is observed with the electron microscope is an artifact. The pertinent question is whether the image reasonably represents the organization of the living cell. Within the limits of resolution of the phase-contrast microscope, one may judge the changes occurring during fixation. Such correlation lends confidence to the interpretation of ultrastructural images reported here.

The structures recognized as typical of *R. solani* are characteristic of basidiomycetous hyphae, with some unique details. Generally, the fine structure of fungal cells is useful in demonstrating protoplasmic similarity within large taxa (i.e. orders and above), but is of dubious value at the generic and specific levels.

Morphology and cytology are blended to derive a full structural characterization of *R. solani*. From each, features are derived that collectively characterize the fungus and serve as a basis for identification. The degree of variation at each level according to environment, clone, or developmental stage indicates the stability of the characters used to identify *R. solani*. Variability is least understood at the ultrastructural level because this is the most recent area of morphological investigation and the least accessible to evaluation. An important consideration, resulting from electron microscope studies of *R. solani*, is that the ultrastructure reflects hyphal age. However tedious the task, variations at all levels must be reconciled if a valid concept of this fungus is to be obtained.

ACKNOWLEDGMENTS. — We thank Dr. Robert K. Webster and Mr. Stanley N. Grove for their able assistance in obtaining some of the original information presented in this paper. We are grateful for the United States Public Health Service Grant RG-5868 and for the National Science Foundation Grant GB-03044.

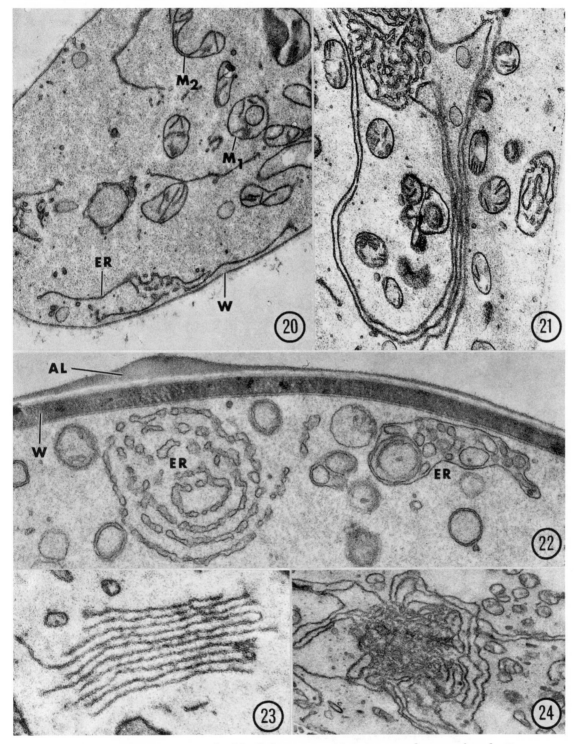

*Fig. 20.* Electron micrograph of hyphal-tip region. $M_1$ is interpreted as cup-shaped mito-chondrion in transverse section; $M_2$, similar mitochondrial form in longitudinal section. Endo-plasmic reticulum, *ER*; hyphal wall, $W \times 20,000$.

*Fig. 21.* Complex configuration of endoplasmic reticulum continuous with simpler continu-ous sheetlike form. $\times 20,000$.

*Fig. 22.* Whorl of endoplasmic reticulum (*ER, left*) showing discontinuities in section. *ER* in face view (*right*) appears a porous sheet. Hyphal wall, *W*; amorphous layer on outer surface of wall, *AL*. $\times 30,000$.

*Fig. 23.* Lamellar configuration of endoplasmic reticulum. $\times 21,000$.

*Fig. 24.* Cytoplasmic membrane complex composed of endoplasmic reticulum. $\times 15,000$.

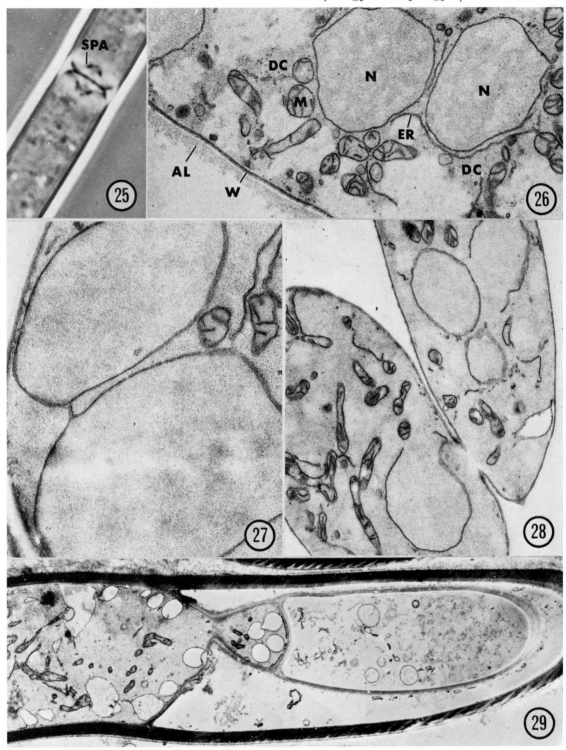

Fig. 25.    Phase-contrast photomicrograph of living mature hypha in septum region. Septal pore apparatus (*SPA*) is apparent. × 2,500.

Fig. 26.    Electron micrograph of part of mature hypha showing dense staining cytoplasm (*DC*) surrounding organelles. Nucleus, *N*; endoplasmic reticulum, *ER*; mitochondrion, *M*; hyphal wall, *W*; amorphous coating, *AL*. × 14,000.

Fig. 27.    Nuclei joined by continuous nuclear envelopes. × 23,000.

Fig. 28.    Electron micrograph showing hyphal fusion in section. × 9,000.

Fig. 29.    Intrahyphal hypha from aged culture. Note thick dark walls (*W*), typical of brown hyphae. × 6,500.

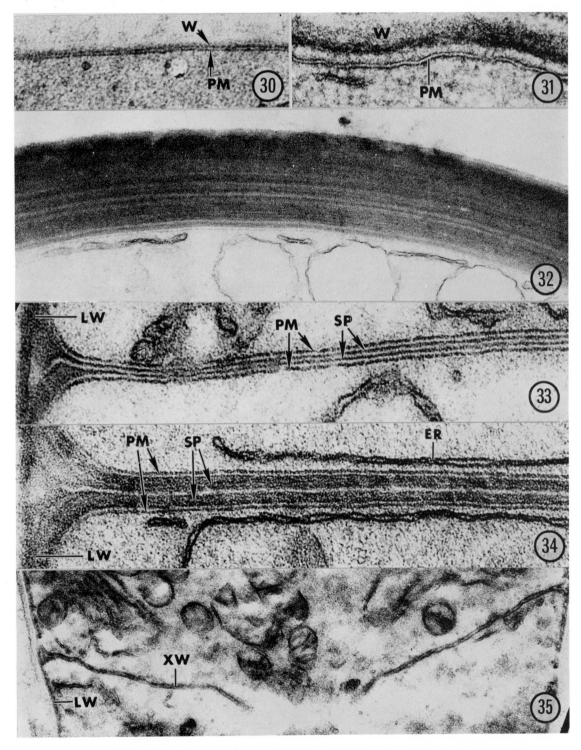

*Fig. 30.* Plasma membrane (*PM*) and hyphal wall (*W*) in hyphal-tip region. × 55,000.

*Fig. 31.* Plasma membrane at high magnification.

*Fig. 32.* Multilayered wall of brown hypha. × 36,000.

*Fig. 33.* Newly formed cross wall in which septal plates (*SP*) appear monolayered. Plasma membrane, *PM*; lateral wall, *LW*. × 77,000.

*Fig. 34.* Mature cross wall showing several lamellae in each septal plate (*SP*). Endoplasmic reticulum, *ER*; plasma membrane, *PM*; lateral wall, *LW*. × 71,000.

*Fig. 35.* A forming cross wall (*XW*). Note absence of septal pore apparatus. Lateral wall, *LW*. × 15,000.

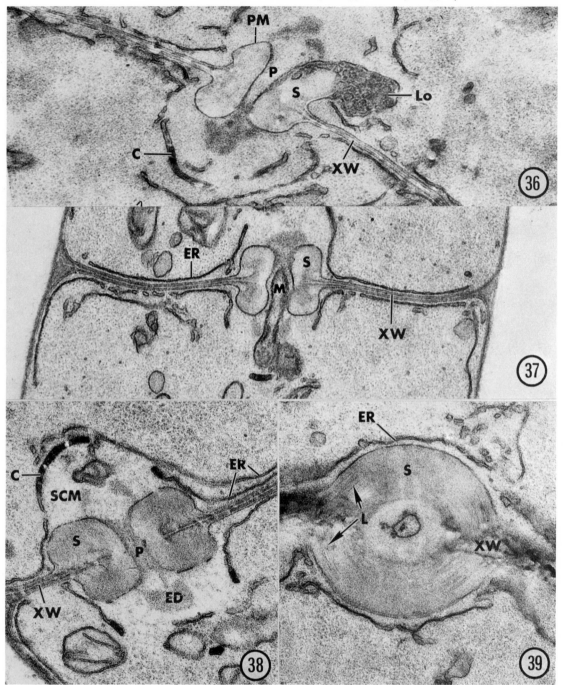

*Fig. 36.* Longitudinal section of septal pore apparatus of very young septum. Lomasome (*LO*) is attached to septal swelling (*S*). Septal pore, *P*; septal pore cap, *C*; plasma membrane, *PM*; cross wall, *XW*. × 38,000.

*Fig. 37.* Longitudinal section of septum in young hypha with mitochondrion (*M*) passing through septal pore. Septal swelling, *S*; cross wall, *XW*; lateral wall, *LW*; endoplasmic reticulum, *ER*. × 27,000.

*Fig. 38.* Longitudinal section through septal pore apparatus in more mature hypha than Fig. 26. Septal pore cap, *C*; subcap matrix, *SCM*; cross wall, *XW*; septal swelling, *S*; endoplasmic reticulum, *ER*; septal pore, *P*. Note clear zone between end of cross wall and septal pore. Curved electron-dense areas (*ED*) occur at each end of pore. × 33,000.

*Fig. 39.* Septal pore apparatus seen in hyphal cross section. Note circular clear zone around pore. This is comparable to clear zone seen in longitudinal section (Fig. 27). Cross wall (*XW*) appears blurred due to oblique plane of section. Septal swelling, *S*; fine lines emanating from cross wall into swelling, *L*; endoplasmic reticulum, *ER*. × 29,000.

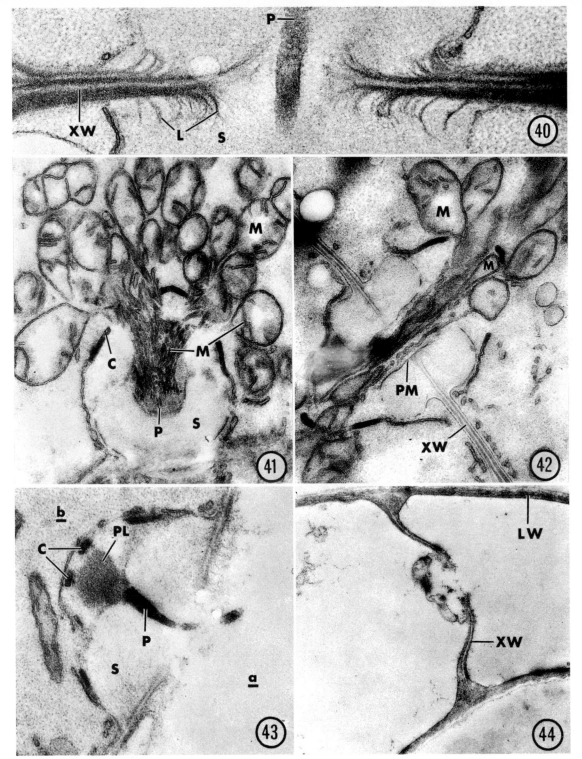

*Fig. 40.* Longitudinal section through mature septal pore apparatus. Lines (*L*) representing portions of cross wall lamellae extend into septal swelling (*S*). Cross wall, *XW*. × 48,000.

*Fig. 41.* Oblique cross section through septal pore apparatus with constricted mitochondria (*M*) moving through pore (*P*). Pore is expanded 0.5μ diam. Septal swelling, S.; septal pore caps, *C*. × 28,000.

*Fig. 42.* Longitudinal section showing septal pore apparatus with mitochondria (*M*) streaming through pore. Plasma membrane (*PM*) is pressed back against cross wall (*XW*) as pore is enlarged. × 27,000.

*Fig. 43.* Portion of hypha in which protoplasm was evacuated from cell *a* but retained in cell *b*. Pore is occluded with electron-dense material and a plugging inclusion (*PL*) lies against pore beneath cap (*C*). Pore apparatus is absent from evacuated cell. × 28,000.

*Fig. 44.* Portion of evacuated hypha from old culture. Only remnants of pore apparatus remain. Lateral wall, *LW*; cross wall, *XW*. × 12,000.

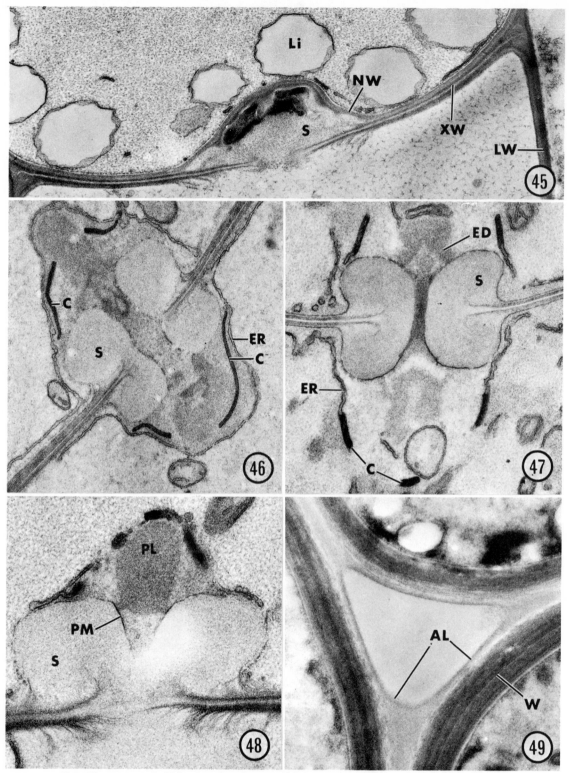

Fig. 45.    Portion of old hypha where lower cell has been evacuated and new wall (NW) has been deposited in intact cell, sealing it off. Cross wall, XW; lateral wall, LW; lipid inclusion (Li). × 23,000.

Fig. 46.    Longitudinal section through septal pore apparatus in old hypha. ER encloses whole apparatus; cytoplasm in pore region stains denser than cytoplasm in rest of hypha. Pore caps (C) are not continuous with ER. Septal swelling, S. × 20,000.

Fig. 47.    Septal pore apparatus in mature hypha. Material in pore stains electron-dense. Electron-dense region (ED) contacts swelling at end of pore, suggesting a plugging precursor. Pore cap, C; endoplasmic reticulum, ER; septal swelling, S × 29,000.

Fig. 48.    Plugging inclusion (PL) over septal pore. Projections from plug contact plasma membrane (PM) by septal swelling (S). × 38,000.

Fig. 49.    Amorphous coating (AL) between monilioid cells of sclerotium. Cell walls, W. × 34,000.

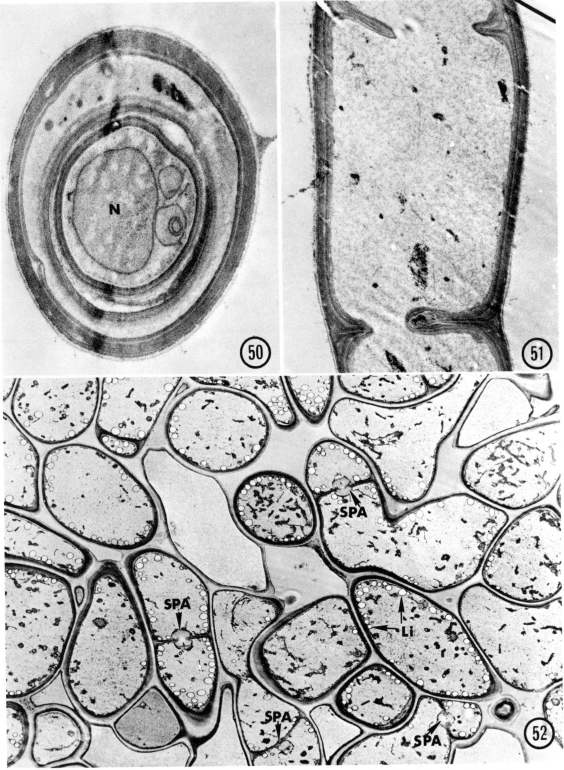

*Fig. 50.* Cross section of hyphae in old colony showing a hypha within a hypha within a hypha. Nucleus, *N.* × 11,500.

*Fig. 51.* Longitudinal section through old hypha showing incomplete, simple septa. × 10,000.

*Fig. 52.* Part of sclerotium showing monilioid cells. Septal pore apparatus, *SPA*; lipid inclusions, *Li.* × 3,000.

# Mechanisms of Variation in Rhizoctonia Solani

N. T. FLENTJE, HELENA M. STRETTON, and A. R. McKENZIE—*Department of Plant Pathology, Waite Agricultural Research Institute, University of Adelaide, South Australia.*

INTRODUCTION.—What is the evidence for variation in *Rhizoctonia solani?*

One basis for assuming that variation exists is the wide range of isolates, differing in a variety of characteristics, but identified as *R. solani* largely on the basis of hyphal morphology. However, Parmeter and Whitney (this vol.) have pointed out that hyphal morphology is unreliable for identification and that not all isolates described as *R. solani* can be accepted as such. Nevertheless, after critical study, they do accept a wide range of isolates as being closely related and correctly described as *R. solani*. Alternative and perhaps more reliable indications of variation are the differences between isolates known to have the same perfect state. A number of workers, including Kotila (1929), Flentje (1956), Whitney and Parmeter (1963), and Papavizas (1965), have induced isolates of *R. solani* to form the perfect state. These isolates, though differing in a wide range of cultural and pathogenic characters, do in fact have the common perfect state, *Thanatephorus cucumeris* (Frank) Donk. But Talbot (1966) suggested that even these isolates may be a collective rather than a natural species. Nevertheless, we believe these latter isolates afford the most reliable evidence of variation and we have therefore restricted our consideration in the following review to those isolates of *R. solani* that are known to have the perfect state, *T. cucumeris,* and to mutants and single-basidiospore cultures derived from the fruiting isolates.

DISTRIBUTION AND BEHAVIOR OF NUCLEI.—An understanding of distribution and behavior of nuclei throughout the life cycle is essential for analyzing mechanisms of variation. The vegetative cells of *R. solani* are multinucleate, usually containing 6-10 haploid nuclei. These nuclei divide simultaneously and conjugately and segregate evenly, half the daughter nuclei moving forward into the new tip cell and the others remaining in what becomes the penultimate cell. Irregularities in the division and segregation of daughter nuclei in a heterokaryon may contribute to variation, but little is known at present about the control of this nuclear division. In older vegetative cells, secondary septa are formed that reduce the number of nuclei to 1-4, depending on the number of septa. Sclerotial cells are also multinucleate and behave as ordinary vegetative hyphal cells (Flentje, Stretton, and Hawn, 1963).

Prior to formation of basidia, the nuclei in the vegetative cells pair and the pairs are separated by septa. The short binucleate cells thus formed proliferate and form basidia. In the basidium, the nuclei fuse to form a diploid and then undergo meiosis to form four haploid nuclei which migrate, one through each of the four sterigmata, to form four uninucleate spores. However, aberrations can .occur. Prebasidial cells may contain three nuclei instead of two, and the extra nucleus apparently may go through the basidium unchanged and be discharged by the formation of an extra sterigma and spore or by inclusion of two nuclei in one spore. The basidium may also form less than four sterigmata (Fig. 1), leading to a regular percentage of binucleate spores (Flentje, Stretton, and Hawn, 1963).

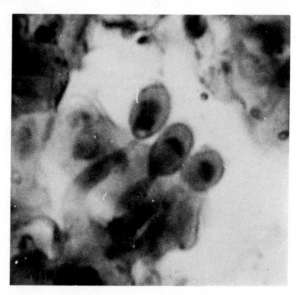

*Fig. 1.* One binucleate spore and 2 uninucleate spores from 1 basidium of *T. cucumeris.*

It should therefore be stressed that not only is general information on distribution and behavior of nuclei essential for the analysis of variation, but in view of the aberrant nuclear behavior in many isolates, detailed cytological information should be obtained for every isolate studied.

NUCLEAR CONDITION OF FIELD ISOLATES. — What is the nuclear condition of *R. solani* growing under natural conditions? The cells of vegetative hyphae and sclerotia are multinucleate, but are they homokaryotic or heterokaryotic? By staining hyphae removed directly from soil or infected seedlings, it has been shown (Flentje, unpub.) that cells are multinucleate, with the same average number of nuclei per cell as when grown on agar. To determine whether these cells are heterokaryotic or otherwise, some method must be used to achieve asexual separation of nuclei, or the isolate must be induced to form basidiospores. To achieve either objective at present, it is necessary to grow the isolate in pure culture, and it could be argued that the change from natural soil conditions to pure culture could bring about changes in nuclear composition. This is a valid criticism. Although field isolates of *R. solani* appear to be very stable in culture, even when grown on different media, there is evidence from laboratory studies that mutation occurs. Furthermore, synthesized heterokaryons can be unstable under culture conditions (McKenzie, 1966). Thus, further study in this area is required before the following evidence on heterokaryosis can be accepted as completely valid for field isolates.

Partial separation of nuclei can be achieved asexually in some isolates by obtaining single cells of older hyphae where secondary septation has occurred and there are only 1-4 nuclei per cell. Such cells, when grown separately, give rise to cultures that differ both from each other and from the parent culture, strongly suggesting there were different types of nuclei in the parent culture (Flentje and Stretton, 1964).

Numerous workers have isolated single basidiospores from a hymenium and shown that the resulting cultures differ from each other in a wide range of characters (Kotila, 1929; Hawn and Vanterpool, 1953; Sims, 1960; Whitney and Parmeter, 1963; Flentje and Stretton, 1964; Papavizas, 1965). Some of this work, however, is inadmissible as evidence of heterokaryosis, as the hymenium may have been formed by a mixture from two or more different isolates as outlined by Papavizas (1965). To be admissible, evidence should be based on isolates grown from hyphal tips and induced to fruit under conditions that exclude subsequent contamination with any other isolate of *R. solani*. A number of cultures of *R. solani* isolated as hyphal tips have been induced to form the basidial stage in pure culture on agar (Whitney and Parmeter, 1963; Papavizas, 1965). Flentje and Stretton (1964) have also induced isolates of *R. solani*, grown originally from hyphal tips, to fruit on soil (Fig. 2) previously treated with aerated steam to free the soil from any

Fig. 2. Fructifications of *T. cucumeris* (root strain) on soil treated with aerated steam at 71°C. for 30 min.

naturally occurring *R. solani*. In each instance where single-basidiospore cultures have been obtained from one field isolate, these cultures have exhibited considerable variation in a wide range of characters, the differences being maintained in successive transfers.

This evidence of genetic segregation through the basidial stage is a strong indication that the parent isolate was heterokaryotic, and unless there has been a change in nuclear composition in culture, it appears that the original field isolates were heterokaryotic. However, as only a few field isolates have been critically examined, it would be unwise to assume from this evidence alone that all field isolates are heterokaryotic.

Nevertheless, there are theoretical grounds for suggesting that field isolates of *R. solani* are likely to be heterokaryotic. In all isolates so far examined, dividing vegetative cells are multinucleate and nuclear division is conjugate followed by an even segregation of daughter nuclei. Evidence will be presented later that, in homokaryotic cultures obtained from single basidiospores, spontaneous mutants occurred during vegetative growth and were retained in the hyphal-tip cells so that homokaryotic cultures became heterokaryotic and remained so because of mutation. Thus, on both theoretical and experimental evidence, it seems that field isolates generally are likely to be heterokaryotic.

MECHANISMS OF VARIATION. — In the following sections of this review, we have not attempted to catalog variation, but rather to analyze the mechanisms of variation and suggest how these mechanisms may operate in nature.

*Sexual recombination.*—Mechanism.—As indicated previously, most field isolates are likely to be heterokaryotic. This allows fusion of unlike nuclei to form the diploid, with consequent recombination during meiosis. Cytological studies (Hawn and Vanterpool, 1953; Saksena, 1961; Flentje, Stretton, and Hawn, 1963) indicate that the behavior of nuclei in the basidium follows the typical pattern with diploidization followed by meiosis. Although, because of the small size of nuclei, the chromosomes are difficult to count, several workers (Hawn and Vanterpool, 1953; Saksena, 1961; Flentje, Stretton, and Hawn, 1963) agree there are six in the haploid nucleus. There are still many questions to be answered regarding the mechanisms involved in recombination. How many different kinds of nuclei can be carried in the heterokaryon? With respect to what factors can an isolate be heterokaryotic? Are all isolates homothallic? Do all types of nuclei participate in diploidization? Is there preferential "selfing" of like nuclei or crossing of unlike nuclei? Is there differential proliferation of prebasidial cells depending on nuclear composition?

From the few genetical studies of *R. solani* to date, answers to some of these questions can be given, but in relation to other questions, tentative suggestions only can be made. Whitney and Parmeter (1963) and Stretton, et al. (1967) have shown that though some isolates of *R. solani* are

homothallic, unlike nuclei in a heterokaryon do fuse and subsequently from recombinants. Whitney and Parmeter (1963) used two homothallic single-spore cultures from one parent, which they designated CDA+ and CDA— because of differences in their ability to grow on Czapek-Dox agar. These two cultures formed a heterokaryon, which was CDA+, and which produced single-spore progeny, half of which were CDA+ and half CDA—. Although tetrad analyses were not carried out, the fact that the progeny differed in cultural characteristics is evidence that the recombination occurred. McKenzie (1966) produced more extensive evidence. He worked with mutant single-spore cultures obtained from a homothallic, homokaryotic parent. The mutant cultures would not fruit alone, but when anastomosed in pairs, readily formed fertile heterokaryons. The mutations behaved as single nonlinked genes, and from each heterokaryon, the nonmutant and double-mutant recombinants were recovered together with the original mutants. Tetrad analyses showed, however, that from some basidia the four spores all gave rise to cultures of one or the other mutant. Thus, nuclear "selfing" frequently occurred, apparently quite at random.

The extent of variation among different single-spore cultures from one parent, described by several authors (Kotila, 1929; Hawn and Vanterpool, 1953; Flentje and Saksena, 1957; Whitney and Parmeter, 1963; Flentje and Stretton, 1964; Papavizas, 1965) indicates that in naturally occurring heterokaryons, dissimilar nuclei paired in the basidia (Fig. 3).

McKenzie (1966) also obtained confirmation by tetrad analysis of the earlier suggestion by Flentje and Stretton (1964) that two dissimilar nuclei could migrate into one spore.

Thus, the basidiospores from a heterokaryon will contain nuclei that are either identical with, or recombinants from the original nuclei, and, in some instances, a spore may contain two dissimilar nuclei.

Extent.—The extent of variation through sexual recombination covers an extremely wide range and can be summarized under the following headings with selected references for the evidence.

Cultural characteristics. These include growth rate and abundance of aerial mycelium (Whitney, 1963; Papavizas, 1965), color of mycelium, zonation, and sclerotial formation (Flentje and Stretton, 1964).

Physiological characteristics. These include ability of spores to germinate and form colonies on common laboratory media (Flentje and Stretton, 1964), enzyme production (Papavizas and Ayers, 1965), tolerance to $CO_2$ (Papavizas, 1964), and ability to grow on Czapek-Dox agar (Whitney and Parmeter, 1963).

Saprophytic behavior. Papavizas (1964) and Olsen, et al. (1967) have shown that single-spore cultures derived from one parent varied considerably in their growth through, and survival ability in soil. Papavizas (1964) showed a correlation between these differences and $CO_2$ tolerance.

Virulence. Several workers have reported differences in virulence among different single-spore cultures obtained from one parent. In some of these investigations (Papavizas and Ayers, 1965), the results are inconclusive because of the techniques used. Tests carried out by growing plants in soil previously inoculated with the fungus investigate a composite process, namely, the ability of the fungus to maintain its population, grow through the soil, infect the host plant, and cause progressive disease. Virulence is concerned only with the last part of

this process. We believe that virulence can be satisfactorily determined only by opposing the host tissue with the fungal inoculum directly.

However, from the results of experiments using techniques that directly oppose fungus and host (Hawn and Vanterpool, 1953; Flentje and Stretton, 1964; Garza-Chapa and Anderson, 1966), it is clear that there are considerable differences in the virulence of single-basidiospore isolates. These differences have not been satisfactorily correlated with any dif-

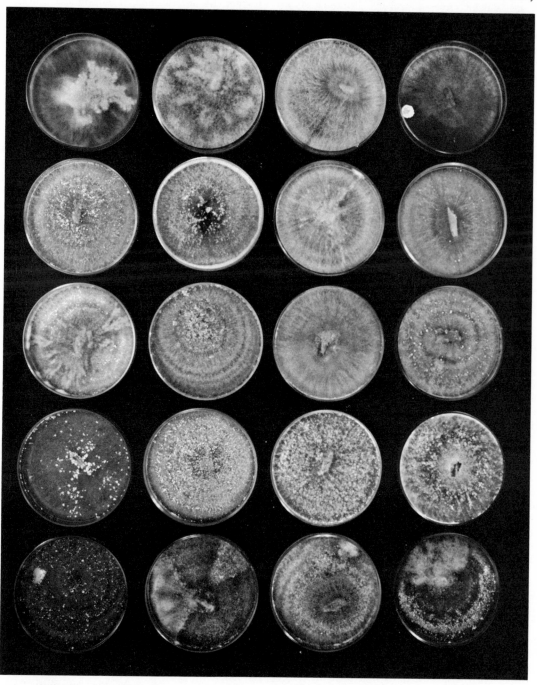

*Fig. 3.*   Single-basidiospore cultures from a stem-strain isolate of *T. cucumeris* showing wide variation in cultural characteristics.

ferences in physiological characters.

Fruiting ability.    First-generation single-basidiospore cultures from one parent vary considerably in
fruiting ability (Kotila, 1929; Hawn and Vanterpool,
1953; Flentje and Stretton, 1964; Papavizas, 1965;
Garza-Chapa and Anderson, 1966). In most instances, the pattern of the results is similar; at first
the majority of single-spore cultures failed to fruit,
a few cultures produced basidia but no spores, and
fewer again produced basidia and spores. These results are inconclusive, however. Stretton, et al.
(1964) showed that environmental conditions influence fruiting and, as the precise conditions are not
known for any isolate, the lack of fruiting in some
single-spore cultures may be due to environmental
rather than genetic factors. Nevertheless, it is most
unlikely that all the results could be explained thus,
especially where replicated tests have been carried
out with isolates that fruit on agar media, thereby
reducing environmental variation.

Stretton, et al. (1967) have shown that some isolates at least are homothallic, but there are apparently a number of sterility factors present that prevent the fruiting of single-spore cultures. Also, mutations can occur that overcome the sterility factors
and induce fruiting in an originally sterile culture.
Flentje, et al. (1967) have shown that five independent mutants from a fertile homothallic homokaryotic line were sterile. From recombination
studies with these mutants (McKenzie, 1966), it
appears that there are a number of stages, similar to
those in the Ascomycetes (Raper, 1960), in the sexual progression from vegetative hypha to the release
of basidiospores, at which blockage may occur. Some
of the mutants investigated apparently fail at an
early stage of this progression, but there is insufficient information for precise definition of the stages
blocked. However, as two of the mutations investigated occurred spontaneously, it is likely that they
represent naturally occurring sterility factors.

Little synthesis can be made at present from this
variation because of its complexity. Many of the
factors listed, such as mycelial characteristics, sclerotial formation, and fruiting also vary with different
environmental conditions. We have no information
as to whether variation in any of the characters
listed is controlled by one or more genes. In any
studies so far carried out, there are so many of the
above characters varying simultaneously that it is
practically impossible to correlate any of the characters with such important features as saprophytic
growth, survival, or virulence. Following the genetical studies of Whitney and Parmeter (1963) and
McKenzie (1966) there are, however, obvious possibilities for future investigations.

It is important to note that so far, though some
single-spore cultures appear to have lost pathogenicity (Hawn and Vanterpool, 1953; Flentje and
Stretton, 1964; Papavizas and Ayers, 1965; Garza-
Chapa and Anderson, 1966), there is no evidence of
any other change in host range between the parent
and progeny cultures. This may be due partly to
the fact that so little work has been carried out along

these lines. Nevertheless, from the studies that have
been made, it is clear that pathogenic specificity is
a much less variable character than those other
characters listed above.

Under mechanism of recombination, we posed the
question, with respect to what factors can an isolate
be heterokaryotic? The answer will have considerable
bearing on the extent of variation by recombination.
The factors apparently include those listed here, but
some attempts have been made to explain these
further by attempting to anastomose single-spore
cultures from different parents. This is further discussed in the later section dealing with heterokaryosis.

*Mutation.*—Mechanism.—Since field isolates of *R.
solani* appear to be so stable in culture, it is commonly believed that mutation is comparatively rare
under ordinary cultural conditions. Theoretically this
is most unlikely. We now know that cultures of *R.
solani* are heterokaryotic, with multinucleate cells
and simultaneous nuclear division which maintains
the full complement of nuclei in the hyphal-tip cells.
Therefore, to alter cultural expression, any mutant
would have to be dominant and of significant advantage over any allele in the group of nuclei present, or it would have to produce an unstable heterokaryon. Alternatively, if cultural appearance depended on nuclear ratios in a heterokaryon, a mutant would have to bring about an alteration in the
nuclear ratio.

Kernkamp, et al. (1952) reported at length on
induced mutation under a range of cultural conditions. While it is likely that mutation accounts for
some of the variation reported, no critical analysis
of the results can be made.

Similarly, in any investigations of variation among
single-basidiospore cultures from one parent culture,
it could be argued (Flentje and Stretton, 1964) that
this variation, arising by recombination through the
sexual stage, is due initially to mutation, but no definitive evidence of mutation has been presented. If
the variation does arise initially through mutation,
then it affects a wide range of cultural characters,
but few if any of these have been sufficiently distinctive to be used as markers for genetic analysis.
However, the frequent occurrence of sectors in some
single-basidiospore cultures (Flentje and Stretton,
1964) is almost certainly due in part to mutation.

Critical information on mutation has been obtained
recently by Flentje, et al. (1967) with one isolate of
*R. solani.* This isolate was fertile, and a selection was
made through three successive generations for a
fertile homokaryotic single-basidiospore culture that
was the same in cultural appearance and pathogenicity as the original field isolate. From the third
generation of spores, a fertile single-spore culture
was obtained that was almost identical with the
original field isolate. This produced a fourth generation of spores that were almost 100% uninucleate,
germinated evenly, and gave rise to cultures that
were identical with each other and with the third-
generation parent culture. This latter culture was

thus apparently homothallic and homokaryotic. From spores shed from this culture and treated immediately with ultraviolet irradiation, four stable mutants, *sparse, stumpy, fleecy,* and *curly,* were obtained. After 6 months of continued subculturing of the third-generation parent culture, it was again fruited and spores collected. A small percentage of these spores germinated unevenly and gave rise to cultures very different from the parent culture. This was apparently due to the occurrence of spontaneous mutations during the 6-months vegetative growth of the parent culture. Two stable spontaneous mutants, *rusty* and *ropy,* were studied in detail.

These six mutants (Fig. 4) were studied in recombination experiments by McKenzie (1966) and five of them behaved as independent single-gene mu-

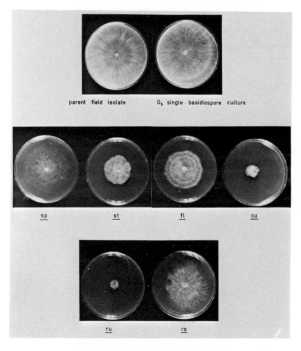

*Fig. 4.* Cultural characteristics of (from top left): parent isolate, of identical G3 single-basidiospore culture, of 4 G4 (fourth generation) induced mutants, and of 2 G4 spontaneous mutants.

tations. Studies with *fleecy* yielded results that do not fit a simple mendelian segregation so the nature of the mutation involved is still uncertain.

Extent. — The extent of variation between the above six mutants and the parent culture is summarized in Table 1.

As all six mutant cultures failed to fruit while some heterokaryons fruited readily under the conditions in which the parent culture fruited, it seems very likely that the mutations caused sterility. However, while heterokaryons between spontaneous and induced mutants fruited readily, heterokaryons between induced or between spontaneous mutants failed to fruit, so it appears that the question of sterility may be quite complex.

Five of the mutants were nonpathogenic, but in each the infection process was interrupted at a different stage. In *stumpy,* growth over the test host was inhibited; in *sparse,* the hyphae failed to adhere to stems; in *ropy,* no infection cushions were formed; in *rusty,* there was apparently no penetration from the few infection cushions formed; and in *curly,* hypersensitive lesions followed infection. The sixth mutant, *fleecy,* was as virulent as the parent isolate. As each of these mutants behaved in recombination experiments as single nonlinked factors, they are possibly on different chromosomes. Thus, there are at least five distinct factors concerned with virulence or pathogenicity and six with sterility. Mutants (Fig. 5) were not tested on other than cruciferous hosts to determine whether host specificity had been altered by mutation.

*Parasexual recombination.*—There is at present no information as to whether parasexual recombination occurs in *R. solani.* The simultaneous division in close proximity of several nuclei in vegetative cells would appear to present easy opportunity for diploidization and genetic exchange. However, because of the absence of asexual spores, there appears to be little possibility at present for investigating such exchange.

*Heterokaryon formation.*—Mechanisms.—1. Through binucleate spores. Flentje and Stretton (1964) showed that a percentage of basidiospores in some isolates of *R. solani* were binucleate and postulated that the condition arose through the migration of two nuclei from the basidium. Saksena (1961a), however, suggested that where there was more than one nucleus in the spore, it was due to mitotic division of the original single nucleus in the spore. McKenzie (1966), using tetrad analysis to study recombination between mutants of *R. solani,* recently produced conclusive evidence that from one basidium, which produced only three sterigmata and three spores, two genetically different nuclei moved into one of the three spores. The frequency with which this occurs has still to be determined, but it will almost certainly be correlated closely with the frequency of occurrence of basidia with less than four sterigmata.

2. By mutation. As described above from the work of Flentje, et al. (1967), spontaneous mutation occurred in a homokaryotic single-spore culture of *R. solani.* The single-spore culture used in this study was similar in growth rate, pathogenicity, and cultural appearance to the parent field isolate from which it was derived. These spontaneous mutants resulted in a change from a homokaryotic to a heterokaryotic condition, the mutant nuclei presumably being carried forward in the multinucleate vegetative cells because of the simultaneous division and even segregation of nuclei. No change from the original cultural appearance, pathogenicity, or fruiting ability was detected over the period the mutations accumulated, though the mutants as shown by spore analysis were slow in growth rate, nonpathogenic to radish, and failed to fruit.

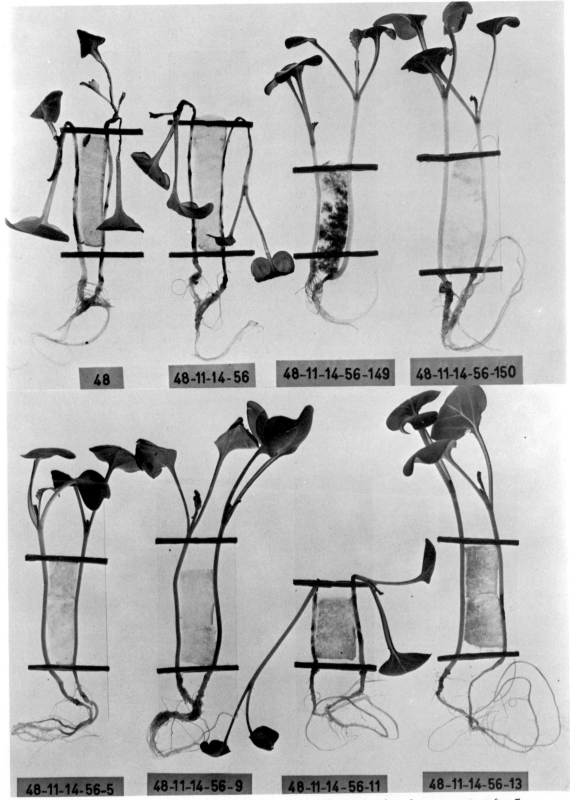

*Fig. 5.* Pathogenicity tests on radish stems showing variation in host-fungus reaction after 7 days of (a) from left: parent field isolate, G3 single-basidiospore culture, and 2 spontaneous mutants derived from G3 spore culture. (b) Four mutant single-basidiospore cultures derived from G3 culture.

TABLE 1. Description of characteristics of parent culture P56 and mutants

| Isolate | Colony diam cm on PMDA after 6 days | Colony color | Cultural appearance | Pathogenicity to radish seedlings | Self-fer-tility |
|---------|------|-------------|---------------------|-----------------------------------|-----------------|
| Parent P56 | 10.0 | Off-white | Wild type | Pathogenic; infection cushions, spreading lesions, seedling death | Yes |
| Sparse | 9.0 | Off-white | Similar to wild type but sparse; few aerial hyphae | Nonpathogenic; no cushions, grows freely over stems without adhering | No |
| Stumpy | 8.5 | Gray | Dense, even radial growth; abundant aerial hyphae | Nonpathogenic; growth inhibited on stems | No |
| Fleecy | 6.3 | White | Dense, even radial growth with dentate periphery; abundant aerial hyphae | Pathogenic; similar to P56 | No |
| Curly | 5.0 | Light brown to brown | Dense, irregular-shaped colony; aerial hyphae and sclerotia in tufts | Hypersensitive reactions; cushions produce localized necrotic flecks | No |
| Rusty | 3.8 | Light brown to red brown | Dense, irregular-shaped colony; aerial hyphae and sclerotia in tufts | Nonpathogenic; few superficial cushions, sparse growth on stems | No |
| Ropy | 6.5 | White to light brown | "Ropy" irregular-shaped colony; few aerial hyphae | Nonpathogenic; no cushions, grows over and adheres to stems | No |

In other single-spore cultures differing from the parent field isolate in growth rate and cultural characters, changes have been observed in cultural appearance and in fruiting ability (Flentje and Stretton, 1964). Where these changes have occurred in single-spore cultures derived from spores more than 99% of which were uninucleate, it is almost certain the changes were due to mutation. The most obvious change in cultural appearance is with faster growing sectors showing out. Some of these faster growing sectors have been fruited and two types of single-spore cultures, one identical with the original slow growing culture and one identical with the faster growing sector, have been obtained, indicating that mutation had occurred.

Similarly, in relation to fruiting, with repeated testing of nonsporing single-spore cultures, occasional flecks of hymenium occur. Cultures obtained by isolating prebasidial cells from these flecks have fruited abundantly (Stretton, et al., 1967) and given rise to single-spore cultures some of which fruited readily and some of which failed to fruit. Heterokaryons formed by mutations of this nature are extremely important in relation to genetic studies with *R. solani*.

3. By anastomosis. Self-anastomosis occurs commonly in cultures of *R. solani* and it is often assumed that cross anastomosis between cultures also occurs commonly. Schultz (1936) and Richter and Schneider (1953) paired a large number of cultures and, on the assumption that successful anastomosis indicated a close genetic relation, they attempted to group their cultures according to their success in anastomosing. However, they did not study the anastomosis reaction in detail, nor did they determine whether heterokaryons were formed following anastomosis.

Whitney and Parmeter (1963) investigated hetero-karyon formation following anastomosis between single-spore cultures derived from one parent. They found that two distinctive single-spore cultures, when paired, regularly gave rise to a culture type different from that of the single-spore cultures but similar to the original parent. Ten other distinctive single-spore isolates from the same parent were then paired with each of the first two single-spore cultures. All 10 gave rise to a new cultural type, resembling the parent, with one or other but not with both of the first two single-spore cultures. It appears from these results that some kind of compatibility factor was controlling heterokaryon formation. These workers also found that, from the original parent, approximately half the single-spore cultures were able to grow satisfactorily on Czapek-Dox agar (CDA+) while the other half were not (CDA−). They opposed CDA+ and CDA− cultures in the three possible combinations, obtaining a new cultural type similar to the parent in each case. The new cultures were then fruited and single-spore cultures obtained. From CDA+ × CDA+ crosses, all progeny were CDA+, from CDA− × CDA− crosses, all progeny were CDA−, and from CDA+ × CDA− crosses, half the progeny were CDA+ and half were CDA−. Their results provide fairly conclusive evidence of heterokaryon formation and it appears that ability to grow on Czapek-Dox agar may segregate as a single factor. With regard to cultural type, however, the progeny of the above crosses varied considerably, indicating that though the initial crosses were between progeny from one parent, the nuclei were heterokaryotic with regard to a number of factors.

Garza-Chapa and Anderson (1966) also investigated heterokaryon formation by anastomosing single-basidiospore cultures derived from several parent isolates. Although they had no specific markers, it appears from general cultural characters that some pairs of single-basidiospore cultures did give rise to heterokaryons.

McKenzie (1966) recently reported extensive investigations of heterokaryon formation using the spontaneous and induced mutants described earlier, namely, *fleecy, curly, stumpy, sparse, rusty,* and *ropy.* These mutants carried morphological, sterility, and virulence markers. Excluding *fleecy,* any two mutants, when paired, anastomosed and produced a cultural type identical in both cultural appearance and virulence on radish seedlings with the parent single-spore culture from which the mutants were obtained. The new growth from paired mutants that fruited yielded four types of single-spore culture (Fig. 6), namely: *a* and *b* identical with the two component mutants; a culture type identical with the parent single spore-culture which fruited readily to give further single-spore cultures identical with itself and each other and was therefore regarded as the recombinant nonmutant (nm); and a very slow growing culture subsequently identified by anastomosis tests as the recombinant double mutant (dm). Although the double mutant when paired with either of the two contributing single mutants failed to give rise to a new growth type, when paired with any of the other noncontributing single mutants, new growth identical with the original parent single-spore cul-

ture resulted. This new combination again fruited and gave rise to eight different culture types identified by anastomosis tests as *a, b,* and *c*(the three original single mutants), nm (recombinant nonmutant), the double mutant combinations (dm 1, dm2, dm 3) of the three original mutants, and the triple mutant (tm). In all the recombinant work, analysis was carried out using cultures from both tetrad and randomly shed single spores. This work thus confirms that of Whitney and Parmeter (1963) concerning the formation of heterokaryons by anastomosis.

The use of distinctive markers by McKenzie (1966) begins to offer critical information on factors affected by mutation and the operation of those factors in a heterokaryon. The parent single-spore culture, from which the mutants were derived, was almost identical with the original field isolate in cultural appearance, pathogenicity, and fruiting ability. The heterokaryons formed from pairs of mutants were indistinguishable from the parent single-spore culture with regard to these characters (Fig. 7), indicating that the mutant nuclei complement each other. With regard to fruiting ability, several important points were established. Among single-spore cultures from most of the heterokaryons, the two mutants were equally represented, suggesting they were both required for meiosis and the subse-

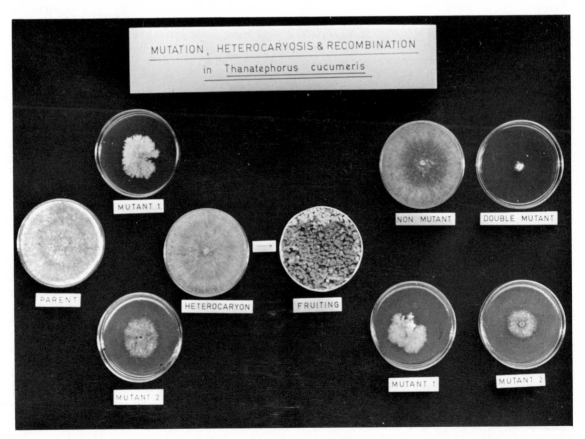

*Fig. 6.* Cultural characteristics of a heterokaryon resulting from anastomosis of 2 mutant basidiospore cultures of *T. cucumeris,* and the 4 types of single-basidiospore progeny resulting from cross.

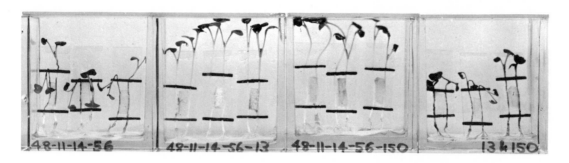

*Fig. 7.* Pathogenicity tests on radish stems showing variation of host-fungus reaction after 7 days of (*from left*): G3 single-basidiospore culture, 2 mutant single-spore cultures derived from G3 culture, and heterokaryon resulting from anastomosis of 2 mutants.

quent production of basidiospores. From one heterokaryon, however, the majority of spores represented only one of the mutants. In isolations made progressively from binucleate prebasidial cells back to multinucleate vegetative cells of this latter heterokaryon, the binucleate cells were predominantly made up of the same mutant as the spores, and only in the multinucleate cells were both mutants commonly represented. Apparently both nuclear types were necessary to initiate fruiting, but then one mutant was able to continue to the formation of spores. These results support the suggestion that several genetic factors influence fruiting ability at different stages.

Although field isolates of *R. solani* are regarded as stable, and Whitney and Parmeter (1963) found their heterokaryons to be stable, there was significant evidence in the investigations of McKenzie (1966) that heterokaryon instability can occur (Fig. 8). When *fleecy* was paired with any of the other five mutants, a heterokaryon with normal growth resulted, but before this had covered a 9 cm petri dish, the growth type had reverted to *fleecy* in each case except the heterokaryon between *fleecy* and *stumpy*, where the growth always reverted to *stumpy*. The heterokaryon formed between *fleecy* and *ropy* was fruited and the single-spore cultures obtained were in appearance predominantly *fleecy* and only occasionally *ropy* or recombinants.

It thus appears that some heterokaryons are quite unstable but the factors responsible for this instability have not yet been investigated.

Although heterokaryons can be formed as a result of anastomosis, successful anastomosis between different cultures may be very restricted. Whitney and Parmeter (1963) found that some single-spore cultures from one parent, when paired, apparently failed to form heterokaryons. Flentje and Stretton (1964) and Stretton, et al. (1967) investigated the anastomosis reaction in some detail. They found that there were several different stages in the reaction. Hyphae of different isolates, growing in close proximity, may have some mutual chemical attraction increasing the frequency of hyphal contact. On contact, the hyphae either may show no reaction to

each other or their walls may fuse. Following wall fusion, a cytoplasmic connection may or may not be formed by a pore through the area of wall fusion. Subsequent to the cytoplasmic connection, a new side branch giving rise to a new colony may be formed from the contributing cells, a successful reaction, or the contributing cells together with some neighboring cells and any new side branch may die within 24 hours of the cytoplasmic connection, a killing reaction (Fig. 9). When field isolates of the same or different geographic origin with different pathogenicity were paired, there was either no reaction, only wall fusion, or cytoplasmic connection followed by death of the contributing cells and up to six cells on each side. No successful anastomoses were observed. When first-generation single-spore

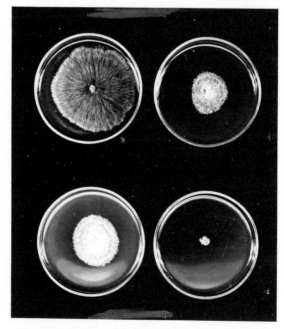

*Fig. 8.* Reversion of heterokaryon formed from 2 morphological mutants, to cultural morphology of 1 of the original mutants. *Top, from left,* mutant A, mutant B. *Bottom, from left,* heterokaryon from A × B, subculture of heterokaryon showing reversion back to mutant A.

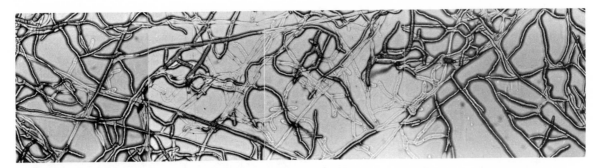

Fig. 9. Death of cells, approximately 72 hr. after anastomosis between 2 F₁ single-basidio-spore cultures *a* and *b*, derived from 1-parent isolate of *T. cucumeris*. Cultures *a* and *b* grown on cellophane overlying water agar.

cultures from one parent were paired, cytoplasmic connections followed by the killing reaction occurred in all pairings. In a small number of these pairings, however, the killing reaction did not occur in every single anastomosis, and where it did occur it appeared more slowly and did not involve cells other than those that anastomosed. It was not possible, because of the lack of genetic markers, to determine whether heterokaryons were formed in any of these instances between the different field isolates or the single-spore cultures.

McKenzie (1966) investigated anastomosis between the mutants described above, all obtained from a single homokaryotic parent. Some pairs of mutants anastomosed without killing, but in other pairs, a killing reaction occurred. If these latter pairs were incubated at 25°C over the anastomosing period, heterokaryons were established (Fig. 10), but if incubated at 5°C, no heterokaryons were established. The affect of temperature appeared to be on the rate of growth of the heterokaryon rather than on the killing reaction, which occurred approximately 24 hours after anastomosis at both temperatures and killed up to six consecutive cells of the heterokaryon. At 25°C, 150-200 new consecutive cells were formed by the heterokaryon and the majority escaped the killing reaction, while at 5°C, only 4-5 new cells were formed and these were all killed.

It is obvious then that the whole anastomosis sequence is of great importance to heterokaryon formation and warrants intensive study. As a working hypothesis from the present results, it is suggested that closely related single-spore isolates may anastomose successfully to form heterokaryons, but in unrelated cultures, either failure to form cytoplasmic connection or cytoplasmic incompatibility inhibits successful heterokaryon formation. The fact that distinct strains of *R. solani* coexist in some soils with no apparent intermediate types (Flentje, 1957) supports this hypothesis.

*Extent.*—Successful laboratory synthesis of heterokaryons so far involves the association of pairs of nuclei that are known to be closely related, being derived from the same source either by segregation or mutation. The extent of variation covers the same characters that were discussed in earlier sections

under mutation and recombination. It is clear that the nuclei in a heterokaryon act in a complementary fashion, a nonmutant allele making the phenotype of the heterokaryon nonmutant for that character despite the presence of a mutant allele. It would appear that the phenotype depends on the presence or absence of the nonmutant allele rather than the ratio of a nonmutant to mutant.

No successful anastomoses have been made between isolates that differ in their host range, and therefore at present it must be assumed that the variation through heterokaryosis does not cover this important character. Thus, the free anastomosis of different pathogenic strains in soil to produce more virulent strains as suggested by Daniels (1963) appears on present evidence most unlikely to occur.

Fig. 10. Heterokaryotic sectors formed from zones of cell death between 2 mutants.

OPERATION AND SIGNIFICANCE OF MECHANISMS OF VARIATION.—Available evidence indicates that there is considerable potential for variation in *R. solani* through a number of mechanisms involving recombination, mutation, and heterokaryosis. The major question to which we seek answers is, how do these mechanisms operate in nature to affect saprophytic growth and survival on one hand and virulence and pathogenicity on the other?

On a worldwide basis, field isolates differ greatly in cultural appearance and pathogenic specificity. This range of variation may appear to be more complex than it really is because there have been no standard reference media on which cultures could be grown for critical comparison and no set of standard reference hosts for which pathogenicity could be described. If agreement could be reached concerning such standard reference media and hosts and methods of testing virulence, it would be a most significant outcome from this symposium. Nevertheless, we ourselves have compared, on standard media and a set of reference hosts, more than 100 isolates from several different countries and the variation among these isolates would support the evidence from the literature that there is a wide range of cultural and pathogenic strains.

Yet, when comprehensive isolations are made from a local area, we usually find very limited variation. Flentje (1957) showed there might be two or three different pathogenic strains in one soil, but that practically all isolates of any one pathogenic strain from that area were the same in general cultural

characters. This work has been repeated recently by de Beer (1965) who used a wider range of isolation techniques with essentially the same results. Papavizas and Davey (1962*b*) isolated from 15 different soils and obtained 16 clones from one soil, 5-9 clones from each of nine other soils, and only 1-3 clones from the remaining five soils.

From this evidence, it appears that there are factors in soil controlling the potential variation and in some cases limiting it to the development of a small number of biologically isolated strains. These controlling factors may be operating along two different lines, one genetic and the other ecological. With regard to the genetic line, on the basis of present evidence, we suggest the following hypothesis.

Assuming one original strain existed in a soil, mutation would have occurred at the normal rate and given rise to variant individuals by segregation through the basidial stage, through unstable heterokaryons, or through secondary septation of cells reducing the number of nuclei in individual cells of hyphal fragments. Some of these variants, being closely related, would have anastomosed to give heterokaryons similar to the original strain, thus tending to restrict the occurrence of individual variants. On the other hand, occasional mutants may have been sufficiently different to be unlikely or unable to form heterokaryons. If these latter variants survived, depending on the ecological conditions, further mutations selected out by the environment could increase the biological isolation of the variants, allowing them to form completely separate strains.

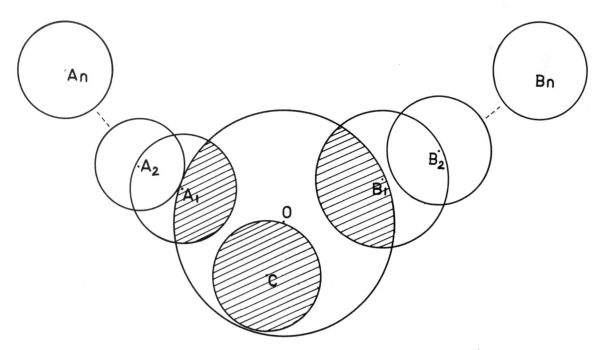

*Fig. 11.* Circle *O* represents gene pool of original isolate. Points A₁ and B₁ represent gene pools of new variants, with reduced possibility of heterokaryon formation with *O* illustrated by the shaded areas. Points and Circles A₂ and B₂ represent next stage of isolation and so on to *An* and *Bn*. *C* represents a segregant completely compatible with *O* and absorbed by heterokaryon formation.

This hypothesis is represented diagrammatically in Fig. 11.

But such a hypothesis immediately raises the questions, what types of mutations and segregation lead to this biological isolation, and what are the underlying mechanisms governing anastomosis and the killing reaction following anastomosis? We do know from the studies of McKenzie (1966) that isolates differing only by single-gene mutations may form unstable heterokaryons and show some killing reaction following anastomosis, and first-generation single-spore cultures from one parent can show killing reaction following anastomosis. Beyond this, however, we have little direct information. With regard to the control of anastomosis, it seems that the ability to form cell-wall fusions, the ability to form cytoplasmic connections, and the killing reaction are all controlled independently. There is some evidence from work with *Neurospora* (Garnjobst 1953, 1955; Holloway, 1955) that a number of different genes may control anastomosis and heterokaryon formation. If this is true also for *R. solani*, it would seem that the genetic factors controlling wall fusion and cytoplasmic connection are widely distributed among different strains and many will form such connections (Richter and Schneider, 1953). The work on cytoplasmic compatibility of *Neurospora* strains does not help us greatly with *R. solani* because the incompatibility in *Neurospora* is connected with mating type behavior. It seems then that a thorough investigation of these factors relating to anastomosis and cytoplasmic compatibility is warranted with *R. solani*. In particular, the compatibility question may have implications far beyond *R. solani* as it may be similar to that between different tissues in higher plants and animals.

With the material now becoming available through genetical studies on *R. solani*, it should be possible to subject the above hypothesis to critical testing using reliable genetic markers. This would also allow for critical testing of the provocative general ideas put forward by Buxton (1960) regarding heterokaryosis.

The second factor controlling variation may be microbiological competition in soil. In presenting a hypothesis, the concept of a wild type offers a useful starting point. The wild type is "a strain approximating to the more or less standard form of the species as found in the wild; without any known abnormalities of mutational origin" (Fincham and Day, 1963). Thus, if we are examining a particular pathogenic strain in any one soil, the wild type represents the sum of all the genetic factors advantageous for growth and survival in this particular situation. The situation will involve all the physical and biological factors of the environment, i.e. for a particular soil over a period of time it will involve soil moisture relations, oxygen and $CO_2$ tensions, pH, food substrates including host and nonhost plants, and competition from other organisms. The wild type will be that type which, from the genetic material available, is best able to cope with these factors. There can be little doubt that mutation will occur in nature, but any mutation will, by virtue of

the definition of the wild type, be deleterious for this situation. Although mutations will accumulate in the wild type, there must be some mechanisms, probably including heterokaryon instability, which limit the extent of accumulation. Such mutants as do accumulate are unlikely to affect either the cultural characters or behavior of the wild type (McKenzie, 1966), provided the wild type complement of genes is retained in the heterokaryon. Such a hypothesis would account for the very limited range of cultural types of any one pathogenic strain that have so far been isolated from any one soil, but would allow for the heterokaryotic condition of isolates. A change in the particular situation so far discussed may occur, either by spread of the organism to different soils or by changes in management and cropping practice in the original soil. This may involve changes in the environmental factors so that the original wild type is no longer the one best able to cope with these factors. It thus in fact ceases to be the wild type. The new selection forces would then operate on mutations that occur to produce a new wild type. This hypothesis allows for the quite different cultural types found in different soils or even from the same soil under different conditions.

To test such a hypothesis, we need much more critical information; firstly, on whether the growth of isolates on laboratory media alters the genetic constitution compared with that in soil; secondly, on what factors in the soil environment comprise the selection forces, if any; and thirdly, on what is the extent of, or limit on, mutation itself.

In the discussion of this hypothesis, attention has been focussed on variation through mutation. Variation through recombination should also be considered, though it is likely that most of the nuclear differences that would be involved in recombination arise initially through mutation. Some forms of disease caused by *R. solani* can be spread aerially by basidiospores (Echandi, 1965). Although other strains of *R. solani* form the perfect state freely in nature on soil or on plant parts, there is no reliable evidence that the single basidiospores are able to survive and form colonies in soil (Flentje and Stretton, 1964). The lack of variant forms in a soil where potatoes had been grown and *R. solani* had fruited freely on the stems (Flentje, unpub.) is perhaps further evidence to support the idea that basidiospores may not survive readily in soil. In some isolates fruited in the laboratory, only a percentage of spores germinated and formed colonies on agar. Even under these conditions, the spores often took 4-5 days at 25°C to establish a total hyphal length of 2-5 mm and under natural soil conditions they may be subject to strong competition for nutrient and the antibiotic effects of the soil flora. Papavizas (1964) and Olsen, et al. (1967) have shown that different single-spore cultures from one parent may differ in survival ability, but further investigation is required before we can accept the idea that the basidiospores will survive and grow in soil. Even if the basidiospores did survive and grow, the resultant colonies would be subject to the same selection forces in soil as outlined above, and the single-spore cultures would all thus be selected toward the wild type parent culture.

While such hypotheses would allow for potential variation, limited actual variation in one soil, and extensive variation on a world wide basis, we would still have one of the most important questions remaining, namely, the origin of different pathogenic strains.

In all the variations obtained under experimental conditions, the only variation in host range (apart from virulence) recorded so far is the apparent loss of pathogenicity due to mutation or recombination. Flentje, Dodman, and Kerr (1963) and McKenzie (1966) have shown there are a number of stages in the host infection process by *R. solani* and some of the earliest appear to be dependent on the nature of the host surface and of host exudates. The only suggestion we can put forward is that a number of successive interdependent mutations may be required to accomplish a change in host range. This orderly succession of mutations would be unlikely to develop under nonselective laboratory conditions, but may develop under natural conditions where the fungus is associated continuously with a potential host plant. The succession may involve a mutation bringing about attachment to the new host surface, a mutation bringing about a response to the new host exudate by the formation of infection structures, and one or more mutations controlling ability to grow through the new host tissue.

To examine these suggestions, it will be necessary to examine in much greater detail the infection process of a number of pathogenic strains, identify what, if any, materials in the host surface and exudates control penetration, and what governs tissue breakdown. A better understanding of these factors would then allow the screening of mutants obtained from genetic studies.

Progress with these studies is likely to be very slow because we have so little information to guide us on the direction of our research. The above hypotheses have been put forward, not with any assurance that they are correct, but in the hope that they will stimulate other workers to improve on them and better guide our work in this field.

Part II.

*Rhizoctonia solani:* the saprophyte

# Physiology of Rhizoctonia Solani

R. T. SHERWOOD—*Crops Research Division, Agricultural Research Service, United States Department of Agriculture, North Carolina State University, Raleigh, North Carolina.*

The complex interrelations of function with environment, gene action, and morphogenesis render any organization of a discussion on the physiology of an organism less than ideal. The outline chosen for this summary of published information on the physiology of *Rhizoctonia solani* Kühn reflects the natural disposition of many workers to study the mycelium, the sclerotium, and the perfect state as separate entities. This choice is not without its fortunate aspects; for, as will become apparent, the development and responses of each of these structures is under different, but undefined, control systems.

Reports on the physiology of *R. solani* must be read with caution. Many isolates referred to in the literature as *R. solani*, or one of its synonyms, undoubtedly do not belong in this species as presently defined (Parmeter and Whitney, this vol.). The present chapter uses the technical fungus names, synonymous with *R. solani*, given in the original reports. In laboratory experiments, conditions other than those being manipulated may also vary from treatment to treatment. Many workers do not discriminate among different aspects of growth and development, such as hyphal elongation, branching, increase in mass, rounding up of cells, nuclear division, vacuolation, senescence, and the like. Cells in many different developmental stages are found within a small sample. With few exceptions, data have not been analyzed statistically. Finally, because this species is highly heterogeneous, results obtained with any one clone usually do not apply to all clones.

THE MYCELIUM.—This discussion of the physiology of the mycelium includes reports based on entire cultures of unspecified composition, because most reports fail to distinguish between purely mycelial cultures and those having additional structures.

*Growth habit.*—Hyphal elongation occurs at the hyphal tip. A lateral branch forms proximal to the septum that separates the ultimate from the penultimate cell (Flentje, Stretton, and Hawn, 1963). Flentje and Saksena (1957) noted that isolates differed in ability to grow from hyphal tips. Some grew out immediately in a straight line, some formed a curling rosette before growing in a straight line, and some did not grow.

Larpent (1962, 1965, 1966) described a kind of apical dominance in *R. solani*. When the fungus was cultured on 3% sucrose, half strength Knop's solution [nitrate replaced by ammonium sulfate], 12% gelatin medium (KG), the principal axis grew well, but lateral branches ceased elongation shortly after their initiation. Decapitation of the main growing point provoked renewed growth of the most recent lateral branches. The principal growing point was regenerated in 1 or 2 hours and resumed rapid growth, but suppression of lateral branch growth was not reestablished (Fig. 1). These results indicated that apical dominance was operative on KG. When transfers were made from young actively

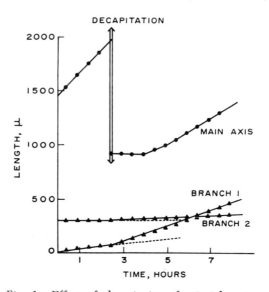

*Fig. 1.* Effect of decapitation of principle axis on growth of branches. Larpent, *C. R. Acad. Sci.* (Paris), 254:1137-1139, (1962).

growing cultures on 10% malt, 12% gelatin medium (MG) to fresh MG, lateral hyphae maintained a rate of elongation approximately equal to that of the main axis, indicating that there was no apical dominance under these conditions. When the fungus was transferred from MG cultures in which the mycelial frontier had nearly reached the edge of the petri dish to fresh MG, apical dominance (measured as the percentage of lateral branches inhibited) was strong while the growth rate was 200-400 $\mu$/hr, but became progressively weaker as the growth rate increased from 400-700 $\mu$/hr. Dominance was also weak when growth was maintained at 150 $\mu$/hr with 8-hydroxyquinoline.

*Growth rate.*—Increase in mass (dry weight) in a liquid medium takes the form of the classical growth curve having an initial period of accelerating growth, a phase of very rapid growth, a decrease in growth rate prior to reaching the maximum mass, and an eventual decrease in mass due to autolysis (Fig. 2; Ikeno, 1933; Townsend, 1957; Israel and Ali, 1964).

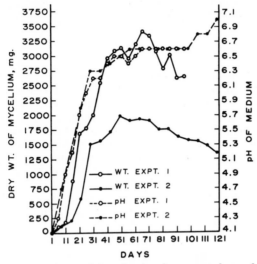

*Fig. 2.* Effect of length of culture period on dry weight of culture produced on 50 ml Richard's solution (Ikeno, *Forsch. Pflanzenkrankh.*, 2:238-256. 1933).

An isolate newly transferred to an agar medium and kept in a uniform environment reaches a uniform rate of linear growth along the agar surface after a period of adjustment to the new environment. The adjustment period is lengthened if transfers are made from inactive cultures or from cultures grown in a markedly different environment.

Isolates vary widely in their maximum rate of linear growth on a suitable agar medium at optimum temperature. Maximum growth rates of different isolates of *R. solani* vary from a trace to about 1.5 mm/hr. Thus, growth rate cannot be used to distinguish *R. solani* from most other species.

Linear-growth capacity is associated to some extent with other characteristics. Usually, isolates from aerial parts of plants, including the "sasakii" type, grow rapidly (Hemmi and Endo, 1933; Exner, 1953; Kontani and Mineo, 1962; Luttrell, 1962); isolates

from near the soil surface, including the "praticola" type, grow quite rapidly; and isolates from subterranean plant parts grow more slowly (Table 1). In a study of 86 clones, Durbin (1959a) found that aerial clones had growth rates ranging from 20-37 mm/day, with a mean of 28 mm; the surface clones grew 13-36 mm/day, with a mean of 22 mm; and the subterranean clones grew 5-22 mm/day, with a mean of 11 mm. Flentje (1956) reported that *Pellicularia praticola* grew 0.5-0.65 mm/hr. According to LeClerg (1941b), the average linear-growth rate of isolates from sugar beets was the most rapid, followed in order by isolates from potato-stolon lesions, potato-stem lesions, and potato-tuber sclerotia. Twenty isolates were studied in each of these four groups, and the differences were significant. El Zarka (1965) found no relation of growth rate to cultural type of 15 isolates from potato.

Growth rate alone cannot be used as a strain specific character. Growth ranges of different strains overlap. Many wild type and single-spore progeny, which on other grounds might be grouped with a particular strain, deviate greatly from the majority of the isolates in the strain. This deviation is usually in the direction of reduced growth rate. Hawn and Vanterpool (1953) reported that growth of single-basidiospore progeny of one clone varied from a trace to 12.5 mm/day compared with 13 mm/day for the parent. Papavizas (1965) reported that growth rate of 60 basidiospore progeny of one isolate of *P. praticola* ranged from 0.03-0.71 mm/hr. Those of one isolate of *P. filamentosa* ranged from 0.10-0.67 mm/hr. The parents grew most rapidly. Exner (1953) reported similar variations in growth rate of single-spore progeny of *P. filamentosa* f. *sasakii*, *P. filamentosa* f. *microsclerotia*, and *P. filamentosa* f. *solani*. Her data indicated that distribution in growth rate among a population from a single hymenium may not be normal; the majority tended to be either rapid or slow. LeClerg (1939a) observed that the growth rate of a sector variant was about one-third to two-thirds that of the parent.

Growth rate of a given isolate at a given temperature can differ on different media, and differential effects in growth rate among isolates times medium can occur (Ullstrup 1936; El Zarka, 1965). According to LeClerg (1934), the growth rate of a particular isolate on potato-dextrose agar (PDA) was not always correlated with that on malt-extract agar;

TABLE 1. Some characteristics of four groups of *R. solani* isolates[1]

| | Group S | Group T | "Praticola" | "Sasakii" |
|---|---|---|---|---|
| No. of isolates | 8 | 10 | 22 | 22 |
| Natural habitat | Potato | Subterranean to surface | Near soil surface | Aerial |
| Virulence | | | | |
| Seedlings | Very low | Low | Moderate to high | Moderate |
| Foliage | Very low | Moderate | High | High |
| Thiamine required | No | Yes | No | No |
| Optimum temp. °C | 24°C | 24-28°C | 28°C | 28°C |
| Growth rate at optimum (mm/hr) | 0.3-0.6 | 0.5-0.9 | 0.8-1.1 | 0.9-1.4 |

[1]Sherwood (unpub.).

some isolates grew more rapidly on one than on the other, while the reverse was true for other isolates.

*Nutrient uptake.*—King and Isaac (1964) investigated nutrient uptake at the hyphal level. Mycelium grown on thin layers of water agar was fed tritiated glucose-6-T and glycine-2-T for various lengths of time at various ages. Distribution of the mass of cytoplasm within cells was measured by interference microscopy, and distribution of the substrate was found by autoradiography. In hyphae 2-3 days old, with cytoplasm throughout, uptake of glucose-6-T or glycine-3-T was relatively uniform throughout with a ½ minute feeding time (Fig. 3). Because translocation in *R. solani* can occur at a rate of 40 $\mu$ per minute, it was concluded that if nutrient uptake were localized at specific sites, these sites must be less than 20 $\mu$ apart. As *R. solani* ages, cytoplasm is withdrawn from older side branches and concen-

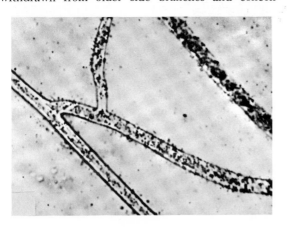

*Fig. 3.* Photomicrograph showing young (2-day-old) hyphae evenly covered by silver grains in the photographic emulsion, indicating uniform distribution of radioactivity following uniform feeding of glucose-6-T (King and Isaac [1964]. Reproduced by permission of the National Research Council of Canada from *Can. J. Botany,* 42:815-820, 1964).

trated in the main trunk hyphae. When mycelia 7-14 days old were fed, uptake was concentrated in the trunk hyphae. There was no detectable correlation between hyphal diameter and autoradiographic grain count (Fig. 4), but grain count was correlated with cell mass. Thus, it appeared that uptake was related to cytoplasmic content rather than to cell-wall surface area. Uptake of labeled glucose could not be detected when mannose, galactose, or fructose were present, but did occur in the presence of sucrose, ribose, or glycine. Glycine-2-T uptake was inhibited by alanine, serine, or sodium glutamate, but not by glucose. These competitive feeding experiments suggested that ribose, sucrose, and glycine entered by metabolic pathways other than those followed by the hexoses, and that glycine entered by the same route as alanine, serine, and sodium glutamate. Further data suggested that in *R. solani,* as in other organisms, sugar uptake may be a two-step process consisting of: (1) energy independent

*Fig. 4.* Portion of old mycelium (10 days) showing uneven distribution of radioactivity after uniform feeding. (King and Isaac [1964]. Reproduced by permission of the National Research Council of Canada from *Can. J. Botany,* 42:815-820, 1964).

transport across the cell membrane, and (2) active accumulation in the cytoplasm. Experimental support for presence of step 1 was provided by blocking uptake with uranyl acetate. In the presence of dinitrophenol, which presumably blocks step 2 but not step 1, there was a significant correlation of grain count with hyphal diameter, but not of grain count with cytoplasmic mass. These results were consistent with the two-step uptake hypothesis.

Grossbard (1958) grew *R. solani* on an agar medium containing $Co^{60}$. Autoradiograms showed a greater grain density within hyphae than in the medium. Thus, it appeared that the isolate accumulated Co against a concentration gradient.

Obrig and Gottlieb (1966) found that on a dry-weight basis, "relatively old" log-phase mycelium absorbed more glucose-U-$C^{14}$, sorbose-6-$C^{14}$, phenylalanine-1-$C^{14}$, and leucine-U-$C^{14}$, and produced more $CO_2$ than did "young" log-phase mycelium. They suggested that permeability of mycelium changes with age. Five-day-old mycelial mats on modified Richard's solution took in lesser amounts of sugars, respired less, and were more efficient in accumulating and increasing carbohydrate reserves than 3- or 7-day-old mats, according to Tolba and Salama (1965).

There is evidence for metabolic specialization within the mycelium in the utilization of cellulosic substrates. Isaac (1964a) grew *R. solani* on cellulose film with low nutrients. Interference microscopy indicated digestion of the film in the area surrounding certain short, branched hyphae in older parts of the mycelium but not in the trunk hyphae bearing the active side branches, or in younger hyphae (Fig. 5). Similar results were reported by Daniels (1963).

Wedding and Kendrick (1959) presented evidence that the effectiveness of living cell membranes in isolating the protoplast from its environment may be impaired by dithiocarbamate. Incubation with dithiocarbamate resulted in an increase in permeability of the mycelium, as measured by loss of dry weight and of $P^{32}$-labeled cell constituents.

*Translocation.* — The double-dish technique has generally been employed in studying translocation of isotopically labeled elements. A watchglass is placed within a petri plate and both are filled with an agar medium (Fig. 6). The medium is seeded

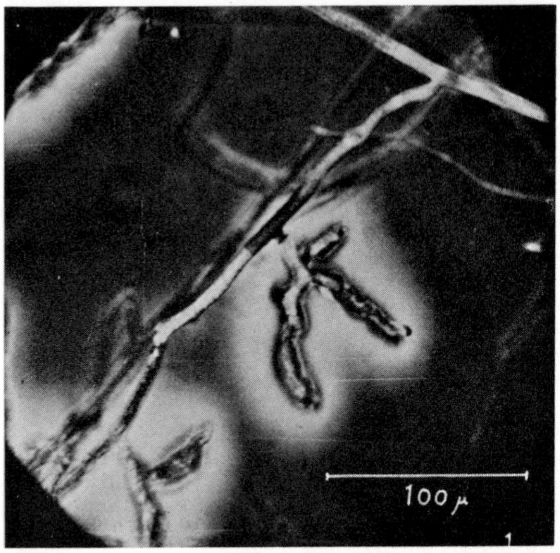

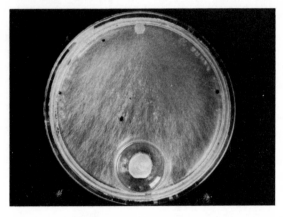

*Fig. 6.* A culture used to study accumulation of $P^{32}$ in sclerotia. Isotope was injected into agar block on watchglass (Littlefield, et al., *Phytopathol.*, 55:536-542, 1965).

isotope applied to older mycelia accumulated in the younger part of the colony, but significantly more isotope was translocated from young to old portions of the colony than from old to young. There was a strong polarity of transport from old, but especially

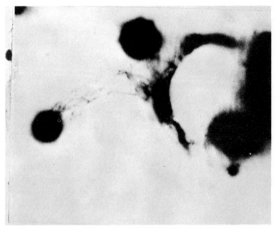

*Fig. 5.* Photomicrograph from Dyson interferometer microscope of portion of mycelium of *R. solani* growing on cellulose film showing areas (light halos) of hydrolysis surrounding some of the short lateral branches (Isaac [1964a]. Reproduced by permission of the National Research Council of Canada from *Can. J. Microbiol.*, 10:621-622, 1964).

in such a manner that the direction of growth is either away from or toward the inner container, and the isotope is added to the inner container. Even if hyphae are killed, isotopes may move along hyphae and over the glass interface by diffusion or capillarity (Monson and Sudia, 1963). However, there may be considerable leakage from hyphae and diffusion through agar. Despite these technical problems, meaningful information on translocation patterns in *R. solani* has been obtained.

Monson (1960) and Monson and Sudia (1963) found that $Zn^{65}$, $S^{35}$, $Sr^{89}$, and $P^{32}$ were apparently translocated from older hyphae to hyphal tips. The amount of nuclide recovered diminished in proportion to the distance from the point of application. $P^{32}$ was translocated in greater quantities but for shorter distances by 9-day-old hyphae than by younger hyphae. Distance by translocation increased daily after application. Greater amounts of $P^{32}$ were translocated from younger to older portions of 8-day-old colonies than in the reverse direction.

Littlefield, et al. (1965) studied translocation of $P^{32}$ by four isolates as a function of temperature. Several watchglasses were placed in a line within a petri dish to reduce diffusion through agar. A larger amount of isotope was moved by diffusion than by translocation; but translocation was more effective than diffusion in moving appreciable quantities for a long distance. As in Monson and Sudia's studies,

from young, hyphae into developing sclerotia (Fig. 7). Radioactivity was detected in individual trunk hyphae connecting the sclerotium with the source. More $P^{32}$ was absorbed and translocated at 27° than at 15°C. There was an interaction of sample location with temperature, because greater proportions were transported for greater distances at 27°C. Greater amounts were translocated at 20°, 25°, and 30° than at 4°, 10°, or 40°C. Light had no apparent effect. Two slow-growing isolates translocated more material than two rapid growers. These observations on temperature effects and accumulation indicate that translocation in *R. solani* may be mediated by an active transport system rather than by a passive process such as diffusion.

Littlefield (personal communication) later found

*Fig. 7.* Radioautograph showing accumulation of $P^{32}$ in sclerotia and in individual hyphae leading to sclerotia (Littlefield, et al., *Phytopathol.*, 55:536-542, 1965).

that translocation of P³² applied at the center of the colony continued throughout the thallus so long as it was growing linearly. On cessation of linear growth, when the agar surface became covered with mycelium, the translocating ability was lost. In cultures forming sclerotia, beginning 1-3 days after cessation of linear growth, the ability to translocate P³² was regained. Transport was directed toward the developing sclerotia. In cultures not forming sclerotia, or those in which sclerotial primordia were removed, the translocating ability was not regained. No consistent correlation existed between the amount accumulated and the age, size, or location of sclerotia. Sclerotia often contained 10,000 - 60,000 times as much P³² per unit weight as nearby mycelium 24 hours after application of the isotope to mycelium as far as 8 cm from the sclerotia.

Flentje, Kerr, Dodman, McKenzie, and Stretton (1963) stated in a preliminary report, that P³² and S³⁵ are not taken up and translocated in older or maturing mycelium. These observations are consistent with those of Littlefield, if one assumes that sclerotia were not being formed.

Grossbard and Stranks (1959) found that Co⁶⁰ is not readily translocated by *R. solani*. Although some movement within masses of mycelium was detected by the double-dish technique, none was found in individual hyphae. Nor could active translocation of radioactive Co or Cs from isotopically labeled inoculum disks of *R. solani* into soil columns be detected. Ui (1966) reported that glucose-C¹⁴ was translocated from old-to-young as well as from young-to-old areas of the culture.

Protoplasmic streaming may provide the vehicle for translocation in *R. solani*. Bracker and Butler (1964) presented visual evidence that the septal pore apparatus does not offer a mechanical barrier to streaming. Buller (1933) described protoplasmic streaming in *R. solani* as a faint gray cloud moving forward in one direction through as many as 12 cells in succession. In older mycelium, the cytoplasm is redistributed from side branches into main hyphae by means of vacuolation and emptying of the branch beginning with the ultimate cell and progressing toward the basal cell. Isaac (1964b) also reported that cytoplasmic movement appeared to be unidirectional rather than cyclical. Experiments on cytoplasmic gushing or withdrawal at points where hyphae were severed with a razor blade failed to demonstrate a positive pressure gradient in the hyphae. Microinterferometric observations demonstrated that there was a greater dry-matter density on the upstream side of the septum than on the downstream side (Figs. 8-9). This would be the expected result if the septum impeded the flow of larger organelles. The proportion of dry material in solution appeared to be greater on the downstream side of the septum than on the upstream side. Isaac suggested that the evidence was consistent with the view that the streaming of the cytoplasm of *R. solani* is due to a counterflow system similar to the type suggested for amoeboid movement.

*Nutrition.*—Nutritional requirements for vegetative growth of most isolates may be completely met by providing an organic carbon source and inorganic salts. A few isolates show an absolute requirement for thiamine. The inclusion of additional vitamins, amino acids, and other natural products in the medium increases the initial growth rate. Eventual dry-

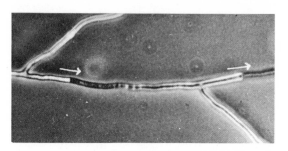

*Fig. 8.* Photomicrograph taken by positive phase contrast of hyphae of *R. solani* showing gradient of cytoplasmic density between adjacent septa. Arrows indicate direction of streaming before fixation in 4% neutral formalin (Isaac [1964b]. Reproduced by permission of the National Research Council of Canada from the *Can. J. Botany,* 42:787-792, 1964).

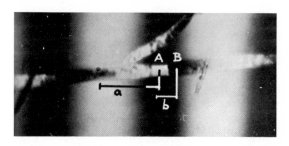

*Fig. 9.* Photomicrograph, taken with Dyson microinterferometer, of hypha showing dry-mass density drop across a septum. Fringe displacement (*a* and *b*) in image is proportional to dry mass of hyphae and is seen to be approximately three times greater on the upstream side of the septum at *A* than it is at *B* (Isaac [1964b]. Reproduced by permission of the National Research Council of Canada from the *Can. J. Botany,* 42:787-792, 1964).

weight yield obtained per gram of carbon source, i.e. the efficiency of utilization of the energy source, is not augmented by vitamins, amino acids, and other natural products (Houston, 1946; Sherwood, unpub.).

*Rhizoctonia solani* utilizes carbon and mineral element sources very efficiently. Steinberg (1950) developed a liquid medium containing 5% sucrose and minimal levels of salts, such that an increase in concentration of any of the salts or prolongation of the culture period gave no increase in yield. The efficiency of utilization by *R. solani* was about 0.5 g dry mycelium per 1.0 g sucrose supplied. All other fungi tested gave lesser yields; *R. solani* also required lower concentrations of ammonium nitrate, dibasic potassium phosphate, and magnesium sulphate to achieve maximal yield. Ross (1960) obtained yields approaching 0.5 g dry mycelium per 1.0 g carbohydrate on several carbohydrates (Table 2). On a glucose-nitrate-salts solution, isolates of the praticola type consistently yield 0.5-0.6 g/1.0 g glucose, while other isolates of *R. solani*, including the sasakii type, yielded 0.4-0.5 g/1.0 g glucose (Sherwood, unpub.). No one has tested the possibility of $CO_2$ fixation.

TABLE 2.   Utilization of carbohydrates by two isolates of *R. solani*[1]

| Substrate | Dry weight (mg/100 mg substrate) | |
|---|---|---|
| | Isolate 6 T | Isolate 7 P |
| D-ribose | 0 | 2 |
| D-xylose | 43 | 42 |
| D-galactose | 52 | 54 |
| D-glucose | 46 | 44 |
| D-cellobiose | 53 | 48 |
| Sucrose | 53 | 48 |
| $\alpha$ - $\alpha'$-D-trehalose | 43 | 42 |
| Hemicellulose | 21 | 16 |
| Inulin | 40 | 5 |
| Pectin | 0 | 36 |
| Starch | 44 | 45 |

[1]Ross (1960).

The fungus grows well on a wide range of natural and semisynthetic media. On any one medium, tissue isolates differ greatly in thickness and color of surface mycelium, abundance and color of aerial mycelium, and zonation (Peltier, 1916; Kernkamp, et al., 1952; Flentje and Saksena, 1957). A similarly wide range of variation may occur among single-basidiospore progeny from a single parent culture (Exner and Chilton, 1943; Kotila, 1947; Flentje and Saksena, 1957; Papavizas, 1965).

Vegetative characters differ when a given isolate is grown on different media. We do not know whether such differences are primarily influenced by nutritive or toxic constituents, osmotic effects, pH, or other unknown effects. Palo (1926) noted that hyphal cells were comparatively long and narrow on water agar and soybean agar but short and wide on cornmeal agar and PDA. Similar observations were reported by Madarang (1941). Variations in pigmentation have been observed by many workers (Monteith and Dahl, 1928; Madarang, 1941). The tendency of hyphae to grow embedded within the substratum or to grow aerially is also influenced by the medium (Weber, 1931). Maier and Staffeldt (1960) noted that most isolates produced predominantly submerged growth on carrot agar but both submerged and aerial growth on PDA.

Carbon sources.—The summary of reports of carbon-source utilization in Table 3 is necessarily qualitative and imprecise, because several authors (Müller, 1924; Forsteneichner, 1931; Akai, et al., 1960) have presented results in relative terms. Moderate to excellent growth is supported by numerous simple sugars and polysaccharides, and to a lesser

extent by alcohols. There is evidence for utilization of cutin (Linskens and Haage, 1963) and streptomycin (Tolba and Salama, 1961a). Compounds utilized very poorly or not at all include ribose, rhamnose, most tricarboxylic acid cycle intermediates, aliphatic acids (Table 3), ferrulic acid, vanillin, p-hydroxybenzoic acid (Ross, 1960), tannic acid, salicylic acid (Carrera, 1951), and methionine (Gottlieb, 1946).

When several isolates are tested on several carbon sources, differential responses are observed. No particular carbon source consistently supports the highest, or lowest, yields of all isolates (Müller, 1924; Forsteneichner, 1931; Bianchini and Wellman, 1958; Akai, et al., 1960; Table 2). Furthermore, the general composition of the medium, in particular the nitrogen source, greatly influences utilization of any given carbon source. Müller (1924) obtained greater growth from nearly all carbohydrates tested in a medium with potassium nitrate than in a medium with ammonium sulfate. Apparent utilization is also related to length of culture period. An isolate tested by Israel and Ali (1964) achieved maximum dry weight more rapidly on maltose, sucrose, and starch media than on glucose and fructose media, though the maximum dry weights achieved were nearly equal. Perombelon and Hadley (1965) reported that maximum yield was achieved more slowly and was less on 1% pectin than on 1% glucose.

Many isolates have shown a marked capacity to decompose cellulose and cellulose derivatives. Variations exist between both isolates and cellulose sources. Daniels (1963) found that all 14 isolates tested grew abundantly on flax fibers with an average 41% decrease in dry weight. The isolates varied in ability to utilize flax fibers, cellulose filter pads, and cellulose film. Natural cellulose from apple wood (Bosch, 1948), sunflower, elder (Kohlmeyer, 1956) and cotton fibers (Bateman, 1964b) have been utilized. Utilization of filter paper was reported by Garrett (1962), Daniels (1963), Pitt (1964b), and Smith (1966). Filter paper supports better growth than apple-wood cellulose (Bosch, 1948) or cotton fibers (Bateman, 1964b). Martinson (1965) and Tribe (1966) reported that cellophane supports growth. Kohlmeyer (1956) found that phriphan (cellulose hydrate) was decomposed, but that collodion (cellulose nitrate) was absolutely resistant to enzymatic degradation. Carboxymethylcellulose (CMC) can be utilized as a sole carbon source (Bateman, 1964b); however, growth on CMC is perhaps slower than on cellulose (Israel and Ali, 1964).

Kohlmeyer (1956) demonstrated that decomposition of cellulose was greatly reduced by the presence of soluble carbohydrates. In the presence of glucose, starch, or maltose, only 0.4-0.5% of the cellulose was utilized, and there was little increase in dry weight above that occurring on the simple sugar alone. With no glucose, about 20% of the cellulose was used, and mycelial dry weight was 31 mg; with 1.5% glucose, 1.5% of the cellulose was decomposed and dry weight was 160 mg.

TABLE 3.   Relative growth of *R. solani* on various carbon sources

| Source | Relative growth and reference | | | |
|---|---|---|---|---|
| | None | Poor | Fair | Good |
| Pentoses | | | | |
| L-arabinose | — | 3, 5 | 5, 14 | 11 |
| D-ribose | 12 | 12 | — | — |
| L-xylose | — | — | — | 11 |
| D-xylose | — | — | — | 3, 12 |
| Hexoses | | | | |
| Glucose | — | 2 | 1, 5, 13 | 1, 2, 3, 5, 7, 8, 9, 11, 12, 13, 14 |
| Galactose | — | — | — | 3, 8, 9, 11, 12 |
| Mannose | — | — | — | 3, 11 |
| L-rhamnose | 1 | 1, 3 | 1 | — |
| Fructose | — | 1 | — | 1, 3, 7, 11, 14 |
| Disaccharides | | | | |
| Lactose | — | — | 2, 3, 8, 9, 14 | 2, 11 |
| Maltose | — | 1 | 2 | 1, 2, 3, 7, 8, 9, 11 |
| Sucrose | — | 5 | 1 | 1, 2, 3, 5, 7, 8, 9, 11, 12, 14 |
| Cellobiose | — | — | — | 3, 11, 12 |
| Trehalose | — | — | — | 11, 12 |
| Higher sugars | | | | |
| Raffinose | — | — | — | 3, 11 |
| Polysaccharides | | | | |
| Amylose | — | — | — | 11 |
| Starch | — | 1 | 1 | 1, 3, 8, 9, 11, 14 |
| Glycogen | — | — | — | 3 |
| Inulin | — | 3, 12, 14 | 11 | 12 |
| Hemicellulose | — | — | 12 | — |
| Pectin | 12 | — | — | 11, 12 |
| Cellulose | — | 5 | — | 4, 5, 9, 11, 12, 13 |
| Sugar acids | | | | |
| Gluconic acid | — | — | — | 11 |
| Saccharic acid | — | — | — | 11 |
| Saccharinic acid | — | — | — | 11 |
| Aliphatic acids | | | | |
| Formic | 11 | — | — | — |
| Acetic | 11, 12 | 12 | 5, 11 | — |
| Butyric | 5, 11 | — | 11 | — |
| Lactic | — | 12 | 11 | — |
| Citric | 11 | 5, 12 | — | — |
| Tartaric | 11 | 12, 13 | 13 | — |
| Oxalic | 11 | — | — | — |
| Fumaric | — | 12 | — | — |
| α-ketaglutaric | — | — | 12 | — |
| Malic | — | 12 | 12 | — |
| Succinic | — | — | 12 | — |
| Alcohols | | | | |
| Methanol | — | 11, 13 | — | — |
| Ethanol | — | 12 | 12, 13 | 11 |
| Ducitol | — | 3 | 11 | — |
| Isoducitol | — | — | — | 11 |
| D-mannitol | — | 12, 14 | 12 | 11 |
| Glycerol | 10 | 5, 13 | 10, 12, 13 | 5, 11 |
| D-sorbitol | — | 2, 3 | — | 2 |
| Others | | | | |
| Alanine | — | 11 | — | — |
| Leucine | — | 11 | — | — |
| Methionine | 6 | — | — | — |
| Arbutin | — | — | — | 11 |
| Salicin | — | — | — | 11 |
| Aesculin | — | — | — | 11 |
| Amygdalin | — | 5, 9 | — | 11 |
| Lignin | — | — | — | 4 |

References:
1.  Akai, et al. (1960).
2.  Bianchini and Wellman (1958).
3.  Butler [E. E.] (1957).
4.  Domsch (1960a, b).
5.  Forsteneichner (1931).
6.  Gottlieb (1946).
7.  Kotila (1929).
8.  Luthra, et al (1940).
9.  Matsumoto (1921).
10. Matsumoto (1923).
11. Müller (1924).
12. Ross (1960).
13. Schultz (1937).
14. Townsend (1957).
————

Sodium polypectate is efficiently utilized. Growth on a pectin medium is slow, principally because the medium becomes quite acid during the hydrolysis of pectin. Eventually, many isolates bring about an increase in pH of the medium and achieve good growth (Sherwood, 1966).

Utilization of cellulose, starch, pectic substances, and probably of many lower molecular weight carbon sources is facilitated by extracellular enzymes (Bateman and Tolmsoff, this vol.).

Nitrogen sources.—The apparent suitability of various nitrogen sources is most often related to the effect of the nitrogen source on changes in pH of the medium during growth. Nitrogen sources promoting increased acidity of unbuffered media give poor growth; however, these generally are satisfactory sources if the pH of the medium is maintained within a suitable range (Table 4). Ammonium chloride, ammonium sulfate, or ammonium nitrate promote acid production to a greater extent than ammonium$_4$-tartrate or ammonium$_4$ oxalate, and accordingly are less favorable for growth in a poorly buffered medium (Townsend, 1957). Rao and Rayudu (1964) noted that nitrogen compounds promoting excessive alkalinity may also limit growth. Sodium nitrate was compared with glutamate. There were equivalent increases in pH and mycelial weight during the first four days, but on the fifth day, pH in the sodium nitrate solution reached 8.2 and growth stopped, while pH of the glutamate solution did not rise steeply and growth continued.

Many nitrogenous compounds have been evaluated by various authors. Interpretation of the results is rendered somewhat uncertain where pH was not controlled or objective measurements were not used. Different isolates vary considerably in capacity to use given nitrogen sources, and no single nitrogen source is consistently superior or inferior for all isolates (Forsteneichner, 1931; Akai, et al., 1960). Potassium nitrate and sodium nitrate support good to excellent growth of all isolates. Potassium nitrite supports no growth to good growth (Forsteneichner, 1931). Various ammonium salts are generally less suitable than nitrate (Müller, 1924). Cyanide is lethal at 0.14% and inhibitory at lower concentrations (Müller, 1955). Urea supports sparse to good growth, depending on the isolate (Forsteneichner, 1931; Akai, et al., 1960). Townsend (1957) considers that urea may be a poor source because of excessive ammonia production. Uric acid gives good growth (Ross, 1960). All of the amino acids tested have supported good growth of at least one isolate; these include alanine, asparagine, leucine, tyrosine (Müller, 1924), phenylalanine, cysteine, methionine, tryptophane (Akai, et al., 1960), and glutamic acid (Rao and Rayudu, 1964). However, some isolates have made little or no growth on these same compounds (Forsteneichner, 1931; Akai, et al., 1960). Casamino acids, casein, peptone, gelatin, gluten, and blood albumin support fair to good growth (Matsumoto, 1921; Müller, 1924; Ross, 1960). Histidine (Akai, et al., 1960), thymus, and adenine (Ross, 1960) support good growth; thymine, guanine, and hypoxanthine give fair growth. No growth occurs in media with cytosine, uracil (Ross, 1960), guanidine carbonate, hippuric acid, solanin (Müller, 1924), or caffeine (Matsumoto, 1921). Wei (1934) and Rusdhi and Sirry (1959) presented data on the effect of nitrogen concentration on growth.

Mineral elements.—The essentiality of phosphorus, sulfur, magnesium, and potassium has not been critically studied. Tyner and Sanford (1935) reported that they obtained substantial mycelial growth in a potassium-free medium, but that magnesium was necessary.

Trace element studies have shown growth responses to iron, zinc, and copper (Houston, 1946; Steinberg, 1950; Ross, 1960; Sherwood, unpub.), manganese (Houston, 1946; Steinberg, 1950; Ross, 1960), molybdenum and gallium (Steinberg, 1950), and calcium (Young and Bennett, 1922; Steinberg, 1948; Sherwood, unpub.). Steinberg (1950) employed the most rigorous and critical methods of purification and obtained exceptionally small deficiency yields (Table 5). The concentration of calcium required for optimum growth (4 ppm) is relatively large for fungal nutrient requirements. Calcium could be partially replaced by strontium (optimum at 6 ppm) but not by sodium, boron, cobalt, vanadium, nickel, aluminum or 10 other metals tested. Effects of deficiencies were usually visible as arrested growth, loss of color, and de-

TABLE 4. Growth[1] on various nitrogen sources in unbuffered and buffered media[2]

| Nitrogen source | Without CaCO$_3$ | | | With 1% CaCO$_3$ | | |
|---|---|---|---|---|---|---|
| | pH | | Dry wt. (mg) | pH | | Dry wt. (mg) |
| | Start | End | | Start | End | |
| Peptone | 5.2 | 5.9 | 172 | 6.7 | 7.7 | 330 |
| Asparagine | 4.5 | 6.2 | 150 | 7.3 | 8.5 | 311 |
| NH$_4$-tartrate | 5.3 | 3.0 | 79 | 6.8 | 7.8 | 195 |
| NH$_4$NO$_3$ | 4.5 | 2.7 | 52 | 6.9 | 7.8 | 258 |
| (NH$_4$)$_2$SO$_4$ | 4.5 | 2.7 | 44 | 8.2 | 7.9 | 227 |
| NH$_4$Cl | 4.7 | 2.6 | 36 | 7.5 | 7.8 | 363 |
| NaNO$_3$ | 4.5 | 7.8 | 99 | 6.8 | 8.4 | 224 |
| Ca(NO$_3$)$_2$ | 4.5 | 7.2 | 122 | 6.8 | 8.2 | 241 |

[1]Ten-days growth on media containing glucose, MgSO$_4$, KH$_2$PO$_4$, and equivalent amounts of nitrogen.
[2]Deshpande (1959*b*).

TABLE 5. Average yields of *R. solani* with micronutrient deficiencies[1]

| Element omitted | Conc. required for max. growth (ppm) | Method of purification and percent yield | |
|---|---|---|---|
| | | Reagent chemicals | Purified with MgO |
| Fe | 0.80 | 18.6 | 5.7 |
| Zn | 0.40 | 10.0 | 3.9 |
| Cu | 0.05 | 92.5 | 30.5 |
| Mn | 0.10 | 76.6 | 9.2 |
| Mo | 0.04 | 10.5 | 7.1 |
| Ga | 0.02 | 94.7 | 17.8 |
| Ca | 4.00 | 28.2 | 12.7 |
| Yield of control (mg) | — | 1,114 | 639 |

[1]Steinberg (1950).

creased sclerotial formation (Steinberg, 1950; Sherwood, unpub.). The failure of Houston, Ross, and Sherwood to demonstrate growth responses with certain elements found to be essential by Steinberg probably resulted from inadequate purification of media by the former workers. An interesting feature of Steinberg's work is the exceptionally poor growth obtained on the omission of molybdenum on a medium containing nitrate. Perhaps molybdenum is required for more than nitrate reduction.

Carrera (1951) showed that an isolate was sensitive to high concentrations of sodium borate, zinc sulphate, and zinc chloride, but tolerant of copper and iron sulphates.

Growth factors. — Many isolates are completely auxoautotrophic (Fries, 1938; Steinberg, 1950; Mathew, 1954*b*), but others appear to have partial or complete requirements for certain exogenously supplied growth factors. Houston (1946) obtained an 880-fold increase in growth with the addition of thiamine or yeast extract to a mineral-sugar solution. Ross (1960) obtained an appreciable increase in growth of one isolate, but not another, with the addition of thiamine. The thiamine-dependent isolate grew equally well when either thiamine, or the pyrimidine moiety of thiamine, or pyrimidine and thiazole were supplied, but grew poorly when thiazole alone was supplied, indicating that it could synthesize the thiazole moiety but not the pyrimidine. Elarosi (1957*b*) reported large gains when thiamine, biotin, or inositol were added singly, but no gains from pyridoxine. He suggested that the isolate had a partial deficiency for the three vitamins and that any one could stimulate formation of the others. Mathew (1952, 1953, 1954*b*) reported thiamine heterotrophy in *Corticium microsclerotia*.

Sherwood (unpub.) studied the growth-factor requirements of several isolates in each of four pathogenically and physiologically distinct groups of *R. solani* on a glucose-salts solution. All the isolates in group T had an absolute requirement for thiamine (see Table 1). The addition of 0.0008 ppm thiamine promoted growth. Below 0.02 ppm, thiamine availability was the principal factor limiting growth, as shown by a linear relation of maximum dry weight attained to logarithm of thiamine concentration.

These isolates did not grow on the basal medium supplemented with pyridoxine, biotin, pantothenic acid, nicotinic acid, riboflavin, p-aminobenzoic acid, and inositol, singly or all together, indicating that thiamine deficiency could not be substituted for by other vitamins. All isolates grew equally well on basal medium plus 0.15 ppm thiamine or on basal medium supplemented with yeast extract, malt extract, and casein hydrolysate, suggesting that other growth factors were not required. No requirements for exogenously supplied growth factors could be detected for any isolates in the three other groups, including typical praticola and sasakii types.

Steinberg (1950) reported that a completely auxoautotrophic isolate lost vigor on successive transfers in potato-dextrose agar (PDA). Mass transfers were made to a synthetic-agar medium containing thiamine, niacine, biotin, calcium pantothenate, pyridoxine, and p-aminobenzoic acid. This led to a rapid regeneration of the original vigor and autotrophy of the isolate.

Three vitamins (not identified) inhibited growth in culture according to Sims (1960). Verona (1952) reported that $5 \times 10^{-3}\%$ vitamin $K_5$ stopped growth.

The pattern of response of certain isolates to gibberellic acid (GA) and the presence of gibberellin-like activity in culture filtrates of one isolate suggest the possible regulating role of gibberellins in *R. solani*. Petersen, et al. (1961) reported that isolate RH-5 from potato showed an increased growth in response to 10 ppm GA in artificial media, whereas two other pathogenic isolates did not respond. Spraying of bean foliage with GA enhanced disease development in beans inoculated with RH-5, but not with the others. According to Petersen, et al. (1963), when sclerotia of two isolates from potato were soaked in 50 ppm potassium GA for 30 minutes and plated on PDA, the linear-growth rate of one isolate was significantly increased. Virulence of the same isolate on bean plants was also increased when the plants were treated with GA. An isolate of *R. solani* that caused the germination rate of peas to increase, in liquid-shake culture produced a substance having gibberellin-like activity in the dwarf-mutant-corn test. Aube and Sackston (1965) also detected gibberellin-like activity in *R. solani* by the dwarf-corn test. Graham (1958) reported that 0, 10, and 40 ppm GA incorporated in PDA did not influence growth rate or appearance.

Indoleacetic acid, or a closely related auxin, was synthesized by isolates grown on Czapek's medium supplemented with tryptophan but not on Czapek's alone. Bean leaves infected by foliage isolates had auxin levels higher than normal, and application of auxins to plants increased disease development (Dodman, et al., 1966).

2,4-dichlorophenoxyacetic acid (2,4-D) may stimulate *R. solani* at low concentrations and inhibit it at high concentrations. Naito and Tani (1952) obtained stimulation of *Hypochnus sasakii* grown on peptone agar with $2 \times 10^{-2}$ to $5 \times 10^{-5}\%$ 2,4-D. Sclerotia were larger at 2 and $8 \times 10^{-2}\%$ than in controls. Numbers of sclerotia were increased by low concentrations and decreased by high concentrations. Kurodani, et al. (1959) found that the linear-growth rate of *H. sasakii* on rice decoction was stimulated by $5 \times 10^{-2}\%$ but not by $5 \times 10^{-3}\%$ 2,4-D. Wagner (1955) tested 2,4-D and 2 methyl, 4-chlorophenoxyacetic acid at 0, 0.16, 1.6, 16.6, 166.6, and 1,666.6 mg/l. An increase in dry weight was obtained with 0.16 mg/l 2,4-D, and a decreased growth rate was obtained with 1,666.6 mg/l of either chemical; but no significant differences from the control occurred with other concentrations. Millikan and Fields (1964) reported that 100 ppm 2,4-D reduced growth by 86%. Simazine at 10 ppm reduced growth 93%. Sato, et al. (1959) reported that mycelial growth was inhibited by 0.3% MH-30 (maleic hydrazide) solution.

Weinhold and Hendrix (1963) found that exposure of PDA, potato-dextrose broth, or V-8 juice agar to light made the media partially inhibitory to *R. solani*. Their evidence supported the hypothesis that the light induced formation of inhibitory peroxides in the media.

Phenolic substances can inhibit growth (Ross, 1960). Isolates may react differentially in bringing about coloration of phenolic compounds in a Bavendamm test (Matsumoto, 1923). Matsumoto, et al. (1932) divided isolates of *H. sasakii* into groups according to color production on tannic acid agar, but these groupings were not correlated with groupings arrived at by other methods. At Raleigh, all isolates in the four groups listed in Table 1 turned tannic acid brown; however, reaction with gallic acid and naphthal was related with group alliance (Sherwood, unpub.).

*Osmotic concentration.*—The growth of *R. solani* is reduced by high osmotic concentration. LeClerg (1939a) found that a sugar beet isolate grew more rapidly on media with 0 and 1% sucrose than with 5%, 10%, or 20% sucrose. A sector variant from the isolates grew more rapidly on 5% and 10% sucrose than at lower concentrations. The variant grew more rapidly than the parent on 10% and 20% glucose. Domsch (1955) found that the inclusion of 5% glucose in an earth substrate reduced relative growth to 34%. Endo (1933) reported that linear growth of *H. sasakii* on potato-decoction agar was more sensitive to sodium chloride at 24° and 28° than at 32°C; 0.1% sodium chloride slightly retarded growth at 24° and 28°C; and 1% sodium chloride decreased growth by 60% at 24°, 40% at 28°, and 33% at 32°C. There was a trace of growth in 2 days on 5% salt at 28° and 32°, but none at 24°C. Müller (1923a,b) observed that linear growth of *Moniliopsis aderholdii* was more severely restricted than that of *R. solani* by increasing concentrations of Van't Hoff solution in potato-leaf agar. Concentrations providing 3% sodium chloride and corresponding amounts of the other constituents virtually stopped growth. Hyphae of *R. solani* looked nearly normal, but *M. aderholdii* formed gnarled and distorted "involution forms." Sherwood (unpub.) grew 20 isolates comprising four strains on PDA with Van't Hoff solution. With Van't Hoff solution containing 4% sodium chloride, growth of all isolates was greatly curtailed, but not stopped. Vegetative cells were parallel-sided, sclerotia were absent, and monilioid cells were rare. Beach (1949) found that total growth on potato-dextrose broth was not impaired by including 1, 2, and $4 \times$ Knop's solution, but was diminished to ½ with $16 \times$ Knop's.

*Atmospheric gases.*—Early investigators noted the poor growth of mycelium submerged in liquid or in sealed culture vessels and considered this an indication that *R. solani* requires an abundant oxygen supply for optimum growth. Briton-Jones (1925) and Luthra, et al. (1940) observed that growth from a seed piece immersed in liquid is much slower than from pieces on the surface of the same medium. Briton-Jones sealed young cultures in test tubes; growth ceased by the second day. The plugs were loosened after 14 days and, except for a few cultures sealed when they were a few hours old, growth resumed. Matsumoto (1921) cultured *R. solani* on a constant volume of medium in sealed flasks of different volumes; growth was proportional to volume of the flask. Balls (1906) removed oxygen from cultures with pyrogallate; in some cultures, air was displaced with nitrogen or $CO_2$. Growth ceased quickly, but resumed when air was readmitted.

Growth is influenced by high $CO_2$ concentrations. Vasudeva (1936) observed that on Richard's agar, growth was greatly reduced in the presence of 20% $CO_2$ and ceased with 50% or more $CO_2$. Luthra, et al. (1940) obtained similar results on synthetic media with various carbon sources; growth was sparse at 30% and nil with 40% $CO_2$. They reported equal growth in cultures incubated in air, 100% oxygen, 50:50 v/v oxygen and nitrogen, and "pure" nitrogen. Blair (1943) obtained progressively smaller colonies on PDA with increase in $CO_2$ concentration from 1-20%. According to Woodcock (1962), no growth occurred on Czapek-Dox yeast-extract broth under pure nitrogen or pure $CO_2$. *Rhizoctonia solani* grew better in pure oxygen than in air.

Durbin (1955, 1958, 1959a) studied the effect of $CO_2$ on the linear-growth rate of 11 isolates each of the aerial, surface, and subterranean types on Czapek's agar in growth tubes with a flowing atmosphere containing 20% $CO_2$, 20% oxygen, and 60% nitrogen, and compared it with that in air. There were highly significant differences among the three types with respect to their tolerance of $CO_2$. Twenty percent $CO_2$ inhibited the growth rate of aerial clones 73-87%, of surface clones 40-62%, and of subterranean clones 20-43%. The $CO_2$ treatment did not cause any gross morphological differences. There was a high negative correlation between the growth rate in air and $CO_2$ tolerance of any of the 33 clones. Evidence was presented favoring the view that $CO_2$ has some metabolic function aside from pH and carbonate-bicarbonate phenomena. Intolerance to $CO_2$ was believed to be one of the factors determining vertical distribution of various clones of *R. solani* in the host environment. Two clones of each of the three types were tested at $CO_2$ concentrations of 0.5%, 1.0%, and 2.5%. The growth rate of an aerial clone and of a surface clone was significantly increased at 0.5%. At 2.5%, the growth rate of the two aerial clones was significantly reduced.

The $CO_2$ tolerance of 21 clones of *R. solani* isolated by trapping from 9 soils was tested by Papavizas and Davey (1962*b*). The growth rates of three clones were inhibited more than 70%, and two were inhibited less than 40% in a $CO_2$-air mixture containing 20% $CO_2$. The other 16 clones, which fell into the intermediate category and would be classified as surface types according to Durbin's scheme, were the most pathogenic of the isolates tested. Papavizas (1964) found that 20 randomly selected single-basidiospore isolates each from single cultures of *R. praticola* and *R. solani* differed considerably in $CO_2$ tolerance. Some single-spore isolates of *R. solani* possessed higher $CO_2$ tolerances than the parent culture.

*Moisture.* — Briton-Jones (1924) reported that slight differences in the moisture content of the medium or in the humidity in the culture vessel can alter the growth rate.

Two studies are available on the effect of relative humidity (RH) on growth. Differences in the methods employed may be the basis for the apparent disagreement of results. Schneider (1953; summarized in Schmiedeknecht, 1960) seeded *R. solani* on thin layers of malt-extract-gelatine on cover slips fastened over small chambers containing solutions of sulfuric acid. The linear-growth rate decreased rapidly with decrease from 100% RH (Fig. 10). Growth was about 50% of maximum at 99%RH and was nearly absent at 96.5% RH. The threshold value for growth was 95.5% RH. *Rhizoctonia solani* was more sensitive to a decrease in vapor pressure than any other fungus tested, and was classified as a hygrophile. Townsend (1934) used salt solutions to control humidity and stated that some growth occurred at 81% RH, but that rapid growth was supported only by higher vapor pressures.

*Temperature.*—Balls (1908) measured elongation of individual hyphae in response to rising tempera-

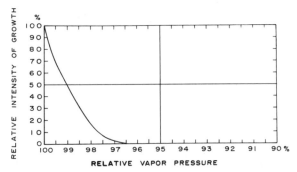

*Fig. 10.* Effect of vapor pressure on growth of *R. solani* (Schneider, *Phytopathol. Z.,* 21:63-78, 1953).

ture over short periods of time. Growth accelerated with rises in temperature up to 30°C, which corresponded fairly well with expectations based on Van't Hoff's law. Above 30°C, the curve flattened out and hyphal growth ceased at 37-38°C. Using longer culture periods, Richards (1923*b*) reported that the rate of increase conformed to the Van't Hoff law only within the range 14-25°C. Newton (1931) found that the shape of the growth curve obtained with closely controlled temperatures was different from that normally seen (Fig. 11). The temperature coefficient was large with low temperatures and declined progressively as the temperature rose. The mycelium was sparsely branched at low temperatures; therefore, at low temperatures, coefficients for linear growth were greater than coefficients for increase in colony mass.

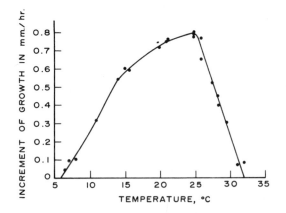

*Fig. 11.* Smoothed growth-temperature curve of *R. solani* (Newton, *Sci. Agr.,* 12:178-182, 1931).

When a culture is brought to one temperature from another, some time is required before a constant growth rate is established at the new temperature. Newton (1931) noted that when growing cultures were transferred from a low temperature to one closed to optimum, the lag period seemed to be a function of the difference in temperature. When they were transferred from 8-25°C, a constant rate was not achieved until the fourth day. According to Monteith and Dahl (1928), initial growth of new transfers placed at 15°C was more rapid if the source culture had also been at 15° than if it had been at 25°C.

The velocity of growth of cultures at temperatures higher than optimum often declines with time. Richards (1923*b*) noted that at 22.4°C or below, the linear-growth rate established in the first 24 hours was maintained for several days; but at constant temperatures from 23.6-32.5°C, the growth rate decreased progressively with time. Balls (1906) also noted that the temperature for sustained growth was lower than that for most rapid initial growth. At 33°C, the fungus grew rapidly for an hour or two but growth soon ceased. If kept at 33°C for 24 hours and then placed at a favorable temperature, growth was not reinitiated for at least another day, indicating that "self-poisoning" occurred at the higher temperature. Balls (1908) presented evidence that decrease and ultimate cessation of growth at high temperature resulted from accumulation of catabolic products, and that these products also formed slowly at lower temperatures, eventually resulting in growth inhibition of the kind generally attributed to staling products. Growth rate of several, but not all, isolates of *H. sasakii* at 36°C decreased from the second to third day according to Matsumoto, et al. (1933). In all of the above studies, drying of the medium at higher temperature could have been a factor in declining growth rate.

Roth (1935) showed that different methods of measuring growth as a function of temperature may yield different growth curves (Fig. 12). Measurement of time required to achieve a given colony diameter on an agar surface gave a curve with steep slopes and a broad "optimum." Measurement

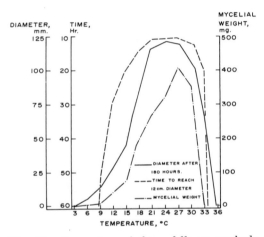

*Fig. 12.* Comparison of three different methods of measuring the effect of temperature on mycelial growth of *R. solani* (Roth, *Phytopathol. Z.*, 8:1-110, 1935).

of dry weight of mycelium formed in a nutrient solution gave a gradual slope from the minimum to the optimum temperatures, and a rather sharp optimum. Intermediate results were given by measurements of increase in colony diameter per unit of time. Bateman and Dimock (1959) reported that the optimum temperature range for growth on PDA was greater when determined by colony diameter than when determined on a dry-weight basis. These differences may be understood by considering the different responses of terminal growth and lateral branch growth to temperature. Below 15-18°C, hyphae are sparsely branched and grow in close contact with, or embedded within, the medium (Richards, 1923b; Weber, 1931). At higher temperatures, mycelia branch more profusely and grow more superficially and aerially. According to Monteith and Dahl (1928), the diameter of the terminal hyphae of one isolate increased with increase of temperature from 15-30°C, while that of another isolate decreased slightly.

Growth curves obtained on liquid media corre-

sponded to those on agar in tests by Roth (1935) and Bosch (1948), but not in tests by Jaarsveldt (1940, 1944).

LeClerg (1939a) found that the growth rate of a sector variant was less than the parent at 20° and 24°C, but greater than the parent at 34°C.

Much of the extensive literature on the effect of various temperatures on the growth of *R. solani* is summarized in Table 6. Nearly all records are in terms of linear growth on agar media. There is extreme variability within the species. The minimum temperatures for growth vary from 2-12°C. Many isolates apparently will not grow at 10°C or less and one isolate reportedly did not grow at 14°C. However, Lauritzen (1929) showed that the initiation of visible growth at low temperatures may be delayed for several days. Some isolates grow poorly at 15-16°C. The so-called optimum temperature varies from 18-33°C, depending on the isolate (Kendrick, 1951), but most frequently is·23-28°C. Some isolates grow poorly at 30°C; the maximum temperature for growth is between 30° and 40°C. The incubators available govern to a large extent the particular temperature intervals tested, therefore the experimentally determined cardinal temperatures are usually only approximate.

Despite the great variation within the species, several investigators have sought, with some success, to correlate growth-temperature characteristics with other characters used to describe strains. LeClerg (1941b) tested 20 isolates in each of four groups at 20°, 25°, and 30°C. With very few exceptions, isolates from sugar beets grew most rapidly at 30°C, and those from potato stolens, potato stems, and potato scurf grew best at 25°C. These findings were confirmed (LeClerg, 1939c; LeClerg, et al., 1942). In several studies, isolates of the sasakii type grew most rapidly at 28-31°C and did not grow at 5-8°C (Matsumoto, et al., 1933; Matsumoto and Yamamoto, 1935; Ito and Kontani, 1952). Houston (1945) found that optimum and maximum temperatures of his type C isolates were 25° and 33°C, respectively, and for types A and B, 28° and 40°C. Richter and Schneider (1953) observed that the minimum, optimum, and maximum temperatures for the "potato" and "crucifer" groups were lower than those of the four other groups studied.

The cardinal temperatures of allied species have been compared with those of *R. solani* (Matsumoto and Yamamoto, 1935; Castellani, 1935; Tims and Bonner, 1942; Mathew, 1954a). Because of the great variability within *R. solani*, such interspecific comparisons seem pointless.

Isolates having relatively low temperature optima may also have low minima and maxima, and those having high optima have high minima and maxima (Thomas, 1925; Richter and Schneider, 1953). Often, isolates from warm habitats have higher cardinal temperatures than those from cold ones. Kharitinova (1958a) noted that isolates of *M. aderholdii* from Azerbaijan had optima of 21-30°C, while those from Leningrad had optima of 13-21°C. Thomas (1925)

TABLE 6. Effect of temperature on growth of *R. solani*[1]

| Temp. associated with indicated growth (°C) | | | | | | |
|---|---|---|---|---|---|---|
| No growth | Min. for growth | Optimum | Max. for growth | No growth | Comments | Reference |
| — | 3 | 27 | 33 | — | Knop's solution | Jaarsveld, 1944 |
| 3.6 | <9 | 24 | — | 36 | Malt-extract solution | Roth, 1935 |
| — | 5 | 25 | >30 | — | PDA | Andreucci, 1964 |
| 9 | <12 | 30 | >36 | — | PDA | Bateman and Dimock, 1959 |
| 4 | — | 24 | — | 37 | PDA | Carrera, 1951 |
| — | 10 | 27 | >32 | — | PDA | Endo, 1963 |
| 4 | <8 | 25, 30 | 40 | — | PDA | Leach, 1947 |
| 1 | 5 | 25 | — | 35 | PDA, 1 isolate | LeClerg, 1934 |
| 5 | 8 | 25, 30 | >35 | 40 | PDA, 4 isolates | LeClerg, 1934 |
| — | 6 | 25 | 32 | — | PDA | Newton, 1931 |
| — | 7 | 27 | 34 | 35 | PDA, 2 pH values | Nightingale and Ramsey, 1936 |
| — | 4 | 23-26 | 32 | — | PDA | Samuel and Garrett, 1932 |
| 5 | <10 | 25, 30 | >35 | 40 | PDA | Smith, 1946 |
| 3 | 7 | 25-28 | 37 | 40 | PDA | Tompkins and Ark, 1946 |
| 3 | 4.5 | 26-30 | 32.5 | 38 | PDA | Townsend, 1934 |
| — | <5 | 20-25 | — | 32 | PDA, spring strain | Ui, et al., 1963 |
| 5 | — | 28 | >32 | 35 | PDA, summer strain | Ui, et al., 1963 |
| — | 5 | 25-28 | 32.5 | — | PDA | Verhoeff, 1963 |
| 8 | — | 26 | — | 36 | PDA | Weber, 1939 |
| 8 | — | 29 | — | 33.5 | PDA | Weber, 1939 |
| — | 8 | 27 | 37 | — | PDA | Wei, 1934 |
| — | 6 | 22-28 | 30 | — | PDA | Yu, 1940 |
| — | 4.6 | 25-27 | >32.6 | 35 | Potato agar | Richards, 1923b |
| — | 5-6 | 25-29 | 35 | — | PDA | Richter, 1936 |
| 10 | — | 28, 32 | 36 | 40 | 3-agar media | Endo, 1940 |
| 0 | 2 | 23 | 31.5 | 34.5 | Carrot agar | Lauritzen, 1929 |
| 10 | — | 28 | >35 | 40 | Czapek's agar | Chowdhury, 1944 |
| 10 | 12 | 26 | 37 | — | Oatmeal agar | McRae, 1934 |
| 3 | 4.5 | 24 | 30.8 | 33 | Leaf-decoction agar | Müller, 1924 |
| — | 7 | 25 | >30 | — | Malt agar | Müller, 1923a |
| 14 | — | 29 | 30-35 | — | Malt agar, *M. aderholdii* | Müller, 1923a |
| — | 5-7 | 23 | 32 | — | Malt agar, *H. solani* | Müller, 1923b |
| 3 | 6 | 24, 27 | 33 | 36 | Malt agar | Roth, 1935 |
| — | 9 | 24-27 | 33 | — | Malt agar | Jaarsveld, 1944 |
| — | 5 | 18-25 | >30 | — | | Frederiksen, et al., 1938 |
| — | 10 | 30 | 40 | — | *H. sasakii* | Hemmi and Yokogi, 1927 |
| — | 10 | 20, 25 | 40 | — | | Rabinovitz-Sereni, 1932 |
| 2-3 | — | 24 | — | 34-35 | | Spokauskiene, 1961 |
| — | 7-11 | 27-29 | >38 | — | | Walker, 1928 |

[1]Additional reports are given by Balls (1906); Gratz (1925); Weber (1931); Wellman (1932); Matsumoto, et al. (1933); Ryker and Gooch (1938); Roth and Riker (1943b); Beach (1949); Kendrick (1951); Ito and Kontani (1952); Singh and Singh (1956); Deshpande (1960a); Fulton and Hanson (1960); Ross (1960); Saksena and Vaartaja (1960); El-Helaly, et al. (1962); Graham, et al. (1962); and Kontani and Mineo (1962).

observed that isolates originating in warm environments did not grow at 3-7°C while those from temperate regions did.

The cardinal temperatures for growth may depend on the medium used in the test. When 15 isolates from potato were grown on two media at 24° and 28°C, one isolate showed a pronounced interaction of medium and temperature optimum (El Zarka, 1965).

*pH.*—Reports on relative growth of cultures on media adjusted to various initial pH values are summarized in Table 7. The data must be interpreted cautiously, because most investigators used unbuffered media, and pH values probably became modified during the culture period. Only one isolate has been reported to grow at pH 2.0 (Richter and Schneider, 1953). Although many isolates will grow at initial values of 2.6-3.0, some do not grow at 3.0. Most rapid growth usually occurs on media initially at 5.0-7.0. In the alkaline range, some isolates reportedly do not grow at 7.8; however, others grow at 8.0-11.0. If the fungus is able to initiate growth on moderately acid or alkaline media, it will usually modify the pH to a range more favorable for itself and will grow well. Thus, most published pH-growth curves, particularly those from unbuffered media, rise steeply from the most acid value, permitting growth, and have a broad optimum. Upper limits are generally not established.

Jackson (1940) buffered a liquid medium with sodium glycerophosphate and renewed the solution every second day. The two isolates tested did not grow at pH 2.5, grew slightly at 3.5, showed a progressive increase up to 5.5 and 6.5, a decrease at

7.5, and very little growth at 8.5. Using buffered media, Roth and Riker (1943*b*) and Bateman (1962) obtained bimodal growth curves. Bateman found that maximum growth of an isolate on buffered PDA occurred at 5.8; less growth occurred at 6.8 than at 5.8 or 7.5. Differences were significant at the 1% level. Similar results were obtained with either potassium phosphate buffer or with McIlvaine's citrate-sodium phosphate buffer on PDA, potato-dextrose broth, or dextrose salts solution. Growth curves for pH were similar on the agar and liquid media with the exception that there was no growth on PDA at 2.5, but there was growth in liquid cultures at 2.2. Roth and Riker found bimodal response with growth peaks at 4.2 and 6.2 for an isolate grown on PDA buffered with sodium phosphate.

Although individual isolates vary in pH responses, there is little evidence that pH response is correlated with other characteristics. Samuel and Garrett (1932) stated that potato strains had a somewhat wider optimum range than wheat strains. In a comprehensive study, Richter and Schneider (1953) evaluated 105 isolates assigned to six strains on the basis of pathogenicity, growth rate, cultural appearance, and anastomosis. Growth on beer-wort agar adjusted with sodium hydroxide or hydrochloric acid to pH values from 2.0-11.6 was measured. Isolates within each strain differed markedly from one another in pH responses, but there were few characteristic differences between strains. Gratz (1925), Tsiang (1947), Kernkamp, et al. (1952), and others present data illustrating differences among isolates.

Changes in pH of the medium during culture are influenced by the isolate used, composition of the medium, initial pH value, and length of culture period. Barker and Walker (1962) reported that most root and stem isolates gave decreases in pH, while two root and all foliage isolates tested gave an increase in pH in 5 days. Castellani (1935) reported that *R. solani* and *R. solani* var. *cedri-deodarae* gave alkaline values, while *R. lupini* gave acid values. Matsumoto, et al. (1932) incorporated indicator dyes in Duggar's agar and found that 17 isolates of *H. sasakii* made the medium more acid than did *R. solani*. The influence of composition of the medium is illustrated by their observation that the same isolates caused an increase in pH of Czapek's agar. The influence of nitrogen and carbon sources on pH changes during culture was discussed in earlier sections. It was also pointed out that under some circumstances, such as with a pectin-salts medium, pH may change first in one direction, then in the opposite, so that the length of the culture period is a factor.

When certain isolates are grown on media adjusted to various initial pH values, the reaction may be altered toward a value characteristic for the particular isolate on that medium. Bosch (1948) tested *R. solani* and *M. aderholdii* on a glucose-nitrate solution with citrate-phosphate buffer. If the initial pH was 3.6-6.2, the values rose to 5.2-6.6 in 18 days; if initial pH was 6.7-7.6, the pH dropped slightly. Weber (1939) found that *C. microsclerotia* grown on PDA at initial pH values of 4, 5, 6, 7, 8, and 9 changed the reaction to pH 6.0-6.8 in 60 hours. Unseeded control medium at pH 8 and 9

TABLE 7. Effect of pH on growth of *R. solani*

| Low pH Growth | | | | High pH Growth | | | | |
|---|---|---|---|---|---|---|---|---|
| None | Poor | Good | Best | Good | Poor | None | Comments | Reference |
| — | 3 | 3 | 5, 6, 7 | 9 | — | — | PDA, 5 isolates | Weber, 1939 |
| — | 3.4 | — | 6-7 | — | 9 | — | PDA, 5 isolates | Samuel and Garret, 1932 |
| — | — | 3.2 | 5.6-6.2 | — | 9.1 | — | PDA, 5 isolates | LeClerg, 1934 |
| 2.8 | 3.5 | — | 6.0-6.2 | — | 10.0 | — | PDA | Carrera, 1951 |
| 2 | 2.4 | 3.5 | — | 7.0 | 9.0 | — | PDA, buffered | Roth and Riker, 1943*b* |
| 2.5 | 3.1 | 3.9 | 5.8 | 7.5 | — | — | PDA, buffered | Bateman, 1962 |
| 2 | 2.4 | varies with isolate | | — | 10.4 | — | Potato agar | Gratz, 1925 |
| 2.47 | 2.57 | 4.6 | 5.4-6.7 | 7.8 | — | 7.85 | Potato broth | Endo, 1940 |
| — | 2.6 | — | 5.5-6.0 | — | 9.9 | — | Potato-dextrose broth | Elarosi, 1957*b* |
| — | — | 3.2 | 6.2 | 7.0 | 9.1 | — | Richard's agar | Chowdhury, 1944 |
| 3.0 | — | 4 | 6-7 | 8 | — | — | Richard's | Ross, 1960 |
| 3.0 | — | 3.6 | — | 6.7 | 7.6 | 7.7 | Richard's, buffered | Bosch, 1948 |
| 3.4 | — | 4.0 | 7.4 | 8.7 | — | — | Czapek's solution | Castellani, 1935 |
| — | — | 3.2 | 6.0 | 8.5 | — | — | Glucose-peptone, buffered | Vasudeva, 1936 |
| 2.4 | — | 3.2 | 6.6 | 9.1 | — | — | Glucose-peptone | Deshpande, 1959*b* |
| 2 | 2 | — | 7.4 | — | 11.6 | — | Beer wort, 104 isolates | Richter and Schneider, 1953 (Result variable with isolate) |
| — | — | 3.0 | 4.0-8.5 | — | 9.5 | — | | Valdez, 1955 |
| — | 2.5 | — | 6.5-7.0 | — | 10.0 | — | | Weber, 1931 |
| 2.9 | 3.2 | — | 6.5 | 8.5 | — | — | Malt-extract soln.-buffered | Roth, 1935 |
| 2.5 | 3.5 | 4.5 | 5.5-6.5 | 7.5 | 8.5 | — | Defined soln.-buffered | Jackson, 1940 |
| 1.8 | 3.2 | — | 5.8-7 | — | 9.8 | — | | Montieth and Dahl, 1932 |
| 2.0 | — | 3.0 | 6.0 | 7.0 | — | 10.8 | Tomato juice | Ragheb and Fabian, 1955 |

declined slightly, presumably due to the effect of atmospheric $CO_2$. Castellani (1935) found that Czapek's solution initially at pH 4.0-8.7 became differentially modified to final pH ranges as follows: *R. solani*, 7.2-7.4; *R. solani* var. *cedri-deodarae*, 7.0-7.1; and *R. lupini*, 5.0-5.2.

Labrousse and Sarejanni (1930) and Matsumoto, et al. (1932) incorporated acid-base indicators and redox potential indicators into media. In studies with several fungal species, including *R. solani*, the observed redox potentials were correlated with abilities of the various species to alter pH.

*Light.*—Butler [E. E.] (1957) found that continuous fluorescent light of 150-400 f.c. suppressed vegetative growth. Endo (1935; 1940) stated that sunlight partially inhibited linear growth of *C. sasakii;* however, his experiments did not seem to rule out temperature increases inside dishes exposed to sunlight as a possible cause of inhibition. Briton-Jones (1924), Rabinovitz-Sereni (1932), and Roger (1942) did not find any reaction to light vs. darkness.

Durbin (1959*b*) compared darkness with fluorescent light of 200 f.c. and energy peaks at 580 and 480 $\mu$. Illumination significantly decreased mycelial dry weight of seven of the nine clones tested (Fig. 13), but did not influence the linear-growth rate. The length and development of lateral hyphae, but not of terminal hyphae, were decreased by illumination.

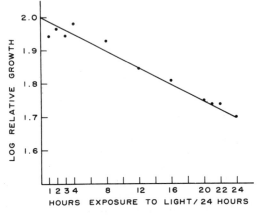

*Fig. 13.* Effect of daily period of illumination on log of mycelial dry weight of clone of *R. solani* after 1 week's growth in shake culture (Durbin, *Phytopathology*, 49:59-60, 1959).

*Tropism.* — Bianchini and Wellman (1958) reported that *P. filamentosa* did not show positive geotropism.

Thigmotropism was observed in two isolates that were highly virulent parasites on fungi (Butler [E. E.], 1957). Tight coils, arising primarily from a single growing point formed around glass fibrils, 8 $\mu$ in diameter, on PDA. Sparse coils originating mainly as side branches from runner hyphae formed on glass tubes 30-45 $\mu$ × 1 cm. The tendency of hyphae to grow along the longitudinal (anticlinal) walls of

epidermal cells prior to penetration could be a tropistic response (Kerr and Flentje, 1957; Khadga, et al., 1963).

*Aversion.* — Chen (1943) paired isolates of *R. solani* on PDA. Pairings were made on the basis of (1) similarity or dissimilarity in cultural appearance, (2) original host, and (3) similarity in hyphal diameters. When a race was paired with itself, a thin line or "barrage" often appeared, but the colonies usually merged after a few days. When different isolates were paired, various types of reactions occurred. In some cases a barrage was formed and persisted; in one colony grew completely over the other. Various intermediate reactions also occurred. The reactions were not correlated with cultural appearance, hyphal diameter, or host origin. One race was antagonistic to four others in one trial but not in another, indicating that the reaction might be influenced by environment. There were no consistent reactions in pairings between cultural variants and their parents, variants and sister variants, and variants from different parents.

Whitney and Parmeter (1963) observed three kinds of reactions when single-spore strains were paired on an agar medium. Some strains grew together to form an indistinct line of contact; some approached each other but failed to mingle (suggesting a barrage); and others formed tufts of aerial mycelium at the line of contact, possibly in conjunction with anastomosis.

Vanterpool (1953) reported that an isolate from flax showed slight aversion to isolates from potato on culture plates. Thirumalachar (1953) claimed to have distinguished two physiologic strains of *Rhizoctonia* by aversion.

*Longevity.*—Reports on the ability of *R. solani* to survive adverse temperature, moisture, or other conditions for various periods of time are summarized in Table 8. Precise comparisons between, and often within, studies are usually not possible, because different investigators used different conditions. The method usually employed to assay viability after treatment has been to place the fungal material in a nutrient medium at a temperature suitable for growth. Although failure to renew growth is usually accepted as an indication of death, obviously this interpretation may be erroneous. A vital staining method for distinguishing living from dead sclerotia of *H. sasakii* was developed by Hemmi and Endo (1928) but has rarely been applied (Endo, 1940). When stained with eosin or acid fuchsin, the cells of living sclerotia were cinnamon buff, but inner cells of dead sclerotia were spinal pink. Georgopoulos and Wilhelm (1962) used disappearance of the vacuolate structure of hyphae and inability to obtain subcultures as criteria of death.

Reports usually lack adequate information concerning the conditions of the test. Adequate attention has not been given to the effects of prior environments, maturity of the fungus, types of cells and structures present, size of sclerotia, changes in temperature, moisture, oxygen content, and other conditions of the test. The principal causes of variable longevity often cannot be distinguished.

An overall perusal of the data suggests certain generalizations. Survival is favored by lower temperature down to 5°C. Temperatures of 45°C or above are injurious. Resistance to high temperatures is greater under dry conditions than under moist ones. Mycelia are less resistant than sclerotia to high temperatures. Under moderately severe stress, the number of viable propagules decreases progressively with time. These relationships are illustrated in Table 9.

According to Nisikado, et al. (1937), agar-slant cultures remained viable at least 13 months when stored at 0°, 5°, 10°, 15°, 20°, and 25°C, but were dead by 4-6 months at 30°C, and 1 month at 35°C. When cultured on rice straw, the longevity was 11 months at 30°C, and 4 months at 35°C. In similar tests, Nisikado and Hirata (1938) found that sclerotia stored in water lost viability more rapidly than air-dried sclerotia at 10°C or higher temperatures. Schneider (1957) reported on storage of slant cultures of *Rhizoctonia*. Thirty-one of 40 cultures sealed with parafilm were viable when transferred 5½-9 years later. Seventy of 76 cultures covered with mineral oil, sealed with parafilm, and stored on the laboratory shelf were viable 6-7½ years later. Two of three strains tested by Park and Bertus (1932. 1934a,b) remained viable more than 600 days in sealed cultures.

Newton (1931) found that actively growing mycelial colonies on PDA were killed by exposure to 50°C for 45-60 minutes but tolerated lower temperatures or shorter exposures. Miller and Stoddard (1956) reported that the approximate thermal death point was 46° for young cultures on sterile stems, 53° for young PDA cultures, and 55°C for older cultures.

Samuel and Garrett (1932) suggested that the strain which infects cereal roots in South Australia oversummers in the form of thin, dark brown, distributive hyphae. Warcup, et al. (1963) also considered that the root-infecting strain, in contrast to other strains, persists mainly as residual hyphae during the dry summer period. Sclerotia could not be found. Strains differ markedly in survival ability in soils, some surviving indefinitely, others dying out in 2-3 months (Flentje, et al., 1964).

MONILIOID CELLS.—Chains of monilioid cells may be born aerially, on the agar surface, on the inner surface of petri dish covers when condensed moisture is available, or they may be immersed in agar. High relative humidity favors monilioid cell formation (Saksena and Vaartaja, 1961). Isolates of the sasakii type do not typically form chains of globose cells (Sherwood, unpublished).

According to Saksena and Vaartaja (1960), size and shape of monilioid cells of an isolate are constant on different substrates. There was good development of chlamydospores on cornmeal agar, without dextrose, in 20-25 days. On asparagine-PDA, many monilioid filaments were formed, but

differentiation of the cells into chlamydospores was slow and irregular. Aggregations of monilioid chains (sporodochia) varied in character and size, depending on substrate and light.

Papavizas (1965) noted that production of monilioid cells by three isolates of *P. praticola* and two isolates of *R. solani* and 60 single-spore progeny of each was a variable character depending on the isolate and medium used. Among the single-spore progeny there was a 1:1 ratio for ability to form monilioid cells on cornmeal agar.

The presence of abundant oil globules and granular materials in the monilioid cells suggests that they are storage cells. The longevity of these structures has not been studied. The monilioid cells, singly or in chains, germinate readily when provided moisture and a fresh food supply. Germination on tap water at 25°C ensues within 5-8 hours. There is usually one germ tube per cell, and it readily forms branched hyphae (Duggar and Stewart, 1901; Duggar, 1915; Saksena and Vaartaja, 1960, 1961).

SCLEROTIA.—*Initiation and development.*—Sclerotial initiation is independent of sclerotial growth and maturation. Thus, number and distribution of sclerotia are not necessarily correlated with size and color. As used in the literature, the terms sclerotial development or sclerotial formation fail to distinguish among these individual aspects but may conveniently indicate whether sclerotia have formed.

Sclerotium initiation is associated with new cell formation. Older elongate cells of hyphae have lost their capacity to become sclerotial initials. The sclerotial-initial cells are broader and shorter than ordinary hyphal cells and proliferate by irregular branching and intertwining to form loosely constructed, undifferentiated sclerotia (Sanford, 1941a, 1956; Townsend and Willets, 1954). The precise events that trigger sclerotial initiation are not known.

Sclerotial formation differs among isolates. Durbin (1959a) reported that 82%, 48%, and 21% of the aerial, surface, and subterranean isolates tested, respectively, formed sclerotia. Exner (1953) found that size of sclerotia among basidiospore isolates from single hymenia of *P. filamentosa* f. *microsclerotia* and of *P. filamentosa* f. *sasakii* varied greatly.

Sclerotia generally are not initiated on agar until the mycelium has grown to the edge of the medium (Matz, 1921; Allington, 1936; Valdez, 1955). Perombelon and Hadley (1965) noted that the first appearance of sclerotial initials of various isolates on two liquid media was associated with the time when dry weight of the colony reached its maximum. Townsend (1957) observed a close correlation in reduction of increased vegetative growth with maturation of sclerotia on liquid media.

Chemical factors.—Certain natural or semisynthetic media are generally more favorable than others for sclerotial formation. However, differential effects among isolates on different media occur. Maier and Staffeldt (1960) showed that one isolate produced sclerotia on carrot agar but not on PDA, while the

TABLE 8. Survival of mycelium, sclerotia, and cultures of *R. solani* at high or low temperatures or under wet or dry conditions

| Mycelium | | Sclerotia | | Cultures | | Treatment | Reference |
|---|---|---|---|---|---|---|---|
| Live | Dead | Live | Dead | Live | Dead | | |
| High temperatures | | | | | | | |
| + | − | − | − | − | − | 50°C, 45 min, PDA | Newton, 1931 |
| − | + | − | − | − | − | 50°, 60 min, PDA | Newton, 1931 |
| − | + | − | − | − | − | 51°, 5 min, suspension | Chowdhury, 1944 |
| − | + | − | − | − | − | 52°, 30 min, seedborne | Baker, 1947 |
| − | + | − | − | − | − | 60°, 5 sec, suspension | Vasudeva, 1936 |
| − | − | + | − | − | − | 40°, 21 days, 50%, 75%, 95% moisture | Tsiang, 1947 |
| − | − | 7/10 | − | − | − | 60°, 7 days, 50%, 75%, 95% moisture | Tsiang, 1947 |
| − | − | 4/10 | − | − | − | 60°, 14 days, 50%, 75% moisture | Tsiang, 1947 |
| − | − | − | + | − | − | 60°, 14 days, 90% moisture | Tsiang, 1947 |
| − | − | − | some | − | − | 65°, 35 days | Tsiang, 1947 |
| − | − | − | 43/47 | − | − | 68°, 35 days | Tsiang, 1947 |
| − | − | − | + | − | − | 71°, 35 days | Tsiang, 1947 |
| − | − | + | − | − | − | 55°, 60 min, on tubers | Newton, 1931 |
| − | − | − | + | − | − | 60°, 60 min, on tubers | Newton, 1931 |
| − | − | − | + | − | − | 59°, 5 min | Chowdhury, 1944 |
| − | − | − | + | − | − | 60°, 5 sec, suspension | Vasudeva, 1936 |
| − | − | − | − | + | − | 34°, 7 days | Nightingale and Ramsey, 1936 |
| − | − | − | − | − | + | 35°, 7 days | Nightingale and Ramsey, 1936 |
| − | − | − | − | − | + | 38°, 3 days | Townsend, 1934 |
| − | − | − | − | − | + | 40°, 12 days | Chowdhury, 1944 |
| − | − | − | − | − | + | 41°, 12 days | McRae, 1934 |
| − | − | − | − | − | 4/16 | 49°, 5 min | Balls, 1906 |
| − | − | − | − | − | + | 50°, 5 min | Balls, 1906 |
| − | − | − | − | − | 4/5 | 50°, 5 min | McRae, 1934 |
| − | − | − | − | − | + | 51°, 5 min | McRae, 1934 |
| Low temperatures | | | | | | | |
| + | − | − | − | − | − | 3°, 20 hr | Kotani and Mineo, 1962 |
| + | − | − | − | − | − | −2°, 15 hr | Kotani and Mineo, 1962 |
| − | + | − | − | − | − | 3°, 25 hr | Kotani and Mineo, 1962 |
| − | + | − | − | − | − | −2°, 20 hr | Kotani and Mineo, 1962 |
| Variable | − | − | − | − | − | 1°, 3.5 mon | Shurtleff, 1953a |
| − | − | + | − | − | − | 1°, 12 mon | Shurtleff, 1953b |
| − | − | − | − | + | − | 1°, 4 wk | LeClerg, 1934 |
| − | − | − | − | + | − | −18° several wk | Gutzevitch, 1934 |
| − | − | − | − | − | + | 2°, 1 wk | Nightingale and Ramsey, 1936 |
| Stored dry | | | | | | | |
| + | − | − | − | − | − | 21 mon, over CaCl$_2$ | Endo, 1940 |
| + | − | − | − | − | − | 123 days, air-dried medium | Müller, 1924 |
| − | + | − | − | − | − | 138 days, air-dried medium | Müller, 1924 |
| − | + | − | − | − | − | 7 mon, dry soil | Endo, 1931 |
| − | − | + | − | − | − | 6 yr, in lab. | Gadd and Bertus, 1928; Park and Bertus, 1932 |
| − | − | + | − | − | − | 5 yr, in sand | Herzog and Wartenburg, 1958 |
| − | − | + | − | − | − | 2½ yr, in sand | Chowdhury, 1944 |
| − | − | + | − | − | − | 21 mon, desiccator | Endo, 1931, 1940 |
| − | − | + | − | − | − | 13 mon, at 26° | Shurtleff, 1953a |
| − | − | + | − | − | − | 19 mon, on potato tuber | Müller, 1924 |
| − | − | 86% | − | − | − | 30 mon, dry | Pitt, 1964b |
| − | − | − | + | − | − | 21 mon, in soil | Endo, 1931 |
| − | − | − | + | − | − | 4 mon, culture | Briton-Jones, 1925 |
| − | − | − | − | + | − | 30 mon, room temp. | Gutzevitch, 1934 |
| − | − | − | − | + | − | 31 mon, in lab | Peltier, 1916 |
| Immersed in water | | | | | | | |
| − | − | + | − | − | − | 571 days, in tap water | Park and Bertus, 1932 |
| − | − | + | − | − | − | 70 days | Wei, 1934 |
| − | − | − | − | + | − | Over 3 mon | Palo, 1926 |

TABLE 9. Effect of temperature and moisture on viability of *C. sasakii*[1]

| | Temp. (°C) | No. of living units after treatment[2] | | | | | |
|---|---|---|---|---|---|---|---|
| | | 0 min | 5 min | 10 min | 20 min | 40 min | 80 min |
| Sclerotia, dry | 60 | 10 | 10 | 10 | 10 | 10 | 10 |
| | 70 | 10 | 7 | 6 | 6 | 5 | 4 |
| | 80 | 10 | 10 | 10 | 5 | 10 | 0 |
| | 90 | 10 | 10 | 10 | 8 | 6 | 0 |
| In water | 45 | 10 | 10 | 10 | 10 | 10 | 10 |
| | 50 | 10 | 10 | 10 | 6 | 2 | 0 |
| | 55 | 10 | 0 | 0 | 0 | 0 | 0 |
| Mycelium, dry | 50 | 10 | 10 | 10 | 10 | 10 | 10 |
| | 60 | 10 | 10 | 10 | 7 | 5 | 0 |
| | 70 | 10 | 0 | 4 | 0 | 0 | 0 |
| In water | 45 | 10 | 10 | 10 | 10 | 10 | 8 |
| | 50 | 10 | 0 | 0 | 0 | 0 | 0 |
| | 55 | 10 | 0 | 0 | 0 | 0 | 0 |

[1]Endo (1940).
[2]Ten sclerotia or pieces of mycelium were tested in each series.

reverse was true for other isolates. Staffeldt and Maier (1959) also noted that some isolates produce submerged as well as aerial sclerotia on certain media. Data on ranges in size of sclerotia on many natural media presented by Palo (1926) and Valdez (1955) indicate that sclerotia are small and infrequent on nutritionally poor media but generally large and frequent on rich media. Matz (1921) and Mathew (1954b) noted that sclerotia of certain isolates are larger in culture than in nature. According to Park and Bertus (1934a), *R. solani* strain A did not form sclerotia under natural conditions, but produced abundant sclerotia in agar. Sclerotia may look considerably different on a synthetic medium than they do on one with natural components (Park and Bertus, 1932; Chowdhury, 1944; Malaguti, 1951; Exner, 1953).

Carbon source and concentration influence number, size, coloration, and distribution of sclerotia (Allington, 1936; Elarosi, 1957b). Allington obtained abundant sclerotia with glucose, sucrose, or starch; moderate numbers with glycerol; and none with lactic acid as the carbon source in a synthetic medium. There were abundant sclerotia with 0.2% glucose and few with 3% glucose. Abundance was correlated with early initiation. Location of sclerotia was influenced by glucose concentration. Using a modified Richard's solution, Townsend (1957) obtained abundant production with glucose, fructose, sucrose, maltose, and starch; moderate numbers with lactose and arabinose; few with inulin; and none with mannitol. Sclerotial production was correlated with mycelial growth. When glucose and sucrose were tested at concentrations from 0.1-5%, sclerotia were initiated earlier with high concentrations, but matured most rapidly with low concentrations. Sclerotia were not mature (as indicated by pigmentation) until sugar concentration had been appreciably reduced, but complete exhaustion of sugar was not essential, indicating that starvation was not essential for sclerotial formation.

Townsend found urea unsatisfactory for sclerotial production in liquid media; but Allington (1936), using an agar medium, claimed that a large number of sclerotia formed with urea. He obtained many small and some large black sclerotia with asparagine; many small, coalesced sclerotia with calcium nitrate, few small sclerotia with sodium nitrate; and few, medium size, hard sclerotia with ammonium nitrate. The pH of the sodium and calcium nitrate media increased during culture, but sclerotial formation was very different on these two media.

From studies on the influence of C/N ratio, Townsend (1957) concluded that nutritional requirements for initiation differ from those for further development and maturation. She tested peptone at 0.1%, 0.3%, 0.6%, and 1.0% in all combinations with glucose at 0.1%, 0.5%, 1.0%, and 3.0%. Initials formed within 5 days. When both glucose and peptone were at high concentration, the greatest number of sclerotia was initiated and maturation was rapid. When a high concentration of either was combined with a low concentration of the other, the number of sclerotia tended to decrease and maturation was delayed. Similar trends occurred when asparagine or potassium nitrate were substituted for peptone.

The effect of mineral element nutrition was studied by Tyner and Sanford (1935). Optimal and minimal amounts of phosphorus for sclerotial formation were 31 and 7.5 ppm. The minimum amount of nitrate nitrogen required was 50-60 ppm and the optimum concentration was about 560 ppm. The minimum requirement for ammonium was slightly lower. Phosphorus or nitrogen content influenced size, color, and distribution. A minimum of 2 ppm potassium was required, but there was no clear optimum. Sclerotial production was somewhat curtailed on media with less than 20 ppm magnesium.

Kotila (1929) obtained sclerotia of *C. praticola* only at the higher concentrations of dibasic potassium phosphate, calcium nitrate, and magnesium sulphate. Some isolates of the praticola type formed sclerotia abundantly on PDA with Van't Hoff solution containing 2% sodium chloride, but did not form sclerotia on the same medium with 0% or 4% sodium chloride (Sherwood, unpub.).

The role of growth factors in sclerotial formation has not been studied.

Biotic factors can influence sclerotial formation. Sanford (1956) found that very few sclerotia were produced in sterilized soil, and few were produced in natural soil. When a small quantity of natural soil was added to sterilized soil, six times as many sclerotia were formed as in the natural soil alone.

Inclusion of Captan, Dichlone, pentachloronitrobenzene (PCNB) or Thiram, but not Maneb, in potato-sucrose agar induced sclerotial · formation (Elsaid and Sinclair, 1964). Weinhold and Hendrix (1963) reported that fewer sclerotia formed on PDA previously exposed to light than on PDA not exposed to light.

*Physical factors.*—Aeration is apparently essential for sclerotial formation (Matsumoto, 1921, 1923). Tyner and Sanford (1935) sealed culture flasks immediately after seeding, or when mycelium covered half of the surface, or when mycelium covered all the surface. In all sealed flasks, mycelium grew fairly well, but no sclerotia formed: They formed in unsealed controls. Briton-Jones (1924, 1925) considered that the better aeration in cultures on steamed potato, carrot, or turnip chunks, as compared with that on agar media, partly accounted for better sclerotial production on the former. Balls (1906) believed that under conditions of good aeration and poor nutrition, the original hypha is unstable and changes to monilioid cells or sclerotia. Conversely, the hypha is the stable form when immersed within culture fluid, and supplied excess food and poor aeration. The two forms coexist in an intermediate environment on the agar surface. However, others have observed that isolates can produce monilioid chains or sclerotia within agar.

High humidity or small areas of condensed moisture seem to be essential for sclerotial formation. Sanford (1956) believed that condensed moisture on the walls of test tubes or in soil favored sclerotial development. Tyner and Sanford (1935) permitted cultures to cover an agar surface, then placed them in chambers at 31%, 60%, and 100% RH. There was a decrease in sclerotial production with decreased humidity. Townsend (1934) obtained sclerotia in cultures at 92% RH but not at 81%. He reported that humidity higher than 92% impaired sclerotial formation. Saksena (1960) and Saksena and Vaartaja (1961) noted that monilioid chains seemed to be stimulated into forming sclerotia by condensed moisture. Briton-Jones (1924, 1925) noted that sclerotia formed better on fresh or rehydrated agar media than on partly dried media. Larger and more numerous sclerotia formed on steamed potatoes than on agar. He attributed this to better aeration and longer retention of moisture. Townsend (1957) found that from 10-37°C, any temperature that would support mycelial growth would permit sclerotial formation. Many authors, however, have stated that sclerotia do not form at all temperatures that support mycelial growth. Kotila (1929) reported that the temperature range for sclerotium formation of *C. praticola* was more restricted than that for mycelial growth. The optimum for sclerotium formation (16-18°C) was lower than for mycelial growth. Balfe (1935) found that one isolate formed sclerotia at 26-28° and 30-32°C, but not at 19° or 24°C, but another isolate formed sclerotia at all these temperatures except the highest. Monteith and Dahl (1928) observed that one isolate produced sclerotia at 30° but not at 15°C, while others formed sclerotia at both temperatures. According to Tyner and Sanford (1935), a shift from higher temperature toward the optimum (18-21°C) for the isolate tested decreased the rate of production but increased the number produced, whereas a change from lower temperature toward the optimum increased both rate and amount.

Hemmi and Endo (1931) and Endo (1940) believed that sclerotium formation was accelerated by a drop in temperature. Sclerotia were five times as abundant in cultures grown at 32°, exposed to 12°C for 4 days, then returned to 32°C than they were in cultures maintained at 32°C. Matsumoto, et al. (1933) stated that time to sclerotial formation was shortened by placing plates at 10°C for 30 minutes daily. Moisture condensation could partly account for these results.

A slightly acid reaction appears to be most favorable for sclerotial formation. Jackson (1940) obtained sclerotia at pH 5.5 and 6.5, but not at 4.5 and 7.5 in buffered media. Townsend (1957) obtained sclerotia on media initially at pH 8.8, but found acid reaction most favorable. Good formation on media initially at 3.5-6.2, but poor formation at 8.4, was observed by Chowdhury (1944). Allington (1936) obtained most sclerotia at pH 7.0 and none at 8.0, but Tyner and Sanford (1935) obtained some at 9.0. Color and type of sclerotia may also be influenced by pH (Monteith and Dahl, 1928).

Isolates react differently to light. No marked influence of light on sclerotial formation was observed by Chowdhury (1944) or Townsend (1957). Hemmi and Endo (1931) and Endo (1940) reported that sclerotia of *H. sasakii* were 3-4 times more abundant in dishes incubated in light than in those in dark. Durbin (1958) grew 96 clones in constant light and constant dark. Twenty percent formed sclerotia only when grown with light, 25% never formed sclerotia, and 55% formed sclerotia in either treatment. There were many differences in morphology and pigmentation of sclerotia between these treatments. Tyner and Sanford (1935) obtained increased numbers of sclerotia by irradiating with uranium oxide.

Matz (1921), Sanford (1956), and others have noted that sclerotia often form more abundantly in certain positions in the petri plate, such as near the seed piece or near the edge. Position effects probably result from variations in moisture, light, staling products, nutrients, and other factors within the culture vessel. The observations of Ui (1966) are of particular interest. He employed the double-dish technique. The inner dish contained 7.5 ml of either Czapek's agar without carbohydrate or of distilled-water agar with carbohydrate. The outer dish contained 15 ml of whichever medium was not used in the inner dish. When *R. solani* was inoculated into the center dish, sclerotia formed only in the area of the dish having the medium deficient in carbohydrate, irrespective of the site in the dish. Use of glucose C$^{14}$ showed that the carbohydrate was translocated into the deficient area. Efficiency of utilization of carbohydrate was greater when the carbohydrate was present in the outer medium.

*Longevity.*—Reports on longevity of sclerotia were given in Tables 8 and 9. Comments on methods used to test viability were presented above. Sclerotia are probably more resistant to adverse conditions than other structures of the fungus. Sclerotia stored dry in the laboratory have remained viable up to 6 years (Park and Bertus, 1932). However, Weber (1939) reported that germination of *C. microsclerotia* sclerotia stored at room temperature declined from 100% at 3 months to 0% at 11 months. Germination of sclerotia stored at 10°C remained 100% at 11 months. Park and Bertus (1932, 1934a,b) exposed sclerotia to direct sunlight (average air temperature, 57°C) for approximately 6 hours each day. Of three strains tested, one was still viable after 284 hours exposure during 57 days, a second strain was unable to germinate after 207 hours during 39 days, and the third strain lost viability after 183 hours during 34 days. Grooshevoy, et al. (1940) reported that sclerotia failed to germinate after being heated at 45°, 50°, and 60°C for 48, 10, and 0.5 hours, respectively, or at 45°C for 3 hours daily on 3 days. Exposure to 144 hours of ultraviolet light had little effect; however, sclerotia exposed to X-rays for 240 minutes required twice as much time to germinate, and for 270 minutes usually did not germinate (Tsiang, 1947).

There are many reports of survival of sclerotia in soil. Sclerotia in soil at 7.7% and 38% moisture remained viable up to 7 and 5 months, respectively (Palo, 1926). Sclerotia in soil at 55% moisture holding capacity for 6 months were 86% viable (Pitt, 1964b), and in sterile soil for 6 months during winter were still living (Endo, 1940). Park and Bertus (1932, 1934a,b) found that two strains were viable at 130 days in air-dry soil or 133 days in soil at about 18.5% moisture-holding capacity. Sclerotia of the third strain studied perished under the same conditions. Sclerotia of this strain also died earlier than the others when stored in tube cultures.

Townsend and Willets (1954) considered that the dense cellular contents, pigmentation, and rather impermeable wall of the sclerotial cells are responsible for their resistance. Matz (1921) believed that within the aerial strains, as contrasted with the mycelial root strains, the sclerotia, and sometimes the mycelial strands, are more distinct and of a harder consist-

ency, and that this is an adaptation to the aerial habitat.

*Germination.*—Initiation of mycelial growth from sclerotia is dependent on suitable moisture and temperature and is not restricted by maturity or dormancy requirements. In the related species *Corticium anceps* (Bres. and Syd.) Greg. both outer and inner cells can germinate, and germination seems to proceed from relatively large groups of cells but not isolated cells (Gregor, 1935).

A single sclerotium of *R. solani* in a fluctuating environment can germinate several successive times. In testing sclerotial regermination, Shurtleff (1953a) surface sterilized sclerotia in each successive test prior to plating them on a germination medium. Sclerotia were harvested at 1-3, 6-8, 15-17, and 90 days of age and were germinated on moist filter paper. The older and larger sclerotia, or sclerotia given a 19-hour presoak in water, produced more abundant growth. Only one of 12 sclerotia 1-3 days old at harvest germinated a second time, but all of the older sclerotia regerminated after being stored dry for 4 months, and again after an additional 13 months storage. Four- and 7-week-old sclerotia were germinated 25 successive times over a 9 month period; mycelial growth rate was not impaired. Sclerotia transferred successively on PDA were 100% viable, but those transferred on water agar showed reduced viability. Pitt (1964b) obtained 10 successive germinations on unsterile soil.

A lack of specificity in response to germination stimuli was observed by Pitt (1964b). Sclerotia that formed against the glass wall of tubes of soil, germinated in response to disturbance with a needle, abrasion with forceps, application of water, or sowing of wheat seeds. Sclerotia formed on potato tubers or unsterile soil gave nearly 100% germination within a few days when placed on distilled water, tap water, sterile soil, unsterile soil, PDA, malt-extract agar, soil-extract agar, and water agar. Sclerotia from beans with web blight gave 100% germination in 12 hours on PDA (Weber, 1939).

Shurtleff (1953a) studied the effect of relative humidity on germination of two isolates. Sclerotia 3, 7, and 20 weeks old were soaked in water and placed in sealed jars with salt solutions at 17-23°C. Sclerotia at 98% and 100% RH gave 100% germination within 36 hours, but sclerotia at 95% RH or less failed to germinate within 12 days. The sclerotia at 98% RH formed secondary sclerotia, while those at 100% RH formed only mycelium. Sclerotia kept at 35%, 42%, 76%, and 95% RH for 2½ weeks and then transferred to 100% RH gave 100% germination. Those from 95% RH germinated poorly.

Müller (1924) obtained germination on distilled water at 8.9° and 30.2° but not at 7° or 31.2°C. The temperature range for sclerotial germination was narrower than for mycelial growth of the same strain. Germination began earlier at temperatures closer to the optimum. Thus, germination began in 4 hours at 23°C but not until 48 hours at 9°C. Shurtleff

(1953a) obtained germination on Czapek's agar in 68 hours at 9.5°C but not after 9 days at 5°C. Dahl (1933) obtained germination of all isolates tested at 12° and 36°, of some at 8°, and none at 40°C. Most rapid germination was at 28° or 32°C. A web-blight isolate germinated in 1-2 days at 25°, 2-4 days at 20°, 4-6 days at 15°, and 8-10 days at 9°C (Kontani and Mineo, 1962). At each temperature, some sclerotia germinated earlier than others. Cardinal temperatures for sclerotial germination of *H. sasakii* are approximately 10°, 28-32°, and 36°C (Matsumoto, 1921; Endo, 1930; Hemmi and Endo, 1933).

Dickinson (1930) suggested that sclerotial germination is enhanced by chilling for a short period. Sclerotia planted on PDA were chilled at 17-20°C then incubated at 20-25°, 25-30°, 25-35°, and 30+°C. After 3½ hours, chilled sclerotia showed a higher percent of germination and more mycelial growth than nonchilled controls; the differences were small at 17½ hours. Dahl (1933), however, found that sclerotia usually germinated less rapidly when temperature was temporarily lowered. Shurtleff (1953a) found no benefit from chilling. He pointed out that chilling might promote germination by inducing condensation of moisture.

THE PERFECT STATE.—*Formation.*—Many isolates have been induced to form the perfect state in the laboratory. We lack sufficient information on physical and chemical prerequisites for fruiting, however, to enable us to bring all isolates to fruit at will. Of the many laboratory conditions tested, those which have been successful have been based on an application of the principle (Kleb's) of transferring well nourished, actively growing vegetative cells to a less nutritive substratum. One isolate may form basidiospores in response to a given treatment but not to another. The response of a second isolate may be the opposite, while a third may not respond to any method. There are differences between laboratories in the ease with which isolates are brought to fruit. Some correlations with strain affiliations have been noted. According to Stretton and coworkers (1964), isolates of *Thanatephorus praticola* fruit if transferred from a rich agar medium to a low nutrient agar. They fruit poorly or not at all when transferred to soil. Parmeter (personal communication) found isolates of this type that did fruit well on soil. Many pathogenic isolates of *T. cucumeris* fruit on soil but not on agar, but some nonpathogenic isolates fruit on both soil and agar.

Kotila (1945a,b; 1947) inoculated sugar beet and rubber leaves with vigorously growing hyphal-tip isolates and kept the plants at approximately 100% RH and 21-25°C. Hymenia formed on the leaves in about 2 weeks.

Many isolates have fruited on water agar or similar nutritionally poor media. Kotila (1947) reported on six hyphal-tip isolates that fruited on 2% distilled-water agar. Hawn and Vanterpool (1953) transferred isolates, apparently of the praticola type, from 30-day-old malt agar or PDA test tube cultures to petri dishes with thin pourings of 2% water

agar. The dishes were kept under a bell jar near a north window, and hymenia developed in 14-21 days. Best results were obtained at 19-20°C and 70-80% RH. Inocula from 4-, 7-, or 14-day-old cultures, or use of 1% water agar gave inferior results. Strains of the solani type would not sporulate (Vanterpool, 1953). Saksena and Vaartaja (1961) reported that 26 isolates of *P. praticola* formed the basidial stage in 15-20 days on cornmeal agar (without dextrose) in plates at 22-25°C in glass cabinets within a room with at least 8 hours of fluorescent light daily. Saksena (1960) transferred from 25-day-old PDA cultures to 2% cornmeal agar and placed the plates over some water in a bell jar in diffused light at about 25°C. Whitney and Parmeter (1963) grew the fungi for 2-3 days on Marmite-PDA, then transferred blocks about 2-3 mm³ containing only a few hyphal tips, to water agar in plates. The plates were incubated at room temperature in diffuse light. Applying the above method, Whitney (1964b) obtained the perfect state on 2.5% water agar, cornmeal agar, soil-extract agar, and a cornmeal-sand mixture. Flentje (1956) grew isolates for 7-8 days on Marmite-PDA at 25°C, transferred them to soil-extract agar, incubated the plates at 25°C for 7 days, then removed the plates to the laboratory bench. Hymenia developed 1-10 days later, and were most prolific in plates exposed to reduced light and marked differences between day and night temperatures. Garza-Chapa and Anderson (1966) found that a combination of the methods of Flentje and of Whitney and Parmeter was most effective. *Thanatephorus cucumeris* was grown on Marmite-PDA 72 hours at 25°C. Disks 3 mm in diameter were transferred from the colony edge to dishes of 2% water agar. The cultures were incubated in the dark at 25°C for 8 days and then transferred to a laboratory where they received diffuse light at 21-25°C. Basidia formed 6-9 days later.

Methods for fruiting on soil have been described as follows. Flentje (1956) mixed cornmeal-sand cultures into unsterilized, Waite Institute loam soil in glass jars at a rate of 2.5%. The jars were kept in the greenhouse, and the soil was watered daily. Hymenia appeared in 5-21 days. They developed both in summer and in winter in light intensities varying from 200 to 2,000 f.c. and were favored by lumpy soil texture. Some strains formed hymenia only if plants were present in the soil, others formed hymenia in the presence or absence of plants. Stretton, et al. (1964) obtained fruiting of *T. cucumeris* on soil surfaces by two methods: (1) Isolates were grown on Marmite-PDA in 9 cm petri dishes for 7-10 days. Tops were removed and cultures were covered to 1 cm depth with steamed soil. The soil was watered 1-3 times daily. Fructifications appeared after 3-14 days. (2) Seven- to 10-day-old Marmite-PDA cultures were mixed with 300 g soil, placed in a drinking cup, and watered 1-3 times daily. The soil was previously treated with aerated steam at 71°C for 30 minutes. Fructifications formed under light periods varying from 8-24 hours, but formed most profusely at 12-16 hours. They formed under light intensities of 4 to 1,450 f.c., but best at 10-440 f.c. Cultures fruited between 20-30°C and fruited better at 40-60% RH than under high humidity. Garza-Chapa and Anderson (1966) fruited many isolates on soil by these methods. The soil was autoclaved 30 minutes at 121°C. Echandi (personal communication) indicated that an acid soil is not suitable.

To induce sporulation, Exner and Chilton (1944) grew isolates in potato-dextrose broth 10 days, washed the mycelial mat in distilled water, and placed it in a small Erlenmeyer flask with a rooted cutting of alligator weed with sufficient water to maintain humidity. Basidial mats formed on the stem surface. Using this method, Sims (1956) found that the concentration of glucose or maltose in the antecedent potato-broth culture influenced sporulation. Echandi (1965) fruited web-blight isolates (*C. microsclerotia*) on soil-rice agar.

Numerous other methods have resulted in sporulation. Müller (1923b) transferred washed mats from 17-day cultures on peptone-starch solution to moist filter paper in flasks. Basidia formed in one of two trials. Schenck (1924) fruited an isolate on blotting paper moistened with a chemically defined medium. Ullstrup (1939) obtained fruiting in 2-6 days at 27-29°C on and around pieces of infected cotton seedlings plated on tap-water agar, and also when mass transfers of basidiospores, but not of mycelium, were made to healthy seedlings on water agar. Sims (1956) listed several additional treatments leading to fruiting of certain isolates. See also Jauch (1947) and Pinckard (1964).

Some authors have reported negative results. Pitt (1964a) applied without success the methods of Flentje, Kotila, Sims, and Hawn and Vanterpool to isolates from sharp eyespot of wheat. Valdez (1955) could not induce sporulation of a rice-sheath isolate by exposure of cut mycelium to sun, alternating light and darkness, addition of toxic compounds, or alternating low and high temperatures. Papavizas (1965) transferred 69 clones (obtained from 15 soils) from rich substrates to poor substrates. All combinations of eight rich media, including a synthetic medium amended with various growth substances, and of five poor media were tested. Eight clones fruited when transferred from yeast-extract-PDA to soil-extract agar. No other combination gave fruiting.

Kotila (1929) studied the influence of environment on fruiting of *C. praticola*. Fruiting was abundant at 65% RH, moderate at 82% RH, meager at 90% RH, and lacking at 32% RH. Optimum temperature for spore formation was 21°C. Temperature range for sporulation was more restricted than that for growth. Spores formed on Coon's medium adjusted to pH values from 4.0-7.6. Spores formed abundantly in cultures kept under a bell jar with forced aeration, but not in cultures without aeration.

Carpenter (1949) found that production and discharge of spores from hymenia was a periodic function. His experiments strongly implicated light in the control of this function; however, they did not rule out the possible effect of temperature or humid-

ity. Flentje, Stretton, and Hawn (1963) reported that sterigmatal growth and spore development began in the last 2-3 hours of the light period, continued throughout the dark period, and declined during the first 2-3 hours of the following light period. Spores began to be shed freely 1 or 2 hours after spore development commenced. Spore discharge continued through the dark period into the first few hours of the next light period. Sterigmata of *T. praticola* and *T. cucumeris* were rare during the light period. Echandi (1965) observed that basidium formation, spore formation, and spore discharge in bean leaves infected by *C. microsclerotia* were mainly nocturnal. Whitney (1964a) found that under well-aerated conditions, many more basidial clusters formed in continuous darkness than in diffuse natural light, and that basidial clusters were very infrequent in natural light supplemented with continuous fluorescent light. The rate of sporulation in continuous darkness was constant with time.

*Basidiospore discharge, longevity, and germination.*—Hawn and Vanterpool (1953) observed that basidiospores are discharged violently by the drop excretion method. Müller (1924) reported on the germination of spores stored on slides for various lengths of time. Germination was 97% after 6 hours of storage, 0-14% after 11 days, and 0% after 44 days.

Basidiospores germinate either by repetition or by germ tube formation. In germination by repetition, the secondary basidiospore develops at the apex of a sterigma, arising usually from the apiculus (Hawn and Vanterpool, 1953). The secondary spore may give rise to tertiary and quarternary basidiospores, called repetition spores or ballistospores (Whitney, 1964b). In germ-tube formation, growth usually starts at the end of the spore opposite to the apiculus (Hawn and Vanterpool, 1953; Ito, 1958). Some spores produce one germ tube at each end of the spore; it is rare for a tube to arise laterally from the spore (Gadd and Bertus, 1928). Hawn and Vanterpool (1953) noted that 68% of sporelings transferred to PDA failed to continue growth.

Substrate and temperature influence germination. Weber (1931) observed 5% germination on potato agar and 20% on cabbage agar. Germination was also more rapid on the latter medium. Gadd and Bertus (1928) obtained more rapid and frequent germination in an aqueous suspension of cornmeal agar than in water. However, Müller (1924) reported that the addition of small amounts of leaf decoction or sugar to distilled water did not improve germination. He noted that the optimum temperature for germination was close to the optimum for mycelial development of the isolate studied. Germination in 5½ hours at 17°, 21°, 25°, and 30°C was 7.1%, 35.9%, 33.2%, and 1.4%, respectively. Carpenter (1951) observed that germination in rain water occurred with 2-6 hours; the spores germinated equally well in darkness or in artificial light.

Hawn and Vanterpool (1953) reported that spores sown on water agar germinated by repetition, but spores sown on rich nutrient agars germinated by germ tubes. Both kinds of germination may occur on soil-extract agar (Flentje, 1956). Comprehensive studies by Whitney (1964b) indicate that a fungal product, rather than the substrate, controls the occurrence of repetition. Basidiospores became multinucleate and germinated by germ tubes when sown on 0.3%, 0.6%, 1.25%, 2.5%, and 5% water agar, on water agar adjusted with hydrochloric acid or sodium hydroxide to pH values from 2.3-11, on potato-Marmite agar, or on Czapek-Dox agar. Germ-tube development was strongly inhibited on water agar at extreme pH values. There was no germination on sterile distilled water at pH 5.8 or on a 10-day-old potato-Marmite agar culture. Spores deposited onto a sporulating water-agar culture or onto cellophane covering the sporulating culture remained uninucleate and germinated by repetition (Fig. 14).

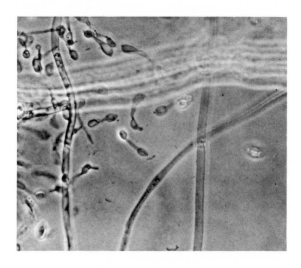

*Fig. 14.* Basidiospore repetition below basidial cluster in fruiting culture (Whitney [1964b]. Reproduced by permission of the National Research Council of Canada from the *Can. J. Botany,* 42:1397-1404, 1964).

Repetition also occurred when spores were sown on certain nonsporulating cultures. When deposited on a culture half covered with mycelium, spores on the unused medium germinated directly, those on an area with mycelium 2-3 days old germinated by repetition, and those in-between germinated in either manner. Agar recovered from a fruiting culture by melting the agar and then resolidifying it, induced repetition. Dilution of this medium with water decreased repetition and increased direct germination. If basidiospores in the initial stages of germination on cellophane on water agar were transferred, while still presumably uninucleate, to a fruiting culture, the germ tubes stopped elongating, and a sterigma bearing a ballistospore formed at the tips of many germ tubes. Older hyphae, even from homokaryotic single-spore strains, did not do this. The chemical factor that induced repetition was extracted from fruiting cultures with ether and localized on chromatograms. Whitney (1964b) suggested that

the substances causing repetition may also prevent nuclear division in the uninucleate young sporelings, and that nuclei in sporelings are functionally different from those in older mycelia.

ASSOCIATIONS AMONG PHYSIOLOGIC PROPERTIES.—Many workers have observed associations among physiologic attributes, morphology, and pathogenicity of tissue isolates from nature. Examples were given above from the work of Schultz (1936), Le Clerg (1941b), Richter and Schneider (1953), Durbin (1958, 1959a), and others. Additional examples are given below. Durbin divided isolates into three groups, based on their original ecological habitat, and found that each group had a rather characteristic growth rate, $CO_2$ sensitivity, and ability to form sclerotia in culture. Richter and Schneider (1953) recognized six major groups on the basis of anastomosis behavior though some isolates could not be assigned to any group. There were correlations among all the following characters: anastomosis behavior, original host plant, hyphal diameter, cultural appearance, pathogenicity, cardinal temperatures, and rate of growth. There was no correlation of these characters with response to pH. There was considerable variation within groups, and overlapping between groups in individual characters.

Houston (1945) divided 56 isolates into three cultural types. Each group had characteristic optimum and maximum temperatures, growth rates, and pathogenicity; there was some overlapping between groups. Three groups were distinguished by Hansen (1963). Group 1 included isolates from wheat and barley that had slow growth rates (0.17-0.31 mm/hr), narrow hyphae (4.3-5.8 $\mu$), and uncolored mycelium. Group 2 included isolates from potato that had medium growth rates (0.43-0.53 mm/hr), broad hyphae (7.5-8.0$\mu$), and dark mycelium. Group 3 included praticola-like isolates that had rapid growth rates (0.76-0.80 mm/hr), intermediate hyphae (6.1-6.3$\mu$), and gray mycelium. Thomas (1925) observed that isolates with hyphae measuring 8.5$\mu$ or more grew about equally well at 17° and 27°C and formed large sclerotia, while isolates with hyphae 7.0$\mu$ or less grew more rapidly at 27°C and formed small sclerotia. Isolates having the most rapid growth rates had high optimum temperatures and did not grow at 5°C.

Sherwood (unpub.) recognized four groups among 62 mycelial isolates (see Table 1). Each group included isolates from several states or countries. When the isolates were paired on water agar, each anastomosed with other isolates within its group, but not with isolates outside its group. On PDA, isolates of groups S and T formed abundant small brown loosely organized sclerotia or no sclerotia; sasakii isolates formed large (3-10 mm diameter) globose or subglobose sclerotia and praticola types formed small crusty dark sclerotia. As shown in Table 1, each group had a characteristic natural habitat, virulence, thiamine requirement, approximate optimum temperature, and growth rate. Aside from anastomosis, no single character was adequate for separating all groups. However, a discontinuity in variation in these natural populations was revealed when these several characters were considered simultaneously. There was strong uniformity within groups for certain characters, such as thiamine requirement. Overlapping occurred between groups in regard to virulence, growth rate, and appearance in culture.

The examples of association between natural habitat and physiologic properties cited above indicate that ecology plays an important role in evolutionary development within *R. solani.* Indeed, habitat may be more important than the host plant as a selection pressure on the physiological responses of the fungus. One is struck by the fact that not only are there apparently discrete physiological types (e.g. the rapidly growing, $CO_2$ intolerant, highly virulent aerial type), but that these adaptations encompass reproductive physiology (e.g. failure of the aerial type to fruit on water agar). This is not to deny that many isolates do not fit neatly into a group.

One would expect to find nonconforming isolates in a highly heterogeneous species that exists in a variety of intergrading habitats. Nevertheless, when tissue isolates from diseased plants in nature are considered, certain combinations of properties recur in a remarkably high frequency. Types recognized in one geographic area are also recognized by other workers in different areas. In each of the three major habitats distinguished by Durbin, particular combinations of physiologic attributes may have a distinct survival advantage over other combinations. Certain aspects of function are expressed as form. The possibility that there are certain major physiological-morphological strains in *R. solani* is therefore inescapable.

ACKNOWLEDGMENTS. — Grateful acknowledgment is extended to the National Research Council of Canada for permission to reproduce Figs. 3, 4, 5, 8, 9, and 14; and to the editor of *Phytopathology* for permission to reproduce Figs. 6, 7, and 13.

# Metabolism of Rhizoctonia Solani

WALTER J. TOLMSOFF—*Crops Research Division, Agricultural Research Service, United States Department of Agriculture, Texas Agricultural and Medical University, College Station, Texas.*

There have been few studies on the metabolism of *Rhizoctonia solani*. Those initiated by Matsumoto in the early 1920's were conducted at a time when information on the metabolism of any organism was meager. For example, it was not until 1927 that Meyerhof gave the name "hexokinase" to an enzyme which, in the presence of magnesium and adenosine-5'-triphosphate, phosphorylates glucose for its entry into various multienzyme pathways of metabolism (Fruton and Simmonds, 1959). Since Matsumoto's work, studies on the metabolism of *Rhizoctonia* have hardly progressed.

In the meantime, the metabolism of many other organisms has been revealed in considerable detail, showing the coordinated activity of multienzyme systems, their intracellular organization and location, factors affecting their activity, the nature of enzyme inheritance and synthesis, etc. The amino acid sequence of several enzymes is known in detail and has been compared in different organisms, revealing that the protein catalysts vary in composition and other properties but perform similar duties under their specific optimum conditions. Biochemists and biophysicists are revealing the tertiary structure of enzymes and the conformational changes associated with their catalytic activity. Hopefully, the knowledge gained from these pioneering efforts in biochemistry will accelerate a better understanding of the metabolism in *R. solani* through future research.

Information on some of the unique metabolic properties of *R. solani* is needed in order to understand how and why this fungus is a pathogen on many plant species and not on others. There is also the problem of great diversity in strains of the fungus. This diversity indicates the need for caution in drawing broad conclusions based on studies with one or two strains of *R. solani*. Still, a starting point is needed from which deviations might grow and help produce a more complete picture of the unique metabolic properties of this unusual pathogen. Sufficient information is not available at this time to describe strain differences in metabolism of *R. solani*. Much research is still needed to develop even a working framework or skeletal picture of the metabolism of *R. solani*.

CARBOHYDRATE METABOLISM.—One of the most common carbohydrates that *R. solani* encounters in nature is D-glucose in oligosaccharide and polymeric forms such as sucrose, starch, cellulose, etc.

*Cellulose.*—This is utilized for growth by many isolates of *R. solani* (Matsumoto, 1921; Kohlmeyer, 1956; Ross, 1960; Garrett, 1962). Cellulose is hydrolyzed in advance of the growing hyphae in culture, presumably by extracellular enzymes. The cellulase enzymes are poorly understood from any biological source. These enzymes are normally absent in cells that have not been exposed to cellulose or one of its decomposition products, e.g. cellobiose.

The basidiomycete *Irpex lacteus* (*Polyporus tulipiferae*) contains at least nine components involved in cellulose degradation that are separable by starch zonalelectrophoresis (Nisizawa, et al., 1962). Although carboxymethylcellulose is commonly used as a substrate in studies of cellulase activity, *Myrothecium verrucaria* contains a separate enzyme system for the decomposition of this substrate as compared to native cellulose (Halliwell, 1961). The cellulase system was strongly adsorbed to cellulose. Agitation of the culture inhibited carboxymethyl cellulase production, but not that of cellulase. Pretreatment of cellulose with phosphoric acid made this substrate more susceptible to attack by the cellulase system.

Although we commonly think of a substrate inducing the production of enzymes for its metabolism, related or unrelated molecules may also possess this capacity in specific instances. For example, sophorose (2-O-$\beta$-D-glucopyranosyl-D-glucose) occurs as an impurity of reagent-grade glucose to the extent of 0.0058% and is 2,500 times more active than cellobiose for induction of cellulase production in *Trichoderma viride* (Mandels, et al., 1962). The induction in *Trichoderma* appears to be highly specific, and sophorose is ineffective for cellulase induction in a number of other organisms.

There is some evidence for cellular specialization in cellulose degradation by *R. solani* (Isaac, 1964a). The short branch hyphae that develop from the main trunk appear active in cellulose degradation while the growing tips and their supporting trunk hyphae appear inactive. The branched hyphae that catalyze the decomposition of cellulose in vitro bear some resemblance to the highly branched infection cushions found in vivo (Dodman and Flentje, this vol.). Since it is known that *R. solani* is stimulated to produce infection cushions by diffusable materials from roots, it should be of interest to determine if the infection cushions of the fungus have an activated cellulase system to aid in penetration of the host.

*Rhizoctonia solani* possesses activity assignable to the B-glucosidase emulsin (Matsumoto, 1923), which is capable of hydrolizing the β-glycosidic bonds of amygdalin and cellobiose to yield free glucose.

*Starch.*—A group of different enzymes catalyzes hydrolysis of starch (amylose, amylopectin, glycogen) to yield dextrins, maltose, free glucose, or glucose-1-phosphate (Fruton and Simmonds, 1959). β-amylase decomposes starch to yield a mixture of products including dextrins, maltotriose, and ultimately maltose and free glucose; phosphorylases yield glucose-1-phosphate. The complete hydrolysis of starch to glucose also requires an enzyme that cleaves the α-(1—6) glycosidic branches of amylopectin or glycogen.

It is reported that *R. solani* possesses amylase activity, based on the appearance of reducing groups during starch hydrolysis (Matsumoto, 1921, 1923; Edwards and Newton, 1937). While *R. solani* probably has the enzymatic mechanisms for degradation of starch to maltose, glucose, or glucose-1-phosphate, the routes by which this might be accomplished are not known. The fungus grows well on starch, indicating that it is capable of converting the polysaccharide to numerous cell constituents. However, utilization of inulin, which is found as a storage polysaccharide of artichoke and dahlia, is questionable (Matsumoto, 1921) and may vary among isolates (Sherwood, this vol.).

Amylase activity of *R. solani* is reported to occur intra- and extracellularly with a pH optimum at 6.2, and a pH range for activity of 3.4-9.4 (Matsumoto, 1923). Greater activity was produced by the fungus in the presence of starch, suggesting enzyme induction by substrate.

The fungus also possesses a mechanism for conversion of maltose, a product of starch hydrolysis, to glucose—possibly by an intracellular α-glucosidase (Matsumoto, 1921, 1923). Cell extraction was required to demonstrate the activity, which was optimum at pH 6.0.

*Sucrose.*—The hydrolysis of sucrose to its monosaccharide constituents, glucose and fructose, is effected by many isolates of *R. solani* (Matsumoto, 1921, 1923; Edwards and Newton, 1937; Tolba and Salama, 1960, 1961*a,b*, 1962*a,b*). It has been suggested that this activity is due to a β-fructofuranosidase type of enzyme located at the outer cytoplasmic cell membrane, rather than a sucrose phosphorylase (Tolba and Salama, 1960, 1961*b*). It appears that the β-fructofuranosidase (invertase) is located in such a position that it is readily available to sucrose in the external medium. The enzyme releases glucose and fructose into the medium as products of sucrose hydrolysis. Although the enzyme is attached to or associated with living cells, it is also found in the culture filtrate of aging cultures, perhaps as a result of autolysis (Matsumoto, 1923). Invertase of

other microorganisms is also believed to be a surface enzyme (Neu and Heppel, 1965).

The designations of β-fructofuranosidase, α-glucosidase, and sucrose phosphorylase refer to various types of sucrose-hydrolyzing enzymes, depending on substrate specificity and method of breaking the glycosidic bond. It appears that the invertase of *R. solani* is of the β-fructofuranosidase type, which hydrolizes a terminal β-D-fructofuranoside from an oligosaccharide such as sucrose or raffinose. The fungus readily hydrolyzes either sucrose or raffinose in the external medium, but not maltose. If *R. solani* instead had an α-glucosidase, sucrose and maltose should be the preferred substrates for hydrolysis, while raffinose would require a separate enzyme, based on current concepts of activity by these enzymes. As previously noted, there may be an intracellular α-glucosidase in *R. solani* active with maltose.

The pH optimum for invertase activity in crude culture filtrates of *R. solani* is reported to be 2.8-3.2 (Matsumoto, 1923). Invertase activity was not affected by inclusion or omission of sucrose from the medium, which suggested that the enzyme is constitutive, rather than adaptive. However, this interpretation is relative because there are various degrees of induction and repression in the synthesis of enzymes.

Based on its utilization of a wide variety of carbohydrates for growth, *R. solani* will probably be found to possess the necessary enzymes to perform the conversions to cellular constituents. It is clear that general observations on carbohydrate utilization and studies on crude culture filtrates or cell homogenates will reveal new information on carbohydrate metabolism at a painfully slow pace. There is a need for detailed studies based on purified enzymes obtained from *R. solani*, showing their substrate specificity, activating and inactivating factors, and mechanisms of action.

*Glucose.*—Organisms incapable of utilizing glucose for growth are exceedingly rare in nature. *Rhizoctonia solani* is not exceptional in this regard. In fact, its efficient utilization of glucose for growth and addition of dry cellular weight in forms other than free glucose may be exceptional (Steinberg, 1948; Ross, 1960). The implication is that the fungus metabolizes glucose efficiently with little loss of energy, and must possess the metabolic pathways necessary for conversion of glucose to numerous cell constituents. Glucose serves as a sole carbon source for the growth of *R. solani* (Sherwood, this vol.).

When sucrose is provided as the sole carbon source to *R. solani*, the hydrolysis products, glucose and fructose, appear in the external medium. From this mixture of sucrose, glucose, and fructose, glucose is preferentially taken up by the fungus, leaving fructose to be concentrated externally (Tolba and Salama, 1960, 1961*b*). The glucose:fructose ratio after 24 hours was about 1:6. Inclusion of sodium fluoride in the growth medium reduced the uptake of glucose without inhibiting the invertase activity. This resulted in larger and more uniform levels of glucose and fructose in the external medium. Silver nitrate at a concentration of 0.1 mM had a similar effect.

The uptake of radioactively labeled glucose-6-tritium occurs primarily in the areas of greatest cytoplasmic density in *R. solani* (King and Isaac, 1964) and is shown to be a competitive process by the inhibition with unlabeled glucose or equimolar mannose, galactose, or fructose. Equivalent weights of unlabeled ribose, sucrose, or glycine did not interfere with the uptake of labeled glucose. Similarly, measurable quantities of glycine-2-tritium were not taken up in the presence of unlabeled glycine or equal weights of alanine or serine, but glucose did not interfere.

These results suggest that there are membrane mechanisms or protein "permeases" for moving substrates across the membranes in a manner that may be similar to bacterial permeases. (Mammals appear to have at least three independent mechanisms for movement of basic, neutral, and acidic amino acids across membranes [Begin and Scholefield, 1965]).

The following list of enzyme activities found in *R. solani* includes, in addition to others, those directly or indirectly associated with glucose catabolism They include most of the enzyme activities studied in the fungus, with the exception of those previously mentioned in carbohydrate metabolism and the pectolytic enzymes discussed by Bateman (this vol.). The following activities have been found in specific or various isolates of *R. solani*:

1. Glycolysis
   *a*. Hexokinase (Molitoris and Van Etten, 1965)
   *b*. Phosphofructokinase (Molitoris and Van Etten, 1965)
   *c*. Glyceraldehyde-3-phosphate dehydrogenase (Dowler, et al., 1963)
   *d*. Pyruvic dehydrogenase (Molitoris, personal communication)
2. Kreb's cycle
   *a*. Malic dehydrogenase (Dowler, et al., 1963; Molitoris and Van Etten, 1965)
   *b*. Aconitase (Molitoris and Van Etten, 1965)
   *c*. Nicotinamide adenine dinucleotide (NAD-specific isocitric dehydrogenase (Molitoris and Van Etten, 1965)
   *d*. Nicotinamide adenine dinucleotide phosphate (NADP)-specific isocitric dehydrogenase (Molitoris, personal communication)
   *e*. Succinic dehydrogenase (Dowler, et al., 1963; Molitoris and Van Etten, 1965; Tolmsoff, 1965)
   *f*. Fumarase (Molitoris and Van Etten, 1965)
3. Pentose phosphate pathway
   *a*. Glucose-6-phosphate dehydrogenase (Dowler, et al., 1963; Molitoris and Van Etten, 1965)
   *b*. 6-phosphogluconic dehydrogenase (Molitoris and Van Etten, 1965)
4. Electron transport
   *a*. NADH-cytochrome c reductase (Dowler et al., 1963)
   *b*. NADH oxidase ($O_2$-uptake) (Dowler, et al., 1963; Molitoris and Van Etten, 1965; Tolmsoff, 1965)
   *c*. Succinic-cytochrome c reductase (Dowler, et al., 1963)
   *d*. Succinoxidase (oxygen uptake) (Tolmsoff, 1965)
   *e*. Cytochrome c oxidase (Molitoris and Van Etten, 1965)
   *f*. Cytochromes a, b, and c, based on absorption spectra of whole cells (Boulter and Derbyshire, 1957)
5. Other enzyme activities
   *a*. $\alpha$-glycerophosphate dehydrogenase (Molitoris and Van Etten, 1965)
   *b*. Isocitritase (Molitoris, personal communication)
   *c*. Glutamic dehydrogenase (Dowler, et al., 1963)
   *d*. L-aspartic-oxaloacetate aminotransferase (Molitoris and Van Etten, 1965)
   *e*. Asparagine-keto acid-amino transferase (Molitoris and Van Etten, 1965)
   *f*. Catalase (Edwards and Newton, 1937)
   *g*. Phosphodiesterase (5′-nucleotide formation) (Hasegawa, et al., 1964*a,b*)
   *h*. Invertase (Tolba and Salama, 1960, 1961*b*)

The list of enzyme activities in *R. solani* should be considered tentative and subject to confirmation in most cases.

Quantitative changes in the level of certain enzymes have been found with aging of *R. solani*. Hexokinase activity increased with age, while phosphofructokinase and one or more of the three glycolytic enzymes between 3-phosphoglycerate and pyruvate decreased (Molitoris and Van Etten, 1965). Other activities that increased with age included the $\alpha$-glycerophosphate and malic dehydrogenases, and cytochrome oxidase. Aconitase, NAD+-specific isocitric dehydrogenase and NADH oxidase activities decreased.

The significance of change in the various enzyme activities with aging of the fungus cannot be estimated at the present time. The changes probably, but not necessarily, reflect quantitative shifts of metabolic traffic along certain pathways. Without knowing if the enzymes are used to maximum capacity, it is impossible to estimate whether a several-fold change in potential activity would alter the traffic. Furthermore, some of the studies are based on cells grown in stationary culture (Molitoris and Van Etten, 1965). The strongly aerobic nature of *R. solani* raises the possibility that certain phases of metabolism in stationary liquid culture may be diverted into abnormal channels.

While many common enzyme activities in the catabolism of glucose have been found in *R. solani*, information is insufficient to connect these pieces of knowledge into functional multienzyme pathways of metabolism. Certain key enzymes in glycolysis, the pentose phosphate pathway, and the Kreb's cycle have not been demonstrated. For example, $\alpha$-ketoglutaric dehydrogenase activity was sought but not found (Dowler, et al., 1963; Molitoris and Van Etten, 1965; Tolmsoff, 1965). This could imply that the Kreb's cycle is not functional in the oxidation of glucose by *R. solani,* though there are other possible explanations. Some isolates are apparently capable of making moderate growth with $\alpha$-keto-glutarate as a sole carbon source (see Sherwood, this vol.).

A fruitful area of research should be the identification of the carbohydrates *R. solani* stores when aging in media containing glucose or sucrose (Tolba and Salama, 1962*a,b*; Van Etten and Molitoris, 1965). These include low molecular weight carbohydrates soluble in trichloroacetic acid and comprise up to 71% of the dry weight of the fungus after 80 hours growth (Molitoris, personal communication). In addition to a relatively high concentration of what are believed to be disaccharides other than sucrose, there are also unidentified polysaccharides. It is important to identify these stored carbohydrates.

RESPIRATORY METABOLISM OF WHOLE CELLS.—*Oxygen.*—Different isolates of *R. solani* possess a persistent high level of $O_2$ consumption in the absence of exogenous substrates when initially grown on a medium containing glucose. Endogenous oxygen uptake with one isolate grown on glucose-salts medium was 902 $\mu$l/hr per g of fresh weight at 30°C (Bateman and Daly, 1967). Another isolate, grown on a more complex medium containing glucose, yeast extract, peptone, and salts, gave an endogenous $O_2$ uptake of 1,320-2,400 $\mu$l/hr per g fresh weight at 25°C (Tolmsoff, 1965).

Unusual features of the oxygen uptake by *R. solani* are the persistence of the elevated rate of oxidation and its lack of influence by the addition of glucose. Starvation of cells about 18 hours old in distilled water for 4-5 hours still resulted in a high level of respiration that was not influenced by external glucose.

In comparison, *Pythium ultimum* grown under identical conditions also has a high rate of endogenous oxygen uptake (1,620 $\mu$l/hr per g at 25°C) when first recovered from the growth medium, but the endogenous rate declines rapidly and is immediately stimulated to 3,600-4,200 $\mu$l/hr per g fresh weight on addition of glucose (Tolmsoff, 1965). Another major difference in glucose oxidation by the two fungi is the pronounced "Crabtree" phenomenon in *Pythium,* and the inability to demonstrate this metabolic control mechanism in *R. solani.* The so-called Crabtree phenomenon results from an inhibition of aerobic respiration by glucose in animal cells that have a strong glycolytic system. That is, glucose initially causes a burst of respiration, but as ATP accumulates, glycolysis and aerobic respiration become inhibited through metabolic control mechanisms. The condition is overcome by addition of 2,4-

dinitrophenol to animal or *Pythium* cells, without which respiration with glucose continues to decline if excess glucose is present. The Crabtree effect in *Pythium* also occurs with other "energy-rich" substrates such as $\alpha$-ketoglutarate or glycine, but not with succinate. However, glutamate behaves unexpectedly by continuously stimulating respiration, acting similarly to combinations of glucose or $\alpha$-keto-glutarate and dinitrophenol.

The lack of respiratory stimulation by glucose in *R. solani* has been observed with cells that were approximately 18-83 hours old. Whereas, a starvation period of 4-5 hours in distilled water had no effect on cells that were about 18 hours old, a similar treatment of cells approximately 83 hours old provided a slightly reduced rate of endogenous oxygen uptake that was increased from 1,320 $\mu$l/hr per gram to a new value of 2,400 $\mu$l by the addition of glucose (Fig. 1). Under these conditions, the increased rate

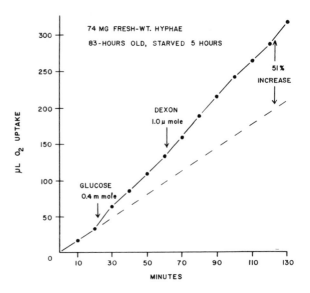

Fig. 1.   Respiration of whole cells of *R. solani.* Medium is 18 g glucose, 3 g Difco yeast extract, 5 g Difco peptone, 1.36 g $KH_2$-$PO_4$, 1.68 g $K_2HPO_4$, 0.5 g $MgSO_4$ × 7$H_2$O, final volume, 1 liter with distilled water. Cells approximately 83 hr. old were washed with 0.01 M K-$PO_4$ buffer, pH 7.0, and starved 5 hr. in buffer. Two ml hyphal suspension was used. Additions were made from sidearm of Warburg vessels in volumes of 1 ml of buffer. Center well of Warburg vessels contained 0.2 ml of 20% KOH for removal of $CO_2$. Temperature of incubation was 25°C.

of respiration does not remain constant. During a 1-hour period following the addition of glucose, two oscillatory cycles of oxygen consumption and repression are apparent. The oscillatory periods of repression resemble initiation of the Crabtree effect, but they do not continue to progress as they do in *Pythium.* This effect is further discussed in relation to $CO_2$ fixation by *R. solani* in the following section.

These observations on respiration of *R. solani* suggest that when grown in a medium containing glucose, the fungus concentrates one or more reserve respiratory substrates that saturate its respiratory system to the exclusion of exogenous glucose oxidation.

Recent information suggests that the high and persistent respiration with low R.Q. values is induced by the methods used in preparing the hyphae of *R. solani* (Tolmsoff, unpub.). Hyphae washed by gently transferring them several times to distilled water without removing the free water have different respiratory properties from those in the same hyphal batch where free water is temporarily removed by vacuum filtration in order to obtain an estimate of fresh weight.

It appears that removal of free water from the hyphae, even for a short period of time, releases an otherwise inactive respiratory pattern to give the high endogenous respiratory rate that is persistent and has a low R.Q. value in the presence or absence of added substrate. In hyphae washed without removal of free water, endogenous respiration declines rapidly; more "normal" R.Q. values with glucose (0.8-0.9 instead of 0.5) are obtained, and the respiratory rate is quickly stimulated by the addition of glucose.

The hyphae prepared by a temporary removal of free water turn dark brown within about 24 hours when incubated in a petri dish on the moist filter paper they were collected on. The cells become rounded and resemble a sclerotial mass. Possibly respiration released by removal of free water is a normal process associated with sclerotial production in *R. solani* rather than an easily induced artifact.

*Patterns of $CO_2$ metabolism.*—It is not surprising that *R. solani* yields $CO_2{}^{14}$ from uniformly labeled glucose-$C^{14}$. This terminal event in glucose catabolism is prevented, through unknown mechanisms, by the well-known fungicides dithiocarbamate and methylisothiocyanate (Wedding and Kendrick, 1959).

There are some uncommon features of $CO_2$ evolution by the fungus in the presence of exogenous glucose. High rates of $O_2$ consumption are in some cases accompanied by low rates of $CO_2$ evolution, providing R.Q. values ($\frac{CO_2}{O_2}$) of 0.50-0.53 over a 1-hour period with cells approximately 18 hours old (Fig. 2). During the second 1-hour period, the R.Q. is seen to increase to 0.71, suggesting that the exogenous glucose slowly displaces an endogenous respiratory substrate. Others, perhaps using different conditions, have observed R.Q. values of 0.8-0.9 in the presence or absence of glucose (Molitoris, personal communication).

Respiratory quotient values can be misleading because they are heavily dependent on numerous intra- and extracellular conditions (Beevers, 1961), and they also tend to obscure details of respiratory patterns that can be of importance (see Fig. 2). However, several observations suggest that $CO_2$ plays a special role in the metabolism of *R. solani*.

An atmosphere containing 20% $CO_2$, 20% oxygen,

and 60% nitrogen was found to inhibit the linear growth of aerial isolates of *R. solani* 73-87%, because of the high concentration of $CO_2$ (Durbin, 1959*a*). Isolates recovered from the soil-surface were inhibited 40-62%, while subterranean isolates were inhibited 20-43%, as compared to growth rates in air (0.03% $CO_2$). A concentration of 1% $CO_2$ was not inhibitory to any of the isolates, while 0.5% stimulated the growth of surface and aerial isolates, but not the subterranean forms. For comparison, several other

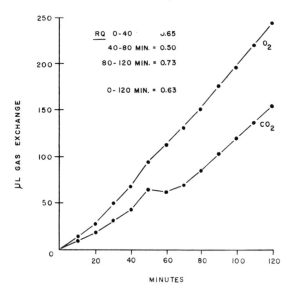

*Fig. 2.* Respiration of whole cells of *R. solani*. Conditions same as Fig. 1, except that cells were about 18 hr. old and were washed and starved 3 hr. in distilled water prior to use. One ml of cell suspension, after rewashing, was added to Warburg vessels (50 mg fresh weight) containing 1.0 ml of 0.02 M $K-PO_4$ buffer, pH 7.0, plus 0.02 M glucose. Results represent average of six vessels.

soil fungi, including *Verticillium albo-atrum, Phytophthora cinnamomi,* and *Fusarium solani* f. *cucurbitae,* were unaffected by 20% $CO_2$, while the growth of *Sclerotium rolfsii* was inhibited 55%.

It is possible that the data of Figs. 1 and 2 suggest a role for $CO_2$ metabolism in *R. solani*. It should be noted that in the aged (83 hour) cells starved 5 hours, the addition of glucose resulted in a temporary increase of oxygen consumption, which was followed by the return to a reduced rate (Fig. 1). A second increase in the rate of oxygen consumption occurred about 50 minutes after the first, followed again by a decline and a third elevation after an additional 50 minutes.

There appears to be oscillatory pattern to oxygen uptake in which there are periods of release and repression of activity. In another hyphal preparation consisting of cells 18 hours old, the oscillatory pattern of oxygen uptake is again evident (Fig. 2). Measurement of $CO_2$ release with the latter preparation showed that the pattern of $CO_2$ evolution paralleled the oscillatory pattern of oxygen uptake. The elevated rate of oxygen uptake accompanied an increased rate of $CO_2$ evolution. A subsequent decline in oxygen uptake was accompanied by a negative $CO_2$ pressure, indicating an uptake of $CO_2$.

Since the fixation of $CO_2$ is an energy-dependent process, it is possible that the alternate periods of respiratory stimulation and repression in *R. solani* are associated with cycles of $CO_2$ fixation. In the case observed, $CO_2$ uptake is initiated when the cells go into respiratory repression. It is possible that the respiratory repression results from a decrease in the ADP level or, conversely, an elevated ATP level. An elevated level of ATP should be most conducive to $CO_2$ fixation.

Some reactions within the living cell, and in cell-free systems, occur in spurts due to the collective action of various endogenous enzyme activators and inactivators (temporary inhibitors). Oscillatory respiration occurs in the yeast *Saccharomyces carlsbergensis*, and there is an oscillation of glycolysis in cell-free preparations (Chance, et al., 1965). The bursts of activity are influenced by several of the reactants and products of glycolysis. An important feature of the oscillatory activity is a cyclic generation of reduced diphosphopyridine nucleotide. An accumulation of ATP enhances reduction of pyridine nucleotide. Conceivably, a cyclic fixation of $CO_2$ by *R. solani* might be linked to the intracellular concentration of ATP, reduced pyridine nucleotides (NADH and NADPH), and other compounds.

Thus, even though *R. solani* fails to demonstrate the Crabtree effect associated with glycolysis during respiration, there is reason to believe that it possesses a highly efficient energy-conserving system. In this system, $CO_2$ fixation might release the respiratory repression by using the energy of ATP and restoring ADP for another cycle of respiratory increase. Coupled with a strong intracellular reducing capacity (see Discussion), the efficient energy-conserving system could result in near-equality of intra- and extracellular $CO_2$ at the atmospheric pressure of $CO_2$. A system of this type might also be related to the observed inhibition of growth by higher $CO_2$ pressures (2.5-20%) and to the known efficiency of *R. solani* in conserving the carbon of glucose during growth. Differences in $CO_2$ or energy metabolism are to be sought in different strains of the fungus, since those insensitive to elevated concentrations of $CO_2$ also have slower growth rates (Durbin, 1959a). Those most sensitive to $CO_2$ have the highest growth rates in air.

There is also an unusual pattern of $CO_2$ evolution from the different carbon atoms of specifically labeled glucose—$C^{14}$ (Bateman and Daly, 1967). With an isolate of *R. solani* grown on glucose-salts medium, the $C_6^{14}/C_1^{14}$ ratio was 0.09 during the first 45 minutes and increased to only 0.33 after 3½ hours. The function of the classical glycolytic, Kreb's cycle, terminal respiration chain, etc., would give a ratio of near 1.0.

The $C_6/C_1$ ratios should be interpreted cautiously in view of the following: (1) endogenous respiratory substrates may largely preclude the entry of exogenous glucose into certain metabolic channels until the endogenous substrate is depleted; (2) the glycolytic system of *R. solani* may be under a state of energy control most of the time with short "flushes" of activity related to oscillatory respiration; and (3) a strong $CO_2$-fixing tendency might negate the significance of $C_6/C_1$-$CO_2$ ratios, especially since the intracellular sites of $CO_2$ generation are not known. In other organisms, the pentose phosphate cycle enzymes that generate $CO_2$ from $C_1$ of glucose are located in the cytoplasm, while pyruvate decarboxylation, which generates $CO_2$ from $C_6$ glucose, is associated with mitochondria. Conceivably, there could be a lopsided reincorporation of released $C_1$ and $C_6$ glucose, depending on where the $CO_2$ release and fixation occurs.

MITOCHONDRIAL METABOLISM. — Information on mitochondrial metabolism of *R. solani* is almost entirely limited to previously unpublished data, and has appeared only in preliminary form (Tolmsoff, 1962a,b, 1965). This section includes original data obtained with an isolate of *R. solani* previously used in studies of fungicidal activity (Eckert, 1957; Hills and Leach, 1962; Tolmsoff, 1965) and ultrastructure (Bracker and Butler, 1963). Methods used in the mitochondrial studies were previously described in detail (Tolmsoff, 1965) and are summarized below.

*Methods.* — Cells approximately 20 hours old were washed with deionized water and suspended in a homogenizing medium at the rate of 1 gram fresh weight per 5 ml of medium. The homogenizing medium consisted of 0.5 M sucrose, 0.02 M potassium phosphate buffer (final pH, 7.2), and 0.005 M ethylene-diaminetetraacetate. Homogenization was accomplished with Ten-Broeck, hand, glass homogenizers, all previous and subsequent operations being performed at 0-4°C. The cell-homogenate was filtered through four layers of cheesecloth and centrifuged at 1,500 g for 5 minutes. The precipitate was discarded and the supernatant recentrifuged at 15,000 g for 15 minutes. The supernatant was then discarded, and the mitochondrial pellets were rinsed and resuspended in a similar medium containing 0.5 M sucrose, 0.005 M potassium phosphate buffer (final pH, 7.1), and 0.001 M ethylenediaminetetraacetate, using the glass homogenizers for resuspension. Centrifugation was repeated at 15,000 g for 15 minutes, the supernatant was discarded, and the mitochondria were resuspended in the same medium.

Solutions of this mitochondrial suspension were added (0.5-1.0 ml) to one of two types of reaction media (2.0 ml) in a 3-ml Oxygraph cell. The reaction media were:

1. Complete reaction medium:
   a. 0.5 M sucrose
   b. 0.01 M glucose
   c. 0.01 M sodium fluoride
   d. 0.01 M potassium phosphate buffer, pH 7.0
   e. 0.003 M magnesium chloride
   f. 0.001 M ethylenediaminetetraacetate

2. Potassium phosphate buffer, 0.025 M, pH 7.5

These two media are hereafter referred to as "complete" and "phosphate," respectively.

Additions to the mitochondrial suspension in the reaction cell were made in volumes of 5-25 $\mu$l and did not disturb the level of oxygen in solution to a significant degree. Oxygen consumption was recorded continuously by the Model K Oxygraph (Gilson Medical Electronics, Madison, Wisc.), which employs a highly sensitive vibrating platinum electrode. Results are expressed as m$\mu$ moles of $O_2$ consumed per min/mg of mitochondrial protein at 25°C. Rates of oxygen uptake represent steady-state conditions established over minimum periods of 1-3 minutes except where noted otherwise.

*Endogenous mitochondrial respiration.*—Mitochondria of *R. solani* had endogenous respiration of 15-17 m$\mu$ moles $O_2$/min per mg protein when suspended in the phosphate medium (Tables 1, 2, 3, 5). This represents a high rate of endogenous respiration in comparison to *P. ultimum* and *Sclerotinia fructicola*, whose mitochondria have no endogenous respiration (Lyda, 1963; Tolmsoff, 1965). However, the high rate of endogenous oxygen consumption was maintained for only 3 minutes by the *R. solani* particles and then gradually declined during 3-6 minutes to a persistent level of 6 m$\mu$ moles $O_2$/min. The persistent respiration continued for at least 15 minutes.

When the mitochondria were suspended in the "complete" reaction medium, their initial endogenous respiration was reduced to 4 m$\mu$ moles $O_2$/min (Tables 7, 8, 9). Apparently there is something about this medium that reduces the oxidation rate of an endogenous substrate. It should be noted that this same medium inhibits succinate oxidation in *Pythium* mitochondria owing to simultaneous combination of magnesium and fluoride, perhaps because of an inhibitory action by magnesium fluoride (Tolmsoff, unpub.). However, the oxidation of NADH is not inhibited by this medium with *Pythium*; it is stimulated.

The results suggest that mitochondria from *R. solani* carry with them a short supply of an endogenous respiratory substrate. When this supply is depleted, they continue to respire at a reduced rate through a slow regeneration of the substrate, or through the oxidation of an entirely different substrate. In future studies on endogenous respiration, it should be interesting to study this phenomenon in

TABLE 2. Respiration of mitochondria from *R. solani*

| Additions | Quantity | $\mu$ moles $O_2$/min per mg protein |
|---|---|---|
| 25 mM K-PO$_4$, pH 7.5 | 2.0 ml | — |
| Mitochondria, 1.0 ml | 0.6 mg protein | 15 |
| $\beta$-hydroxybutyrate | 25.0 $\mu$ moles | 6 |
| NAD | 0.25 $\mu$ moles | 6 |
| Cytochrome c | 0.02 $\mu$ moles | 6 |
| $\alpha$-glycerophosphate | 25.0 $\mu$ moles | 6 |
| L-lactate | 2.5 $\mu$ moles | 6 |
| Succinate | 25.0 $\mu$ moles | 28 |
| NADH | 0.6 $\mu$ moles | 81 |
| Dexon | 0.3 $\mu$ moles | 20 |
| Antimycin A | 0.027 $\mu$ moles | 9 |

TABLE 3. Respiration of cytoplasmic supernatant from *R. solani*

| Additions | Quantity | $\mu$ moles $O_2$/min per mg protein |
|---|---|---|
| Sucrose-PO$_4$-EDTA-supt | 3.0 ml | 8 |
| NADH | 1.25 $\mu$ moles | 8 |
| Cytochrome c | 0.02 $\mu$ moles | 8 |
| Succinate | 25.0 $\mu$ moles | 8 |

TABLE 4. Respiration of mitochondria from *R. solani*

| Additions | Quantity | $\mu$ moles $O_2$/min per mg protein |
|---|---|---|
| 25 mM K-PO$_4$, pH 7.5 | 2.0 ml | — |
| Mitochondria, 1.0 ml | 0.6 mg protein | 15 |
| NADH | 0.6 $\mu$ moles | 43 |
| Cytochrome c | 0.02 $\mu$ moles | 102 |
| (None) | (NADH exhausted) | 5 |
| Succinate | 25.0 $\mu$ moles | 34 |
| ATP | 2.5 $\mu$ moles | 30 |
| MgSO$_4$ | 2.5 $\mu$ moles | 28 |
| ADP | 2.5 $\mu$ moles | 23 |
| Malonate | 25.0 $\mu$ moles | — |

relation to the oscillatory pattern of respiration in whole cells. That is, mitochondria might be isolated with a higher content of endogenous substrate if recovered from the cells immediately following the period of $CO_2$ fixation. On the other hand, the endogenous substrate might be depleted during certain phases of oscillation.

*Oxidation of NADH, NADPH, and succinate.*— NADH and succinate provide two of the three primary respiratory substrates for mitochondrial respiration within a wide biological spectrum, a third being L-$\alpha$-glycerophosphate. NADPH and many other mitochondrial substrates commonly generate one of the primary respiratory substrates before their electrons are acceptable to the energy-conserving respiratory chain. It is not surprising that mitochondria of

TABLE 1. Respiration of mitochondria from *R. solani*[1]

| Additions | Quantity | $\mu$ moles $O_2$/min per mg protein |
|---|---|---|
| 25 mM K-PO$_4$, pH 7.5 | 2.0 ml | |
| Mitochondria, 1.0 ml | 0.6 mg protein | 15 |
| Cytochrome c | 0.02 $\mu$ moles | 15 |
| (None) | (Decline of endogenous respiration after 3-6 min) | — |
| | | 6 |
| Pyruvate | 2.5 $\mu$ moles | 6 |
| NAD | 0.25 $\mu$ moles | 6 |
| DL-isocitrate | 10.0 $\mu$ moles | 6 |
| Coenzyme A | 0.03 $\mu$ moles | 6 |
| Thiamine diphosphate | 0.21 $\mu$ moles | 6 |
| NADP | 0.25 $\mu$ moles | 6 |
| $\alpha$-ketoglutarate | 2.5 $\mu$ moles | 6 |
| NADH | 2.5 $\mu$ moles | 100 |

[1]Methods of mitochondrial preparation and study are described in the text.

TABLE 5. Respiration of mitochondria from *R. solani*

| Additions | Quantity | $\mu$ moles $O_2$/min per mg protein |
|---|---|---|
| 25 mM K-PO$_4$, pH 7.5 | 2.0 ml | — |
| Mitochondria, 1.0 ml | 0.6 mg protein | 17 |
| NADH | 2.5 $\mu$ moles | 60 |
| Cytochrome c | 0.02 $\mu$ moles | 119 |
| Dexon | 0.12 $\mu$ moles | 5 |
| MgSO$_4$ | 2.5 $\mu$ moles | 14 |
| NADH | 2.5 $\mu$ moles | 13 |
| ZnSO$_4$ | 2.5 $\mu$ moles | 13 |
| Dexon | 0.12 $\mu$ moles | 11 |
| Succinate | 25.0 $\mu$ moles | 33 |

TABLE 6. Respiration of mitochondria from *R. solani*

| Additions | Quantity | $\mu$ moles $O_2$/min per mg protein |
|---|---|---|
| 25 mM K-PO$_4$, pH 7.5 | 2.0 ml | — |
| Mitochondria, 1.0 ml | 0.6 mg protein | 9 |
| NADPH | 2.5 $\mu$ moles | 28 |
| Cytochrome c | 0.02 $\mu$ moles | 51 |
| ADP | 2.5 $\mu$ moles | 34 |
| MgSO$_4$ | 2.5 $\mu$ moles | 34 |
| Hexokinase | 0.5 mg | 25 |
| Dexon | 0.3 $\mu$ moles | — |
| Succinate | 25.0 $\mu$ moles | 38 |
| Antimycin A | 0.21 $\mu$ moles | 15 |

TABLE 7. Respiration of mitochondria from *R. solani*

| Additions | Quantity | $\mu$ moles $O_2$/min per mg protein |
|---|---|---|
| Reaction medium, pH 7.0[1] | 2.0 ml | — |
| Mitochondria, 1.0 ml | 0.6 mg protein | 4 |
| Succinate | 25.0 $\mu$ moles | 21 |
| ATP | 2.5 $\mu$ moles | 7 |
| Lipoate (oxidized) | 0.18 $\mu$ moles | 51 |
| (None) | — — — | 11 |
| NAD | 0.25 $\mu$ moles | 11 |
| Lipoate (oxidized) | 0.18 $\mu$ moles | 11 |

[1]The reaction medium used is described in the text.

TABLE 8. Respiration of mitochondria from *R. solani*

| Additions | Quantity | $\mu$ moles $O_2$/min per mg protein |
|---|---|---|
| Reaction medium, pH 7.0 | 2.0 ml | — |
| Mitochondria, 1.0 ml | 0.6 mg protein | 4 |
| NADH | 2.5 $\mu$ moles | 17 |
| ADP | 2.5 $\mu$ moles | 11 |
| Succinate | 25.0 $\mu$ moles | 21 |

TABLE 9. Respiration of mitochondria from *R. solani*

| Additions | Quantity | $\mu$ moles $O_2$/min per mg protein |
|---|---|---|
| Reaction medium, pH 7.0 | 2.0 ml | — |
| Mitochondria, 1.0 ml | 0.6 mg protein | 4 |
| NADPH | 2.5 $\mu$ moles | 15 |
| Cytochrome c | 0.02 $\mu$ moles | 30 |
| ADP | 2.5 $\mu$ moles | 23 |
| DL-isocitrate | 15.0 $\mu$ moles | 21 |
| Dexon | 0.6 $\mu$ moles | — |

*R. solani* fit the common metabolic scheme by the oxidation of NADH, NADPH, and succinate (Tables 1-9). Nevertheless, in comparison to other organisms, the mitochondria of *R. solani* respond differently to these common substrates.

In the complete reaction medium, mitochondria of *R. solani* consumed only 17 m$\mu$ moles of $O_2$/min in the presence of excess NADH (Table 8), which is not significanctly higher than initial endogenous rates of respiration in the phosphate medium (15-17 m$\mu$ moles/min). The respiratory rate with NADH was increased to 43-60 $\mu$ moles $O_2$/min when the mitochondria were suspended in "phosphate" medium (Tables 4, 5). The hypotonic phosphate medium stimulated the rate of respiration with NADH by 2.5-3.5 times compared to the rate observed in the complete reaction medium.

A comparison is perhaps pertinent here to demonstrate a later point. Mitochondria of *P. ultimum* and *S. fructicola* do not show the stimulatory response to hypotonic phosphate medium in the oxidation of NADH; instead, both of these fungal systems are inhibited by these conditions, perhaps largely because of the omission of magnesium from the medium (Tolmsoff, 1965). In the complete reaction medium, where NADH oxidation by *R. solani* mitochondria is retarded, the corresponding particles of *Pythium* or *Slerotinia* oxidize NADH about tenfold, or more, faster on a protein basis.

A possible explanation for these differences in response to NADH between the fungi might lie in the results with NADPH and succinate in the *R. solani* particles. In the complete reaction medium, NADPH and succinate provided respiratory rates of 15 and 21 m$\mu$ moles $O_2$/min, respectively (Tables 7, 8, 9). In other words, succinate provided the highest respiratory rate, with NADH and NADPH being almost equal but repressed. The suspension of mitochondria in the phosphate medium increased the oxidation rate with succinate and NADPH 1.8 and 1.9 times, respectively, but 2.5 times with NADH as compared to the complete reaction medium.

If similar respiratory mechanisms are functional in the mitochondria of *R. solani* as compared to other organisms, the results are probably underestimated by the data. That is, the stimulatory effect of the phosphate medium on NADH oxidation may be underestimated because of the absence of magnesium, and a stimulatory effect on succinate oxidation may be exaggerated because of the absence of magnesium fluoride.

Interpreted on the basis of information available from other organisms, these results suggest that the mitochondria of *R. solani* are highly impermeable to penetration by reduced pyridine nucleotides (NADH and NADPH), but more permeable to succinate. Mitochondrial membrane damage caused by hypotonicity appears to increase the permeability to the reduced pyridine nucleotides, while possibly having less effect on succinate penetration. The impermeability of *R. solani* mitochondria to external reduced pyridine nucleotides is consistent with other phases of metabolism in this fungus (see Discussion).

It should be noted that during mitochondrial isolation from *R. solani* by the procedure outlined, all of the NADH- and succinate-oxidase activities are recovered with the mitochondrial fraction, and none is lost to the supernatant (Table 3). Similar conditions of mitochondrial isolation from *P. ultimum* are known to result in mitochondrial damage with a substantial loss of these activities to a submitochondrial unit released to the supernatant (Tolmsoff, 1965).

*Other mitochondrial substrates.* — Mitochondria from *R. solani* have not provided oxygen uptake with typical mitochondrial substrates such as pyruvate, $\alpha$-ketoglutarate, isocitrate, L-$\alpha$-glycerophosphate, $\beta$-hydroxybutyrate, glutamate, and lactate (Tables 1, 2). Recognized coenzymes and cofactors have also been ineffective in stimulating the oxidation of these substrates. Failure to oxidize these substrates is due to an absence, or inactivity, of the dehydrogenases from mitochondria, as demonstrated by the ability of the particles to oxidize NADH or succinate following the addition of various potential, but inactive, substrates and cofactors (Tables 1, 2).

The inactivity (or absence) of numerous typical mitochondrial dehydrogenases from *R. solani* particles deserves further comment. In view of other metabolic properties of the fungus, it is possible that the mitochondria are deficient in those activities that could not be shown. However, the physiological and morphological integrity of the isolated particles has not been determined. Their impermeability to external NADH implies a high degree of integrity, but this should be confirmed by other methods.

It is of interest that cell homogenates of *R. solani* have also failed to reveal the presence of $\alpha$-ketoglutarate dehydrogenase activity (Molitoris and Van Etten, 1965). Without this key enzyme system, the complete Kreb's cycle cannot function because of the strong exergonic nature of the complex series of reactions it catalyzes. This provides additional support for the suggestion that vegetative hyphae of *R. solani* may lack Kreb's cycle activity.

*Respiratory stimulation by horse-heart cytochrome c.*—Mitochondrial respiration of *R. solani* was approximately doubled by low levels of horse-heart cytochrome c in the oxidation of NADH and NADPH in a phosphate medium (Tables 4-6). The oxidation of NADPH was also doubled by cytochrome c in the complete reaction medium (Table 9).

The stimulation of NADH oxidation by cytochrome c suggests that mitochondria of *R. solani* possess certain respiratory features in common with other mitochondrial organisms. Succinate oxidation was not stimulated by cytochrome c, and this too is a common result.

In the presence of cytochrome c, *R. solani* mitochondria consumed oxygen with NADH at a rate of 119 m$\mu$ moles/min per mg protein in the phosphate medium (Table 5). For comparison, the best preparations from *P. ultimum* have given respiratory rates of 152 m$\mu$ moles O$_2$/min in the absence of added cytochrome c, and 369 in its presence, when tested in the complete reaction medium (Tolmsoff, 1965).

The stimulatory effect of cytochrome c on respiration should be interpreted cautiously. The conventional concept of added cytochrome c fitting into a shuttle site in the electron transport chain, which might be vacated through loss of the endogenous enzyme, leaves important observations unexplained. For example, the added enzyme might permit respiratory electrons to bypass endogenous cytochrome c in mammalian mitochondria (Camerino and Smith, 1964). There is also the problem of magnesium producing the same degree of stimulation in mitochondria of *P. ultimum* competitively with added cytochrome c (Tolmsoff, unpub.). This type of response suggests that the added enzyme has some role other than, or in addition to, the simple filling of a vacant electron-shuttle site.

The data of Tables 4-6 also suggest another role for the added cytochrome c. The oxidation of NADH without cytochrome c occurred at 60 m$\mu$ moles O$_2$/min and was increased to 119 on addition of the enzyme. Under identical conditions, NADPH gave oxygen uptake of 28 and 51 m$\mu$ moles without and with cytochrome c, respectively. It is apparent that the endogenous respiratory chain was capable of carrying electrons to reduce 60 m$\mu$ moles O$_2$/min, but cytochrome c still increased the respiratory rate with NADPH from 28 m$\mu$ moles to the new level of 51 m$\mu$ moles. The inhibition of respiration with both substrates by Dexon suggests that their electrons pass through the sensitive NADH-linked flavoprotein-cytochrome b site (Tolmsoff, 1965). Thus, it appears that added cytochrome c has an effect other than filling a vacant site on the respiratory chain.

Future studies on the role of cytochrome c in stimulating the respiration of fungal mitochondria should consider the possibility that the added enzyme might alter the structure and permeability of mitochondrial membranes. The selective effect of cytochrome c in stimulating NADH-linked respiration, but not that of succinate, and the competitive role of magnesium in *Pythium* point also to a possibility of affecting the catalytic action of an enzyme specifically associated with NADH oxidation.

It should be noted that beef heart and *Pythium* mitochondria possess two types of respiratory units (Green, 1958; Tolmsoff, 1965). The one released from mitochondria as a small electron-transport particle is not stimulated by cytochrome c or magnesium, whereas the respiratory system more firmly attached to mitochondria responds to these factors in the oxidation of NADH.

It remains to be determined whether the two respiratory units represent different stages in the biosynthesis of a complete respiratory system in which cytochrome c is incorporated last. Studies with *Pythium* mitochondria isolated at different pH levels (6.0-7.5) and from different ages of the cells suggest that an incomplete respiratory unit, which is tightly bound to the membranes, is the one that responds to cytochrome c (Tolmsoff, unpub.).

*Effects of adenine nucleotides on mitochondrial respiration.*—Respiratory rates with NADH and NADPH in mitochondria of *R. solani* were reduced 23-35% by ADP in the complete reaction medium (Tables 8, 9) or the phosphate medium (Table 6). ADP had little or no effect on succinate oxidation. On the other hand, ATP depressed succinate oxidation by 67% in the complete medium (Table 7) and by 12% in the phosphate buffer.

These apparent anomalous responses to the adenine nucleotides in respiration could again reflect the permeability properties of *R. solani* mitochondria. ADP (Packer, 1960) and ATP (Neubert and Lehninger, 1962) induce mitochondrial contraction in animal mitochondria. Although the adenine nucleotides increase respiratory activities with certain substrates, their inhibition of choline oxidation by rat-liver mitochondria is believed to result from a decreased permeability to choline (Kagawa, et al., 1965; Wilken, et al. 1965).

The mitochondria of *R. solani* again demonstrate some unique responses to common metabolites. Perhaps under in vivo conditions, its mitochondria are highly impervious to the reduced pyridine nucleotides and many other cytoplasmic components. The fact that the mitochondria possess a relatively high potential for the oxidation of NADH and succinate suggests that these respiratory substrates are generated within the mitochondria by endogenous substrates that remain to be identified, and that may not be the common Kreb's cycle acids.

Future studies on mitochondrial swelling and shrinking, as influenced by ADP, ATP, AMP, calcium, magnesium, respiratory substrates, and other factors should help to explain the unusual mitochondrial behavior of *R. solani*. It is known that in the complete reaction medium described, mitochondria of *P. ultimum* and *S. fructicola* shrink in response to succinate or ATP, and expand during NADH oxidation in the absence of ADP (Tolmsoff, unpub.). There is reason to suspect that the mitochondria of *R. solani* will also respond by swelling or shrinking.

*Lipid oxidation by mitochondria from R. solani.*—Two common coenzyme lipids, oxidized lipoic acid and coenzyme $Q_{10}$ (co-$Q_{10}$), were found to provide a high degree of respiratory stimulation in *R. solani* mitochondria. The lipids were added to mitochondrial suspensions in $10 \mu l$ quantities of acetone. Acetone by itself had no effect on respiration.

The initial rates of respiration and total oxygen consumed were usually proportional to the amount of lipid added (except under the conditions shown in Table 7). With either lipid, oxygen uptake was initially rapid for about 1 minute or less and then declined to lower rates. The reduced rate of respiration, following the initial burst, persisted much longer with co-$Q_{10}$ than with lipoate. Highest rates of oxygen uptake with either lipid were obtained in phosphate buffer (pH 7.0) containing 0.1 mM

MnSO$_4$. In this medium, repeated additions of small quantities of lipid gave repeated bursts of respiration.

There were striking differences in the way that the two lipids were oxidized. Over the range of concentrations of lipoate employed (0.18-1.8 $\mu$ moles), the $\frac{1}{2}O_2$/lipoate ratio was 0.3-0.5, showing that lipoate was utilized inefficiently for respiration. However, the $\frac{1}{2}O_2$/co-$Q_{10}$ ratio was consistently 6.5 and greater when this lipid was employed in quantities of 0.05-0.12 $\mu$ moles. Thus, a minimum of several oxygen atoms was consumed per molecule of co-$Q_{10}$ added to the mitochondria. Considering the insolubility of co-$Q_{10}$ in aqueous solution, it is possible that the observed $O_2$ ratios are underestimated.

Another difference in the oxidation of the two lipids was their response to Dexon, which is known to be a selective inhibitor of NADH oxidation (Tolmsoff, 1965). Dexon stimulated the respiratory rate with lipoate 7-10 fold, but not the quantity of oxygen consumed. On the other hand, Dexon prevented oxygen uptake with co-$Q_{10}$, suggesting that the latter was oxidized through NADH generation.

It is unfortunate that other common lipids such as the fatty acids were not tested with the *R. solani* mitochondria. The results do not show whether the aromatic nucleus or polyisoprene chain of coenzyme Q contributes to respiration. It is also possible that co-$Q_{10}$ exerts its effect through releasing endogenous compounds for respiration, perhaps displacing other mitochondrial lipids. Future tests using $C^{14}$-labeled substrates and measuring the release of $CO_2^{14}$ would be helpful in determining if *R. solani* oxidizes the lipid isoprene chain of co-$Q_{10}$. It should be noted that the low R.Q. values obtained with whole cells (Fig. 2) might also suggest the endogenous oxidation of lipids or other highly reduced compounds.

*Effects of enzyme inhibitors on mitochondrial respiration.*—

1. *Malonic acid.* The oxidation of succinate is prevented by malonic acid at an equal molar concentration (Table 4).

2. *Antimycin A.* At a concentration of 7.5 $\mu$M, antimycin A provided 43% inhibition of NADH-cytochrome c reductase activity in cell-free extracts of *R. solani* (Dowler, et al., 1963). Mitochondrial oxidation of succinate was inhibited 39% by 70 $\mu$M antimycin in the presence of added cytochrome c (Table 6).

3. *Azide and cyanide.* These two inhibitors act on a similar area of respiration to prevent cytochrome c-oxidase (cytochromes a + a$_3$) from passing electrons to oxygen in the terminal step of respiration. Sodium azide inhibited the NADH oxidase (−340 m$\mu$) 86% at 10 mM in cell-free extracts of *R. solani* (Dowler, et al., 1963). Potassium cyanide (1 mM) prevented the mitochondrial oxidation of lipoic acid and co-$Q_{10}$, but its effects on succinate and NADH oxidation are not known.

4. *Pentachloronitrobenzene (PCNB).* In preliminary studies with isolated mitochondria from *R. so-*

*lani*, PCNB (50 $\mu$M) inhibited the endogenous respiration by 77%, while the oxidation of succinate was stimulated 1.7 times (Tolmsoff, unpub.). If these results are confirmed, PCNB might prove useful in identifying the endogenous mitochondrial respiratory substrate of this fungus.

5. Dexon (p-dimethylaminobenzenediazo sodium sulfonate). Dexon is a highly selective enzyme inhibitor that prevents respiratory electron flow between the NADH-linked dehydrogenase flavoprotein and cytochrome b (Tolmsoff, 1965). The original Dexon molecule is not the active respiratory inhibitor, but is the starting material for an enzymatic synthesis of a potent NADH oxidase inhibitor. These conversions are made on a light-activated molecule and require a respiratory substrate (NADH or succinate) and probably ADP for completion. The respiratory reducing equivalents probably intervene prior to ADP in the formation of the inhibitor.

The *R. solani* mitochondria apparently contain the mechanisms for activating Dexon, and they also contain the site inhibited by the activated molecule (Tables 2, 5). In a hypotonic phosphate medium used to obtain reasonable rates of NADH oxidation, mitochondria of *R. solani* react to Dexon similarly to those from *Pythium*. The oxidation of NADPH by mitochondria is also inhibited by Dexon in either the complete reaction medium or the hypotonic phosphate buffer (Tables 6,9). This could suggest that NADPH is oxidized via NADH generation within the *R. solani* mitochondria.

The similarity in response to Dexon with mitochondria from *Pythium* and *R. solani* presents a curious situation. In the presence of glucose, whole-cell respiration of *Pythium* is prevented by Dexon, whereas that of *R. solani* is unaffected (Tolmsoff, 1965). *Rhizoctonia solani* catalyzes the decomposition of Dexon, possibly requiring NADPH or NADH for the reaction, and the decomposing system may be distributed in the mitochondria and cytoplasm. Both NADH and pyruvate stimulate the mitochondrial decomposition of Dexon, even though pyruvate does not contribute to respiration in the absence of Dexon. This could indicate that the pyruvic dehydrogenase of *R. solani* is a mitochondrial-surface enzyme, if indeed the pyruvate-linked decomposition of Dexon is due to NADH generation by pyruvate.

It was previously considered likely that the basis for resistance to Dexon by *R. solani* might reside in mitochondrial impermeability, such as that known to occur in isolated mitochondria of *S. fructicola* (Tolmsoff, 1965). NADH oxidation by the latter is highly resistant to Dexon in the presence or absence of ADP until the mitochondria are damaged by hypotonicity or one of the natural steroid detergents such as deoxycholate or digitonin. When injured, the *Sclerotinia* mitochondria also react to Dexon similarly to the *Pythium* system.

There are at least three possible reasons for the failure of Dexon to inhibit whole-cell respiration of *R. solani* in the presence of glucose: (1) the impermeability of mitochondrial membranes to Dexon; (2) the endogenous respiration of a substrate such as succinate, which bypasses the Dexon-sensitive site; and (3) the NADH (or NADPH?)-linked destruction of Dexon by cytoplasm and mitochondria of *R. solani*.

The impermeability factor in *R. solani* may not be important because, as noted, Dexon inhibits the oxidation of NADPH in the presence of ADP. The ADP represses NADPH oxidation in the absence of Dexon, but, in its presence, there is an immediate and total inhibition of respiration similar to that observed with NADH oxidation in *Pythium* mitochondria, but very unlike the response of *Sclerotinia* mitochondria. This suggests that the *R. solani* mitochondria are not impermeable to Dexon in the reactions required for its activation and inhibition in spite of possible restrictions on NADH or NADPH permeability caused by ADP.

It is considered unlikely that the endogenous respiration bypasses the NADH pathway because this is the most efficient system known for the conservation of respiratory energy in most biological systems. The fact that *R. solani* possesses an active NADH oxidase argues against the inactivity of this system in respiration because inactive enzymes are usually eliminated by proteolysis.

This leads to the third possibility of enzymatic destruction of Dexon before it can reach the mitochondria in whole cells. Previously, this interpretation was considered unlikely because of the relatively slow decomposition of Dexon by whole cells or isolated cell fractions of *R. solani* as observed by total loss of Dexon color. However, this interpretation may no longer be tenable in view of a probable strong reductive potential of the cytoplasm in *R. solani* (discussed later). The possibility should be considered that the resistance of *R. solani* to Dexon might reflect a strongly reductive cytoplasm capable of enzymatically destroying Dexon before it can reach the mitochondria for activation to an inhibitor.

It should be noted that *Pythium* lacks the system for decomposition of Dexon, *Sclerotinia* appears to possess a membrane system that eliminates the inhibitor, and *R. solani* might possess an efficient mechanism for the reductive destruction of the compound before it can do damage. These observations on a single fungistatic inhibitor are easily extended to possible significance in other fungi. For example, Dexon is also destroyed by whole cells of *Rhizopus stolonifer* and *Penicillium* sp. (Ramsey [R.], personal communication). In the case of *Rhizopus*, which like *R. solani* has an uncommonly high rate of growth, one might look for a reductive destruction of Dexon. In *Penicillium*, which has a much slower growth rate, one might look for mitochondrial exclusion of Dexon, a system for enzymatic destruction of the inhibitor, or both. Thus, future studies on the action of Dexon, as on numerous other inhibitors, might reveal general metabolic properties of an organism that are otherwise difficult to study.

NONRESPIRATORY OXIDASES.—*Ascorbic acid oxidase.* — Isolated mitochondria of *R. solani* have given variable degrees of oxygen uptake with ascorbate as a substrate, from high rates of activity to negative reaction (Tolmsoff, unpub.). The reason for variable activity is not known, and could be related to age of the fungus, or various degrees of mitochondrial contamination by other cellular fractions. For example, *M. verrucaria* has ascorbic acid oxidase activity associated with the cell wall.

*Polyphenol oxidase, tyrosinase, laccase, catalase.*— The phenol oxidase of *Corticium sasakii* (*T. cucumeris*) is active in oxidizing gallic acid, hydroquinone, catechol, and guaicum, (the activity decreasing in that order [Nagata, 1960]). Tyrosine and p-cresol were not acted on. Maximum phenol oxidase activity was found after 6-15 days of growth, with very little activity after 3 days.

Maximum laccase activity was found after 3-9 days, while tyrosinase was maximum after 10-12 days. Optimum activities were observed at pH values of 5.6 for laccase and 4.6 for tyrosinase with relatively higher laccase activity. The type of medium on which the fungus was grown influenced both the total activities and the age at which maximum activities were developed. Sclerotia contained both tyrosinase and laccase activities, and the laccase activity of mycelial mats was confined intracellularly.

Laccase activity was inhibited 80-85% by 0.1 mM sodium azide and 100% by 1 mM. Potassium cyanide inhibited laccase activity 30-50% at 0.1 mM and 83-100% at 1 mM, whereas diethyldithiocarbamate was not inhibitory at 2 mM.

The type of medium on which *R. solani* is grown has a pronounced effect on catalase released to the external medium. Gelatin provided a high level of growth with an absence of catalase activity, whereas peptone gave less growth and more catalase activity (Edwards and Newton, 1937). It was suggested that the peptone of the medium protected the released catalase from inactivation.

The physiological significance of the various nonrespiratory oxidases is uncertain. Their production is highly dependent on the nature of the substrate on which the fungus is grown. It is likely that some of these oxidases are involved in detoxification reactions, and in reactions that prepare aromatic compounds for degradative pathways of metabolism. They may play an important role in senescent metabolism, preparing the organism for hibernation by removal of certain molecules that interfere or are no longer needed. Others, such as catalase, are required for various flavin-linked, nonrespiratory oxidations. In general, nonrespiratory oxidases require a high level of oxygen for catalytic activity, and their action may be somewhat subdued by intracellular conditions. For example, the potato tuber contains phenol oxidases, but these are activated by cell disruption (Beevers, 1961).

STEROID METABOLISM.—Ergosterol is a primary steroid of numerous fungi (Lavate and Bentley, 1964), and also occurs in *R. solani* (Van Etten and Molitoris, 1965). When expressed in terms of dry weight of *R. solani*, ergosterol was found to decrease

with age, as did also soluble amino nitrogen, RNA, DNA, and protein, while total lipids remained constant and carbohydrates increased. When expressed on a DNA basis, ergosterol again decreased, while RNA, protein, and soluble amino nitrogen remained constant, and total lipids, fatty acids, and carbohydrates increased. Thus, the metabolism of ergosterol follows a pattern on aging different from several other common constituents. It appears that ergosterol is present at highest levels during juvenility with a decrease during aging.

In view of the importance of steroids in membrane properties and possible enzyme-control mechanisms, and considering some unusual mitochondrial membrane properties in *R. solani,* it should be of special interest to study the role of steroids in the metabolism of this fungus.

During the past few years, a group of Japanese workers has been attempting to find microbial methods for the conversion of certain readily available steroids into mammalian steroid hormones on a commercial basis. The unusual steroid interconversions carried out by *C. sasakii* (*T. cucumeris*) have led to extensive studies with this fungus (Hasegawa and Takahashi, 1958, 1959; Takahashi, 1961, 1963, 1964). Although it is not possible to summarize all of the steroid interconversions catalyzed by *C. sasakii* and related fungi, a few should be mentioned.

An uncommon oxidation of 17-$\alpha$-hydroxydeoxycorticosterone (Cortexolone or Richstein's compound S) by *R. solani* and related fungi involves the simultaneous hydroxylation of $C_{11}$ to the sterioisomeric 11-$\alpha$ and 11-$\beta$ hydroxy-derivatives (Hasegawa and Takahashi, 1958; Greenspan, 1960; Takahashi, 1961, 1964). This same position is subject to dehydrogenation (Takahashi, 1964), and to formation of a $C_{11}$-keto derivative (Greenspan, 1960).

Other oxidations of the steroid molecule by *R. solani* occur at $C_1\beta$, $C_2\beta$, $C_6\beta$, $C_7\varepsilon$, $C_{15}\varepsilon$, $C_{15}$keto, and many others (Greenspan, 1960). Some of the predominant products formed by *C. sasakii* from cortexolone include 19-hydroxycortexolone; the two hydrocortisone epimers mentioned above ($C_{11}$-$\alpha$ and $\beta$-hydroxy derivatives); $\Delta^4$-pregnene-6$\beta$-, 17-$\alpha$, 21-triol-3,20-dione; and, $\delta$1,4-pregnadiene-17-$\alpha$,21-diol-3,20-diene (Hasegawa and Takahashi, 1958, 1959; Takahashi 1961, 1963).

Different isolates of *Pellicularia filamentosa* (*T. cucumeris*) differ in their ability to convert cortexolone to other steroids (Takahashi, 1961).

POTENTIAL USE OF FUNGICIDES IN STUDIES OF METABOLISM.—It was previously pointed out that if the mechanism of PCNB action in *R. solani* were known, or if the basis for lack of respiratory inhibition by Dexon were known, more rapid progress could be made in understanding the unusual metabolism of this fungus. There are many other compounds that have a high degree of activity in preventing the growth of *R. solani*. Studies on the action of some of these compounds should also be revealing.

Methylarsine sulfide is one of the most active compounds available for preventing seedling diseases caused by *R. solani* (Leach and Tolmsoff, unpub.); it has less activity against *P. ultimum*. Other arsine compounds are active against *C. sasakii*, particularly certain arsine xanthates (Oda, et al., 1961). The most effective have the empirical formula: RAs (SC-(:S)-OR')$_2$, where R is methyl and R' is propyl or R is propyl and R' is hexyl.

Methylisothiocyanate and dithiocarbamate prevent the growth and respiration ($-CO_2$) of *R. solani* (Wedding and Kendrick, 1959). These compounds cause a loss of P$^{32}$ from cells, and it has been suggested that dithiocarbamate affects the cell membranes. The loss of phosphorus from cells need not be an indication of a direct action on membranes, since inhibition of respiration and oxidative phosphorylation would reduce the ATP level required to maintain membrane integrity. Activation of ATPase or phosphatases might also increase the cellular level of inorganic phosphate with loss to the external medium. More detailed studies are needed on the action of these fungicides.

It should be noted that measurements of $CO_2$ release can be helpful in the interpretation of action by inhibitors, but they can also be misleading. For example, the inhibition of $CO_2$ evolution in *Pythium* by Dexon follows closely behind the inhibition of oxygen consumption (Tolmsoff, 1965). An effect of this type on $CO_2$ should not be as surprising as one that is secondary, where the prevention of NADH oxidation is likely to deplete NAD+, the latter being required in many decarboxylation reactions.

Captan [N-(trichloromethylthio)-4-cyclohexene-1,2-dicarboximide] and Dichlone (2,3-dichloro-1,4-napthoquinone) are effective inhibitors of respiration in *P. ultimum* (Tolmsoff, unpub.). These fungicides have varying degrees of activity against different isolates of *R. solani*, as does also pentachloronitrobenzene when measured by fungicidal activity (Sinclair, 1960).

Also active in growth inhibition of *R. solani* are 6-chloro- and 6-bromobenzoxazolone-2 (Eckstein and Zukowski, 1958).

Studies with methylmercuridicyandiamide have shown that *Aspergillus niger* concentrates intracellular nonprotein thiols from the medium, which may help to protect the fungus against the action of this fungicide (Ashworth and Amin, 1964). *Rhizoctonia solani* and *P. ultimum* did not concentrate the thiols as extensively, and it was suggested that this might account for a greater activity of mercurials against these fungi. This same study revealed a general difference in the properties of the proteins between the pathogenic fungi. The proteins of *R. solani* were precipitated by either the sodium chloride-phosphate mixture or 80% acetone, whereas the *Pythium* proteins were not precipitated by the salt-phosphate mixture.

The toxicity of allyl alcohol to *R. solani* is believed due to an enzymatic oxidation of the alcohol to acrolein, which inhibits urease and certain sulfyhydryl-dependent enzymes (Legator and Racusen, 1959).

Filipin (a tetraene antibiotic) inhibits the growth of *R. solani* (Gattani, 1957) and numerous other fungi (Gottlieb, 1961). It selectively affects those organisms that produce steroids, or require steroids for growth (Gottlieb, et al., 1961; Lampen, et al., 1963; Demel, et al., 1965). The polyene antibiotics should prove useful to studies of membrane properties and the integrity of membrane systems.

A material produced by *Endothia parasitica,* and known as diaporthin, inhibits the growth of *R. solani* (Gauman, 1957), as does also a polypeptide antibiotic known as duramycin (Lindenfelser, et al., 1958). Pimaricin is similarly effective (Hine, 1963).

Sulfanilamide is reported to inhibit the growth of *R. solani*, with the inhibition reversible by p-aminobenzoate (Tolba and Salama, 1962*a*). This suggests that the action of sulfanilamide in the fungus is similar to that of other organisms in preventing the synthesis of essential folic acid compounds that have numerous metabolic functions.

The action of these various compounds against *R. solani* is mentioned because they and many others might serve as tools for investigating the metabolism of this fungus. In the past, metabolic inhibitors have served a useful function in helping to identify vital reactions, their sites of occurrence, and their mechanism of reaction. A number of chemical inhibitors are now available for such studies with *R. solani*.

METABOLIC PRODUCTS OF R. SOLANI.—It is difficult to draw a line between compounds synthesized in a cell and retained and used there, and those formed and released to the external medium. The release of compounds by living cells is influenced by many factors, including composition of the medium, other environmental factors, time of investigation in relation to aging and autolysis, etc. However, some of the common metabolites of *R. solani* deserve mention because of the possible reflection on metabolic pathways involved in their synthesis. These metabolites are of special interest in that most arise where glucose is a sole carbon source.

Among the aromatic compounds found in culture filtrates of *R. solani* are phenylacetic acid, m-hydroxyphenylacetic acid, p-hydroxyphenylacetic acid, and 3-nitro-4-hydroxyphenylacetic acid (Aoki, et al., 1963). Nonaromatic compounds included $\beta$-furoic acid, succinic acid, and lactic acid.

When sucrose is fed as the sole carbon source with sodium nitrate as the nitrogen source, the following sugars appear in the external medium in addition to sucrose: glucose, fructose, unidentified pentose, maltose, and two unidentified carbohydrates (Ogura, et al., 1961). With the exception of maltose, these same sugars are found in the hyphae. Several common amino acids are also found in the medium and hyphae. Perhaps it is significant that variations in the presence or absence of nearly all of the common metabolites occurred between different isolates of this fungus.

With the efficient conservation of glucose carbon by *R. solani*, it should not be surprising if this fungus is conservative in the release of metabolites. The release of some of the aromatic compounds is of special interest in that phenylacetic acid is a plant growth regulator.

DISCUSSION.—*General metabolic properties of R. solani.*—The information on metabolism of *R. solani*, though limited, falls into a pattern compatible with itself and with some of the physiological and ecological properties of this fungus. The following is a general picture of this metabolism. The picture is not intended to include all strains of *R. solani*, but it may suggest more fruitful points of entry for future research on a more detailed level, which hopefully might reveal strain differences.

Two important metabolic features required for the picture include: a high level of cytoplasmic-reducing equivalents in the form of NADH, NADPH, or both, and an efficient energy-conserving system.

*Cytoplasmic reducing potential.*—Several pieces of information suggest that *R. solani* will be found to possess a strongly reductive cytoplasm. These pieces of information are: (1) an unusually high rate of growth; (2) an impermeability of its mitochondria to reduced NADH and NADPH; (3) an apparent absence of common electron-shuttle systems in the mitochondria, i.e., $\alpha$-glycerophosphate and $\beta$-hydroxybutyric dehydrogenases; (4) uncommonly low $C_6/C_1$ ratios with glucose (0.09); (5) a relatively high degree of sensitivity to $CO_2$ with an apparent cyclic fixation of the gas; (6) efficient utilization of nitrate for growth in comparison to ammonium; (7) sensitivity to acid conditions; and (8) a reductive decomposition of Dexon.

The high rate of growth of mammalian tumors is associated with a loss of the mitochondrial electron-shuttle enzymes, $\alpha$-glycerophosphate and $\beta$-hydroxybutyrate dehydrogenases, and an elevated level of cytoplasmic-reduced pyridine nucleotides (Boxer and Develin, 1961). The tumor cells have a high rate of glycolysis and an elevated transhydrogenase activity (Foster and Taylor, 1966), which may make it possible for the NADH generated in glycolysis to equate with NADPH in reactions of synthesis. The tumor cells retain an efficient conservation of respiratory energy (Chance and Hess, 1959). The synthesis of the steroid cholesterol in the tumor lacks the normal feedback control mechanism and pyridine nucleotide specificity (Foster and Taylor, 1966.)

A high level of cytoplasmic-reductive potential, which may be associated with rapid growth, demands an absence of electron-shuttle systems and a sharp partitioning of the reduced pyridine nucleotides between the cytoplasm and mitochondria. Both of these properties appear to be fulfilled by the single isolate of *R. solani* examined. Its mitochondria are relatively impervious to the reduced pyridine nucleotides and further exclude them in the presence of adenine nucleotides found in the living cell. In the cell, the mitochondria might be highly impervious to the reduced pyridine nucleotides in the cytoplasm. This could lead to a situation conducive for reactions of synthesis and rapid growth.

The low $C_6/C_1$ ratios obtained with glucose suggest that this substrate is oxidized preferentially, if not exclusively, by a nonglycolytic pathway (Bateman and Daly, 1967). Glucose-6-phosphate and 6-phosphogluconate dehydrogenases are active in *R. solani* (Dowler, et al., 1963; Molitoris and Van Etten, 1965) and are to be suspected in helping to provide the low $C_6/C_1$ ratio from glucose. For each carbon of glucose released by these enzymes, two molecules of NADPH are generated. Thus, function of the pentose phosphate cycle might simultaneously maintain a high level of cytoplasmic-reduced pyridine nucleotide and provide intermediates required for rapid growth. Perhaps 5-bromouracil, an inhibitor of glucose-6-phosphate dehydrogenase (Hochster, 1961), could be used to study the contribution of different pathways to respiration in *R. solani*.

The low R.Q. obtained in the presence of glucose (Fig. 2) is indicative of the conservation of glucose carbons. This observation is complimentary to the known efficiency of glucose conversion to cellular constituents in *R. solani* (Sherwood, this vol.).

Other basidiomycetes share a resemblance to certain phases of metabolism in *R. solani*. *Lactarius torminosus* lacks certain glycolytic enzyme activities and is believed to possess a strong pentose phosphate cycle (Meloche, 1962). *Schizophyllum commune* has a strong endogenous respiration unaffected by glucose in the vegetative stage (Niederpruem and Hackett, 1961). However, basidiospores of *Schizophyllum* show respiratory stimulation with glucose and a lack of inhibition by fluoride, fluoroacetate, or malonate (Niederpruem, 1964). The lack of inhibition by these compounds might suggest that glucose is oxidized by a pathway that does not involve glycolysis or the Kreb's cycle. Some of the smut fungi are also implicated in a strong pentose phosphate pathway of glucose catabolis at the expense of glycolysis (Meloche, 1962).

The metabolism of nitrogen by *R. solani* is consistent with a strongly reductive cytoplasm. Nitrates are effectively used for growth (Sherwood, this vol.). They serve as a "sponge" of free protons and the reducing equivalents of pyridine nucleotide during formation of ammonium. Thus, in an unbuffered medium, the pH rises as the fungus uses nitrate. If ammonium is instead supplied as the nitrogen source, the pH falls to levels that inhibit growth unless the medium is well buffered (Deshpande, 1959*b*). If both ammonium and nitrate are present, the fungus responds as though only ammonium was present, and the pH falls. This suggests that ammonium might repress the formation or inhibit the activity of nitrate reductase in *R. solani*.

Most fungi tolerate acid conditions better than *R. solani* does, but growth at acid pH varies with the substrate supplied. This phenomenon might be related to the generation of excess reducing equivalents and protons by the oxidation of glucose but not when more highly oxidized substrates are provided (i.e. $AH_2 + NADP^+ \leftrightarrows A + NADPH + H^+$). The level of hydrogen ions might reach repressing levels in the cytoplasm. Inhibition of growth at acid pH could also be related to $CO_2$ metabolism since the solubility of $CO_2$ (bicarbonate) decreases with increasing acidity.

*Energy conservation.*—The efficiency with which *R. solani* uses glucose for growth while simultaneously confronted with energy-draining reactions associated with nitrate reduction, $CO_2$ fixation, and synthesis of numerous molecules required for growth indicates that the fungus must possess an efficient energy-conserving system. The most efficient system known for the conservation of respiratory energy is the oxidative phosphorylation process associated with mitochondria. Identification of the endogenous mitochondrial respiratory substrate in *R. solani* should be of significant aid in studies of its conservation of energy.

*Carbon dioxide metabolism.*— Carbon dioxide was once thought to be a dead-end product of metabolism in various organisms that lack chlorophyll. It is now apparent that $CO_2$ is involved in numerous synthetic and degradative pathways. The synthesis of pyrimidines, purines, pteridines, riboflavin, steroids, oxaloacetate, fatty acids, and a number of other molecules requires $CO_2$ as a substrate or catalyst (Fruton and Simmonds, 1959). Carbon dioxide can be pictured as a catalyst in the synthesis of fatty acids where its addition to acetyl-CoA yields malonyl-CoA. The latter is the immediate precursor of the $C_2$ unit added to an existing fatty acid molecule for extension of the carbon chain. As the $C_2$ unit is added to the fatty acid molecule, the original $CO_2$ is released (Bressler and Wakil, 1961).

It is interesting that the subdivision of ecological strains of *R. solani* into subterranean, soil-surface, and aerial types appears to be related to their $CO_2$ tolerance or stimulation by $CO_2$ (Durbin, 1959a). This suggests that the more rapidly growing surface and aerial strains might require the replenishment of an essential metabolite (e.g. oxaloacetate, fatty acids, carbamyl phosphate, etc.) through periodic $CO_2$ fixation. It would also appear possible that their inhibition by excess $CO_2$ might result from an uncontrolled synthesis of a metabolite required in low concentrations and inhibitory at higher levels. For example, oxaloacetate and palmityl-CoA (Ontko and Jackson, 1964) inhibit the mitochondrial oxidation of succinate, and palmityl-CoA induces mitochondrial expansion. Palmityl-CoA is extremely inhibitory to certain enzymes such as glucose-6-phosphate dehydrogenase and glutamic dehydrogenase (Taketa and Pogell, 1966).

It remains to be determined how $CO_2$ influences the metabolism of different strains of *R. solani*. The results of Fig. 2 suggest that the fixation of $CO_2$ is turned on and off abruptly. The value of cells that are uniform in age for the study of this phenomenon

should be apparent. A mixture of different ages might yield averaged responses in gas exchange and obscure important metabolic patterns that could help to identify the $CO_2$ response.

A brief consideration of $CO_2$ metabolism in other organisms might suggest similarities for which to look in *R. solani*.

The virulence of *Pasterurella pestis* is lost when it is grown at 37°C, at which temperature the solubility of $CO_2$ is reduced, and avirulent strains develop (Baugh, et al., 1964a). However, the virulent strain of the pathogen develops at 26° or at 37° if bicarbonate or a spent medium is added to the culture. Orotic acid, cytosine, uracil, or citrulline can replace the $CO_2$ requirements for growth of the virulent strain, and each of these compounds is known to be linked to $CO_2$ metabolism via carbamyl phosphate (Baugh, et. al., 1964b). In this bacterium, a deficiency in the ability to utilize $CO_2$ for the synthesis of carbamyl phosphate appears linked to its loss of virulence.

*Cytophaga succinicans* requires $CO_2$ for the fermentation of glucose, and the $CO_2$ is found in succinate as an end product of metabolism (Anderson and Ordal, 1961). Succinate is believed to be formed from oxaloacetate in this organism.

The stimulatory effect of $CO_2$ on growth and reproduction of fungi was previously reviewed (Cochrane, 1958). It appears that $CO_2$ stimulates the accumulation of succinate and other organic acids in fungi, and that an available nitrogen source accentuates the fixation of $CO_2$.

In the basidiomycete *S. commune*, the absence of $CO_2$ prevents germination of basidiospores (Hafiz and Niederpruem, 1963). On the other hand, $CO_2$ prevents fruiting of the fungus when it is exposed to the gas prior to formation of fruiting primordia, but $CO_2$ is without effect after the primordia are formed (Niederpruem, 1963).

Carbon dioxide also affects the autotropic response of germinating conidia in *Botrytis* (Jaffe, 1966). The germ tubes of nearby conidia commonly grow toward each other or parallel, but $CO_2$ induces them to grow away from each other.

In view of the importance of $CO_2$ metabolism in other microorganisms, and the effects it has on growth and development of *R. solani*, it would appear that a great deal might be learned about the metabolism of this fungus by following the fixation of radioactively labeled $CO_2$ during various stages of growth and differentiation.

Another fruitful avenue for gaining insight into the metabolism of *R. solani* might be through investigations on the precise sites and modes of action by various fungicides and growth inhibitors. By revealing the enzymes affected (or other mechanisms of inhibition), the more important metabolic pathways or properties unique to *R. solani* might become apparent.

# Colonization and Growth of Rhizoctonia Solani in Soil

GEORGE C. PAPAVIZAS—*Crops Research Division, Agricultural Research Service, United States Department of Agriculture, Beltsville, Maryland.*

During the past 15 years, considerable knowledge has become available on growth and saprophytic behavior of several soil-borne microorganisms associated with root rots, wilts, and seedling diseases of crop plants. Extensive studies have been made of the saprophytic behavior of some cereal root-rot fungi by the Cambridge Botany School (Garrett, 1956). These studies, published in a series of elegant papers (Butler [F.C.], 1953*a,b,c;* Lucas, 1955) are now well known and require little further comment at this time, except to say that they were responsible for stimulating further research along the same lines with many root-infecting fungi.

For a little over a century, students of the widespread, predominantly soil-borne fungi of the genus *Rhizoctonia* have inquired into the occurrence, distribution, and pathogenicity of its species, into epidemiology of Rhizoctonia diseases and the principles of their control, and into their behavioral characteristics. Much has been learned of the species, their potential, and their limitations insofar as they affect pathogenesis. Until the past few years, however, very little effort has been expended in studying growth and saprophytic behavior of *Rhizoctonia* spp. in soil. In the past decade, interest in the microecological behavior of *R. solani* Kühn under competitive conconditions has markedly increased as a result of the development of a considerable amount of detailed knowledge of the behavior of this parasite under conditions of nonantagonism and the recent development of rapid and easy techniques for measuring growth and saprophytic colonization in natural soils (Kendrick and Jackson, 1958), and saprophytic colonization in natural soils (Warcup, 1955; Thornton, 1956; Kendrick and Jackson, 1958; Papavizas and Davey, 1959*b*, 1961, 1962*b;* Martinson, 1963).

I shall attempt to summarize the available knowledge on saprophytic colonization and growth of *R. solani* in soil as affected by several biological and nonbiological determinants of the soil ecosystem.

AUTECOLOGICAL BEHAVIOR OF R. SOLANI IN SOIL. —*Growth of mycelium in soil.*—In seeking precise and detailed fundamental information about the behavior of *R. solani* in the complex environment of the soil ecosystem, I may be in disagreement with deep-rooted concepts that experimental conditions in vivo are too complex for the proper analysis of events. Since the pioneering research of Blair (1943) on the growth and behavior of *R. solani* in soil, however, new interest has developed in the functioning of the complex soil system in relation to the study of *R. solani* behavior. Although most of the relevant problems are still unresolved, more plant pathologists and microbiologists are now more interested in the actual behavior of this parasite in nature than in its performance under highly artificial conditions, however commendable and profitable the interest in that performance may be.

Blair (1943) was the first to introduce the Rossi-Cholodny "soil-plate method" to study growth of *R. solani* in unsterilized soils. Since the mycelium of this parasite is characterized by particular and recognizable morphological peculiarities, it was easy to follow its linear extension through soil along the glass surface of the slides from the inoculum to the tips of the advanced mycelium. With this technique, Blair demonstrated that *R. solani* could grow in unsterilized soil for relatively long distances without any energy sources other than those present in natural soils and quite independently of the inoculum from which growth was initiated.

Perhaps the most illuminating experiment on growth and spread of *R. solani* in natural soil dealt with infection of radish seeds placed in a ring at various distances from an inoculum disk, the radii of the rings being 1-9 cm (Blair, 1943). The closer to the disk the seeds were placed, the earlier and more extensive was the damping-off that occurred. At a distance of 9 cm from the inoculum, about 40% of the seedlings were eventually parasitized by the fungus in each of two soils tested.

The fact that *R. solani* can be isolated from natural soils by screened-immersion plates (Thornton, 1956), by immersion tubes (Chesters, 1948; Martinson, 1963), by colonization of organic substrate segments (Papavizas and Davey, 1959*b*, 1962*b;* Davey and Papavizas, 1962; El Zarka, 1963) or of seeds (Messiaen, 1957; Kendrick and Jackson, 1958), or by picking individual hyphae existing in natural soils (Warcup, 1955, 1957) substantiates the contention that the parasite exists as active mycelium in soil. Warcup (1957) obtained *R. solani* from wheat fields by means of his hyphal-isolation method and showed that the parasite is a frequent and early colonizer of buried leaf tissues of wheat and grasses. Several other investigators (Herzog and Wartenberg, 1958; Boosalis and Scharen, 1959; Herzog, 1961) showed that *R. solani* can exist in soil and in organic debris as vegetative mycelium and also as dormant sclerotia.

There remain, after all these fairly specific and relevant instances, very few doubts about the ability of *R. solani* to make free and independent growth in the natural and complex soil microenvironment. Some doubts were expressed by Radha and Menon (1957) who found that *R. solani* from coconut roots grew profusely in sterilized soil but very poorly in natural soil, and by Winter (1950) who showed that *R. solani* invaded unsterile soil only from an already colonized substrate. Differences, possibly connected with inherent characteristics of the isolates used by Blair (1943) on the one hand and Winter (1950) and Radha and Menon (1957) on the other, may explain some of the contrasting views on growth of this parasite under natural conditions.

*The importance of substrate availability on growth.* —It is now common knowledge that food bases in the soil ecosystem may influence not only survival of soil-borne fungi (Garrett, 1938; Park, 1956), but also the nutritional status of fungal propagules and, consequently, the inoculum potential *sensu* Garret (1956). To understand in particular the mechanisms of survival of *R. solani* in soil, however, and to design cogent biological tests in the future, precise information on the role of soil substrates on the growth of the parasite is required.

Several reports already show the importance of available substrates on the growth of *R. solani* in soil (Blair, 1943; Winter, 1950; Boyle, 1956*a;* Das and Western, 1959; Garrett, 1962). These reports, in addition to giving information relevant to the present thesis, stimulated a number of questions in our minds. Is a substrate needed for initiation or maintenance of growth or for both? Does *R. solani* draw nutrients from soil solutions during its growth? If so, what is the precise nature of the metabolized nutrients? The last question cannot be discussed because of lack of pertinent data.

The first question was discussed by Blair (1943) who showed that growth of *R. solani* was diminished more on sand than on soil on removal of the inoculum disk. Growth ceased 13 days after the inoculum plug was placed on the sand in glass tubes. At this stage, the hyphae had traveled about 5 mm from the inoculum. In three natural soils, however, the parasite grew about 12-17 cm more than it did on sand; growth continued to an appreciable extent when he removed the inoculum disks after 2 days or substituted nutrient-free mycelial strands for the regular inoculum disks. Blair concluded that a certain "food potential" may be necessary to initiate, though not to maintain, saprophytic growth of *R. solani* in soil, and that additional growth on natural soil was supported by available nutrients in the soil

ecosystem. The "food base" concept had originally been proposed by Garrett (1938).

Winter (1950) presented evidence that not only initiation, but also sustenance of growth of *R. solani* in soil is dependent on the presence of a food base. In his experiments, growth depended entirely on nutrients derived from the food base (inoculum). The parasite grew well from inoculum placed on pure sand moistened with tap water, the extent of growth depending on the size of the food base. The parasite could also grow from infected potato sprouts to a depth of 30-40 cm and contaminate the soil with sclerotia. Later, Winter (1951) reiterated his view that *R. solani* was virtually unable to develop as a saprophyte without a food base. When he supplemented soil with straw dust (with or without calcium nitrate) the parasite grew equally well in unsterilized compost, garden soil, or sand.

Considerable support for Winter's views may be derived from Boyle's (1956*a*) concept of a food base being important in mediating not only saprophytism, but also parasitism, by a *Rhizoctonia* sp. under natural soil conditions. In his experiments (Boyle, 1956*b*), crop refuse served as a food base enabling a *Rhizoctonia* sp. to attack peanut plants. Cropping practices designed to keep organic matter out of the upper level of soil gave excellent control of peanut root and pod rot caused by *Sclerotium rolfsii* Sacc. and a *Rhizoctonia* sp. Further support for Winter's views comes from the work of Das and Western (1959) who found that in natural soils *R. solani* made vigorous growth for 6 days, probably deriving nutrients from the food base. However, after 6 days very little growth occurred. On the other hand, removal of the food base from sterilized soil after 5 days did not affect subsequent rate of growth. Complete and phosphate fertilizers and moderate applications of nitrogen and potassium increased growth of the parasite in sterilized soil but, in natural soils, the nutritional effects were largely masked by antagonism of the associated soil microflora.

As yet it is difficult to assess the role that substrate availability may play in the growth of *R. solani*. More investigations are needed, particularly of the possibility of storage and translocation of nutrients by the saprophytic mycelium, which can absorbs and translocate readily, throughout its length, various mineral ions (Monson and Sudia, 1963; Littlefield, et al., 1965). The ability of the *R. solani* mycelium to absorb, translocate, and store nutrients would not minimize the significance of a food base in initiating growth, but the phenomenon itself would reduce continuous nutritional dependence on food bases for sustenance and growth. Between nutritional exhaustion of a food base and incidental or regular appearance of another food base, *R. solani* would remain dormant as thick-walled mycelium and sclerotia in the exhausted food base (Boosalis and Scharen, 1959). New nutritional contacts may be made when fresh organic matter is added to soil, by mycelial migration supported by stored and translocated nutrients, or by nutrients derived directly from the soil itself. Ephemeral substrates would not only support growth of *R. solani*, but also limit its spread around them. The radius of mycelial migration and the duration of growth would depend on the kind of substrate nutrients, on the differential ability of clones to absorb and translocate nutrients, on nutrients avalable in the soil matrix itself, and on other biochemical and ecological determinants of the soil microenvironment.

*Characteristics and morphology of growth.*—It is evident from Garrett's book (1956) and from a number of review articles (Kendrick and Zentmyer, 1957; Menzies, 1963*a*) that characteristics and morphology of growth (growth habit *sensu* Garrett [1951]) are among the least known aspects of behavior of this organism in soil. This stems from lack of proper and precise laboratory and field techniques whereby growth habit and behavior can be examined in a strictly controlled environment.

Garrett (1956) named the continuous external growth of soil fungi on roots "ectotrophic growth habit." Samuel and Garrett (1932) and Hynes (1937) showed that *R. solani* behaved like a typical soil-inhabiting parasite, with no ectotrophic growth habit, and with the ability to infect only the immature apical region of cereal roots. Characteristic red-brown external mycelium of *R. solani* abounded both on diseased roots and in a ramifying condition in the soil matrix. The parasite evidently grew indiscriminately through soil and over any roots with which it came in contact, invading susceptible root tips but not mature regions of roots. Hyphae of the parasite were usually distributed externally on the root surfaces in an irregular and sparse fashion, unlike that of the ectotrophic mycelium of root-infecting fungi, which usually develop quite a uniform sheath around the host root (Garrett, 1956).

The degree of mycelial organization of *R. solani* into sclerotia has been studied almost exclusively under artificial conditions. Any suggestions based on the work in vitro about factors that may influence initiation and development of sclerotia in vivo must therefore be largely speculative and cannot be critically assessed at the present time. Nevertheless, the exercise should be made since it will at least indicate some of the aspects for future scrutiny under natural conditions. High nitrogen and carbon in agar media, for instance, favored sclerotial development in some experiments (Tyner and Sanford, 1935), whereas in other experiments (Allington, 1936), relatively low carbon and nitrogen favored sclerotial production on agar media. Initiation of sclerotia in vitro differed in nutritional requirements from the process of further sclerotial development and particularly from that of maturation (Townsend, 1957). More carbon and nitrogen were needed for maturation than initiation. Maturation of sclerotia began only after mycelial growth stopped. Townsend (1957) postulated that sclerotial maturation, and therefore survival was prevented by competition for food supply, but she did not specify the competitors.

Sanford (1941*a*) presented some evidence that sclerotial formation was a result of sudden and vigorous growth of new *R. solani* hyphae and that hyphae a few days old lost their ability for sclerotial initiation and development. An adequate supply of nitrate nitrogen, a temperature range of 15-20°C, and especially a high relative humidity in the soil interstices were the most critical determinants for sclerotial initiation in steam-sterilized soil. Later, Sanford (1956) showed that growth and the process of sclerotial formation were related to condensation of soil-air moisture on the glass of the containers rather than to total water content of soil and thus by implication, to the relative humidity of the soil interstices. Sclerotial formation by mycelia growing through soil in glass tubes reached a maximum within 6 days of inoculation and within 6 cm from the inoculum (Blair, 1943). Since no sclerotia developed farther than 6 cm from the inoculum, Blair thought that sclerotial formation depended on an abundance of nutrients provided by the inoculum. Sclerotia were produced sparsely on the sand, presumably because of scarcity of inorganic nutrients. According to Das and Western (1959), sclerotia were formed in unsterile soil only.

The studies of the latter workers on sclerotial formation by *R. solani* illustrate the complementary nature of observations made under contrived experimental conditions. They also serve to remind us that though we can often make valid deductions from observations in vitro, we must carefully assess the influence of variables such as concentration of nutrients, moisture content and relative humidity, temperature, and other determinants that may be unaccounted for in artificial experiments. Of the many variable involved, perhaps the most important are presence or absence of food supply and microbial competition for that supply; little is known about these determinants in relation to sclerotial initiation and development.

SAPROPHYTIC COLONIZATION AND SUBSTRATE EXPLOITATION BY R. SOLANI.—*Saprophytism and competitive saprophytic ability.*—In this section, I would like to examine briefly the terms "competitive saprophytic colonization," "competitive saprophytic ability," and "competitive saprophytic activity" as various expressions of saprophytism encompassing the overall ability of *R. solani* to colonize competitively and exploit nutrient substrates in soil. From an ecological viewpoint, I regard both "competitive saprophytic colonization" and saprophytism as having the same meaning. Competitive saprophytic ability *sensu* Garrett (1950, 1956) is "the summation of physiological characteristics that make for success in competitive colonization of dead organic substrates." Garrett (1956) suggested that competitive saprophytic ability of a fungus in soil would depend not only on growth rate, production of antibiotics or toxins, enzymatic ability, and tolerance of toxins or antibiotics produced by other microorganisms, but also on other microdeterminants of the soil ecosystem, including number and variety of antagonists exploiting a substrate.

Since the competitive saprophytic ability of *R. solani*, an intrinsic characteristic, is regarded as crucial to substrate exploitation and to survival of the parasite in soil (Garrett, 1956), it is important that we have an accurate and simple means of measuring it and of determining the influence of determinants—both intrinsic and extrinsic—mediating saprophytism. For *R. solani*, Papavizas and Davey (1961) used the term "competitive saprophytic activity" as a percentage expression of Garrett's (1956) "competitive saprophytic ability," a genetic characteristic mediated by extrinsic determinants of the soil microenvironment. Papavizas and Davey (1959b, 1961) determined the percentage of colonization by incubating stem segments of mature buckwheat plants or other plant materials (1962b) in soil for certain time intervals, then by recovering the segments from soil, washing and culturing them on a minimal agar medium, and determining the percentage of colonization by *R. solani*. They showed (1962b) that *R. solani* existed saprophytically in soil, in the absence of hosts, as morphologically distinct isolates. These isolates differed not only in their saprophytic and parasitic abilities, but also in their tolerance of $CO_2$ and other antimicrobial agents.

There are several other reports strongly in favor of competitive saprophytism in *R. solani*. El Zarka (1963) used dry mature stems of Jew's mallow plant (*Corchorus olitorius* L.) to assay the vertical distribution of the parasite in soils. Martinson (1963), using soil microbiological sampling tubes, confirmed previous reports (Papavizas and Davey, 1961) that the rapid saprophytic invasion phase of a substrate buried in soil takes place from 1-2 days, though, occasionally, it may keep increasing up to 3 or 4 days. Saprophytism of *R. solani* has also been used successfully as a basis for isolation by burying maize kernels in soils (Messiaen, 1957; Kendrick and Jackson, 1958). Boyle (1965b) found that inoculum of an unidentified species of *Rhizoctonia* was ineffective on peanuts in sandy soil from which all organic matter had been removed. The pathogen remained active if stems and leaves of peanuts, cotton, soybean, or corn were incorporated in the upper few inches of soil.

*Intrinsic growth rate and success in saprophytic colonization.*—Intrinsic growth rate was recognized (Garrett, 1956) as one of the most important factors determining the outcome of any struggle between a particular soil-inhibiting fungus and associated microflora in soil for colonization and exploitation of substrates. Considerable evidence has now arisen with *R. solani* confirming or contradicting the suggested relationship. Wastie (1961) suggested that not only intrinsic growth rate, but also tolerance of competition from other microbes, may influence success or failure of soil-inhibiting fungi in the struggle for successful organic matter exploitation. One would therefore expect strict correlation between innate growth rate and extent of saprophytism only in the absence of antagonism in its general sense.

In the case of *R. solani*, however, where tolerance to antagonism was marked (Lockwood, 1959; Wastie, 1961) a close correlation is expected between growth rate and degree of saprophytism. Actually, Wastie (1961) showed that the degree of success in competitive colonization by *R. solani* was correlated better with its growth rate in pure culture than with its degree of tolerance to antibiotics.

Further observations were brought forward by Wastie (1961) that while *R. solani* was the most antibiotic-tolerant fungus, it was one of the three poorest colonizers of organic matter out of 14 soil-inhabiting fungi tested. Rao (1959) also considered *R. solani* a poor saprophyte and placed it in the same category with several other soil-borne fungi of low saprophytism. According to Rao, there may not be any correlation between intrinsic growth rate measured in pure cultures on agar or the tolerance of antifungal substances with saprophytic activity of *R. solani*. He arrived at his conclusions by using the agar plate method for estimating competitive saprophytic ability, and attributed the rather unexpectedly low saprophytism to more favorable conditions of antagonism prevailing on the plates than could ever be expected in soil. Rao admitted, however, that *R. solani* is a successful saprophyte under conditions of reduced competition in natural soils, and that the agar plate method may not be reliable for determining saprophytism.

Our recent data, obtained with wild-type isolates of *R. solani* and single-basidiospore isolates of *Thanatephorus cucumeris* (Frank) Donk (Papavizas, 1964, 1965), lend considerable support to Garrett's concept (1956) that saprophytism and intrinsic growth rate are positively correlated. Single-spore isolates possessing high growth rate on agar exhibited in natural soils not only high tolerance of antimicrobial agents and high saprophytic activity, but also prolonged survival in the colonized substrates (Table 1). To further understand the saprophytic behavior of *R. solani*, perhaps, most of all, this symposium should encourage the design of more realistic laboratory and greenhouse experiments to simulate the behavior of this parasite in nature.

*Effect of substrate and soil nutrients on saprophytism.*—At any one time, *R. solani* is under the continuous influence of several primary and secondary physicochemical and biological determinants that may be responsible for success or failure in the struggle for substrate exploitation in the soil ecosystem. The presence or absence of energy materials, their kind and quantity, and their availability are are the primary factors directly determining saprophytic colonization and growth of *R. solani*, and indirectly determining these by influencing other antagonists.

In one of the original experiments, Papavizas and Davey (1961) studied the effect of time of substrate incubation in soil on the saprophytic activity of *R. solani*. Substrate segments incubated in soils infested with *R. solani* for 2-4 days had the highest percentage of colonization (Fig. 1). After this brief

TABLE 1.  Growth rate, saprophytic activity, and saprophytic survival of single-basidiospore isolates of *T. cucumeris*

| Single-basidiospore isolate | Rate of growth[1] (mm/hr) | Colonization[2] % | Residual colonization[3] % | Inhibition of colonization by 20% $CO_2$ % |
|---|---|---|---|---|
| R118-11 | 0.72 | 100 | 90 | 0 |
| R118-13 | 0.58 | 100 | 89 | 18 |
| R118 (parent) | 0.71 | 100 | 67 | 30 |
| R118-1 | 0.52 | 96 | 66 | 29 |
| R118-32 | 0.50 | 84 | 46 | 50 |
| R118-43 | 0.51 | 72 | 26 | 67 |
| R118-14 | 0.42 | 64 | 17 | 76 |
| R118-6 | 0.33 | 60 | 16 | 88 |
| R118-48 | 0.29 | 46 | 6 | 92 |
| R118-42 | 0.16 | 40 | 4 | 91 |

[1]On Czapek-Dox agar containing 0.5 g yeast extract/l.
[2]Assayed by use of buried mature buckwheat stem segments, colonized by the fungus.
[3]One hundred percent minus percentage of residual colonization equals decolonization.

initial and intensive activity, a period of declining activity (beginning at about the third or fourth day of incubation) was observed, followed by a period of relative equilibrium (for a number of weeks). Saprophytic survival within the colonized segments decreased with prolonged incubation, but the para-

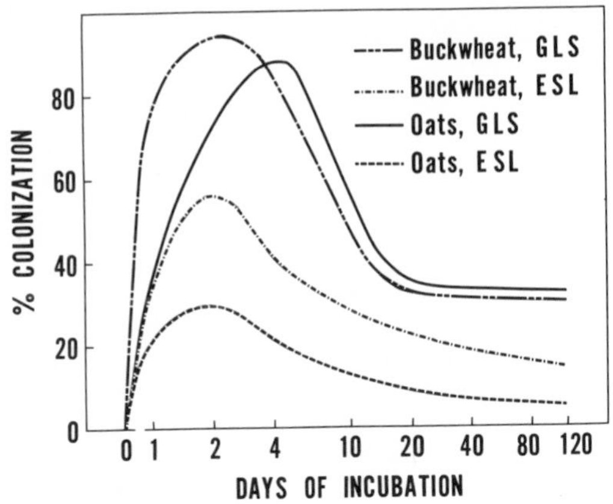

Fig. 1.  Competitive saprophytic colonization by *R. solani* of mature buckwheat and oat-stem segments buried for various lengths of time in greenhouse loamy sand (*GLS*) and Elsinboro sandy loam (*ESL*) naturally infested with the pathogen.

site was still recoverable from the colonized substrates after 120 days in soil. It was suggested that availability of rapidly assimilable nutrients in the substrate may explain the intensive saprophytism in the initial phases of colonization. As quickly as food reserves decreased, saprophytic activity declined. Absence of an appreciable decline of the saprophytic activity after the tenth day of incubation might have been due to sclerotial formation, to the presence of a limited food supply released by the decomposition of organic matter, or to slow utilization of cellulose by *R. solani* (Garrett, 1962). These results were confirmed recently by Sneh, et al. (1966) who found that the highest number of colonized stem segments of various crop plants was attained

within 2-4 days of their exposure in soil naturally infested with *R. solani*.

Supplemental phosphorus and calcium added to substrates were not important in affecting saprophytic activity of *R. solani* (Papavizas and Davey, 1961). Substrate enrichment with sodium nitrate, however, produced a marked increase of saprophytism. When carbon:nitrogen (C/N) ratios of substrate segments had been modified by addition of glucose or sodium nitrate to produce a range of C/N ratios, the carbon-to-nitrogen balance determined the extent of *R. solani* saprophytism (Papavizas and Davey, 1961). A decrease of the C/N ratio of the substrate, by the addition of increasing concentrations of nitrogen, produced a marked increase in saprophytic activity and an increase in the longevity of active fungal mycelium in colonized substrate segments (Fig. 2). Reduction of saprophytism was hastened by supplementing the substrate with glucose, which raised substrate C/N ratios. In these experiments, an increase in numbers

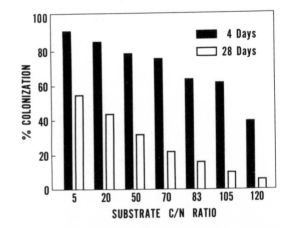

Fig. 2.  Effect of C/N ratios of buckwheat-stem segments on their competitive saprophytic colonization by *R. solani*. C/N ratios below and above 83 were obtained by addition to segments of increasing concentrations of $NaNO_3$ and glucose, respectively.

and activities of organic-matter decomposers may have resulted in nitrogen immobilization and consequently in nitrogen deficiency for *R. solani*. Nitrogen depletion decreased not only the initial colonization of the substrate, but also its complete exploitation by accelerating the rate of decline of the *R. solani* mycelium in the colonized substrate.

The indirect effect of substrate availability on the competitive saprophytic activity of *R. solani* was shown in experiments (Davey and Papavizas, 1963) in which cellulose powder, oat straw, and soybean hay were adjusted to a range of C/N ratios through addition of calculated amounts of nitrogen from ammonium nitrate. With cellulose, maximum inhibition of saprophytism occurred in soils receiving amendment adjusted to C/N ratios within the range of 40/80 a week after incorporation, 40-100 after 3 weeks, and 40-200 after 5 weeks (Fig. 3). Inhibi-

C/N ratios 10-20, nitrogen stimulated saprophytic activity of the pathogen, the stimulation being more pronounced after 1 week than after 3 and 5 weeks. At C/N ratio 5, there was stimulation after 1 week, a mild depression after 3 weeks, and a deep depression after 5 weeks (Fig. 3). A concomitant decrease in soil pH (from original pH 6.6 to 4.8) and a sharp increase in soil fungi (Fig. 4) were observed in soil at C/N ratio. Saprophytic activity of *R. solani*, however, is not affected to an appreciable extent by soil reactions as low as pH 4.5 (Papavizas and Davey, 1961).

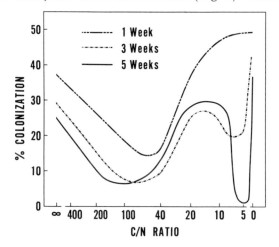

*Fig. 3.* Effect of C/N ratio of N-enriched cellulose powder incorporated in soil naturally infested with *R. solani* and time after incorporation on the competitive saprophytic activity of the pathogen.

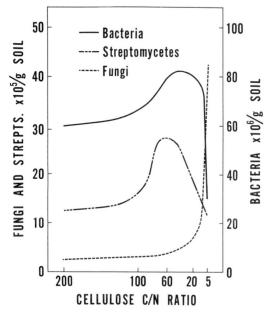

*Fig. 4.* Numbers of soil microorganisms as affected by C/N ratio of N-enriched cellulose powder incorporated in soil 3 weeks prior to sampling.

tion was obtained also by cellulose-nitrogen combinations having either higher (200-400) or lower (5, 10, and 20) C/N ratios. In these experiments, nitrogen exerted two distinct but opposite effects on saprophytism, the first observed at C/N ratios of 40-100, and the second at 20 and below. The inhibitory effect at 40-100 was perhaps due to intensified microbial activity, especially that of bacteria and actinomycetes, associated with cellulose decomposition stimulated within that particular C/N ratio range and predetermined by inherent microbial characteristics and environmental factors (Fig. 4). Nitrogen at C/N ratios 40-100 undoubtedly increased the decomposition rate of organic matter and promoted not only a rapid output of $CO_2$, which may be fungistatic to *R. solani* at certain concentrations (see section on $CO_2$), but also antimicrobial substances of specific or unspecific nature. At C/N ratio 0 (400 ppm nitrogen only), the process of absorption and assimilation of an abundance of nitrogen prevented nitrogen starvation and stimulated considerable saprophytism by *R. solani*. Also, at

That the conclusions drawn in the foregoing paragraphs may not be more than an oversimplification of microecological interactions of the soil ecosystem was shown by the fact that the same kind of depressing or stimulatory effects of organic matter on saprophytism were not always produced by the same C/N ratio ranges (Davey and Papavizas, 1963). The effect depended on the kind of substrate used. Thus, saprophytic activity of *R. solani* was significantly suppressed by an oat-straw-nitrogen combination with C/N ratios 5-85 (Fig. 5). However, saprophytic activity was suppressed by a soybean hay-N combination at a C/N ratio range lower than that of cellulose and oat straw (Fig. 6). Chandra and Bollen (1960) observed that the C/N ratio of an organic material may not be a good indicator of the rate of its decomposition in soil. To understand the mechanisms of suppression or stimulation of saprophytism of *R. solani* by available substrates in the soil microenvironment, we should know not only the C/N ratios of the substrates, but also the relative availabilities of carbon and nitrogen, and, undoubtedly, of other organic and inorganic nutrients. We should also bear in mind that numerous inconsistencies in the literature concerning the effects of carbon and nitrogen on the saprophytic activity of *R. solani* may be related to its wide range of variation.

*Cellulose decomposition.*—The problem of cellulose decomposition by *R. solani* in soil is of particular importance, not only in the context of stimulation and maintenance of saprophytic growth of the parasite by cellulose, but also because cellulose is perhaps the most abundant organic compound in soil (Alexander, 1964). Work on cellulose decomposition by *R. solani* has been concentrated for the most part on utilization of cellulose as a sole source of carbon by *R. solani* isolates in vitro. There has been less attention to utilization of cellulose or its by-products of degradation in natural soils.

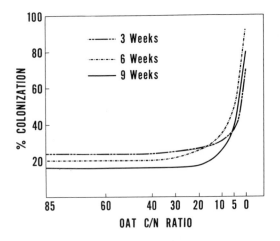

*Fig. 5.* Effect of C/N ratio of N-enriched mature oat straw incorporated in soil naturally infested with *R. solani* and time after incorporation on the competitive saprophytic activity of the pathogen.

Isolates of *R. solani* can utilize cellulose as the sole source of carbon in vitro, grow well on cellulose fibers, and produce considerable amounts of soluble cellulolytic enzymes in the culture filtrates (Ross, 1960). On the other hand, though wheat-stem isolates decomposed cellulose in pure culture and competitively colonized certain types of cellulose in unsterile soil, they were unable to colonize mature cereal straws in soil (Pitt, 1964b). In another study (Kohlmeyer, 1956), *R. solani* was the only member of a group of 11 soil-inhabiting fungi characterized by a rather poor cellulose-decomposing ability. The *R. solani* mycelium was attached loosely to cellulose membranes, neither penetrating nor forming grooves.

The foregoing results suggest that it is difficult to assess the role that cellulose plays in the competitive saprophytism and carbon nutrition of *R. solani* particularly of the possibility of adaptive utilization in natural soil. More investigations need to be made, particularly of the possibility of adaptice utilization of cellulose and the by-products of microbial degradation. *Rhizoctonia solani* was placed (Garrett,

1951) in the saprophytic "sugar fungi," a term first used by Burges (1939) to denote soil-inhabiting fungi that are normally unable to decompose cellulose or lignin. Garrett's ecological classification of *R. solani* among the saprophytic sugar fungi was based mainly on Blair's report (1943) that several clones

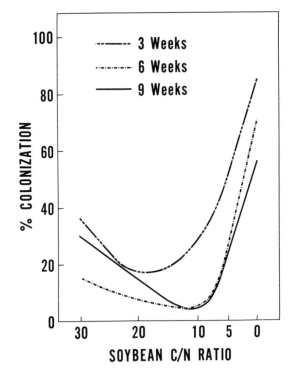

*Fig. 6.* Effect of C/N ratio of N-enriched mature soybean hay incorporated in soil naturally infested with *R. solani* and time after incorporation on the competitive saprophytic activity of the pathogen.

of *R. solani* were unable to grow over filter paper. The isolates could decompose cellulose only slightly and much more slowly than typical cellulose decomposers. The ecological disadvantage of *R. solani* in natural conditions in the presence of decomposing cellulosic materials was attributed by Blair to its inability to live on cellulose as a substrate in soil, and, by inference, to its inability to compete with strong cellulose decomposers.

If saprophytism of *R. solani* is suppressed by strong cellulose decomposers (Davey and Papavizas, 1963), this does not preclude the possibility, at least under some conditions, that the parasite can decompose cellulose. A relevant and convincing instance of cellulose colonization and degradation by *R. solani* in natural soil was provided by Garrett himself (1962). He found that isolates of the parasite from swede, lettuce, and potato could not only utilize cellulose as a sole source of carbon, but also were able to competitively colonize and destroy cellulose in natural soil. Saprophytic growth in soil in the absence of living organic matter was attributed by Garrett (1962) to the ability of the parasite to colonize and decompose cellulose. Tribe (1960a) also showed that *Rhizoctonia* spp. were among the primary colonizers of boiled cellulose films buried in soil. *Rhizoctonia solani* alone reduced a piece of cellulose film to a mushy condition in only 3 weeks. In an extension of this work, Tribe (1960b) showed *R. solani* mycelium to be dominant on cellulose film buried in two natural soils. In a mixture of equal parts of a *Rhizoctonia*-free soil with a *Rhizoctonia*-infested soil, cellulose was colonized thoroughly by *Rhizoctonia*. Similar results have been reported by Daniels (1963) who demonstrated unequivocally the cellulolytic activity of several isolates tested, both in unsterilized soil and in pure cultures, and on both processed and native forms of cellulosic materials. In unsterilized soil, mycelial growth of 14 isolates on cellulosic materials was rated from poor to abundant, depending on the isolate and the kind of material used as a substrate. Of particular relevance was the ability of several isolates to degrade native cellulose similar to that from flax fibers. Daniels concluded that "cellulolytic ability was a normal characteristic of *R. solani,* whatever its natural habitat, and that it was able to grow on cellulosic substances in unsterile soil, although here it was somewhat restricted by soil mycostasis."

A relevant point was provided by Papavizas (1964) in studies on colonization of cellulosic materials in soil by several single-basidiospore isolates of *T. cucumeris.* The isolates differed greatly in their ability not only to colonize an organic substrate, but also in the length of time they remained active and exploited the substrate. Addition of mature oat straw to soil either increased saprophytic activity of some single-spore isolates (and, by inference, the ability to utilize cellulosic materials) or reduced it. Single-basidiospore isolates whose saprophytism was enhanced by oat straw were those also possessing high survival ability in soil and high tolerance to antimicrobial agents.

From the foregoing discussions on cellulose decomposition in vivo and in vitro, it is rather difficult to classify *R. solani* among cellulose decomposers or among sugar fungi. The difficulty stems from the fact that in different strains or in single-spore isolates the capacity for cellulose decomposition may vary from very feeble to very strong; also, *R. solani* behaves as a primitive sugar fungus in all other respects. If we accept that *R. solani* decomposes cellulose, as appears to be the case (Tribe, 1960a,b; Garrett, 1962; Daniels, 1963; Papavizas, 1964), we cannot by definition classify it as a sugar fungus. And yet it behaves like a saprophytic sugar fungus: It possesses the ability to utilize easily available sugars, pentosans, and possibly hemicelluloses; it is considered among the first invaders of injured, moribund, or dead plant tissues in soil; it has a high intrinsic mycelial growth rate that may be related to the exploitation and exhaustion of soil food bases; it forms dormant and resistant structures under some conditions; and it may infect living roots only occasionally and according to existing opportunities.

Therefore, from the standpoint of microecological grouping, *R. solani* should perhaps still be classified as a primitive sugar fungus, but be regarded somewhat exceptional in this respect, as Garrett indicated (1956).

*Saprophytism and survival.*—It is extremely difficult to determine which characteristics of *R. solani* are the most important for its survival or predominance in the soil ecosystem. Garrett (1956) introduced the concept that strong saprophytes are usually able to survive in soil for a longer period of time than weak saprophytes. He even postulated that small differences in saprophytic ability, acting over a wide span of time and space in the natural microenvironment, could be of great importance in saprophytic survival.

There is now considerable experimental evidence to support Garrett's concept that degree of saprophytism and longevity of *R. solani* are strongly interrelated. There is also evidence that *R. solani* persists in soil primarily as a saprophyte in tissues infected during parasitism, or by saprophytically colonizing dead plant tissues in which it can remain dormant or active for long periods of time (Boosalis and Scharen, 1959).

*Parasitic survival.*—Certain isolates of *R. solani* may depend on parasitism for survival. *Rhizoctonia solani* for example did not survive in the absence of a susceptible host in eastern Kansas when temperatures during the growing season were too high for survival of soil-borne mycelium or for production of sclerotia (Elmer, 1942). Several investigators believe that the parasitic phase may be important in the survival of *R. solani* in soil (Daniels, 1963; Pitt, 1964b). Pitt (1964b) found that *R. solani* declined rapidly after artificial introduction into unsterile soil. His wheat-stem isolates were incapable of prolonged saprophytic survival in bare soil following an initially healthy infestation, though they were capable of parasitic survival on a number of susceptible crops. Pitt also observed a limited survival in naturally infected cereal straws buried in soil. He concluded that saprophytic survival of *R. solani* clones from wheat stems is not a major factor in the persistence and survival of the sharp eyespot disease. Sanford (1952) found that susceptible host plants were more important for survival in soil than were dead or living roots of nonsusceptible hosts. According to Sanford, *R. solani* "disappeared" from heavily infested soils in less than 4 months in the absence of a susceptible crop, but survived up to 8 months under soils planted to susceptible crops. He concluded that parasitism is more important than saprophytism in the survival of *R. solani.*

*Saprophytic survival.*—If *R. solani* disappeared from soil lacking suitable hosts (Sanford, 1946, 1947, 1952), sclerotia and resting mycelia such as those observed by Boosalis and Scharen (1959) would undoubtedly become the only means of survival and would be very important in disease control. It is

116          *Colonization and Growth of Rhizoctonia Solani in Soil*

now realized, however, that this may not account entirely for survival. It may seem pedantic, but it is not really so, to indicate at this point that a fungus so variable in its ability to colonize substrates in soil (Papavizas and Davey, 1961; Martinson, 1963) would not be likely to depend on parasitic nutrition alone for its survival in soil. Relevant evidence, obtained by Papavizas and Davey (1960, 1961) in the absence of a susceptible host, was compared with evidence obtained in the presence of a susceptible host, and interpreted to mean that the *R. solani* clones present in the naturally infested soils used in that investigation did not depend entirely on parasitic nutrition for at least 20 weeks.

Several soil factors and treatments may exercise a pronounced effect on the saprophytic survival of *R. solani* in soil. The surviving mycelium of *R. solani* within precolonized substrates, for instance, appeared to be less sensitive to adverse microenvironmental conditions created by decomposing amendments than the active saprophytic phase of this fungus (Papavizas, et al., 1962). Mature oat straw and nitrogen were combined to produce a wide range of amendment C/N ratios, and precolonized substrate segments were incubated in soils amended with the oat straw. After certain intervals, the segments were recovered, and the percentage of originally colonized segments still containing viable *R. solani* was determined by culturing the segments on water agar. The C/N ratios of oat straw were critical in determining the percentage of residual colonization (and therefore reduction of saprophytic survival) of the substrate (Fig. 7). Oat straw with C/N ratios of 30 and 85 was more effective in reducing saprophytic survival of *R. solani* in colonized substrate segments than oat straw with C/N ratio 10, or supplemental nitrogen only. The sensitivity of *R. solani* to amendment decomposition and fungicidal effects was greater during the process of saprophytic colonization of the substrate than during the saprophytic survival period of the mycelium within the substrate tissues.

Saprophytic survival of *R. solani* may be influenced not only by the carbon-to-nitrogen balance of the substrate and of the soil environment (Papavizas and Davey, 1961; Papavizas, et al., 1962; Davey and Papavizas, 1963), but also by the innate constitution of isolates. Several single-basidiospore isolates of *T. cucumeris* differed considerably in their tolerance to $CO_2$, competitive saprophytic activity, and ability to survive within colonized substrate segments (Papavizas, 1964). Although most of the single-spore isolates tested were weak saprophytes, several isolates were strong saprophytes with long survival in organic matter buried in natural soil and with low sensitivity to decomposing oat straw and antifungal agents. With a few exceptions, isolates possessing high saprophytic activities also possessed high tolerance to $CO_2$ and high survival in precolonized substrate segments.

Single-basiodiospore isolates differed not only in their active saprophytism and saprophytic survival, but also in their virulence and pectolytic enzyme production (Papavizas, 1964; Papavizas and Ayers, (1965). Survival and saprophytism of isolates, even of virulent and strong saprophytes, were reduced with time in unsterile soil (Fig. 8). Reduction, the rate of which depended on the isolate, occurred in unamended soil even with some isolates possessing high saprophytic activity, despite the fact that these demonstrated an initial high saprophytic activity and virulence, suggesting a successful establishment in soil. Although most isolates ceased to be parasitic on a number of host plants within 6-9 weeks, they maintained at least some of their saprophytic activity. Papavizas and Ayers (1965) suggested that reduction of survival (indicated by decline of parasitism and saprophytism with time in natural soil) was perhaps due to a decline in inoculum potential of the parasite brought about by depletion of certain nutrients in ephemeral food bases of the soil ecosystem. Since a higher inoculum potential is required for infection than for colonization (Garrett, 1956), parasitic activity appeared to decline more rapidly than saprophytic activity. These findings may now help us understand why Sanford (1952) thought that parasitism was more important for survival than saprophytism.

*Intensity of saprophytic growth vs. virulence.—* Valuable information about intensity of *R. solani* parasitism would be obtained by direct comparison with intensity of saprophytism. Most of the root-infecting fungi are more virulent in sterilized soil where good saprophytic growth occurs than in natural soil (Garrett, 1956). According to Sanford (1941a), however, *R. solani* may be an exception to this general concept. Conditions favoring its maximum saprophytism may not be optimum for maximum parasitic activity. The extent and intensity of infection of potato sprouts was much less when the mycelium of the pathogen grew copiously on steam-sterilized soil than when it grew sparsely on

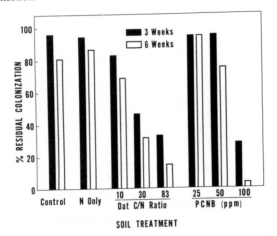

*Fig. 7.* Effect of N, oat straw, and PCNB on survival of *R. solani* in precolonized buckwheat-stem segments.

unsterile soil (Sanford, 1941*a*). When inoculum on steam-sterilized soil was mixed with unsterile soil in various proportions ranging from 1/16 to 3/4 of the mixtures, virulence of the parasite was greatest at the smallest concentrations of inoculum. His abundantly growing isolates of *R. solani* were among the least pathogenic to potato sprouts.

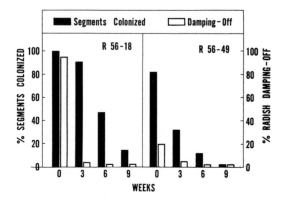

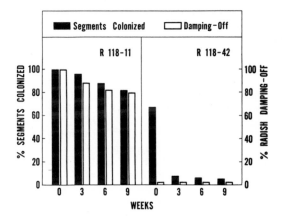

*Fig. 8.* Saprophytic activity (percentage of substrate segments colonized) and virulence (percentage of radish damping-off) of 4 single-basidiospore isolates of *T. cucumeris* at intervals after addition of inocula to soil.

To interpret the observation that virulence of *R. solani* was greater in natural soils (where growth is presumably retarded) than in sterilized soils, Sanford suggested that autotoxic staling products, detrimental to the parasitic action of *R. solani*, may develop as a result of abundant mycelial growth in steam-sterilized soil. In view of results with $CO_2$ (Papavizas and Davey, 1962*a*), it is also possible that accumulation of this gas as a result of increased growth of the mycelium may suppress the parasitic activity more rapidly than the saprophytic activity. Also, enough nutritional changes may take place during soil autoclaving to account for this phenomenon. Available nitrogen content may increase in autoclaved soil, and this increase may result in disease reduction on potato sprouts; Sanford (1947) found this to be the case with supplemental nitrogen added to soil. This view, however, may not be

consistent with the fact, observed by others, that *R. solani* was increased and not decreased by supplemental nitrogen (Anderson, 1939; Couch and Bloom, 1958; Das and Western, 1959; Davey and Papavizas, 1960).

Flentje and Stretton (1964) found that lesions caused by *R. solani* isolates on wheat plants grown in sterilized soil had a more limited spread and a smaller stunting effect than those on host plants grown in unsterile soil. In partially sterilized soil (heated to 160°F for 30 minutes), *R. solani* was mildly virulent whereas in untreated soil the parasite was extremely virulent (Flentje, 1965). In the unsterile soil, many other organisms were present along with *R. solani,* whereas only a few were present with the primary parasite in the heated soil, implying that weak pathogens or even saprophytes may become synergistic with the primary invader in the development of disease symptoms.

Additional evidence has now been produced (Papavizas, 1964) that is not in agreement with Sanford's (1941*a*) findings. Single-basidiospore isolates of *T. cucumeris* that made strong saprophytic growth in natural soil were also considered among the most virulent isolates (Papavizas and Ayers, 1965). In one experiment with three inoculum concentrations (Papavizas, unpub.), virulence of *R. solani* remained extremely high for at least 80 weeks, despite the fact that saprophytic growth was excellent and saprophytic activity very high in sterilized soil. Invariably, we observed less virulence in natural than in sterilized soil. Kernkamp, et al. (1952) also reported that there was more root rot caused by *R. solani* in sterilized than in natural soil.

That there is a direct relationship between saprophytic activity and parasitism of *R. solani* was also suggested by Martinson's experiments (1963), which showed a significant linear relationship between frequency of emergence of radishes and saprophytic activity measured by the frequency of holes invaded per sampling microbiological tube by *R. solani*. Recently, Sneh, et al. (1966) observed a high degree of correlation between the soil infestation level and the saprophytic activity of *R. solani* as assessed with the plant-segment colonization method (Papavizas and Davey, 1961), the plant debris particle isolation method (Boosalis and Scharen, 1959), or the immersion-tube method (Martinson, 1963). Sneh, et al. (1966) also found a good correlation between inoculum density (estimated from saprophytic activity on plant segments) and the disease-severity index of bean seedlings, whereas a lesser correlation was observed between degree of soil infestation and percentage of diseased seedlings. The plant-segment colonization method was the most sensitive and accurate method, and its superiority over other methods was established by statistical analysis. In the experiments by Martinson (1963) and in those by Sneh, et al. (1966), the degree of parasitism then could be predicted from the degree of saprophytism, a correlation that holds true only for primitive parasites such as *R. solani* (Garrett, 1956).

SOIL MICROECOLOGICAL DETERMINANTS AFFECTING GROWTH AND SAPROPHYTIC COLONIZATION.—Growth response, saprophytic activity, and saprophytic survival of *R. solani* are mediated by a number of determinants provided by plant roots, by decomposing plant materials, and by the soil itself. Specifically, growth and saprophytism may be influenced by nonbiotic factors such as temperature, moisture, and soil reaction, and factors of biotic significance such as antibiosis, competition, lysis, and available soil nutrients.

*Nonbiotic soil determinants.*—Temperature.—The relation of temperature to disease development and growth of the pathogen in cultures or in sterilized soils has been extensively studied in the past. It will be covered in other papers of this symposium. The effect of soil temperature on saprophytism in unsterilized soil in the absence of a host has received little attention. Papavizas and Davey (1961) noted that optimum soil temperature for saprophytic activity differed from one soil to another. For instance, significantly greater saprophytic activity of *R. solani* in greenhouse loamy sand occurred at 20°C than at other temperatures tested. In this soil, activity decreased markedly at 30°C. In Immokalee, fine sand (a Florida soil naturally infested with the parasite) *R. solani* was most active saprophytically at 26-30°C and significanctly less active above and below these temperatures. The discrepancies observed in the optimum temperatures of colonization in a Maryland soil and a Florida soil can best be explained by assuming a past natural selection of isolates adapted respectively to moderate and high temperatures. A comprehensive review of the heterogenity of *R. solani* with respect to the behavior of its isolates at various temperatures has been published (Bateman and Dimock, 1959).

In Martinson's experiments (1963) with artificially infested soil, the percentage of holes invaded saprophytically by *R. solani* increased with temperature increases from 15-25°C, but there were no differences between 25° and 30°. The effect of temperature on saprophytism became most pronounced at inoculum densities of more than 8,000 ppm inoculum in soil.

As far as can be determined, the foregoing experiments on temperature are the only ones dealing with the effect of temperature on competitive saprophytism of *R. solani* in soil. More research is needed to resolve numerous questions on the role that soil temperature plays in organic substrate colonization and exploitation and in saprophytic survival of this parasite in natural soils.

Soil reaction.—It is very difficult to distinguish between direct effects of soil pH on growth and saprophytic colonization of *R. solani* in soil and the indirect effects mediated by a changed physicochemical and biological environment as a result of pH changes of microloci in the soil itself, in organic debris particles, and, possibly, in plant rhizospheres.

Although the effect of soil reaction on growth and virulence of *R. solani* was extensively studied in the past, only a few studies dealt directly with the effect of soil pH on growth in natural soil. Blair (1943) studied the effect of soil reaction on growth of *R. solani* on Rossi-Cholodny slides buried in natural soils. Good growth was made by the isolates of *R. solani* over a reaction range of pH 5.8-8.1. A neutral reaction was optimum for growth. Papavizas and Davey (1961) placed emphasis on soil reaction influencing the ability of a number of isolates to colonize competitively a substrate in unsterile soil. In two soils, saprophytic growth was good at a reaction range of pH 4.5-8.1. Colonization declined to less than 10% at pH values of approximately 4.0. Optimum saprophytism occurred at a neutral or slightly alkaline reaction.

Soil moisture and aeration.—Considerable attention has been given to soil moisture and its role as a determinant of growth of root-infecting fungi. The recent review by Griffin (1963) has adequately covered the subject and very little needs to be said here. With respect to *R. solani*, Papavizas and Davey (1961) found that its saprophytism was significantly higher when the soil moisture was maintained at 20-60% of the moisture-holding capacity (MHC) than when it was maintained at moisture contents higher than 60%. At 70 and 80%, there was an appreciable reduction of colonization: at 90%, saprophytic colonization was almost eliminated. These results agree with those of Blair (1943), who studied saprophytic growth with glass slides in unsterilized soil and with those of Das and Western (1959), who studied growth in tubes containing sterilized, artificially infested soil. These investigators showed that growth of the parasite was best in soils of relatively low moisture content, in the range of 33-60% MHC, and that growth in wetter soils was greatly restricted or even suppressed completely.

Papavizas and Davey (1961) found that maximum saprophytic activity of *R. solani* in unsterilized soils occurred within approximately the same soil moisture range in which the maximum parasitic activity was reported by others (Roth and Riker, 1943*b*; Beach, 1949; Das and Western, 1959). Sanford (1938*b*), for example, found that, even in a fairly dry sterilized soil at 16°C, the parasite made 5 cm of linear growth in 10 days, an interval Sanford considered adequate for infection of young potato sprouts from a set bearing viable sclerotia.

The reduction of growth at high soil moisture content was attributed by Blair (1943) to a decline in soil aeration with an increase in moisture content. He was able to increase growth of *R. solani* significantly on the slides by forced aeration of the soil at 50% saturation or by absorption of $CO_2$ from soil by means of saturated alkalies. Inhibition of saprophytism at high soil moisture levels may have been due not only to poor aeration per se, but also to a consequent increase in $CO_2$ concentration. Results with $CO_2$ (Papavizas and Davey, 1962*a*) may also provide a possible explanation for the demonstration (Papavizas and Davey, 1961) that saprophytic activity is greater in soils at 20-60% than at 70-90% MHC where $CO_2$ concentrations are expected to be high. High moisture content would also stimulate bacterial activity, which in turn may result in lysis of the fungal mycelium. Kovoor (1954) found that at a moisture level of 30% saturation, there was no bacterial action against the hyphae of *R. bataticola* (*Macrophomina phaseoli*) on glass slides buried in soil. At 50% saturation, bacterial attack was perceptible; at 80%, considerable bacterial attack was followed by complete lysis of mycelium.

The concept of microbial stimulation by high soil moisture mediating reduction of survival of *R. solani* was also supported by Radha and Menon's results (1957). They believed that mycelial growth and survival of *R. solani* was related more to the activities of associated microorganisms than to soil moisture per se. In their work, high soil moisture had no adverse effect on *R. solani* in sterilized soil, but in unsterilized soil, growth and survival was considerably affected. Thus, at 25% and 50% MHC, *R. solani* remained viable in 100% and 98% of precolonized coconut-root segments after 24 weeks of incubation, respectively. At 100% MHC, the parasite survived in 40% of the segments only after 12 weeks of incubation.

More information is needed on the quantitative and qualitative aspects of soil aeration in relation to soil texture and compaction, moisture and organic content, and on the relationship of these determinants to saprophytism and survival ability. Future research along these lines, for the mechanisms that may lead to specific relationships between soil microflora and either soil moisture and aeration or *R. solani* saprophytism, should offer ample scope for realistic experiments.

*Biotic soil determinants.*—Antagonism encompassing all possible microbial associations and interactions of soil and rhizosphere microorganisms may influence the saprophytic growth and survival of *R. solani* in at least three ways: by affecting initiation of substrate colonization; by curtailing saprophytic substrate exploitation and duration of saprophytic survival in colonized substrates; and by reducing the ability of the fungus to spread out of the colonized substrate. Interference of associated microflora with the foregoing activities of *R. solani* may take place in the soil itself or at the surface of plant roots and organic matter particles.

Investigations concerned with the effect of antagonism on *R. solani* in vitro are numerous, but they are outside the scope of this paper except where they pertain to results in soil. The reader is referred to several comprehensive treatments of the subject of antagonism that have appeared in the past few years and that discuss *R. solani* among other root-infecting fungi (Wood and Tveit, 1955; Garrett, 1956, 1965; Kendrick and Zentmyer, 1957; Boosalis, 1964; Papavizas, 1966).

*Antibiosis.*—Despite the scarcity of information on antibiosis affecting growth of *R. solani* at the plant root surfaces, some idea of rhizosphere antibiosis can be gained from quantitative and qualitative differences in microbial populations in the rhizosphere and their possible effect on *R. solani* (Papavizas, 1963). High applications of quickly available nitrogen to natural soils with a wide range in soil reaction were detrimental to rhizosphere microorganisms antagonistic to *R. solani* (Fig. 9). In acid soils, the majority of the bean-rhizosphere antagonists adverse-

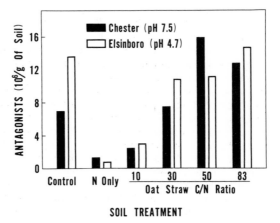

*Fig. 9.* Numbers of microorganisms isolated as antagonists to *R. solani* from rhizosphere of bean plants as affected by oat straw and supplemental N added to Chester loam (pH 7.5) and Elsinboro sandy loam (pH 4.7).

ly affected by supplemental nitrogen were true bacteria. More than 80% of the total antagonists from the rhizosphere of beans grown in very acid soils were violet-pigmented pseudomonads that produced a potent antifungal antibiotic (Ayers and Papavizas, 1963). The antagonistic bacteria or the crude antibiotic applied to substrate segments buried in soils with natural *R. solani* populations suppressed initiation of substrate colonization by the fungus. The amount of suppression, however, was not great, and indicated little or no potential as a control agent for this fungus.

When carbohydrates are decomposed by soil microorganisms, as much as 50-80% of the total carbon is liberated as $CO_2$ (Waksman, 1952). Carbon dioxide, the product not only of microbial decomposition but also of microbial respiration and root respiration, is an important microecological determinant in the life cycle, behavior, and competitive saprophytic activity of *R. solani*. Carbon dioxide is an important determinant not only because of its potential role as a differential fungistatic (Durbin, 1959a; Papavizas and Davey, 1962a) and as a possible fungicidal agent, but also because of its potential role in changing microsite reactions in soils and the development of bicarbonates (Alexander, 1964).

The inhibitory effect of $CO_2$ on growth and saprophytic activity of *R. solani* was first illustrated by Blair (1943). He found that the presence of a saturated alkali solution improved linear extension of *R. solani* on glass slides infested with the pathogen and buried in soil containing straw meal. In his experiments in large glazed clay crocks sealed with cellophane, the increased growth on the glass surface was presumed to be due to reduction of the $CO_2$ content in the soil atmosphere. Growth of the fungus was promoted by forced aeration. Durbin (1959a) provided evidence that isolates of *R. solani* differed in their sensitivity to $CO_2$ concentration, and that subterranean isolates were more tolerant to $CO_2$ than aerial isolates were.

To study the effect of $CO_2$ on *R. solani*, Papavizas and Davey (1962a) followed the alternative approach of enriching the $CO_2$ content of the soil atmosphere by continuous passage of gas mixtures through columns of natural soil, artificially or naturally infested with *R. solani*. Considerable inhibition of the saprophytic and parasitic activities of *R. solani* by $CO_2$ was observed. The depressing effect of $CO_2$ depended on its concentration, on the kind of soil, and on the inoculum potential of the pathogen. The inhibitory action of the gas on the saprophytic activity of *R. solani* was shown by a straight-line dosage-response curve obtained from the values for percentage of inhibition of colonization by increasing $CO_2$ concentrations. *Rhizoctonia solani* was more sensitive to $CO_2$ in the parasitic phase than in the saprophytic phase, and the pathogen in the latter phase was more sensitive to $CO_2$ than in the saprophytic survival phase within precolonized substrate segments. The differential depressing effect of $CO_2$ on the parasitic and saprophytic activities of *R. solani* may be due to higher accumulation of the gas in the microhabitat of the host rhizosphere than in that of the colonizable dead substrate. Both root respiration and the greatly enhanced microbial population in the rhizosphere may be responsible for the increase. Waksman (1952) showed, for instance, that of a total quantity of $CO_2$ liberated by roots and soils, about 30% was due to root respiration. Also, it is possible that the greater inhibition of parasitism as compared to saprophytism is due to longer exposure to $CO_2$ during the parasitic phase than during the saprophytic phase. Hyphae of *R. solani* not only must reach the plant surface, but also must develop infection structures on the plant surface prior to penetration (Flentje, 1965).

Herzog and Wartenberg (1958) and Herzog (1961) showed that growth and sclerotial formation of *R. solani* were affected by several soil and rhizosphere determinants. Growth was retarded by antifungal "antibionts" and favored by unknown substances released by roots of higher plants. These substances may have a double function: they may stimulate saprophytic growth by "stimulating" resistance of the *R. solani* mycelium to the antibionts, or they may protect the mycelium from its own antibionts. According to Herzog and Wartenberg (1958), decomposing plant materials inhibited growth of *R. solani* because they were good substrates for fungistatic antibiont production. The antibionts may prevent

sclerotial germination and therefore preserve and enhance their survival value. These investigators, however, did not express any views on the possible nature or the origin of the antifungal antibionts.

Fungistasis and lysis.—The widespread occurrence of *R. solani* in natural soils (Thornton, 1956; Warcup, 1957; Papavizas and Davey, 1962b) away from plant roots, the extensive mycelial growth of some clones in natural unamended soil (Blair, 1943), and its tolerance of specific or nonspecific agents of microbial origin (Wastie, 1961; Papavizas, 1964) suggest that, unlike the great majority of soil-inhabiting fungi, *R. solani* hyphae and sclerotia possess considerable resistance to those soil-fungistatic principles described by Dobbs and Hinson (1953). Daniels (1963) could not demonstrate any appreciable fungistatic effect on 10 isolates of *R. solani* inoculated on agar disks and placed in intimate contact with moist loam. This is somewhat contradicted by the findings of Herzog and Wartenberg (1958) that sclerotia of *R. solani* did not germinate in the presence of antifungal antibionts.

The foregoing results are all that are available from the literature on the effect of fungistasis on *R. solani*. Since susceptibility or resistance of *R. solani* hyphae and propagules to fungistasis in soil, and in colonized debris particles, is likely critical for survival, it is surprising how little we know about fungistasis. Numerous problems remain to be studied, especially problems related to sclerotial germination in soil and within plant debris particles.

It is generally recognized that *R. solani* mycelium is resistant to attack by lytic soil microorganisms. Lockwood (1959, 1960) found, for instance, that of several fungi tested, *R. solani* was the only fungus resistant to lysis brought about by the action of natural soils placed on top of agar cultures. This observation is supported by the finding that *R. solani* could persist in sugar beet residues in soil as viable mycelia and sclerotia for at least 6-10 months (Boosalis and Scharen, 1959). A relevant instance to support this view was provided by Potgieter (1965), who attributed the resistance to lysis of the *R. solani* mycelium to phenolic polymers naturally present in the cell walls.

There is some evidence in the literature to support the notion that lysis of mycelium of *R. solani* may play a significant role in the success or failure of this fungus as a saprophyte in soil under some specific conditions. Andersen and Huber (1962) postulated that "bacterial necrosis" (lysis?) of *R. solani* is perhaps the most important mechanism for the control of this pathogen under bean culture. Papavizas (1963) observed that the C/N balance of an amendment was of great importance in determining the amount of lysis of this parasite and of three other root-infecting fungi. A considerable amount of lysis occurred in natural soil amended with oat straw of C/N ratios of 30, 50, and 85 (Fig. 10). The amount of lysis increased with an increase of C/N ratio of amendment. The least destruction of mycelium occurred in unamended soil, and in soil supplemented with nitrogen only.

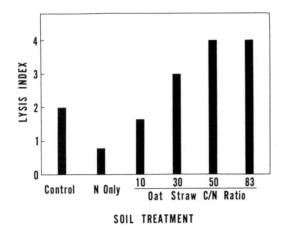

*Fig. 10.* Lysis of mycelium of *R. solani* in cultures covered for 14 days with Elsinboro sandy loam amended with oat straw and supplemental N 9 weeks prior to sampling. 0 = 91-100% mycelium present; 4 = no mycelium present.

Hyperparasitism.—Boosalis (1964) pointed out that it is very difficult, if not impossible, to evaluate the importance of hyperparasitism on the growth and survival of root-infecting fungi in nature. With respect to *R. solani*, it is impossible to state whether hyperparasites in natural soils actually mediate growth (saprophytic colonization, and survival of the pathogen, since most of the work on hyperparasitism by soil saprophytes has been done in vitro. Microecological investigations on soil-borne mycoparasites of *R. solani* have been neglected primarily because methods for this type of research have been lacking.

*Trichoderma viride* (Weindling, 1932; Boosalis, 1956), *Papulospora* sp. (Warren, 1948), and *Penicillium vermiculatum* (Boosalis, 1956) have been reported as parasites of *R. solani*. Boosalis (1956) found that *R. solani* introduced into natural soil amended with green plant materials and kept at 28°C was parasitized by a *Trichoderma* sp. and *P. vermiculatum*. Even under such favorable conditions for parasitism, however, only about 18% of the *R. solani* hyphae were actually parasitized. The importance of hyperparasitism in the microecological behavior of *R. solani* in natural soils was questioned by Boosalis (1964), who was unable to detect any appreciable amount of hyperparasitism of *R. solani* hyphae screened from several infested fields. Thornton (1953), however, observed actinomyces mycelium forming an intimate hyphal contact association with Rhizoctonia-like hyphae. From a total number of 1,765 microscope fields on previously buried glass slides exhibiting Rhizoctonia-like hyphae, 1,127 fields showed the actinomyces-fungus hyphal growth association. However, no destruction of fungus or of actinomyces hyphae was observed.

A reversed but important condition of hyperpara-

sitism, which may play an important role in the survival of *R. solani* in soil, exists. This was shown by E. E. Butler's elegant studies (1957) of the ability of *R. solani* to become itself a hyperparasite feeding on other common soil saprophytes and destroying them. But questions remain unanswered. What significance has the ability of *R. solani* to survive as a hyperparasite? How much is this advantage offset by the fact that *R. solani* itself may not only be a hyperparasite, but may also provide a food base for other hyperparasites in soil?

Competition.—The struggle for essential or stimulatory nutrients of the soil microenvironment, which are in limited enough supply perhaps to affect the growth and saprophytic colonization of substrates by *R. solani* in soil, and also at the soil-root interfaces, has not been studied as extensively as antibiosis or hyperparasitism. This is rather surprising because, if growth and saprophytism of *R. solani* are related to competition for nutrients, then artificially inducing the appropriate deficiency would, theoretically, result in a reduction of the inoculum potential of the fungus. Increasing energy sources and other kinds of food sources would increase inoculum potential and therefore disease. Increases in food supply, however, would not necessarily benefit *R. solani* under all conditions. Essential or stimulatory nutrients may enhance activities of associated microorganisms to a greater extent than those of *R. solani* on a proportional basis. *Rhizoctonia solani* may then suffer deleterious effects, not only as a result of competition in a strict and direct sense, but also as a result of associated microbial activity mediating the production of specific or nonspecific antimicrobial toxins or antibiotics, or production of $CO_2$ and other fungistatic substances. In other words, saprophytism of *R. solani* can be masked by competitive and antagonistic interactions of associated microbial populations possessing greater abilities for successful exploitation of the substrate.

The foregoing concepts may explain why *R. solani* is suppressed by several organic amendments of particular C/N ratios (Snyder, et al., 1959; Davey and Papavizas, 1960), despite its ability to colonize competitively and exploit organic amendment particles. The effectiveness of some organic amendments in controlling *R. solani* diseases or in suppressing saprophytic activities may be associated with modifications of the soil ecosystem, resulting in a severe nitrogen deficiency. Actual control in such cases may be attributed to a poor competitiveness by this pathogen for nitrogen (Snyder, et al., 1959; Davey and Papavizas, 1963). Substrate enrichment with N favored saprophytic activity and survival of *R. solani*, whereas reduction in nitrogen content had the reverse effect (Papavizas and Davey, 1961). Papavizas and Davey attributed the suppression at low nitrogen content to competition for available nitrogen by associated microorganisms decomposing the organic food bases. In this context, reference has already been made (Papavizas, et al., 1962; Papa-

vizas, et al., 1964) to the fact that nitrogen depletion would not only decrease initial saprophytism but
would accelerate the rate of decline of the *R. solani*
mycelium in colonized substrates.

Blair (1943) also showed that saprophytism of *R.
solani* was substantially reduced by dried green
amendments decomposing in soil. Decomposition of
the amendments apparently favored growth and
multiplication of competing microorganisms rather
than *R. solani*. The depressive effect of the amendments could be attributed in part to nitrogen starvation of the mycelium mediated by the saprophytes
multiplying on the decomposing residues.

The ability, or rather the inability, of a root-infecting fungus such as *R. solani* to compete with other
microbes inhabiting the same microsites of the soil
ecosystem is only one of many ecological and physiological characteristics that may be exploited to obtain
biological control of the pathogen. Before this can
be achieved, however we have to know more about
competition by *R. solani* for energy-yielding materials and other substances in the bulk of soil and at
the soil-root interfaces, where competitive interactions may be immensely more intricate. We have to
know more about the ways competition among associated microbes regulates their number. We have
to know more about the biology and autoecological
behavior of *R. solani* that determine success or fail-

ure in the competitive establishment in and exploitation of food bases. Last but not least, we must
understand all specific and nonspecific effects on
competitive interactions exercised by the host plant,
especially through excretion of various kinds of
growth-promoting substances in the general sense,
which are either essential for, or stimulatory to, other
microbes in the immediate vicinity of roots.

CONCLUSIONS.—This discussion has been largely
and intentionally limited to examples that demonstrate the saprophytic abilities of *R. solani* in the soil
ecosystem, rather than in vitro. This emphasis on in
vivo behavior has been deliberate, for it is firmly believed that the importance of microecological factors
in soil influencing saprophytic growth needs to be
thoroughly explored as a necessary prelude to successful control measures for diseases caused by soil-
inhabiting fungi such as *R. solani*.

Conditions that will affect saprophytic existence
and growth of an organism such as *R. solani* (which,
during most of its life cycle, exists in the absence of
a host in the soil ecosystem as a saprophyte in close
association with an antagonistic microflora) will be
of extraordinary significance in the life cycle of the
pathogen, and will profoundly influence its saprophytic survival and its subsequent phytopathogenic
potentialities.

Part III.

*Rhizoctonia solani:* the pathogen

# Types of Rhizoctonia Diseases and Their Occurrence

KENNETH F. BAKER—*Department of Plant Pathology, University of California, Berkeley.*

*Rhizoctonia solani*, as presently understood, probably causes more different types of diseases to a wider variety of plants, over a larger part of the world, and under more diverse environmental conditions, than any other plant-pathogenic species. For example, one strain of this fungus causes an important root rot of wheat under semiarid conditions, another has destroyed large areas of submerged aquatic plants, and still others produce web blight of aerial parts in the humid tropics and subtropics. The species is able to produce such varied effects as disfiguring superficial black sclerotia on potato tubers, brown patch of turf grasses, and storage rots of roots and fruits. Although best known as a prime cause of damping-off of seedlings, it also attacks aerial parts of mature trees during wet periods. It is, furthermore, a potent parasite of other fungi (Butler, E. E. 1957).

This adaptability is more apparent than real, for it is largely a product of the enormously diverse characteristics of the numerous strains that presently constitute this species. Fortunately for man, this adaptability does not apply to individual strains encountered in any given situation. The coincidence of an adapted strain, a favorable environment (including the absence of effective antagonists), and a susceptible host in a suitable stage of development does occur sufficiently often, however, to make this one of the most important plant pathogens. The types and severity of disease produced can be best understood in terms of the coincidence of the favorable stage in the sequence or rhythm of each of these factors. An attempt is therefore made here to relate symptomatology to the orderly body of knowledge on the ecology of *R. solani*.

RHYTHMS INVOLVED IN RHIZOCTONIA DISEASES.—Scharfetter (1922) introduced into ecology consideration of the seasonal climatic cycle (*Klimarhytmik*), the seasonal development of the individual plant (*Vegetationsrhytmik*), and the annual development of the whole groups of plants (*Formationsrhytmik*). An extension of this concept is utilized here—hence the title of this paper might well be "Rhythm in the Rhizocsphere."

*Host rhythm.*—Higher plants characteristically pass through a life sequence: seed→germination→seedling → growth → mature plant→flowering →fruit→ seed. The type of germination influences the length of exposure of the nutrient-rich cotyledons to invasion by parasites. In hypogeal germination, exemplified by the garden pea, the cotyledons remain underground and susceptible to attack. In epigeal germination, exemplified by the common bean, the cotyledons are carried above the soil, and may thereby escape decay. Factors (such as deep planting [Singh, 1955; Ruppel, et al., 1964], reduced vitality of seed [Germ, 1960], or excessively cold, hot, moist, or saline soil [Beach, 1949]) that delay seedling emergence may increase seed decay and preemergence damping-off.

The seedling is composed of juvenile tissue that is primary, meristematic, and susceptible to *Rhizoctonia*. As the tissue matures, the basal stem becomes increasingly resistant to the fungus, and the juvenile tissue then occurs at the apex of stems and in axial buds, higher on the plant. It is for this reason that much of the injury by *R. solani* to mature plants is from invasion of aerial parts, and requires very moist conditions. The idea that old plants are relatively nonsusceptible seems largely to have arisen from the fact that the susceptible juvenile tissue has moved into an unfavorable dry environment. By comparison, *Pythium* spp. generally attack rootlet tips, which are continuously susceptible and in a favorable environment—the soil. Thus, the type of injury from *Pythium* is not changed with host maturation as much as it is for *Rhizoctonia*.

Perennial plants also pass through a yearly sequence: dormancy → new growth → maturity → flowering → fruit → seed → dormancy. The new shoots in this case present relatively juvenile tissue close to the ground, and the plant is therefore periodically susceptible to attack by *R. solani*.

There is another cycle, of the dying (early summer) and replacement (early winter) of transient roots in strawberry and probably other plants, which is important in root invasion by low-grade pathogens (Wilhelm, 1959).

*Pathogen rhythm.*—In soils which are moist for most of the year, *R. solani* remains in more or less continuous vegetative growth; in soils which become extremely dry for several months (e.g. unirrigated soils in South Australia and southern California), the fungus necessarily becomes dormant as thick-walled mycelium or sclerotia. The aerial web-blight forms are characterized by rapid vegetative growth (taking maximum advantage of favorable climatic periods) and production of abundant sclerotia (protection against dry periods, as well as providing a means of dissemination). Both soil and aerial forms develop the basidial state *(Thanatephorus cucumeris)* following wet weather and a period of parasitic activity, usually on healthy tissue adjacent to that invaded (Figs. 41, 43). The importance of the basidiospores in dissemination of the fungus to soil still requires demonstration, but they may serve to disseminate aerial forms to other plants (Fig. 29; Echandi, 1965). It is possible, but not demonstrated, that basidiospores of soil strains may disseminate to soil via a plant infection or by anastomosis of the germ tube with mycelium of an established saprophytic strain of *R. solani*. No other spore forms are known for the fungus.

*Environment rhythm.*—Although the most obvious climatic rhythm is the sequence, winter-spring-summer-fall, there are others more significant for *Rhizoctonia*. Temperature tends rather generally to rise from a winter minimum to a summer maximum, and then to decline. Soil temperatures tend to slowly follow the air-temperature cycles. It is possible to reduce the Rhizoctonia seed decay, damping-off, and hypocotyl decay of some warm-season plants by delaying planting until the soil warms in the spring. Planting beans in the Salinas Valley, California, after the soil warms to 18.4°C thus reduces plant loss (Snyder, unpub.). This takes advantage of the environment rhythm.

Rainfall, however, may be quite variable throughout the year. On the Pacific Coast of the United States, in southern Australia, and in the Mediterranean area, precipitation is greatest in the winter months and declines to a minimum in the summer. On the East Coast of the United States, northern Australia, Central Europe, and Central America, rainfall tends to be more uniformly distributed throughout the year, or to be somewhat greater in summer than in winter.

TYPES OF DISEASES PRODUCED BY R. SOLANI.—The types of disease produced in various crops by *R. solani* will now be considered in relation to the three general types of rhythms just described. The diseases are arranged according to the host sequence of development.

*Seed decay.*—*Rhizoctonia* may invade the seed while still in the fruit, decaying it there or merely infecting it. The decay process is then resumed after the seed is planted and before germination (Fig. 1; Baker [K.], 1947; Neergaard, 1958) (Host rhythm). The seed may also be invaded by growth of the fungus from infested soil in which it is planted; the greater the amount of inoculum in the soil, the more certainly and the more rapidly this occurs. In either case, the invaded seed serves as a food base, enabling the pathogen to reach adjacent seedlings. It is thus common to find seedlings with preemergence damping-off in the vicinity of a rotted seed. The greater the distance seeds are separated, the less the probability of such spread (Singh and Singh, 1955). Thus, losses are usually greater in seed flats and nursery beds than in the wild (Pathogen rhythm).

The importance of this seed decay transcends the loss of the infected seed. Infected seed introduces, with maximum probability of success, a strain of the pathogen virulent to that host, and preselected for it by two events: In commercial seed production (1) an appropriate strain of the fungus in the soil is favored over other strains by the presence of the host, and (2) the strain is further selected in many cases by invasion of the fruit (and thus the seed) from the soil. The significance of this chain of events has been too often neglected by plant pathologists. Such seed transmission has been reported for 38 crop species (Baker [K.], 1947; Neergaard, 1958; Noble, et al., 1958).

*Damping-off of seedlings.*—Preemergence phase.—The preemergence phase of this disease is an extension of seed decay; the two together are often designated as "poor stand" (Fig. 3). This delayed attack may result from unfavorable environment (temperature, moisture) or from insufficient inoculum for faster action, decay only occurring following germination (Fig. 2). The longer that seedling emergence is delayed, the greater the opportunity for invasion and, other conditions being favorable, the greater is the preemergence damping-off. Singh (1955) thus found that, in soil infested with *R. solani*, guar seed sowed at 2.5 cm depth gave 85% emergence of seedlings, whereas at 5.0 cm it was 83%, at 7.8 cm it was 77%, at 11.5 cm it was 75%, and at 14.0 cm it was 68%. Ruppel, et al. (1964) found that when *Tephrosia* seeds were planted 1.3 cm deep there was 75.2% survival, at 2.5 cm there was 59.2%, at 5.0 cm, 56.7%, and at 7.6 cm, 12.1% (Host rhythm). Planting seed of low vitality (Hartley, et al., 1918; Germ, 1960; Sinclair, 1965) decreases the velocity of germination and prolongs subterranean exposure under conditions favorable to infection. Similarly, planting seed in soil so cold, warm, moist, or saline as to delay germination will increase damping-off if the condition is not equally unfavorable to *Rhizoctonia* (Leach, 1947; Beach, 1949; Peace, 1962) (Environment rhythm). Plants with hypogeal germination (pea) may remain in a zone of high infection risk for a longer time than those with epigeal germination (bean), and may require chemical protection against cotyledon decay (Host rhythm).

Postemergence phase.—Postemergence damping-off represents still a further delay in attack, expression of symptoms, or both. Symptoms may develop any time after emergence through the soil surface, until the seedling is past the most susceptible juvenile stage (Figs. 10, 11). The susceptibility of the seedling to attack by *R. solani* declines with maturation and lignification of tissues (Host rhythm). This increased resistance may be due to conversion of pectin to calcium pectate, rendering the tissues resistant to the polygalacturonase of the fungus (Bateman and Lumsden, 1965). Flentje, Dodman, and Kerr (1963) suggested that the resistance developed by older seedlings is due to impermeable waxes deposited on the surface; if these are abraded or removed, resistance of the seedling may be lost. The transition to a mature plant is a continuous process, but may fairly be said to have been effected when the seedling is no longer dependent on the cotyledons, or when they have been shed.

Damping-off refers to the decay of the stem at about soil level, causing it to fall over because it has as yet no thickened supporting tissue. The name may be derived from either the moist conditions under which the disease occurs, or the moist nature of the decay produced. It is generally possible to differentiate macroscopically between postemergence damping-off caused by *Rhizoctonia* and by *Pythium* spp. (Roth and Riker, 1943*a*; Wright, 1944; Ellis and Cox, 1951; Baker [K.], 1957). The former produces stem decay near soil level, but may later advance downward into the roots; *Pythium* spp. generally infect at root tips, in small roots, or root hairs, and advance upward through the plant to the soil surface. Small bits of soil or organic matter dangle from the coarse tough mycelium of *Rhizoctonia* attached to the infected seedlings when removed from the soil (Duggar, 1916); this does not occur with *Pythium*. Examination with a hand lens will distinguish the coarse, hyaline to brown mycelium of *Rhizoctonia* (Fig. 35) from the fine hyaline hyphae of *Pythium*. This method of examination also applies to seed decay and preemergence damping-off. In critical studies, such diagnoses should be confirmed by microscopic or cultural examination.

All of the preceding types of disease tend to occur in circular or irregular patches (Fig. 5) when seed is randomly sown, or in linear strips (Fig. 6) when it is sown in rows. In a solid mat of plants (e.g. turf) the diseased areas are also circular (Fig. 4). These represent foci of survival or of introduction of the pathogen. In disinfested soil commonly used in nursery seedbeds, these areas obviously represent introduction of the pathogen with seed or transplants, contamination from bits of soil or plant tissue splashed in by water, carried in by wind or workmen, or survival on the containers (Baker [K.], 1957). The more nearly sterile the soil (Environment rhythm), the more rapid will be the spread, and the larger the area of spread (Baker [K.], 1962). For this reason, treatment of soil with aerated steam and reduced temperatures has been devised (Baker [K.] and Olsen, 1960; Baker [K.], 1962; Olsen and Baker, 1968) in order to leave as many antagonistic microorganisms as possible.

It is sometimes claimed, especially by forest nurserymen, that damping-off selects the strongest seedlings for transplanting. That this is a fallacy is shown by the total destruction of seedlings in invaded areas, frequently in a whole bed. Furthermore, selection

against Rhizoctonia damping-off is not necessarily based on nursery-seedling characteristics determining survival or growth in the field. Rhizoctonia-induced damping-off seems to have no mitigating virtue such as may obtain for *Phytophthora cinnamomi* in permitting selection among seedlings for field resistance.

It is a mistake to regard Rhizoctonia damping-off as a disease restricted to juvenile plants. This idea has led many growers and pathologists to suppress damping-off in the seed bed by chemical inhibitors added to the soil, or by environmental or cultural modifications such as reduced watering, increased light intensity, soil acidification, temperature control, or chemical seed treatment. The undesirability of such suppression, as opposed to growing seedlings free from *Rhizoctonia* is emphasized by two facts: (1) Some transplant crops (e.g. cabbage, pepper, aster, stock, *Nicotiana* [Fig. 16]) may sustain serious losses from *R. solani* in the adult form, as explained later. Suppression under the relatively controlled conditions of the seedbed may be mere postponement of loss to the largely uncontrollable field environment where the investment loss is even greater. (2) Transplants are the most efficient means of infesting field soils, as explained above. For example, *Rhizoctonia* strains thus host-selected, may be spread from the forest nursery to the reforestation site, where they may be important in preventing regeneration of seedlings (Hartley, et al., 1918; Ness, 1927; Thrupp, 1927; Haig, 1936; Wilde and White, 1939; Smith [L.F.], 1940; Smith [D.M.], 1951; Vaartaja, 1952; Duncan, 1954). *The aim should be to produce transplants free of the pathogen, not merely free of symptoms.*

*Wirestem and soreshin.*—As the plant matures, it becomes increasingly resistant to *Rhizoctonia* at the soil level (Host rhythm). Some crop plants, however, are attacked for a considerable period after the seedling stage. For example, cabbage and other crucifers are frequently attacked after plants are 10-15 cm high (Gratz, 1925). If the weather becomes dry after infection, the cortex decays in sharply defined areas encircling the stem and collapses, but since the stiffened stele provides support, the plant remains erect. The decay extends through the stem in damping-off; in wirestem, only the cortex is decayed. The stem is wiry and slender at the point of the lesion, giving rise to the name "wirestem" (Fig. 10). Such plants usually succumb, and certainly are non-productive. Gratz (1925) showed that the transplanting of such stock to the field leads to further loss. Slightly diseased plants gave 12% loss in number of heads and 10% in pounds produced, over that of healthy transplants; severely diseased plants gave 36% fewer heads and 30% less yield. In addition, harvest was delayed. Similar lesions are produced on pepper (Baker [K.], 1947) and tomato (Felix, 1955). On tomato, the lesions may have alternating light and dark brown bands, which are probably comparable to the diurnal bands produced by the fungus when growing in fruit (Fig. 36) or in culture (Peltier, 1916). Similar symptoms on cotton were called "soreshin" by southern planters before 1900 (Atkinson, 1892), and the name has continued for these stem-girdling lesions (Fig. 11). Partial recovery may be made when the soil warms up, if new root growth develops above the lesion.

Fulton, et al. (1956) showed that planting cotton seed in cool soil increased losses from *R. solani*. In an April 30 planting, the fungus was recovered from 100% of the diseased seedlings, 25% in a May 20 sowing, and 50% in a June 4 sowing. This effect results from prolonging the susceptible period in the host rhythm. Neal and Newsom (1951) related the incidence of soreshin to thrips injury, which was so severe that it reduced leaf area by 70%. In a plot in which thrips were controlled, 23.5% of the plants developed soreshin; in the thrips-injured plot, 85% had the disease. This effect was apparently produced through reduction of photosynthesis, which affected the host rhythm.

*Root rot.*—Erection of the genus *Rhizoctonia* was originally based (De Candolle, 1815) on sterile mycelium attacking alfalfa roots; the name means "root killer," and the German name for *R. solani* has adopted this name as Wurzeltöter. While root rot is not generally the most important type of disease caused by this fungus, severe losses can be produced on some crops. Injurious cankers of the tap root of alfalfa, "for which no control is known," are caused by *R. solani* during the hot summer under irrigated conditions in the southwestern United States (Smith [O.F.], 1943, 1945; Erwin, 1956). Dark, circular to oblong, sunken cankers with brown borders develop at the point of origin of secondary roots (Fig. 12). Secondary roots may be invaded directly, or may be cut off by the basal lesion. Only a few of many isolates of *R. solani* tested proved pathogenic to alfalfa roots, but many attacked the seedlings. As Durbin (1957) pointed out, subterranean types of *R. solani* (such as the alfalfa isolate) are more tolerant of $CO_2$ than are surface or aerial isolates. This may well explain the avirulence of surface strains to deeply-lying alfalfa roots (Environment rhythm). *Rhizoctonia solani* may also attack the main and fine roots of clover, causing stunting of the plants (Benedict, 1954). Roots of sweet pea seedlings are more susceptible to decay by *Thielaviopsis, Rhizoctonia,* and *Pythium* if the plants are suffering shock effect from infection by pea-enation mosaic (K. F. Baker and W. C. Snyder, unpublished data). Roots of lettuce are attacked in Florida, as part of the bottom-rot syndrome (Weber and Foster, 1928).

Cereals in some areas (England, South Australia, Canada, India) sustain serious losses from root rot caused by *R. solani* (Fig. 5). Affected spots in the fields may be 1.8 m in diameter, and recur in successive years. Roots may be "killed back to brown stumps," causing plants to become stunted and spindly (Samuel and Garrett, 1932; Blair, 1942; Moore, 1959). In England, this disease is called purple patch.

Underground fleshy roots are often affected by cankers, as well as being attacked at the crown. On sugar beet (Fig. 14), lesions up to 2.5 cm in diameter, with concentric light and dark bands, are produced on the side of the root in early season; later, the lesions may be large, rough, dry, and pithy (Richards, 1921a). Turnips have similar lesions in the field, as well as in storage (Dana, 1925b; Lauritzen, 1929). Sclerotia may develop on the surface of the cankers on either host. Carrots may be similarly affected (Fig. 13). Tulip bulbs may be russetted by *R. solani* (Moore, 1959). Potato tubers may exhibit russetted scabby areas, said to result from *Rhizoctonia* invasion. The fungus may invade flea-beetle injuries and produce deep pits (Ramsey, 1917), cracks, or furrows (Ramsey, et al., 1949).

*Rhizoctonia solani* may act as an endophytic mycorrhizal fungus on orchid seedlings. Strains parasitic to other plants may be noninjurious to orchids, although recoverable from roots of mature plants. It may even form a basidial collar on the stem (Downie, 1957; Harley, 1965). I have observed orchids in southern California, much weakened from growing under unfavorable poorly drained conditions, in which *R. solani* had become actively parasitic (Environment rhythm).

*Rhizoctonia solani* has been shown to attack the roots of conifers (Hartley, 1921) and coffee trees (Crandall and Arillaga, 1955) under some conditions, but the importance of this trouble under field conditions has not been demonstrated.

*Hypocotyl and stem cankers.*—Stem lesions may develop on fully mature plants with well-developed secondary tissue (Fig. 16). Cankers may occur on mature tomato plants at the base of stem branches; these often have alternate light and dark bands. Infection may occur through a leaf in contact with the soil, and spread to the cortex and stele (Conover, 1949). Basal cankers or "foot rot" may also be produced on tomato plants in glasshouses (Small, 1927). Similar stem rot occurs on cabbage (Fig. 10; Weber, 1932b), stock (Fig. 24; Dimock, 1941), and celery "crater spot," (Fig. 15; Houston and Kendrick, 1949). The fungus causes a destructive "neck rot" of gladiolus, infecting at soil level through the leaf bases, and producing brown shredded lesions from decay of parenchymatous tissue (Fig. 7). Lesions may develop on corms at the point of attachment of a leaf. Infection of plants tends to spread linearly down the row (Fig. 6; Creager, 1945). Infection arises from soil infestation or from an infected cormel or corm.

The stem rot of carnation is very destructive where this plant is grown in soil infested by *R. solani,* or where infected cuttings are used. A soft moist rot starts at the soil surface and advances into the stem; the decayed cortex easily rubs off, leaving the stele beneath. Strands of mycelium, and sometimes sclerotia, appear on the surface. Roots remain intact until late in the disease, but the tops wilt, turn brown, and die (Baker [K.] and Sciaroni, 1952).

*Rhizoctonia* produces a brown cortical stem canker of lima bean, showing variable depths of penetration (Fig. 9). The appearance of the canker on lima bean would lead one to expect severe plant injury. When PCNB and Captan were applied in the field furrows, a reduction of 71.5% in number of infected plants and 82.4% in disease severity resulted (Kendrick, et al., 1957). No increase in yield was obtained, however, by this control (Kendrick, 1956, 1957), suggesting that the disease was more disfiguring than injurious. Perhaps some other factor became limiting, once the disease was controlled.

The well-known stem or sprout rot of potato provides an excellent example of the importance of the host rhythm in the occurrence of the disease. The meristematic tip of tuber sprouts is very susceptible to *R. solani*, and may be killed before emergence from the soil (Fig. 18). A dormant bud on the stem below then starts growth, and may be killed in turn; this has been repeated up to 11 times. Conditions belowground may be optimal for infection, and great loss to the crop then results. Shallow planting or favorable growing conditions may reduce the time of exposure, and thus decrease the losses. As the plant stems mature and become more resistant, the lesions formed become smaller and less injurious (Figs. 17, 19). Such cankers on emerged stems produce a different type of symptom. They may restrict downward movement of photosynthates, which then accumulate in the tops. This accumulation may produce stunting and rosetting of the top, the resultant production of anthocyanin may cause purple pigmentation, and aerial tubers may form in the axils of branches and petioles (Fig. 20; Walker, 1957; Chupp and Sherf, 1960) (Host rhythm). Infection may result from mycelia either in the soil or produced from sclerotia on the planting piece (Fig. 37). Fortunately, only a small percentage of the sclerotia on tubers are strongly pathogenic (Sanford, 1938a; Person, 1945). The underground tuber-bearing stolons may be attacked, with resultant reduction in their length (Fig. 21) and in the number and size of tubers produced. Similar lesions are produced on stolons of white clover following flowering (Garren, 1955). A white hymenial layer is often formed by *R. solani* on stems just above the soil line (Fig. 41).

Stem lesions of shoots in hop crowns, similar to those of potato, may result when new shoots are covered with soil. Such white shoots were susceptible, but noncovered green ones were not (Jones, 1952) (Host rhythm).

A collar rot of mature tea shrubs (*Camellia sinensis*) was said to be caused by *R. solani* in India when plants were set deeper than they were in the nursery. The bark, which had been injured at ground level, was disintegrated and girdled below ground (Venkataramani and Venkata Ram, 1959).

A more unusual type of stem rot was produced on aquatic plants in the coastal waters of Virginia and North Carolina following the opening of a canal. Dark lesions occurred at the soil surface, regardless of the depth and concentration of sea water, producing great destruction to four genera of plants. This strain of *R. solani* was able to attack potatoes, and that from potatoes to attack the aquatic plants (Bourn and Jenkins, 1928b).

*Bottom rot or head rot.*—After the cabbage or lettuce plant has formed a head, it may be invaded from below by *R. solani*. Leaves in contact with the soil may be invaded, and the fungus spread to the stem and to the leaves above. A white hymenium may be formed on the stem (Fig. 41) or basal leaves. Initially, the lesions may be sharply defined and brown; they may dry out if the weather becomes drier, becoming papery and brown. The fungus may thus be confined to the lower leaves. Usually, however, it continues to grow upward inside the head, and may mummify it in 10 days. Sclerotia may be abundantly formed in the head. Under some conditions the leaves may be shed, leaving a naked stalk capped with a small head (Fig. 22; Weber, 1931; Townsend, 1934; Walker, 1957). Losses are increased when soil is thrown against the plants in ridging. Other plants with a basal rosette of leaves (e.g. aster [Fig. 23], may sustain leaf decay.

*Crown and bud rot.*—Crown rot of sugar beet is one of the historic diseases caused by *R. solani*. Infection apparently occurs in the young leaflets or in leaf bases, and causes petiole decay (Environment and Host rhythms). As the crown leaves die out, lateral young ones appear. The crown of the plant may eventually be killed, and the fungus advance into the top of the fleshy root, causing a dry brown decay (Fig. 26; Edson, 1915; Walker, 1957).

Alfalfa is also similarly invaded in many parts of the United States, producing cankers of the crown bud (Fig. 8). At first water-soaked, the tissue becomes necrotic and brown, and discoloration may extend above and below the invaded area. The fungus may invade the vessels. The plant may be killed or productivity sharply reduced (Cherewick, 1948; Erwin, 1954).

During the rainy months in California, the terminal bud (apex or crown) of strawberry plants may be destroyed by *R. solani* in large areas in the field. The flower buds, which are the first to arise from the crown, are killed. Lateral dormant buds may later grow out (Fig. 27) sometimes giving a witches'-broom effect (Fig. 28; Wilhelm, 1957). A preplanting soak of the plant in gibberellin (10 ppm) accelerates the rate of emergence of shoots from the soil (Host rhythm) and thus decreases Rhizoctonia bud rot (S. Wilhelm, unpub.).

*Aerial (web, leaf, thread) blights.*—When the environment rhythm in the tropics and subtropics is in a warm, rainy, humid cycle, *R. solani* may spread through the tops of plants. In some cases, these strains lead an aerial existence independent of the soil. Durbin (1959a) has characterized these strains as rapid growing, $CO_2$ sensitive, and producing abundant sclerotia (Fig. 31; Table 1). The first characteristic is advantageous in utilizing fully any short periods favorable to growth, the second reflects the constant low $CO_2$ content of their environment, the third pertains to the need for means of survival during unfavorable periods. The extremes of moisture are far greater in the aerial than in the subterranean habitat. The closer the seedlings are together, the greater the spread of aerial blight (Singh, 1955). Seeding of *Cyamopsis psoralioides* in the field at 5, 10, 15, and 20 seers per acre gave a 3-year average of leaf blight (caused by *R. solani*) of 3.2, 6.5, 11.3, and 23.1%, respectively. If plants were thinned to 15, 30, 45, 60, and 90 cm, Rhizoctonia leaf blight was 3.1, 2.8, 0.9, 0.2, and 0.03%, respectively, whereas the unthinned check was 4.9%.

TABLE 1. Characterization of strains of *R. solani* by point of invasion of host[1]

| Character | Aerial | At soil surface | Subterranean |
|---|---|---|---|
| Average percentage of growth inhibition by 20% $CO_2$ | 80±1.4 | 52±2.2 | 31±2.6 |
| Average growth rate in air (cm/24 hr) | 29±1.4 | 23±0.8 | 11±1.0 |
| Percentage of clones producing sclerotia | 82 | 42 | 21 |

[1]Durbin (1959a).

An aerial strain that occurs destructively on bean in Costa Rica (Echandi, 1965) and other areas in the humid tropics produces small, brown, necrotic spots on leaves (Fig. 29) from basidiospore infections (Pathogen and Environment rhythms). The spots coalesce and produce mycelia that spread to other leaves and plants. Basidia are produced on healthy tissue adjacent to lesions (Fig. 42); spores are shed at night. Pre- and postemergence damping-off can also be produced, but it is not clear that the fungus can survive in competition with soil microorganisms. Strains that attack trees may be even less dependent on the soil. One strain causes leaf spots of rubber trees in Peru (Kotila, 1945a); it could also attack leaves of sugar beet, but did not cause damping-off of seedlings nor rot the roots. The areolate leaf spot of citrus caused by the fungus in Surinam (Stahel, 1940) is interesting in that only very young leaves and twigs are infected. If new shoots come out in dry weather they remain healthy and presumably resistant (Environment and Host rhythms). The small brown spots are concentrically brown-ringed. Infections again result from basidiospores as well as mycelium. *Rhizoctonia solani* lives over in Louisiana in bark cracks of mature fig trees, and spreads short distances from tree to tree by mycelial growth and by air-borne sclerotia. Leaves develop ragged brown spots that may decay and drop out (shot hole). Sclerotia form on infected tissues, basidia on nearby healthy tissue. Fruit may be rotted, and twigs slightly invaded. Basidiospores are not important in spread. *Rhizoctonia solani* from bean seedlings can infect fig leaves, but evidence is lacking that this strain can persist in the aerial habitat (Matz, 1917;

Tims and Mills, 1943). Aerial invasion of Maranta (Ramakrishnan and Ramakrishnan, 1948) and ginger (Sundaram, 1953) in India seems to be of the same type.

The strains that cause brown patch of turf grasses appear to act in every way as the truly aerial forms just discussed, though growing so close to the ground that soil-surface types might be expected. The diseased areas of grass are usually 30-90 cm in diameter, but may reach 6-15 m. A dark purplish-green advancing margin 1.3-5.0 cm wide, in which the mycelium is webbed, is visible in the mornings (Fig. 4), but soon dries, and the central leaves die and turn light brown. The crowns and roots are only rarely invaded; leaves are infected through stomata and through mowing wounds. Sclerotia are formed near the base of the plant. Because grass is close to the moist soil, there may be a good deal of dew and guttation fluid on the leaves at night; the environment is therefore similar to that in trees in the humid tropics (Environment rhythm). "Poling" the grass in early morning to knock off the water droplets reduces the rate of spread of the fungus (Dickinson, 1930; Monteith and Dahl, 1932; Howard, et al., 1951; Couch, 1962). The fungus strains involved are so sensitive to temperature and humidity that only occasionally are conditions favorable; Dickinson (1930) found only 5% of the days during the brown-patch season in Massachusetts to be favorable. Apparently the typical disease does not occur in England (Sampson and Western, 1954; Moore, 1959; Smith [J.D.], 1959), possibly because of low temperatures. However, a "sharp eyespot" of wheat leaves is caused there by *R. solani* (Sampson and Western, 1954; Moore, 1959; Pitt, 1964a). A disease of rice in the Philippines, and of rice, Bermuda grass, sugar cane, etc. in Louisiana (Palo, 1926; Ryker, 1939) also appears to be similar. The leaf spots of sugar cane and other Gramineae, and those on *Maranta arundaceae*, holly, and citrus may be banded or with brown zones (Ryker, 1939; Stahel, 1940; Cooley, 1942; Ramakrishnan and Ramakrishnan, 1948; Exner, 1953). The leaf spot of *Tephrosia vogelii* is limited by veins, and is angular in shape (Ruppel, et al., 1965). Leaf spots of *Fittonia verschaffeltii* var. *argyroneura* (Baker [K.], 1957), as well as of fig, cotton, Maranta, and *T. vogelii* may decay and drop out, producing a shot-hole effect.

It would be instructive to determine whether the brown-patch or other true aerial strains of *R. solani* can survive in the soil (as distinguished from host tissue in soil), and infect the host from it.

There are numerous other cases of attack of aerial parts by *R. solani*, but the fungus strains seem to be less specialized for that habitat. Aerial parts of soybeans in Louisiana (Atkins and Lewis, 1954), beans (Figs. 30, 31) and lima beans in Florida (Weber, 1939), *T. vogelii* in Puerto Rico (Ruppel, et al., 1965), *Lespedeza* in Louisiana and North Carolina (Allison and Wells, 1951; Stroube, 1954a), coffee seedlings in El Salvador (Weber and Abrego, 1958), larch in Japan (Itô, et al., 1955), tulip in Washington (MacLean, 1948), cotton in Louisiana (Neal, 1944; Kotila, 1945*b;* Pinckard and Luke, 1967), sugar beet in Virginia and the north central states (Kotila, 1947), and celery in California (Houston and Kendrick, 1949), are attacked by *R. solani* under favorable conditions. It may also infect at the soil surface and survive in soil. It should be noted that alfalfa exhibits aerial infection (Allison and Wells, 1951), damping-off and crown rot (invasion at soil surface) (Erwin, 1954), and deep root rot (Smith [O.F.], 1943, 1945) (Pathogen and Environment rhythms).

*Soil rot of parts in contact with soil.*—Apparently any aboveground plant part that comes into contact with soil heavily infested with *R. solani* may become infected. Fruits of various types provide the commonest examples of this. Tomato fruits, either green or ripe, develop small, firm, brown spots on the area in contact with soil. These later enlarge, become soft, and may have concentric zones (Fig. 36). Mycelia (Fig. 40) and sclerotia may develop on the surface. A white basidial hymenium may develop on the surface of the fruit, surrounding the area in contact with the soil (Fig. 43). If the soil remains moist for a few days, or the fruit is borne under heavy vine growth that keeps the soil moist, infection may occur (Host and Environment rhythms). Less frequently, infection may occur during moist weather from soil splashed onto fruit as high as 35-45 cm above the soil. Infection is through intact epidermis or through wounds. Similar soil-rot lesions are produced on pepper, eggplant, cucumber, and bean (Fig. 38) and pea pods (Fulton, 1908; Wolf, 1914; Ramsey and Bailey, 1929; Weber, 1932*a;* Ramsey and Wiant, 1941; Ramsey and Smith, 1961). These infections may be only in the incipient stage at time of harvest, and later develop in storage (see below). Soil rot in fruits to be used for seed leads to seed transmission (see Seed decay).

Leaves in contact with the soil are quite commonly infected by *R. solani;* examples are poinsettia (Fig. 25; Tompkins, 1959), China aster (Fig. 23; Ullstrup, 1936), gardenia (Fig. 32), lettuce (Townsend, 1934), and cabbage (Fig. 22; Weber, 1931). Cutting rot in propagation beds frequently spreads through the foliage as top rot. Cooley (1942) found that the fungus spread to the top of holly cuttings from leaves in contact with the soil. Stems in contact with soil, or on which soil is splashed, may be similarly infected (Figs. 33, 34; Weber and Ramsey, 1926; Davis, 1941; Conover, 1949; Whitney, 1959). Zinnia flower heads, hand picked and piled on canvas in the field, may absorb so much moisture from the soil that *R. solani* grows through the canvas, rots the dead petals, and invades the seed. Use of preservatives to prevent decay of the canvas also prevents this type of *Rhizoctonia* infection (Baker [K.], 1947).

*Storage rots and blemishes.*—The best-known symptom of *R. solani* is the appearance of black

sclerotia on potato tubers (Fig. 37); this is disfiguring, but otherwise of little importance. It is called "dirt that won't wash off," black scurf, black speck, scurf, black scab, and black-speck scab. It is of interest that, in the tropics, the sclerotia on tubers may germinate in storage and grow through the cloth bags, rotting them (Chupp and Sherf, 1960).

Soil rot of tomatoes may spread by mycelial growth (Fig. 35) from an infected fruit in the packed box in storage, through the paper wraps, and cause "nest rot" (Figs. 39, 40; Rosenbaum, 1918; Ramsey and Bailey, 1929). Infected beans may produce a similar condition (Ramsey and Wiant, 1941). Root rot of turnips and carrot (Lauritzen, 1929), beet (Walker, 1957), and radish (Ramsey and Smith, 1961) may also develop in storage.

HOST RANGE AND STRAINS OF THE FUNGUS.—The species, *R. solani,* has a very wide host range that has been enumerated by a number of authors (Duggar and Stewart, 1901; Peltier, 1916; Dana, 1925*b;* Braun, 1930; Viennot-Bourgin, 1949; Stroube, 1954*b;* Walker, 1957; Chupp and Sherf, 1960). No claims apparently have been made that any plant species is immune to the fungus, though many plants have been found not to be attacked by given strains of it (e.g. Williams and Walker, 1966). Tabulations of the hosts of some specialized strains are available (Matsumoto, 1921; Gratz, 1925; Storey, 1941; Person, 1944*b;* Durbin, 1957).

The strains of *R. solani* may differ in: the hosts they are able to attack; the virulence of attack, ranging for a given host from nonpathogenic to highly virulent (Storey, 1941); the temperature at which attack will occur (Kendrick, 1951); the ability to develop in the lower soil levels, the soil surface, or in the aerial habitat (Durbin, 1959*a*); the ability to tolerate $CO_2$ (Durbin, 1959*a*); the ability to form sclerotia (Durbin, 1959*a*); the growth rate (Durbin, 1959*a*); the survival ability in a given soil (Olsen, Flentje, and Baker, 1967). Durbin (1957) has shown that some strains caused preemergence, but almost no postemergence damping-off of pepper seedlings, some post-, but not preemergence damping-off, and some caused both.

It is known that strains are not as sharply host-limited as in the rusts or *Fusarium oxysporum,* for example. One form may attack hosts A to F, another hosts C to J, a third B to G, and a fourth may be nonpathogenic to any hosts tested. The data on cross-inoculation studies with strains of *R. solani* are still too meager to permit estimation of either the actual number of strains or their characterization. At least two natural groupings are already available. Durbin (1959*a*) provided a broad basis for separation of strains of *R. solani* by the source habitat: (1) true aerial forms; (2) strains active only in or near the soil surface; (3) subterranean isolates. Some of the identifying characteristics of each were provided. Flentje and Saksena (1957) placed a number of isolates from South Australia soils in "a strain specialized to Cruciferae"; recognition of this strain by other investigators has proved feasible. Flentje, et al. (this vol.) found that none of the mutations studied altered the host range of the parent strain. If other broad groupings could be similarly defined, clarification of the strain situation in the species would be facilitated. At present, one can only conjecture as to their number, and this is more likely to result in an over- rather than an underestimate.

For progress in understanding the biology of *R. solani*, reduction of the strains to a clearly recognizable status is the first order of business. Once the strain complex is defined, the accumulation of data in an orderly manner becomes possible. At present, by contrast, it is impossible to correlate data on host range, geographical distribution, economic importance, ecology, or control procedures from different workers or areas.

It must be obvious from all that has been said in this symposium that to erect or maintain quarantines against the species *R. solani* is unjustified. In the present state of our knowledge, to base quarantines on strains of the fungus is still impossible, but this may be desirable in the future. Similarly, practical agriculturists have come to think of *R. solani* as a single entity. Since the species is widespread, the conclusion is often reached that the introduction of more of the fungus to a given soil is unimportant. This ignores the significance of strain differences, the importance of the quantity of inoculum in instituting disease, and the likelihood that the new strain will become permanently established in the soil. Evaluation of the relative disease potential of two strains is a specialist's job, and others can with safety only assume them to be different and dangerous until proved otherwise. In such a situation there is no such thing as a safe soil-infesting pathogenic *R. solani!*

The taxonomy and nomenclature of the sexual and mycelial forms have been considered at length in this symposium. Of necessity, little has been said about the strains. Let us hope that, by the time of the next symposium on *R. solani*, a workable scheme for arranging the strains of this fungus will be available. Future emphasis should be on the recognition of strains, their distribution, host range, ecology, and economic importance.

GEOGRAPHICAL DISTRIBUTION.—The species *R. solani* has been reported from many areas on many crops. It is a reasonable conclusion from the literature that the fungus occurs in all arable land in the world, whether cultivated or not.

With reference to any given strain of the fungus, the distribution is unknown, but it certainly is much more restricted than for the species. Since strains have not been adequately defined, mapping of their distribution is necessarily uncertain and has not been attempted. If the geographical distribution of certain defined strains was known, it would be helpful in planning crop-rotation programs, in quarantine administration, and in understanding the ecology of the fungus.

ECONOMIC IMPORTANCE.—Because of the extraordinary host range, geographical distribution, and environmental adaptability of *R. solani*, and the variety of diseases it produces, it is one of the really important plant pathogens. Weber (1932a), for example, states that in Florida, *R. solani* "attacks a large number of truck crop plants and is generally considered the most destructive fungus found in the state. It is also probably the most difficult to control."

The versatility of the fungus makes difficult the collection of data on its economic importance, since information on so many crops and distinct diseases must be compiled. Data on the total losses caused by *R. solani* on crop plants are, without question, excessively low, because the figures include other fungi under the general headings, root rots, damping-off, etc., rather than being individually tabulated.

The losses in the United States (Anon. 1965b) specifically mentioned for *R. solani* in 1964 on cotton (half of collective loss assigned to *R. solani*), dry beans, green snap beans, potatoes, and tomato transplants totaled nearly $6 million. There are many more categories of collective losses in which those attributed to *R. solani* cannot be determined. The losses in California in 1963 (Anon. 1965a) specifically mentioned for *R. solani* (half of collective loss assigned to *R. solani*): bedding plants, miscellaneous bulbs, cotton, cowpeas, foliage plants, orchids, strawberries (a third of the collective loss assigned to *R. solani*: conifer seedlings, flower seed crops, lilies, miscellaneous potted plants, poinsettias, stocks, sugar beets; for the total specified loss) on alfalfa, China asters, azaleas, beans (dry, green snap, and lima), carnations, flax, potatoes—more than $13 million, or more than 33,000 acres of crops lost. The higher figure for California than the United States in these surveys reflects the more detailed reporting in California. A similar type of survey in 1958 (O'Reilly, et al., 1958) estimated that 4 million acres of crop land were infected by *R. solani* in California, Washington, and Oregon.

EPILOGUE.—*Rhizoctonia solani* is an ancient species; just how ancient may eventually be revealed directly by the fossil record. The present indirect evidence for this conclusion is of several kinds:

1. The species is worldwide in distribution in crop plants, and even occurs commonly in uncultivated areas such as forests, uncultivated desert land in Idaho, and mallee land in South Australia. This wide distribution clearly indicates that the fungus has been present a long time.

2. The species consists of a large number of strains, distinguishable on the basis of: host range; the virulence of attack on a given host; the type of attack on the host; the temperature at which such attack will occur; the ability to grow and survive in the lower levels of soil, at the soil surface, or as true aerial pathogens; the ability to tolerate appreciable concentrations of $CO_2$; and probably many other

environmental features as yet unstudied. To develop such variability, and to stabilize, select, and establish the strains in appropriate ecological niches, would require a very long period of time.

3. The species has evolved rather complicated relationships with other soil microorganisms in its ability to survive and grow through the soil quite successfully under a wide range of conditions, despite the large variety of antagonists involved (Baker [K.], et al., 1968; Olsen and Baker, 1967). Such adjustments require a great deal of time.

4. Higher agarics and polypores are known in fossil form at least from the Oligocene Epoch, 30 million years ago (Conwentz, 1890; Pia, 1927) and clamp connections were present in mycelium for at least that long (Conwentz, 1890). Conceivably, then, *R. solani* may well be several million years old, and have had ample time to develop the characteristics mentioned.

Finally, it is suggested that the basic pattern for plant pathogenic *R. solani* is represented in the strains that populate the surface soil. These are quite variable in sclerotial formation, $CO_2$ tolerance, growth rate, and pathogenicity. From this parental material probably evolved the strains able to successfully establish and persist parasitically in the aerial habitat, and those that were able to develop, survive, and parasitize plants 30 cm, more or less, under the soil surface. Each of these derivatives is necessarily more specialized than the strains inhabiting the soil

surface, and these parasitic strains in turn are more specialized than the saprophytic strains residing in the soil surface. The evolutionary sequence might thus be:

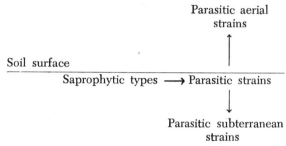

*Rhizoctonia solani* thus emerges as a fungus of ancient lineage, great versatility, and enormous capacity for destruction of man's crops. The relatively simple structure and life cycle of this fungus, evidence of continuing evolutionary plasticity, may well be basic to its marked success in both time and space.

ACKNOWLEDGMENTS. — I am greatly indebted to the many persons who supplied photographs of diseases caused by *R. solani*. Space limitation unfortunately prevented the use of a number of the photographs. The prints were prepared for publication by W. H. Fuller.

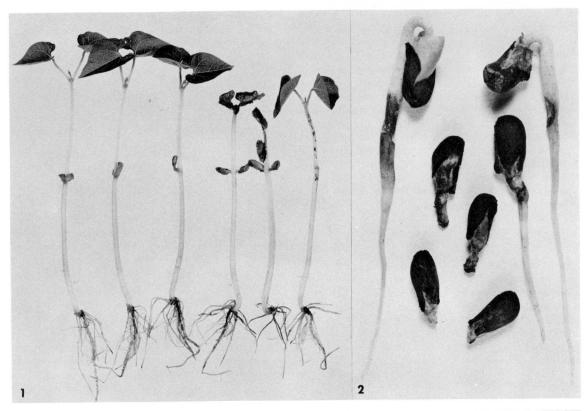

Fig. 1.    Bean (*Phaseolus vulgaris* L.) seedlings grown from seed infected by *R. solani*, show-ing post-emergence effect on cotyledons, stems, and first leaves. Seedlings from uninfected seed at left. (Photo by B. R. Houston, Univ. Calif., Davis.)

Fig. 2.    Pre-emergence damping-off of cotton (*Gossypium hirsutum* L. var. Acala) seedlings caused by *R. solani*. (Photo by D. S. Hayman, Univ. Calif., Berkeley.)

Fig. 3.    Stand reduction of cotton (*Gossypium hirsutum* L.) from seed decay and pre- and post-emergence damping-off caused by *R. solani* in Arizona. (Photo by J. T. Presley, Agri. Res. Service, U.S. Dept. Agri., Beltsville, Md.)

*Fig. 4.* Brown-patch disease of colonial bentgrass (*Agrostis tenuis* Sibth.) putting-green turf caused by *R. solani.* Dark purplish-green advancing margin is shown. (Photo by H. B. Couch, Va. Polytechnic Inst., Blacksburg.)

*Fig. 5.* Circular patches of wheat (*Triticum vulgare* Vill.) with root rot caused by *R. solani* in field in Eyre Peninsula, South Australia. (Photo by N. T. Flentje, Waite Agr. Res. Inst., Adelaide, S. Australia. *In* Baker, Flentje, Olsen, and Stretton, 1967.)

*Fig. 6.* Neck rot of gladiolus (*Gladiolus* sp.) plants grown from cormels in the field, caused by *R. solani.* Linear spread in row is shown. (Photo by D. B. Creager, Ill. Nat. Hist. Survey, Urbana. *In* Forsberg, 1946.)

*Fig. 7.* Shredded condition ("neck rot") of base of gladiolus plants shown in Fig. 6. (Photo credit as in Fig. 6.)

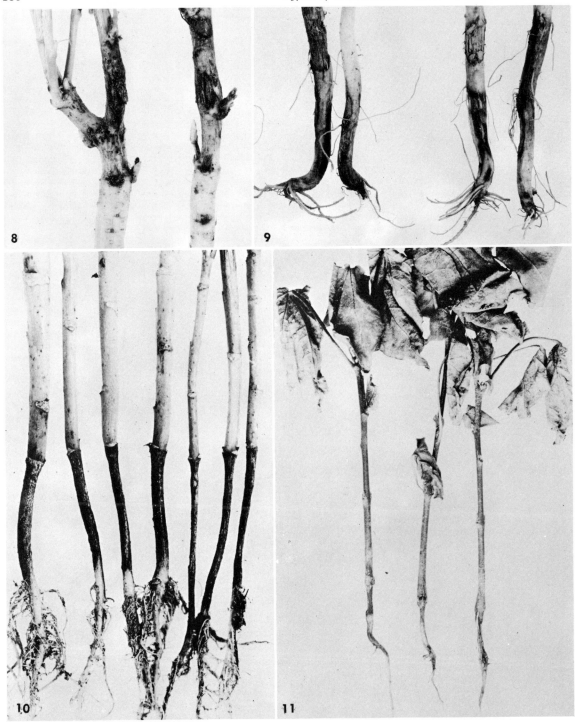

*Fig. 8.*   Stem lesions of crowns of alfalfa (*Medicago sativa* L.) caused by *R. solani*; field infection. (Photo by D. C. Erwin, Univ. Calif., Riverside.)

*Fig. 9.*   Hypocotyl infections of bean (*Phaseolus vulgaris* L.) plants by *R. solani*. Field infections, showing types of cortical invasion. (Photo by Dept. Plant Pathol., Univ. Calif., Davis. Courtesy of L. D. Leach.)

*Fig. 10.*   Wire stem of cabbage (*Brassica oleracea* L. var. *capitata* L.) plants, showing dry firm cortical infections by *R. solani*. (Photo by G. F. Weber, Univ. Fla., Gainesville, *In* Weber. 1931.)

*Fig. 11.*   Sore shin of cotton (*Gossypium hirsutum* L.) seedlings caused by *R. solani*, showing cortical decay of hypocotyls. (Photo by J. T. Presley, Agr. Res. Service, U.S. Dept. of Agr., Beltsville, Md.)

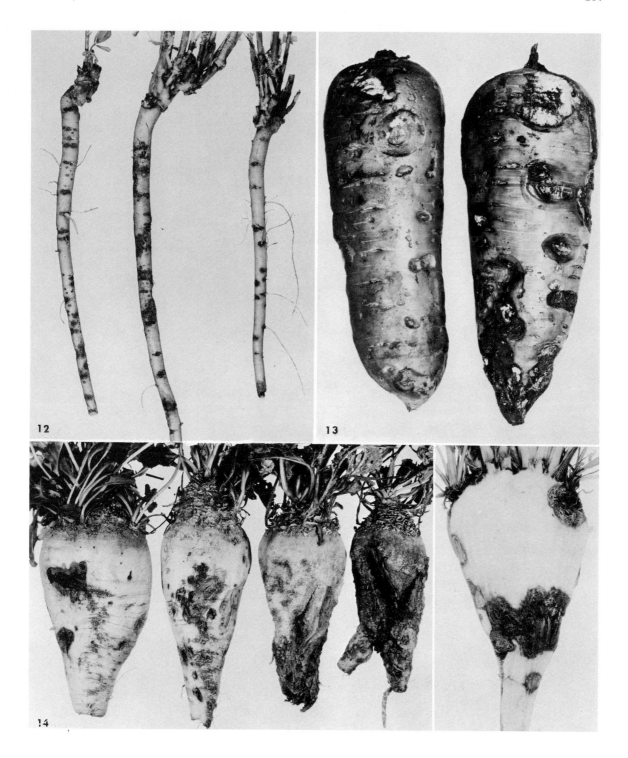

*Fig. 12.* Cankers of alfalfa (*Medicago sativa* L.) tap roots caused by *R. solani*. Infections occur at points of emerging secondary roots, and extend several in. below soil surface. (Photo by D. C. Erwin, Univ. of Calif., Riverside.)

*Fig. 13.* Crater rot of carrot (*Daucus carota* L. var *sativa*. DC.) roots caused by *R. solani*; from Ill. storage house. (Photo by G. B. Ramsey, Agr. Res. Service, U.S. Dept. of Agr., Chicago, Ill. Courtesy of W. L. Smith, Jr.)

*Fig. 14.* Cankers of fleshy roots of sugar beets (*Beta vulgaris* L.) caused by *R. solani*. Longitudinal section of infected root at right. (Photo by B. R. Houston, Univ. of Calif., Davis.)

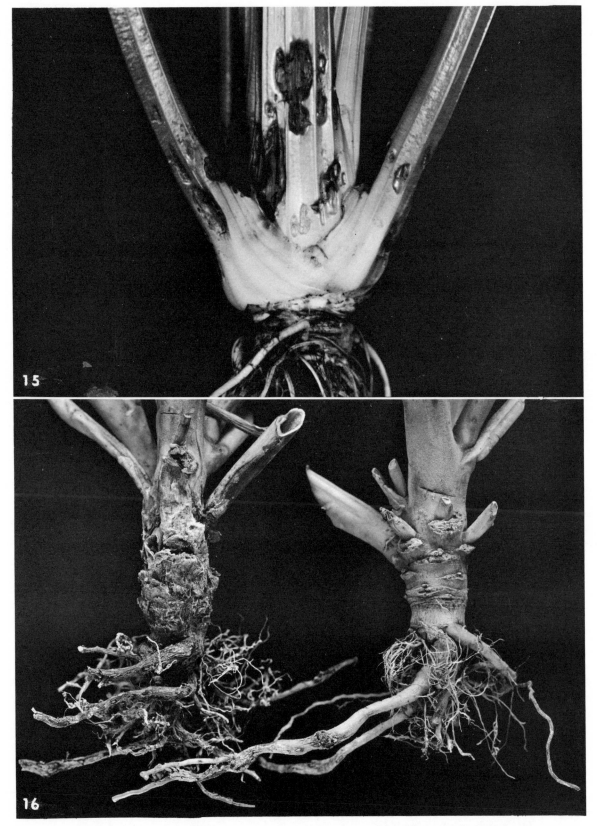

*Fig. 15.* Crater spot of celery (*Apium graveolens* L. var. *dulce* DC.) stems caused by *R. solani*; field infection. (Photo by B. R. Houston, Univ. of Calif., Davis. *In* Houston and Kendrick, 1949.)

*Fig. 16.* Stem rot of *Nicotiana suaveolens* Lehm. var. *undulata* Comes caused by *R. solani*; field infections. Severe infection (*left*), mild (*right*). (Photo by J. T. Middleton, Univ. of Calif., Berkeley.)

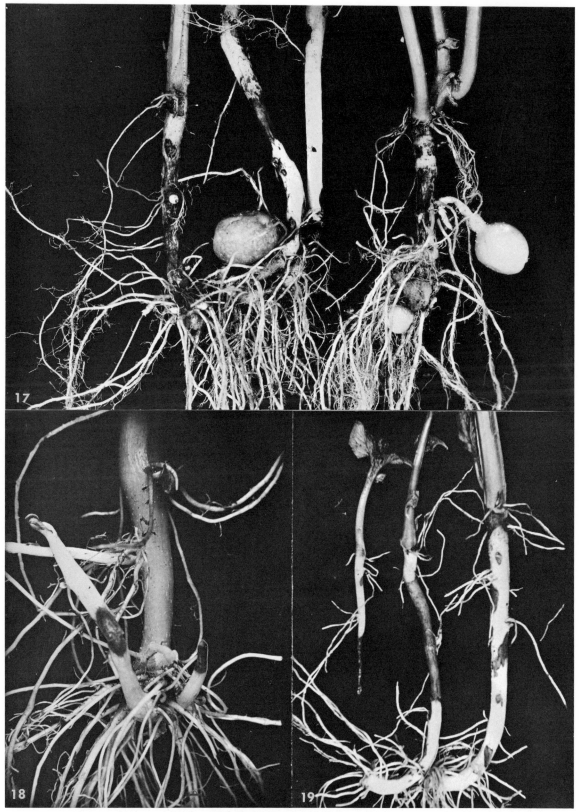

*Fig. 17.* Stem rot of potato (*Solanum tuberosum* L.) caused by *R. solani*. Cankers develop on subterranean white stems. (Photo by B. R. Houston, Univ. of Calif., Davis.)

*Fig. 18.* Sprout rot of potato, showing lateral infection (*left*) and terminal infection (*right*). New shoots often develop below infection, and may be infected in turn. (Photo by Dept. Plant Pathol., Wash. State Univ., Pullman. Courtesy of S. B. Locke.)

*Fig. 19.* Stem rot of potato. (Photo credit as in Fig. 17.)

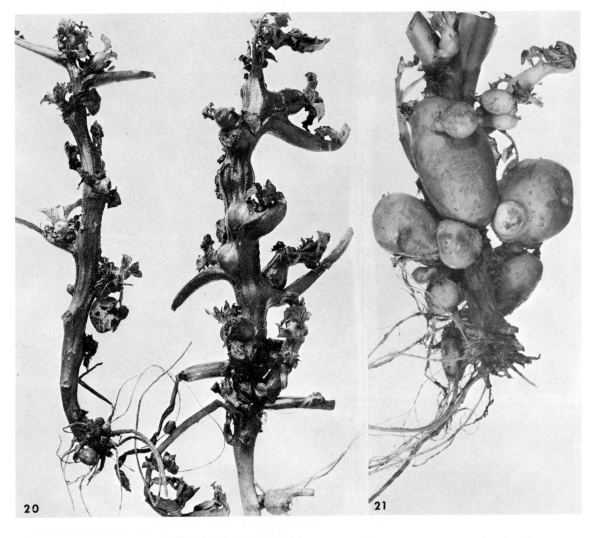

*Fig. 20.* Aerial tubers and swollen leafy stems of potato (*Solanum tuberosum* L.) induced by photosynthates accumulated there because stem cankers produced by *R. solani* prevented downward translocation. (Photo by R. E. Smith, Univ. of Calif., Berkeley.)

*Fig. 21.* Aerial tubers (*top*) and tubers on shortened stolons (*bottom*) on potato with basal stem cankers. (Photo by Howard Lyon, Dept. Plant Pathol., Cornell Univ., Ithaca, N.Y.)

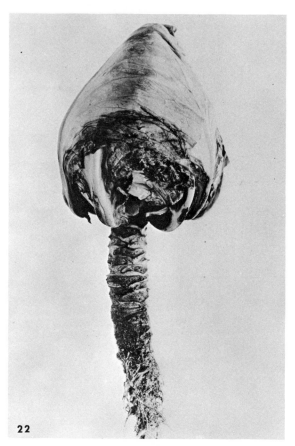

*Fig. 22.* Bottom rot of cabbage (*Brassica oleracea* L. var. *capitata* L.) plants caused by *R. solani;* basal leaves had decayed and fallen away. (Photo by G. F. Weber, Univ. of Fla., Gainesville. *In* Weber, 1931.)

Fig. 23.   Decay of basal leaves of China aster (*Callistephus chinensis* Nees.) resulting from contact infection by *R. solani* from soil. Healthy plant at right. (Photo by C. M. Tompkins, Univ. of Calif., Berkeley.)

Fig. 24.   Basal stem rot of annual stock (*Mathiola incana* R. Br.) seedlings 12 days after being transplanted to soil infested with *R. solani*. (Photo by A. W. Dimock, Cornell Univ., Ithaca, N.Y. *In* Dimock, 1941.)

Fig. 25.   Stem rot of poinsettia (*Euphorbia pulcherrima* Willd.) cuttings (*right*) caused by *R. solani*; healthy cuttings at left. (Photo by A. W. Dimock, Cornell Univ., Ithaca, N.Y.)

*Fig. 26.* Crown rot of sugar beet (*Beta vulgaris* L.) caused by *R. solani;* longitudinal section of advanced stage of disease. (Photo by B. R. Houston, Univ. of Calif., Davis.)

*Fig. 27.* Bud rot of Shasta strawberry. (*Fragaria* sp.) Early stage of field infection, showing formation of new white shoots replacing dark buds infected by *R. solani*. (Photo by S. Wilhelm, Univ. of Calif., Berkeley. *In* Wilhelm, 1957.)

*Fig. 28.* Witches'-broom effect of Shasta strawberry, an advanced stage of the disease produced by successive infection of buds as they grow out. (Photo by S. Wilhelm, Univ. of Calif., Berkeley.)

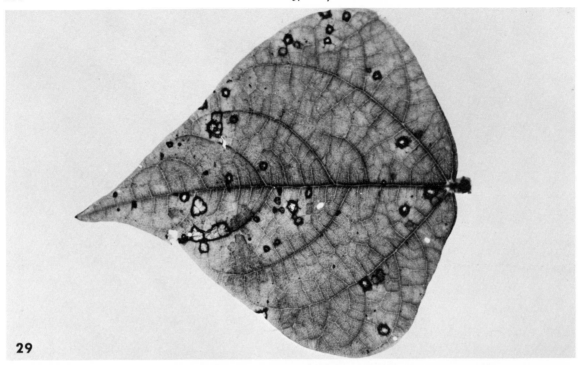

*Fig. 29.* Leaf spots of bean (*Phaseolus vulgaris* L.) resulting from aerial infections by basidiospores of *T. cucumeris* (Frank) Donk; field infections. (Photo by E. Echandi, Inst. Interamericano de Ciemcias Agricoles, Turrialba, Costa Rica. *In* Echandi, 1965.)

*Fig. 30.* Web blight of bean, showing defoliation produced by *R. solani* (*R. microsclerotia* Matz) in a Florida field. (Photo by G. F. Weber, Univ. of Fla., Gainesville. *In* Weber, 1939.)

*Fig. 31.* Naturally infected bean stems, showing abundant surface development of sclerotia of *R. solani*. (Photo by G. F. Weber, Univ. of Fla., Gainesville. *In* Weber, 1939.)

*Fig. 32.* Leafspots of gardenia (*Gardenia jasminoides* Ellis) caused by *R. solani* in cutting bed. (Photo by C. M. Tompkins, Univ. of Calif., Berkeley.)

*Fig. 33, 34.* Red pine (*Pinus resinosa* Ait.) seedlings, showing bud killing and top rot (*center plant of Fig. 32; Fig. 33*) resulting from splashing soil infested by *R. solani* onto the growing point. (Photo by H. S. Whitney, Forest Res. Lab., Can. Dept. Forestry, Calgary, Alberta.)

*Fig. 35.* Bean (*Phaseolus vulgaris* L.) stem held under moist conditions, showing development of mycelia of *R. solani*. (Photo by Dept. Plant Pathol., Univ. of Calif., Davis. Courtesy of L. D. Leach.)

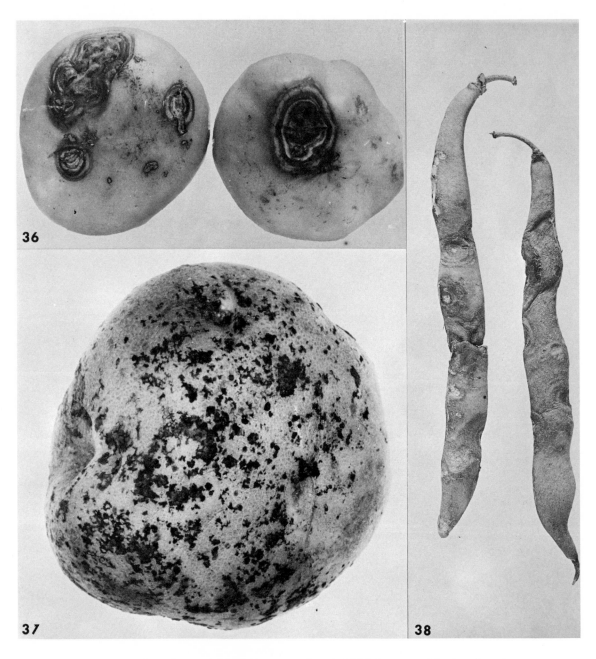

*Fig. 36.* Soil rot of tomato (*Lycopersicon esculentum* Mill.) fruit caused by *R. solani*, showing concentric rings of light and dark brown frequently produced; field infections. (Photo by G. F. Weber, Univ. of Fla., Gainesville. *In* Weber and Ramsey, 1926.)

*Fig. 37.* Sclerotia (black scurf, "dirt that won't wash off") of *R. solani* on potato (*Solanum tuberosum* L.) tuber. (Photo by Howard Lyon, Dept. Plant Pathol., Cornell Univ., Ithaca, N.Y.)

*Fig. 38.* Soil rot of bean (*Phaseolus vulgaris* L.) pods from La. caused by *R. solani*. (Photo by Agr. Res. Service, U.S. Dept. of Agr., Beltsville, Md. Courtesy of R. W. Goth.)

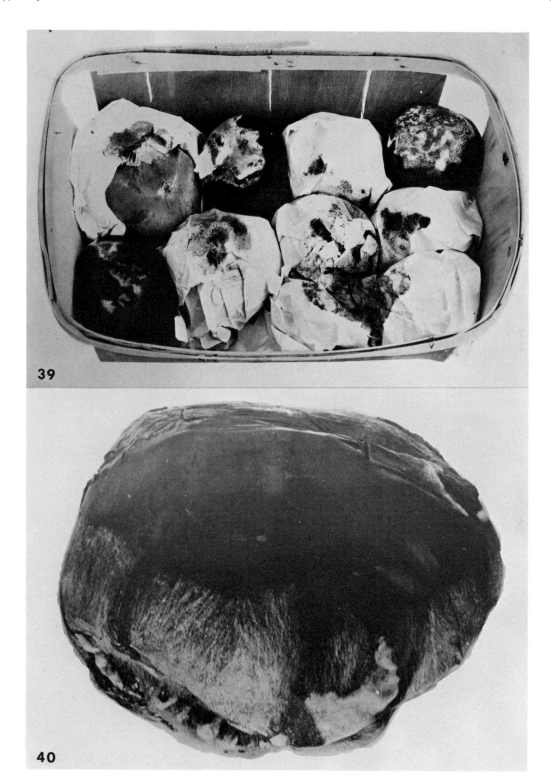

*Fig. 39.* Nest rot of greenwrap tomatoes (*Lycopersicon esculentum* Mill.) during shipment, caused by *R. solani*. Mycelia grow from fruit to fruit through wraps, as shown in Fig. 40. (Photo by G. B. Ramsey, Agr. Res. Service, U.S. Dept. of Agr., Chicago, Ill. Courtesy of L. Beraha. *In* Ramsey and Link, 1932.)

*Fig. 40.* Advanced stage of soil rot of tomato fruit caused by *R. solani,* showing mycelial overgrowth. (Photo by G. F. Weber, Univ. of Fla., Gainesville.)

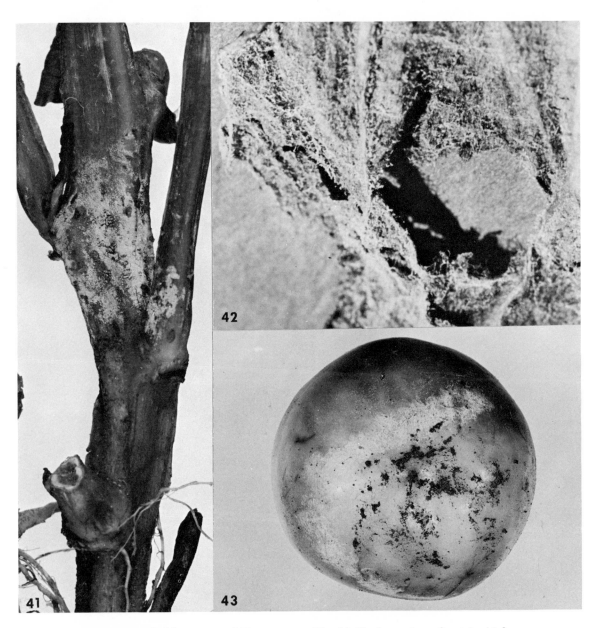

Fig. 41.   Basidial hymenium of *T. cucumeris* (Frank) Donk on stem of potato (*Solanum tuberosum* L.) just above soil surface on uninfected tissue. (Photo by Howard Lyon, Dept. Plant Pathol., Cornell Univ., Ithaca, N.Y.)

Fig. 42.   Basidial stage of *T. cucumeris* formed on infected bean (*Phaseolus vulgaris* L.) leaf. (Photo by E. Echandi, Inst. Interamericano de Ciencias Agricoles, Turrialba, Costa Rica. *In* Echandi, 1965.)

Fig. 43.   Basidial hymenium of *T. cucumeris* on tomato (*Lycopersicon esculentum* Mill.) fruit surrounding area of contact with soil. (Photo by B. R. Houston, Univ. of Calif., Davis.)

# The Mechanism and Physiology of Plant Penetration by Rhizoctonia Solani

R. L. DODMAN and N. T. FLENTJE—*Departmentof Plant Pathology, Waite Agricultural Research Institute, University of Adelaide, South Australia.*

INTRODUCTION.—This review is concerned with the initial stages of pathogenesis by *Rhizoctonia solani* Kühn (*Thanatephorus cucumeris* [Frank] Donk). It deals with the growth of the fungus around and on the plant surface prior to any organization for penetration, the influence of the plant on such growth, the means whereby the fungus gains entry into plants, the factors controlling penetration, and the relation between these factors and specialization. Since there is little critical information on the influence of environmental factors on the initiation of disease, the main emphasis here is on host and fungal factors contributing to infection.

GROWTH AROUND THE PLANT PRIOR TO PENETRATION.—The fungus may exist in the soil in a number of forms. If present in a resting form, presumably some type of germination and growth must occur before penetration; if actively growing, some further growth will probably precede penetration. It may be of value to consider how the host influences these events and whether this fungal activity plays an important part in the establishment of parasitism. It should be pointed out at this stage that an attempt is being made to separate growth toward, around, and on the plant surface prior to any organization for penetration from the growth and development of specialized structures associated with penetration.

There is now abundant evidence that plant exudates influence germination and growth of many soil microorganisms. Some observations on the effects of plant exudates on *R. solani* have been made, but less has been done with this fungus than with others, perhaps because of the difficulties associated with measuring responses by this fungus.

Kerr (1956) demonstrated the effect of root exudates on the growth of *R. solani* by enclosing seedlings in cellophane bags and placing these bags in soil infested with the fungus. When the bags were removed, the hyphae were aggregated on the outside of the bags directly opposite the enclosed seedling roots. Kerr proposed that exudates from the roots diffused through the cellophane and induced this increased growth. Other studies, by Wyllie (1962) and Martinson (1965), have provided similar evidence with a number of different hosts and isolates; Martinson also found that seeds were capable of stimulating the growth of *R. solani*.

A number of workers have demonstrated that the growth of *R. solani* is stimulated by exudates collected from plants and bioassayed both in vitro and in soil. Kerr and Flentje (1957), Flentje, Dodman, and Kerr (1963), and De Silva and Wood (1964) collected exudates from seedlings and showed that these stimulated fungal growth on cellophane in vitro. Other in vitro studies by Nour El Dein and Sharkas (1964b) demonstrated that root exudates from tomato seedlings provided a good source of nutrients for growth of this fungus in liquid culture. Assays in soil were carried out by Martinson and Baker (1962), who added exudates from seedlings to agar in soil microbiological sampling tubes and increased the frequency of isolation of *Rhizoctonia* from soil, presumably by stimulating fungal growth. In later studies, Martinson (1965) showed that exudates collected from seeds and placed in cellophane bags caused a stimulation of fungal growth on the outside of the bags when they were placed in infested soil.

In a number of instances, stimulation of growth of microorganisms by exudates decreases with the age of seedlings. This effect has been correlated with a decrease in the amount of exudate released by seedlings as they become older. De Silva and Wood (1964) found that exudates from younger seedlings caused a greater stimulation of growth of *R. solani* than did exudates from older seedlings, and it is probable that this is due to a decline in materials exuded.

Although little is known about the nature of the materials stimulating growth of *R. solani*, these are most probably carbohydrates and amino acids commonly found in large quantities in the exudates from the seeds and roots of many plants. Nour El Dein and Sharkas (1964b) identified a number of amino acids in exudates from tomato roots and correlated the level of stimulation of growth of *R. solani* with the amounts of amino acids liberated. Martinson (1965) found that certain carbohydrates placed in cellophane bags induced a stimulation of growth of *R. solani*. Further studies of the stimulatory materials are required.

However, it is clear that plant exudates do influence the development of *R. solani*. It may be suggested that the infuences discussed above are largely on saprophytic rather than parasitic growth, but these phases cannot be regarded as wholly independent. The support of survival and growth on and near plant roots as they grow in soil will increase the population of the fungus and probably provide better opportunities for penetration and infection. Much has yet to be learned about these interactions among plant materials, saprophytic growth and survival, and disease. Exudates may also be concerned with host specificity by stimulating the growth of isolates pathogenic to specific hosts. This possibility will be discussed later.

It is clear that the effects of exudates on fungal growth may be of considerable importance in diseases caused by *R. solani*.

THE MODE OF PENETRATION.—*The process of penetration.*—Penetration into plants by *R. solani* may be accomplished in a number of ways. Entry may occur through the intact cuticle and epidermis either from complex organized infection structures or without defined morphological structures. The fungus may also penetrate through natural openings and through wounds. Although isolates of the fungus often penetrate in only one way, it is not uncommon for some isolates to penetrate the same host in several different ways.

The processes of penetration and also the mechanism of each type of penetration are now considered.

Penetration through the intact plant surface.— Penetration of the intact cuticle and epidermis by *R. solani* has been reported by many workers, but in very few cases have these studies provided detailed observations on the actual means of entry. However, it now seems reasonably clear that this type of penetration is commonly effected by means of complex infection structures, and an understanding of the various stages in the development of these structures is now emerging. It is also apparent that different types of infection structures, each characteristic of different isolates, may be formed to effect penetration. The most frequently reported type of structure is the dome-shaped infection cushion; the formation of these structures will be discussed first and will be followed by a consideration of the mode of development of other structures.

The formation of infection cushions by *R. solani* was first described by Abdel-Salam (1933) from studies of damping-off of lettuce and tomato. Prior to this, Duggar (1915) used the term infection cushion for structures formed by *R. crocorum* on the roots of a number of hosts, but found no evidence that these structures were formed by *R. solani*. Abdel-Salam (1933) provided little detailed information on the development and structure of infection cushions, merely indicating that they were formed by the branching of hyphae to form short, swollen cells, which aggregated into a "wartlike, spongy microsclerotium." Subsequently, Ullstrup (1936) briefly reported the formation of infection cushions on the leaves of China aster, and Nakayama (1940) described the development of infection cushions on

cotton seedlings. Nakayama provided good illustrations of cushions, but did not contribute to the understanding of how the structures are formed. However, he did observe that hyphae of the fungus growing over the host surface tended to follow the lines of junction of the underlying epidermal cells; this phenomenon has been observed by a number of workers (Flentje, 1957; Christou, 1962; Khadga, et al., 1963; Dodman, 1965).

Detailed studies of the development of infection cushions by *R. solani* have been carried out by Flentje (1957), Christou (1962), Gonzalez and Owen (1963), Khadga, et al., (1963), Chi and Childers (1964), and Dodman (1965). Flentje and Hagedorn (1964) have briefly reported the formation of infection cushions by *R. practicola* (= *R. solani*). The generalized account given below of the development of infection cushions has been constructed from the descriptions given by these workers.

It has been frequently reported that hyphae of *R. solani* grow in close contact with the plant surface, though Gonzalez and Owen (1963) described only loose attachment to the surface of tomato fruits. Flentje (1957) showed that hyphae are attached to the plant surface by mucilaginous material, but Nakayama (1940) found no evidence of attachment by this means; it appears that isolates vary in this respect.

In most cases, it has been found that hyphae, in growing over the plant surface, tend to follow the lines of junction of the underlying anticlinal cell walls (Fig. 1). Khadga, et. al., (1963) suggested that growth in this manner depends on the density of hyphae on the plant surface. Gonzalez and Owen (1963) reported that hyphae do not follow the junctions of epidermal cells on tomato fruits. Dodman (unpub.) also observed that hyphae do not grow in this way on radish cotyledons. It seems that this pattern of growth may be influenced by the nature of the surface on which the fungus is growing.

During growth over the plant surface, the hyphae or branches formed from them may change direction of growth either by growing along the junction of epidermal cell walls that are at an angle to the original direction of growth, or by growing in an oblique or transverse direction over epidermal cells.

The branches formed from the hyphae extending over the plant surface may continue growth and become indistinguishable from the elongating hyphae or they may initiate infection cushions. The branches from which cushions are formed show an unusual and characteristic type of growth which is markedly different from the normal pattern of extension and branching. The characteristic pattern of vegetative growth on a surface such as agar has been described by Flentje, Stretton, and Hawn (1963). This pattern (Fig. 2) is due to the continued elongation of hyphal tips and the production of primary branches at fairly regular intervals behind the growing tips. The distance between branches usually decreases

*Fig. 1.* Growth of hyphae along lines of junction of underlying anticlinal cell walls and formation of short, stubby, swollen side branches. × 370 (Christou, 1962.)

with increasing order of branching. However, on the host surface a different pattern of growth occurs because the primary and secondary branches usually do not elongate and branch at regular intervals. Instead, the branches either cease to elongate and form many side branches in close proximity to each other or they continue to grow but curl back on themselves and finally branch repeatedly. The branches thus become short, stubby, swollen, curved or coiled, and much branched (Figs. 1, 3). Gonzalez and Owen (1963) report that cells forming infection cushions often resemble sclerotial cells, whereas other workers have found them to be as described above.

Dome-shaped infection cushions may be formed in several different ways; they are commonly formed by the aggregation of short, stubby side branches from adjacent hyphae (Fig. 4a), but they may also be formed by the aggregation of such branches from only one hypha (Fig. 4b), or by terminal branching

*Fig. 2.* Characteristic pattern of growth of *R. solani* on agar. × 50.

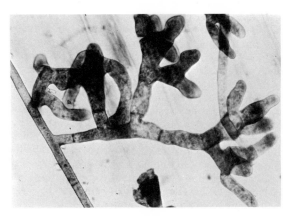

*Fig. 3.* Short, curled, irregularly swollen side branches × 350.

or a single hypha (Khadga, et. al., 1963). Such structures usually consist of tightly compacted hyphae formed into a discrete cushion that may vary considerably in size. Flentje (1957) found that cushions on cruciferous hosts measured up to 300 $\mu$ in diameter and 150-170 $\mu$ high, while Dodman (in press) reported that cushions may be up to 500 $\mu$ in diameter and 120 $\mu$ in height. The long axis of hyphal cells at the base of these cushions in contact with the plant surface is usually orientated at right angles to the plant surface and the cells develop swollen, flattened tips (Fig. 5). It is not known how hyphae become orientated in this way. Structures that are not as clearly defined as dome-shaped cushions, but that apparently are able to effect penetration, may be formed by the intermingling of a tangled mass of hyphae.

At present there is little evidence to indicate why side branches aggregate to form complex infection cushions. An association of different hyphae in single cushions might allow anastomosis and heterocaryon formation, but it is unlikely that this is of significance as it occurs freely at any stage of vegetative growth. It is possible that aggregation is due to an attraction between hyphae, as there is evidence that hyphae in close proximity appear to attract each other (Flentje and Stretton, 1964). Thus, excessive branching in a localized area and attraction between hyphae may result in the formation of tightly compacted infection cushions.

Infection cushions are usually closely appressed to the host surface and are apparently attached to it in some way. Flentje (1957) has suggested that individual hyphae become attached to the host by means of a mucilaginous sheath, but the presence of this material around cushions was not mentioned. Ullstrup (1936) found no evidence of any cementing material around cushions. Christou (1962) and Khadga, et. al. (1963) consider that attachment is due to the flattened hyphal tips at the base of each cushion. It appears that each hyphal tip acts as a single appressorium, but this does not explain how attachment is effected.

It is most commonly reported that penetration from dome-shaped infection cushions occurs by the simultaneous development of numerous fine infection pegs from the swollen, flattened cells at the base of the structure and in contact with the plant surface (Fig. 5). However, this is not the only means of penetration from cushions. Nakayma (1940) reported that only one infection peg was formed from the center of each cushion while Dodman (in press) showed that hyphal tips beneath cushions may penetrate through the cuticle and between epidermal cell walls without forming infection pegs. Where infection pegs are formed, they may penetrate the cuticle and then enlarge and grow between the cuticle and epidermal wall, finally penetrating into or between cells. They may also penetrate both the cuticle and epidermal wall and enlarge in the cell lumen. Usually several cells are penetrated from each infection cushion, with more than one infection peg per cell. Bateman (this vol.) discussed the fur-

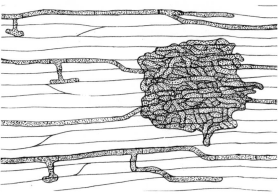

Fig. 4a. Formation of infection cushion by aggregation of side branches from adjacent hyphae. × 130. (Flentje, 1957.) (b) Formation of infection cushion by aggregation of branches from single hypha. × 600.

ther growth of hyphae in the tissue.

Some isolates of *R. solani* penetrate the intact plant surface from lobate appressoria, which are characteristic complex infection structures that differ markedly from infection cushions. The development of these structures has been described by Abdel-Salam (1933), Townsend (1934), and Flentje (1957) and illustrated by Verhoeff (1963, personal communication). Recently, Dodman (in press) has observed the formation of these structures by a number of isolates from widely differing localities.

The development of these structures is similar in early stages to the formation of dome-shaped cushions, in that hyphae adhere to the plant surface, grow along the lines of junction of underlying epidermal cells, and form short, swollen side branches. However, these branches do not continue to form further side branches that aggregate into cushions; rather, the side branches swell at their tips and each branch becomes, in effect, an appressorium from which penetration occurs (Fig. 6).

Although these lobate appressoria differ morphologically from dome-shaped cushions, functionally they are very similar since they are capable of effecting multiple penetration of the plant surface. It is not clear at present whether penetration pegs are

*Fig. 5.* Longitudinal section of infection cushion showing orientation and swelling of hyphal branches and penetration pegs. × 710. (Christou, 1962.)

formed simultaneously from the several lobes constituting this type of infection structure, or whether each lobe functions independently. This is probably not of great significance as the lobes most probably develop at a similar time and thus will probably also penetrate at a similar time, resulting in an effective multiple invasion.

Penetration through natural openings. — *Rhizoctonia solani* may penetrate into plants via stomata on the stems, cotyledons, and leaves. No appressorium is

formed over the stoma; the hypha simply grows through the stomatal opening into the substomatal cavity, decreasing in diameter if the size of the stoma requires it (Fig. 7). This hypha branches and fills the substomatal cavity, and then the branch hyphae invade the tissue beneath the stoma. Monteith (1926), Townsend (1934), Ullstrup (1936), and Dodman (in press) have reported penetration through stomata, usually on leaves, but also on cotyledons and stems. Ullstrup (1936) reports that some

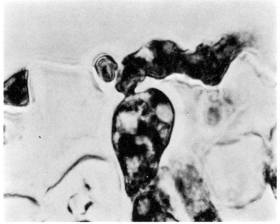

*Fig. 6a.* Development of side branches to form lobate appressoria. × 500. (*b*) Transverse section of lobate appressorium showing penetration peg. × 1,200.

isolates are able to penetrate only through stomata, and that on leaves with stomata on their lower surface only, penetration occurs solely through the lower side of the leaf.

Other natural openings penetrated by *R. solani* are the lenticels on potato tubers. Ramsey (1917) reported that certain isolates entered tubers via lenticels without the formation of any infection structures; penetration of the intact surfaces of tubers also occurred.

Penetration through wounds. — *Rhizoctonia solani* may penetrate into plants through wounds, which may be either associated with plant development or induced by injurious agents. In most cases, penetration through wounds is not the only way in which the fungus enters the plant.

Wester and Goth (1965) found that cracks in the seed coat of beans provided a ready means of penetration for *R. solani*, resulting in the destruction of the plumule and radicle. Another type of wound penetration has been reported by Nakayama (1940), Boosalis (1950), Kernkamp, et al. (1952), and Wyllie (1959), who showed that penetration into roots commonly occurs at the point of development of lateral roots, where the root cortex and epidermis are ruptured by the emerging root tip.

Ullstrup (1936) reported that isolates not able to penetrate leaves in other ways, were able to penetrate through wounds, whereas Khadga, et al. (1963) showed that isolates that usually penetrated cotton hypocotyls from infection cushions easily penetrated seedlings through wounds. On the other hand, Ramsey and Bailey (1929) and Schroeder and Provvidenti (1961) found that rotting of tomato fruits by *R. solani* developed only from penetration through surface wounds.

Wounds may also be induced by other agents, including materials produced by the fungus itself: penetration may occur through such wounds (Boosalis, 1950).

In penetrating through wounds, the fungal hyphae enter the plant without the formation of any infection structures or mats of hyphae; penetration of exposed cell walls probably occurs in the same way as the fungus penetrates cell walls within the host tissue.

*The mechanism of penetration.*—It is probable that the intact plant surface represents a physical barrier to fungal growth and therefore force is required to penetrate that barrier by mechanical means alone. However, stomata and wounds provide no physical barrier, and it may thus be expected that the mechanisms of penetration will differ for the different types of penetration.

To penetrate a physical barrier such as the intact plant surface, some method of exerting force is needed, and it appears that this may be obtained by the attachment of hyphae and infection structures to the plant surface. Furthermore, the force required for penetration is usually minimized by the formation of very fine infection pegs. The fungus is therefore well adapted for penetration by mechanical means alone. However, proof that penetration is entirely mechanical is not easily obtained. Flentje, Dodman, and Kerr (1963) found that penetration of inert collodion membranes occurred from infection cushions, suggesting that in this case, penetration was mechanical.

There appears to be very little additional direct evidence on this point, though there have been several reports that there have been no observable effects on the plant tissue beneath infection cushions prior to penetration. It has been suggested from these observations that the function of the cushion is mechanical rather than chemical. On the other hand, there are numerous reports of the toxic materials produced by this fungus (Bateman and Sher-

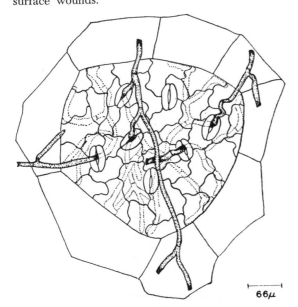

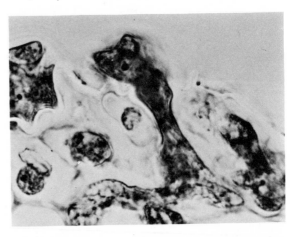

*Fig. 7a.* Penetration through stomata. (Ullstrup, 1936.) (*b*) Transverse section of bean hypocotyl showing slightly constricted hypha penetrating through stoma and then branching and penetrating cortical cells. × 1,200. (Dodman, in press.)

wood, this vol.); these materials could enter the intact host and cause considerable damage. These toxins have not been shown to be associated with infection structures, but this possibility should not be ignored. In addition, the possibility that the cuticle is degraded beneath infection structures should not be overlooked since *R. solani* produces cutinolytic enzymes (Linskens and Haage, 1963). The importance of pectic and cellulolytic enzymes in penetration also needs consideration; it is unlikely that these play a part in penetration of parts of the plant covered by a cuticle, which cannot be degraded by these enzymes, at least not until the cuticle is penetrated or destroyed. However, this may not apply to roots, where it appears that the cuticle is either lacking or different from that on stems and leaves.

Although it is clear that some types of infection structures are well suited for either mechanical or chemical penetration, or both, it is not clear how penetration of the intact plant surface is effected in the absence of clearly defined infection structures. It is possible that a hyphal mat or even an individual hypha firmly attached to the plant can exert the force required to penetrate the plant surface; this may be aided by the softening of the barrier by chemical action prior to penetration. It is also possible that barriers penetrated in this way provide less resistance than those where infection cushions are formed. The rare penetration of plants by individual hyphae, where the most common method of penetration is from cushions, may be due to localized areas where the cuticle and epidermis are weaker than the other areas.

Little is known about the mechanism of penetration through stomata and through wounds. In both cases, it seems that there is no physical barrier to penetration and that the fungus is able to enter the plant tissue without any obstruction. There appears to be no information on the influence of stomatal aperture on penetration.

It is not clear at present whether chemical action prior to penetration (Boosalis, 1950) represents a case of penetration through the intact plant surface aided by chemical action or an example of wound penetration. If the chemical action does not remove the physical barrier, it would be penetration of the intact plant, but if the physical barrier were destroyed, then it would be wound penetration. Thus, the type of chemical activity is important. Are the cells killed by toxins and the physical barrier unaltered, or are the cuticle and epidermis degraded and ruptured by enzyme action? Further investigations of these aspects are needed.

THE PHYSIOLOGY OF THE INITIATION OF INFECTION.—*Factors controlling penetration.*—It has already been pointed out that the characteristic reactions in the process of penetration are attachment to the host surface, growth along the lines of junction of epidermal cell walls, the formation of short, swollen, side branches, and, in some cases, the aggregation

of these branches to form infection cushions, from which penetration occurs. It is of considerable importance to understand what initiates each of these stages in the process of penetration.

At present, little is known of the factors controlling attachment and growth along the host surface. However, two hypotheses have been put forward to explain the branching and aggregation leading to infection-cushion formation. The first suggests that materials from the host initiate the formation of infection cushions, whereas the second proposes that infection-cushion formation is induced by contact with the plant surface.

The first hypothesis was put forward by Flentje (1957) following studies of the reactions of several isolates of *Rhizoctonia* on different plant species. He found that isolates of *R. solani* pathogenic to crucifers formed infection cushions on and penetrated the hypocotyls of cruciferous plants, whereas the same isolates failed to show these reactions on the hypocotyls of solanaceous hosts. Similarly, isolates pathogenic to solanaceous plants reacted on the hypocotyls of solanaceous plants, but not on the hypocotyls of crucifers. Flentje concluded from these observations that the hypothesis of a contact stimulus controlling cushion formation and penetration was inadequate and, as an alternative explanation, he suggested that a chemical stimulus might be involved.

Evidence supporting the hypothesis that exudates initiate infection-cushion formation was provided by Kerr and Flentje (1957) who found that exudates from radish seedling roots and stems induced a crucifer-attacking isolate to form typical infection cushions on pieces of washed epidermis from radish stems, whereas no such structures were formed when the exudate was replaced with water. The same isolate failed to react on pieces of epidermis removed from other plants or on inert surfaces such as glass, cellophane, cotton, and collodion (Flentje, 1957). On the other hand, though clumps of hyphae were formed when root and stem exudates were placed under cellophane and the crucifer-attacking isolate grown over the cellophane, these clumps were not identical with infection cushions. Thus, the evidence that exudates induced cushion development could not be regarded as conclusive since typical cushions were formed only when the cuticle and epidermis of a susceptible host was present.

Further investigations by Flentje, Dodman, and Kerr (1963) showed that structures morphologically identical to infection cushions were formed by crucifer-attacking isolates on collodion membranes covering the stems of radish seedlings and also covering pieces of radish stem and cotyledon tissue. Also, an identical reaction was obtained when the plant tissue was replaced with blocks of agar containing exudate collected specifically from either radish stems or cotyledons (Fig. 8a, b, c). Infection cushions were only formed on collodion where there was contact between the membrane and the tissue or agar containing exudate. No cushions were formed on collodion over agar that did not contain exudate (Fig. 8d). Further studies with a wide range of hosts and isolates have confirmed these results (Dodman, unpub.).

In addition to the evidence already mentioned, other workers have found indications that infection-cushion formation by *R. solani* is induced by host materials. Wyllie (1962) observed that structures resembling the early stages of cushion development were formed on the outside of cellophane tubing adjacent to enclosed seedling roots. More recently, Martinson (1965) found that infection cushions develop on the surface of a number of synthetic films contiguous to radish and bean seed or seedlings. Materials from seed and seedlings were able to diffuse through cellophane and nylon film and stimulate the development of cushions. Concentrated exudates from seeds and also some carbohydrates induced similar, though not identical, reactions. There are thus several reports that infection-cushion formation is initiated by host materials.

As previously mentioned, a second hypothesis has been put forward by De Silva and Wood (1964),

who suggested that the nature of the epidermis determines the development of infection cushions. They formed this hypothesis on the basis of experiments with an isolate of *R. praticola* that attacked both lettuce and cabbage seedlings severely, and with two isolates of *R. solani,* one that attacked cabbage seedlings only and another that attacked lettuce severely and cabbage less severely. Their main evidence for the hypothesis was that, when tested on washed epidermal strips from the different hosts, each isolate formed infection cushions on the epidermis from the host to which it was pathogenic and the addition of root exudates from the different hosts made no significant difference to cushion formation.

There are obvious differences between the results of De Silva and Wood and of Flentje, Dodman, and Kerr, and it is important to examine these differences to advance our understanding of the infection process. The differences involve three aspects of the work namely, the formation of infection cushions, the source of exudates, and specificity. It is apparent that

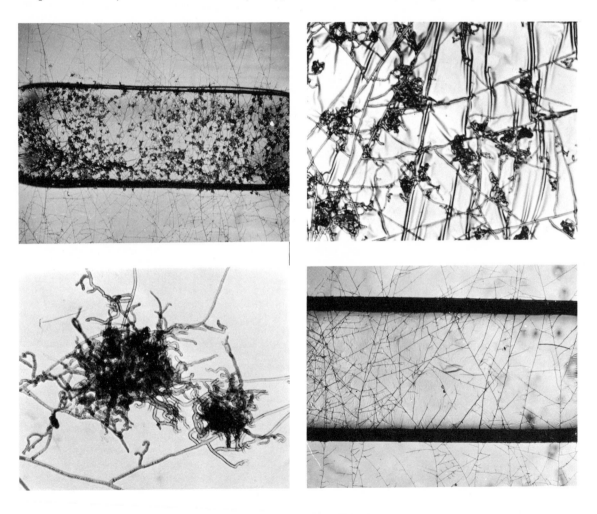

*Fig. 8a, b, c.* Infection cushions formed on collodion membranes over agar containing radish exudate (*a* × 8, *b* × 50, *c* × 90). *Fig. 8d.* Growth on collodion over distilled-water agar.

there is some confusion about the importance and interrelationship of these aspects and that it would be desirable to try to clarify the situation regarding cushion formation. Specificity will be dealt with later.

The hypothesis that exudates stimulate cushion formation was originally based on the specificity of various isolates to different hosts. This has led to an overemphasis of the importance of specificity, when, in fact, the hypothesis could have been developed without involving specificity. Similarly, the stimulation of growth by exudates and the relationship with specificity was not essential for the hypothesis. Thus, the important points to consider are the factors that can be obtained from the part of the plant where penetration normally occurs and the role that these factors play in inducing the specific morphological changes associated with the formation of infection structures.

Flentje, Dodman, and Kerr (1963) obtained exudates from radish hypocotyls and showed that these stimulate the formation of structures morphologically identical to infection cushions on both epidermal strips and collodion membranes. No such structures were formed without exudate. Although a suitable surface was required for cushion formation in these experiments, the surface alone did not induce any morphogenetic response. The only conclusion that can be reached from these results is that some material obtained from the host stem initiated the fungal response.

The main evidence supporting the hypothesis of De Silva and Wood is that different fungal isolates formed cushions on washed strips of host cuticle and epidermis without the addition of exudate. However, they point out that under these conditions the cushions are not identical with those formed on the intact host. In the absence of illustrations, it is difficult to assess the exact nature of the structures formed on the epidermal strips, since other structures formed by the fungus may show some superficial resemblance to cushions, especially in the early stages of development. The difficulty in assessment is further increased by the fact that De Silva and Wood recorded "infection cushions" on washed strips from nonsusceptible hosts, e.g., isolates S and L formed cushions on wheat epidermis. Thus, the specificity obtained in pathogenicity tests was not duplicated on epidermal strips.

In summary, then, there is strong evidence that exudates induce some isolates to form infection cushions whereas the evidence for a contact stimulus is not so convincing. It appears that further experiments are required with inert replicas of the stem surface or some other system that excludes chemical action before it can be concluded that the physical nature of the surface alone initiates infection-cushion formation. However, it must be stressed that a wider range of isolates should be studied before any generalizations can be made.

At the present, stimuli for penetration have only been investigated with isolates that form infection cushions. There is no evidence that there are factors controlling other types of direct penetration, as there is little information on aspects such as host specificity and reactions on inert surfaces. Even less is known about factors controlling penetration through stomata and wounds and it seems that this area should be investigated to determine whether the host exerts any influence on the pathogen in the prepenetration phase of infection.

*The relation between prepenetration reactions and specificity.*—Different hosts.—One of the outstanding features of *R. solani* is its ability to parasitize a wide range of different plant species. This has been adequately pointed out elsewhere in this symposium. Many workers have studied the host range of isolates of *R. solani* in attempts to characterize strains on the basis of pathogenicity. It is now accepted that some isolates have a limited host range, while others attack many different plants. It is of considerable importance to plant pathologists to understand why hosts are susceptible or resistant to isolates of *R. solani* or, conversely, why isolates attack some hosts but not others.

Most of the information on this aspect is concerned with isolates that penetrate directly through the intact plant surface by means of infection cushions. A considerable widening of our knowledge in this area is needed.

There are many stages in the process of infection where the ability of an isolate to cause disease may be influenced. Perhaps the earliest stage in the process of infection where the plant may influence the fungus is during the germination of sclerotia or resting hyphae in soil and the continued growth of mycelium through the soil and in the rhizosphere. These events may be operative through exudates released from seeds, roots, stems, or any part of the plant in the soil. The findings of Kerr (1956) with seedlings enclosed in cellophane bags suggest that exudates may selectively influence isolates of *R. solani* at this early stage of infection. Herzog (1961) has also suggested that stimulation of growth of *R. solani* by exudates from different plants increases with the susceptibility of the plants. The significance of this as far as disease is concerned was not discussed. However, Nour El Dein and Sharkas (1964b) have reported that the level of disease caused by *R. solani* in three tomato varieties is directly correlated with the stimulation of growth induced by exudates from the different varieties. Schroth and Cook (1964) have reported similar findings with bean varieties in relation to damping-off caused by *Pythium* and *Rhizoctonia*. On the other hand, Flentje, Dodman, and Kerr (1963) and De Silva and Wood (1964) have shown that there is no correlation between in vitro stimulation of growth by root exudates and pathogenic specificity of isolates of *R. solani*. De Silva and Wood concluded that their results refuted the hypothesis that exudates are an important factor in the specificity of parasitism of different isolates. This conclusion is unjustified since available evidence indicates that specificity is determined by the ability of an isolate to form infection cushions and since it is probable that stimulation of growth by root exudates does not bear any relation to cushion development. The relation between stimulation of growth by root exudates and specificity requires further investigation.

Flentje (1957) indicated several stages in the establishment of infection where successful infection may be halted. These are:

(1) failure of hyphae to attach to plant surface
(2) failure to form infection structures
(3) failure of infection pegs to penetrate
(4) failure of penetrating hyphae to continue invasion of tissue due to a hypersensitive reaction

It should also be pointed out that effective invasion may be prevented in other ways after penetration has occurred, e.g. resistance to fungal toxins and enzymes, thus localizing infection in a limited area. These postpenetration aspects are covered elsewhere and will not be discussed further here.

Considering the prepenetration phase of infection, there is an increasing amount of evidence that successful penetration is dependent on a series of important steps, which may be analogous in many ways to the pathways involved in biochemical metabolism. This idea is supported by the recent findings of Flentje, et al. (1967), who have been investigating natural and induced mutants from parent isolates of *T. cucumeris*. From one parent isolate, which forms dome-shaped infection cushions on and penetrates the stems of crucifer seedlings, they obtained a number of nonpathogenic mutants. Investigations have shown that among these mutants there are some that will not grow on the plant surface, some that fail to attach to the plant, some that attach but form no infection structures, and some that attach and form infection structures but fail to penetrate.

Although the reactions of these mutants indicate that there is a series of stages involved in penetration, there is still little evidence on the nature of the factors controlling the various stages. At present, the full significance of the failure of some mutants to grow on the plant surface has not been determined since this same phenomenon has not been observed in nature. However, the next stage in the sequence, i.e. attachment to the host, was considered by Flentje (1957) to be necessary for successful penetration. The factors that determine attachment are unknown, though it has been observed that attachment often does not occur on a resistant plant. It appears unlikely that exudates control attachment, since hyphae may become firmly attached to washed epidermal strips, and in some cases to collodion and cellophane membranes, and glass slides. The significance of this part of the process of penetration needs to be determined.

There is now good evidence that the ability of an isolate to establish infection is often determined at the stage of infection-cushion formation and penetration. On susceptible hosts, infection cushions are formed and penetration occurs, but this does not take place on resistant hosts. There is evidence (Flentje, Dodman, and Kerr, 1963) that exudate from radish seedlings stimulates infection-cushion formation by crucifer-attacking isolates, but further studies are required before it can be concluded that each host exudes a specific material which induces cushion formation by isolates attacking those hosts. Until the nature of the active material in radish exudate is determined, it is difficult to demonstrate conclusively that the same material is not exuded by other plants. It is possible that plants resistant to crucifer-attacking isolates exude the same active material as radish, but also exude other materials that suppress infection-cushion formation by crucifer isolates.

If chemical stimuli control the formation of infection cushions by isolates specific to certain hosts, it would appear that isolates that have a wide host range either do not respond to these stimuli at all or that they are able to react to the range of stimulating materials exuded by different plants. Alternatively, the same stimulating material could be exuded by all plants, and specific materials suppressing cushion formation exuded by different plants; isolates with a wide host range would be unaffected by the latter type of material.

De Silva and Wood (1964) suggested that the nature of the epidermis may determine specificity, though their evidence in favor of this hypothesis is limited. However, there is other evidence to suggest that the surface plays some part in cushion development and may therefore be involved in specificity. Flentje, Dodman, and Kerr (1963) observed differences in cushion formation on cellophane and collodion, though the source of exudate was similar in both cases. More recently, it has been found that cushions form in drops of exudate on slides, but only in contact with the glass. Finally, some isolates of *R. solani* have been found that form cushions on collodion membranes containing a trace of paraffin, but do not respond in the absence of paraffin (Dodman, unpub.). These results suggest that the specific type of surface may be important in cushion formation and that differences in the surfaces of different hosts may be involved in specificity.

The failure of infection pegs to penetrate after an infection cushion has been formed was observed by Flentje (1957) and is a possible means of host resistance. However, the mechanism of resistance in this case is not understood.

Little is known about the specificity of isolates that penetrate by other means, either directly or through natural openings and wounds. It is not known whether the host is able to prevent penetration by isolates that enter through stomata. These aspects offer considerable scope for investigation.

*Hosts of different ages.*—It is commonly found that plants become less susceptible or even resistant to attack by *R. solani* as they become older. This is especially true with damping-off, which is characteristically a seedling disease. There are a number of possible reasons why seedlings become less susceptible as they grow older.

Barker (1961) found that certain isolates attacking beans cause almost 100% preemergence damping-off if seed is sown in infested soil. However, if seedlings are transplanted into the same soil, infection is very slight and usually no seedlings are killed. Other workers (Schroth and Snyder, 1961; Schroth and Cook, 1964; Cook and Snyder, 1965) have shown that bean seeds exude large amounts of materials capable of stimulating fungal growth, whereas bean hypocotyls exude relatively little of this nutrient material. It is thus possible that materials liberated by seeds cause a marked stimulation of growth around the seed and emerging roots and shoots, leading to a high incidence of preemergence damping-off. On the other hand, once the cotyledons have emerged from the soil, there is little stimulation of growth by the small amount of material released from the hypocotyl, and, consequently, little infection.

Other explanations for changes in susceptibility have been put forward. Flentje, Dodman, and Kerr (1963) observed that the number of infection cushions formed on the hypocotyls of radish seedlings decreased with seedling age. They suggested that the availability of the material stimulating cushion formation declined as seedlings became older. This was supported by the finding that seedlings that had become resistant could be made susceptible again by rubbing them with absorbent tissue. It was suggested that this treatment removed some of the impermeable waxes deposited onto the cuticle surface, thereby allowing renewed exudation of the stimulatory materials. It is equally possible that these cuticular waxes change the nature of the cuticle surface and thus alter the reaction of isolates that respond to specific cuticular configurations, as suggested by De Silva and Wood (1964).

Finally, it should be stressed that resistance may develop after penetration; this is covered by Bateman (this vol.).

Different parts of plants.—There are a number of reports that some isolates of *R. solani* attack only certain parts of plants. Flentje (1957) showed that different isolates may attack stems and cotyledons, but not roots, and vice versa. He has also shown that one isolate will infect the petioles of cabbage, but not the stems. Recently, Dodman (unpub.) observed that a number of isolates, which cause severe preemergence damping-off, tend to form infection cushions at the top of the stem near the growing apex but not at the base of the stem. It appears that this may be due to the lack of exudation from the lower part of the stem of the materials that stimulate cushion formation.

The reasons for this type of specificity may be similar to those determining host specificity, in that isolates may respond to specific chemicals exuded by one part of the plant, but not another. In addition, differences in the cuticular surface, such as the difference between roots and stems, may be of importance. Again, postpenetration resistance must also be considered. Further studies of these aspects should be made.

DISCUSSION.—Although it has been stated earlier in this symposium that there is a large volume of literature on *R. solani* and that the amount is increasing rapidly, it is also very clear from a comprehensive study of this literature that we still know very little about this fungus and diseases caused by it. Our understanding of the early stages of infection and the factors controlling these stages is no exception; many of the areas where our knowledge is deficient have already been pointed out and will not be repeated here.

It is of interest to speculate why a fungus such as *R. solani*, which is an extremely good soil saprophyte, should have developed parasitic capabilities. It is also interesting to consider how these capabilities evolved. At present, it does not seem possible to rationalize the development of parasitic abilities by this fungus, but it should be one of the things we consider at times. It may be interesting to make comparisons of saprophytic and parasitic abilities to determine if there are any relationships between these aspects or any detectable trends in the development of isolates that are better adapted to parasitism than saprophytism.

It is not readily apparent why *R. solani* has developed complex infection structures when the majority of pathogens penetrating directly do so by means of simple appressoria. The complex structures appear to be similar in many ways to simple appressoria, in that they are the sites of mechanical or chemical action on the host surface. However, the large complex structures, the formation of numerous infection pegs, and the simultaneous penetration into several host cells may provide a more efficient means of penetration and overcoming host resistance than that given by a small appressorium with a single infection peg.

It is possible that other factors have also influenced the development of complex infection structures. Growth over the host surface and the formation of infection cushions may take a period of days, whereas the germination of spores and the formation of appressoria by other fungi may occur in a matter of hours. It is thus tempting to suggest that the development of complex infection structures is associated with a fungus that inhabits environments such as the surfaces of plants in and close to soil, and areas of moist, dense vegetation, where moisture is available for long periods.

The formation of complex infection structures may also be associated with fungi that do not readily produce spores. It is possible that multiple penetration from a single complex structure may achieve successful penetration, whereas fungi that sporulate freely achieve the same result by the action of a number of spores. On the other hand, the formation of complex infection structures may be an accident of morphogenesis and may have no evolutionary significance. In connection with this aspect, it would be of considerable importance to know what is the mechanism of penetration when infection occurs from basidiospores. Carpenter (1951) and Echandi (1965) have reported that basidiospores are the means of dispersal and infection in foliar diseases of *Hevea* and bean, but no observations on the mode of penetration were made. It is possible that, in these cases, basidiospores germinate to form germ tubes and appressoria and thus react as typical air-borne pathogens.

The hypothesis of a chemical stimulus controlling the formation of infection structures was first put forward for *R. solani*. Recently, reports that chemical stimuli are involved in the formation of infection structures by other fungi have been presented. Flentje (1959, 1965) speculated on the development of this interrelationship with *R. solani* and host plants. He suggested that the development of antheridia in *Achlya* and their attraction to oogonia may be analogous to the development of infection cushions by *R. solani* and the penetration of the host surface. Since it is known that the developments in *Achlya* are controlled by fungal hormones, it is possible that the same may be true for *R. solani* and that the response of the fungus to plant materials developed as a result of "mistaken identity." Such ideas offer scope for further investigations.

Since the development of infection structures is the result of marked morphogenesis, it is not unrealistic to think of fungal hormones being involved in the formation of these structures. Furthermore, some of the fungal cellular changes that occur are in many ways analogous to the reactions in plants induced by plant hormones. During cushion formation, the rate of nuclear division in the fungal side branches appears to be increased, while the rate of cell-wall formation and cell elongation is apparently suppressed. Further cytological investigations of this type, together with studies of cell physiology, should aid greatly in determining the nature of the materials inducing cushion formation. In addition, further studies of the active materials in exudates should enable the identification of the specific factors controlling infection-cushion development and penetration.

When this is achieved, and if it is found that these materials act as a type of fungal hormone, it may be possible to synthesize analogues of the active materials and interfere with the growth of the fungus in much the same way as analogues of plant hormones interfere with plant growth. If suitable analogues could be found and applied to plants, these might prevent the fungus from responding to the material that induces cushion formation, and thereby provide a means of controlling diseases caused by isolates of *R. solani* that penetrate from dome-shaped infection cushions.

This type of control would not be effective for isolates penetrating in other ways, but an examination of the factors involved in the early stages of the different types of penetration could lead to the development of control measures.

# Pathogenesis and Disease

D. F. BATEMAN—*Department of Plant Pathology, Cornell University, Ithaca, New York.*

The diseases of higher plants caused by *Rhizoctonia solani* Kühn are many and varied. This is not unexpected since this binomial has come to represent a vast array of fungal strains, having similar morphological features, and exhibiting a broad spectrum of physiological and biochemical characteristics. This property of variation of the fungus as well as of the Rhizoctonia diseases is well established and becomes obvious almost immediately to anyone who examines the literature relating to almost any aspect of the biology and pathology of this fungus.

A given strain of *R. solani* may exist in nature as a saprophyte, a symbiont, a parasite, and a plant pathogen. The adaptability of this organism to its environment and its ability to survive under a multitude of conditions make it one of the most striking examples of flexibility in the biological world. This ability to live and grow in such a variety of environments suggests that it must have an exceedingly broad and diverse genetic make-up and that it is able inductively to call into play a large variety of physiological and biochemical processes.

This paper is limited to a resumé of the available information on pathogenesis and disease processes associated with the *R. solani* diseases. The physiology and behavior of this organism in other environments are covered in the other papers of this symposium.

Information about the processes associated with pathogenesis and disease among the *R. solani* diseases is sparse, and our detailed knowledge about any of the processes involved is exceedingly limited. Since the results available have been obtained with diverse strains of the pathogen in association with a large number of host species, the present coverage is directed toward those diseases caused by *R. solani* that exhibit one or more of the following symptoms: damping-off, stem or hypocotyl cankers, root rot, storage-organ decay, and foliar blight. The mechanisms involved in production of each of the above symptoms may differ considerably in detail. On the other hand, there are perhaps mechanisms involved in each that are common to the entire group. Our state of knowledge about these diseases does not presently permit distinctions among them at this level. Thus, this general treatment of the entire group of diseases associated with *R. solani* is undertaken with an awareness of a possible broad range of differences that may exist among the various types of diseases with respect to pathogenesis, but it is hoped that an exposure of our ignorance about some of the fundamentals of this group of diseases will stimulate more basic research within this area.

PATHOLOGICAL HISTOLOGY.—Histological studies have been made on a number of *Rhizoctonia*-infected plant tissues, and considerable information is available on certain host-pathogen combinations. If one considers, however, the extensive host range of *R. solani* and variation existing among the Rhizoctonia diseases, it must be realized that more work of a histological nature is needed if we are to obtain a thorough understanding of the various relationships of this pathogen to its hosts at the cellular level. Knowledge of the pathological histology within the *R. solani* diseases is not only a requisite to understanding the physical host-parasite relationships, but it should be an essential prelude to physiological and biochemical studies of these diseases. The latter holds the promise of providing knowledge that can be used in formulating control measures based on an understanding of host-pathogen relationship at the molecular level.

*Penetration and ramification in the host.*—Factors associated with initiation and development of infection cushions and other structures and/or mechanisms involved with ingress of *R. solani* are thoroughly discussed by Dodman and Flentje (this vol.). Therefore it should be sufficient merely to reemphasize that *R. solani* most frequently enters its hosts by means of infection pegs that arise from the undersurfaces of the characteristic infection cushions, a phenomenon first described by Prillieux (1891). Penetration may also occur through natural openings such as stomates or lenticels (Ramsey, 1917; Townsend, 1934; Ullstrup, 1936; Shurtleff, 1953b; Valdez, 1955), or wounds (Boosalis, 1950), or by direct penetration by individual hyphae (Matsumoto, 1921). These latter mechanisms of entrance into host tissues appear to be rare, but any of these mechanisms may be common with certain *R. solani* strains (Ullstrup, 1936; Kerkamp, et al., 1952; Christou, 1962).

It has not been established whether direct penetration by infection pegs from infection cushions or individual hyphae is effected primarily by mechanical pressure or by enzymatic destruction of host constituents or a combination of both. Several investigators have favored the mechanical-pressure hypothesis of ingress (Chowdhury, 1946; Christou, 1962; Gonzales and Owen, 1963). Matsumoto (1921, 1923), on the other hand, has expressed the opinion that cell-wall degrading enzymes in addition to mechanical pressure exerted by penetrating hyphae are important for ingress. Adequate evidence to support either view is lacking, though *R. solani* is known to produce a variety of enzymes capable of destroying various components of the host. Recent discovery of an inducible cutinase produced by *R. solani*, as well as localization of enzyme production by specialized branched hyphae of this pathogen, strongly suggests that enzymatic mechanisms may be involved in ingress (Linskens and Haage, 1963; Isaac, 1964a). A thorough analysis of ingress associated with infection cushions and related structures from the biochemical point of view and in combination with histological studies of the process would make a valuable contribution to our knowledge.

There are many similarities as well as some distinct differences among the various strains of *R. solani* with respect to ramification within host tissues. This spectrum of differences is illustrated by the studies of Abdel-Salam (1933), who made a comparative histological examination of a lettuce strain and a tomato strain of *R. solani* on lettuce. The lettuce strain penetrated host tissues intercellularly, and intracellular penetration was rarely observed. Disorganization of host tissues was observed only in the areas occupied by the fungus, i.e. there was no marked **evidence of** damage in advance of hyphal growth. In contrast, the tomato strain of *R. solani* penetrated primarily in an intracellular manner, and infected tissues exhibited evidence of lethal action in advance of hyphal penetration in the form of collapsed, disorganized cells. It appears that the *R. solani* strain may be the primary determinant with respect to whether tissue is damaged in advance of penetrating hyphae. The majority of reported histological studies fall somewhere between these contrasting examples, i.e. tissues may be invaded both intercellularly and intracellularly by a given fungal strain and prepenetration injury to the host may vary considerably (Flentje, 1957; Wyllie, 1961; Christou, 1962; Van Etten, et al., 1967). Most *R. solani* infections apparently are characterized by both inter- and intracellular penetration of host tissues, with intercellular penetration often preceding intracellular development; also, where this type of development is noted, there is generally little marked evidence of host injury in advance of the pathogen (Townsend, 1934; Christou, 1962; Gonzalez and Owen, 1963; Khadga, et al., 1963; Chi and Childers, 1964). During ramification in the host, the penetrating hyphae may or may not exhibit a constriction at the point where host cell walls are penetrated (Bourn and Jenkins, 1928a; Townsend, 1934; Gonzalez and Owen, 1963; Van Etten, et al., 1967).

The studies of Christou (1962) on *R. solani* infections in bean hypocotyls have shown that entrance into the host may occur intercellularly, between the epidermal cells, or by direct penetration through the epidermal cell wall into the cell lumen. Once the fungus enters the cortical tissues, initial penetration proceeds intercellularly, followed by intracellular development of mycelium. Ramification of penetrating hyphae may be checked within one to four cell layers of the endodermis or it may proceed into the vascular tissues (Van Etten, et al., 1967). Young host plants are generally more susceptible than older plants to *R. solani* (Hedgcock, 1904; Sharma, 1960; Pitt, 1964a; Bateman and Lumsden, 1965). In cases where detailed studies have been made on cotton, alfalfa, and other host plants, *R. solani* has been found capable of invading vascular as well as cortical tissues (Drayton, 1915; Nakayama, 1940; Erwin, 1954; Wyllie, 1962; Khadga, et al., 1963; Chi and Childers, 1964).

*Rhizoctonia solani* may infect seeds of a number of plants at the time of or just prior to maturation (Baker [K.], 1947; Neergaard, 1958). This type of association serves as an important means of dissemination of this pathogen (Hedgcock, 1904; Crosier, 1936; Baker [K.], 1947; Leach and Pierpoint, 1958). Baker ([K.], 1947) has made a careful histological study of infected seed of pepper and of other plants. He observed that *R. solani* can penetrate pericarp tissue in contact with the soil and grow into the placenta. From the placenta, the fungus may extend into the funiculus; at this point, the mycelium may be stopped by the endosperm cuticle. If a crack appears over the radical apex, the radical and endosperm may be invaded. Many seeds dry before there is enough decay to prevent germination and in such cases developing seedlings become infected during and following germination.

*Alteration of host cells and tissue collapse.*—One of the first symptoms of cellular injury of *R. solani* is a definite browning of the host cell walls (Boosalis, 1950; Christou, 1962). This appears to be true whether injury precedes or accompanies hyphal penetration. *Rhizoctonia solani* infections generally develop rapidly, and the period between hyphal penetration of tissue and tissue collapse is transient in many instances (Khagda, et al., 1963; Chi and Childers, 1964). Cellular collapse of host cells is generally preceded by intracellular penetration by the fungus and is accompanied by a browning of the host protoplast. The pathogen may proliferate profusely, filling the host cells with densely packed hyphae (Christou, 1962; Chi and Childers, 1964). Apparently, with certain root-infecting strains, the bulk of the proliferating hyphae are destroyed with disappearance of the decaying roots, and only distributive type hyphae remain (Samuel and Garrett, 1932). With other strains of this fungus, the proliferating hyphae are transformed into sclerotial-like masses within the dead host tissues (Drayton, 1915; Townsend, 1934; Sharma, 1960; Chi and Childers, 1964; Christou, 1966). Several investigators who have worked with *R. solani* strains that induce host injury in advance of penetration have attributed the necrotic response of the host to diffusable fungal metabolites or toxins (Boosalis, 1950; Kernkamp, et al., 1952; Wyllie, 1962). Boosalis (1950) demonstrated that various *R. solani* isolates may cause varying amounts of injury to a given soybean variety; however, different soybean varieties did not show appreciable variation in susceptibility to a given fungal isolate. Wyllie (1962) showed that metabolic by-products of certain *R. solani* isolates are toxic to soybean roots and cause considerable damage in the absence of the pathogen. Other pathogenic isolates of this fungus produce very little or varying amounts of metabolites injurious to host tissues (Boosalis, 1950; Carpenter, 1951; Hawn, 1959; Gonzalez and Owen, 1963). Toxic metabolites, excluding enzymes, apparently are not requisite for pathogenesis (Gonzalez and Owen, 1963), though such substances may account for many of the symptoms observed in certain *R. solani* diseases.

Much of the damage to host tissues may be accounted for by the nature of the host's response, i.e. the host cells may produce toxic metabolites that are not only injurious to the invading pathogen but also cause death of the host tissue. Flentje (1957) has observed successful ingress of lettuce stems by several *R. solani* isolates that failed to result in progressive invasion due to a hypersensitive response of the host immediately following penetration. The invaded cells and their immediate neighbors were killed and exhibited a yellow-brown color. Since the pathogen could not be isolated from such infections, it was concluded that it had been killed as a result of the host response. The significance of the host response was demonstrated by subjecting the host to narcotics; these delayed or prevented the host reaction and permitted progressive invasion of lettuce tissues by the same isolates that were restricted prior to the treatment.

The phenomenon of cell and tissue collapse is characteristic of many Rhizoctonia diseases and is generally associated with extensive intracellular development by the pathogen. Although many factors may contribute to cell wall and tissue collapse, the enzymes that degrade the structural components of the host cell walls are probably of paramount importance in this respect (Bateman, 1964b, 1965, 1967). Host cell walls in bean hypocotyl lesions, particularly in the older portions of lesions, lose their birefringent properties. This indicates that the crystalline cellulose, the primary structural component of the host cell wall, has been destroyed by cellulase. Studies of the role of cellulolytic enzymes in pathogenesis have not indicated that these enzymes are essential to the early phases of pathogenesis by *R. solani*, though their activity may be associated with development of certain characteristic symptoms such as cellular collapse during later phases of pathogenesis. The concave nature of lesions or cankers on stems and hypocotyls of various hosts may be attributed to the death, collapse, and shrinkage of invaded host cells (Christou, 1962; Gonzalez and Owen, 1963; Bateman, 1964b).

*Sclerotium formation in infected tissues.*—Sclerotia generally are formed within the infected host following death of the invaded cells, and they are usually limited to the tissues that support the greatest fungal growth during pathogenesis. For example, in potato-

stem infections, the pathogen may be observed in the cortex, vascular bundles, and pith, but hyphal proliferation is greatest in the cortical tissues, and sclerotial-like masses are limited primarily to the cortical tissue (Drayton, 1915). The sclerotia originate from barrel-shaped cells that aggregate in cortical and sometimes in pith cells of the host plant; they generally are not found in stelar tissues (Townsend, 1934; Christou, 1962; Chi and Childers, 1964). The sclerotia produced in diseased or dead plant tissues exhibit little or no organization of mycelium into morphological regions of rind and medulla, and they appear to be similar to those produced by the fungus in vitro.

*Host response and recovery.*—Aside from the host responses mentioned above, which relate primarily to the susceptible reaction, there is relatively little information concerning resistance and host recovery from a histological viewpoint. It is perhaps significant that most authors who have examined *R. solani*-infected tissues histologically have not mentioned formation of phellogen layers or development of mechanical barriers by the host in advance of the invading pathogen (Drayton, 1915; Boosalis, 1950; Hawn, 1959; Christou, 1962; Wyllie, 1962; Khadga, et al., 1963; Chi and Childers, 1964; Van Etten, et al., 1967). Published photographs of tissue sections made through lesion areas of several host plants (Drayton, 1915; Boosalis, 1950; Christou, 1962; Chi and Childers, 1964; Van Etten, et al., 1967) do not reveal any marked changes in host cells within the boundary areas between lesions and apparently healthy tissues. Yet, in many instances the pathogen is limited to well-defined lesions, particularly in older plant tissues. The hypersensitive response described by Flentje (1957) for lettuce indicates that this host responded by producing a substance(s) lethal both to the fungus and the host cells and thus the host was resistant to progressive invasion. In cases where enlargement of stem and hypocotyl lesions are arrested and progressive invasion ceases, it seems plausible that chemical changes induced in the host may have restricted progressive invasion of uninfected tissues by the pathogen. Phytoalexin type of resistance mechanisms (Pierre and Bateman, 1967) have been investigated recently in the *R. solani* disease of bean. Further studies in this area with a number of hosts may furnish explanations for many of the unexplained phenomena relating to lesion size limitation and apparently induced resistance and/or host recovery.

Although the delimitation of many types of *R. solani* infections cannot be explained on the basis of the host's developing phellogens that give rise to layers of suberized cells in advance of the pathogen, others can. For example, the studies of Thatcher (1942) clearly indicate the protective role of phellogen or wound-periderm formation in restricting *R. solani* infection in potato tubers. He states that "ultimate progress of the fungus is restricted by extensive phellogen formation in the healthy tissue just beyond the margin of the necrotic zone. The phellogen and the suberized cells developing from it are not always effective as a barrier, in which case the lethal action of the fungus continues until another phellogen layer surrounds the secondary necrotic region." Suberization of a three- or four-cell layer in advance of *R. solani* has also been reported to have a protective function in another potato-tuber disease, "dry core" (Ramsey, 1917). In addition, cotton plants attacked by *R. solani* at high temperatures are apparently induced to produce suberized cell layers below the invaded tissue and these have been considered to function in limiting pathogen movement and aiding host recovery (Fahmy, 1931). Plants with only cortical infections generally recover, but those in which a progressive vascular invasion takes place are often killed outright. Between these extremes certain plants may survive but fail to recover completely (Neal, 1942; Stewart and Whitehead, 1955).

PATHOGENIC MECHANISMS AND PATHOGENESIS.—The processes and mechanisms associated with disease development are a function of both the host and the pathogen, and disease may be considered the sum of their interactions. The continuing injurious abnormal physiological activity resulting from the host-pathogen interaction represents anabolic and catabolic activities that, strictly speaking, may not be characteristic of either the host or the pathogen individually but of the host-pathogen complex. During the early phases of pathogenic attack, the pathogen may be regarded as the aggressive member of the complex and, initially at least, may be expected to elaborate certain "attacking mechanisms" that favor host invasion. If the host fails to respond or responds too slowly in a defensive manner, death results. Generally, the host's defensive mechanisms are operative and to a large degree they are effective in limiting the pathogen to a defined area or areas within the host, even in many cases where we consider the host response to be a susceptible one.

*Attacking mechanisms.*—The pathogenic mechanisms of *R. solani* are varied, complex, and, to a large degree, inadequately explored. Some of the more obvious characteristics that may be associated with disease production, such as the ability to produce cell-wall-degrading and tissue-macerating enzymes and the production of phytotoxic metabolites, have been investigated. Much of the data regarding these systems have been obtained by studying the pathogen in culture rather than examining the enzymes and metabolic by-products produced by the fungus in diseased plant tissues. Since *R. solani* is an extremely adaptable organism with broad genetic make-up, it seems reasonable to assume that enzymes and metabolites produced in one environment may differ considerably from those produced in another.

The differences in pectic enzyme production by *R. solani* in culture and in *R. solani*-infected bean tissues can be used to illustrate this point (Bateman, 1963a, 1967). The pectic enzymes that degraded the $\alpha$-1,4 bonds of pectic substances obtained from diseased plant tissues differed from those produced by the fungus in vitro with respect to stability in relation to pH and temperature, the amounts of reducing groups liberated in enzyme-substrate reaction mixtures, and the

number and quantity of reaction products when incubated with pectate. Although culture studies may not yield the most desirable information with respect to elucidation of pathogenic mechanisms, they do provide valuable data on the potential of the pathogen to utilize various substrates known to be present within host tissues. Caution must be exercised, however, in attributing a role in pathogenesis to enzymes and metabolites that have been studied only from in vitro sources when attempting to explain disease phenomena.

Enzymes.—The ability of phytopathogenic fungi to degrade cutin represents a relatively unexplored area of research. Few investigators have seriously considered this problem, though it would appear to be of great significance in the initiation of ingress. Matsumoto and Hirane (1933) noted that in no case was the cuticular layer of mature camphor leaves attacked by *Hypochnus sasakii* (= *Thanatephorus cucumeris*), but in the case of young soft leaves, penetration of the cuticular layer was readily accomplished by peg-like hyphae. Our thinking about enzymatic involvement in cuticular penetration has perhaps been dampened by the general acceptance of the hypothesis that contact stimuli are solely responsible for induction of appressorial-like or infection structures and that mechanical force is the primary mechanism of entrance of penetration pegs into host tissues (Brown and Harvey, 1927; Brown, et al., 1948). The work of Linskens and Haage (1963) strongly suggests that cutinase may be involved in cuticular penetrations by *R. solani*. The inducible cutinase produced by this fungus was shown to be more reactive with host cutin than with nonhost cutin. The whole problem area of the enzymatic ability of phytopathogenic fungi to utilize cutin and cutin-like substances is in need of more extensive investigation.

Pectic enzymes.—*Rhizoctonia solani* is capable of producing a variety of pectic enzymes when grown on pectic substances in culture. Examinations of diseased tissues have revealed evidence of pectic enzyme action within host tissues (Christou, 1962; Bateman, 1963a; Papavizas and Ayers, 1965). Barker and Walker (1962) were able to correlate the pectolytic activities in culture of a number of *R. solani* isolates with their pathogenicity on their respective host plants. Similar studies by others with several *R. solani* isolates as well as single-basidiospore cultures failed to show any correlation between the pectolytic activities in the culture filtrates of a given isolate or single-basidiospore culture and its pathogenic capabilities (Papaviz and Ayers, 1965; Perombelon and Hadley, 1965). Numerous investigators have demonstrated the pectin-degrading ability of *R. solani* in vitro (Ragheb and Fabian, 1955; Deshpande, 1960b; Deshpande, 1961; Barker and Walker, 1962; Bateman, 1963a; Gupta, 1963; Hadley and Perombelon, 1963; Nour El Dein and Sharkas, 1963). Studies of pectic enzymes in Rhizoctonia-infected tissues are considerably more limited and most of our information has been derived from Rhizoctonia-infected bean hypocotyl tissues

(Bateman, 1963*a*; 1964*a*; Bateman and Lumsden, 1965; Bateman and Rogowicz, 1965; Papavizas and Ayers, 1965). Pectin-degrading enzymes would appear to function primarily in aiding the pathogen to spread through host tissues, particularly where intercellular penetration is involved.

Water extracts of young or mature *R. solani* lesions consistently yield active preparations of a polygalacturonase and pectin methylesterase (Bateman, 1963*a*). More recent studies have demonstrated weak pectin *trans*-eliminase activities in extracts of infected plant tissues (Sherwood, 1966; Bateman, 1967). The polygalacturonase component associated with diseased tissues has a pH optimum for activity between pH 4.5-5, and this enzyme is more reactive with pectate than with pectin. *Rhizoctonia solani* is known to produce pectin methylesterase, and this enzyme is also known to be a component of most higher plant tissues. In the Rhizoctonia-disease of bean, the pectin methylesterase content of diseased tissues may be several fold higher than that of healthy tissue. Evidence indicates that a major portion of the increase in pectin methylesterase in the infected host is of plant rather than pathogen origin (Bateman, 1963*a*). Polygalacturonase and pectin methylesterase are considered to work together in destroying the highly methylated pectic substances in plant tissues. Studies with a highly purified polygalacturonase from diseased bean tissues indicate that this enzyme, in the absence of significant amounts of cellulase, pectin methylesterase, protease, and other enzymes, readily macerated plant tissues (Bateman, 1963*c*, unpub.). Byrde and Fielding (1965) have obtained evidence that an arabinase from *Sclerotinia fructigena* is capable of macerating plant tissues and that *R. solani* is capable of producing a similar enzyme. Recent studies with a pectin *trans*-eliminase from *Fusarium solani* f. *phaseoli* have demonstrated that this enzyme also macerates plant tissues (Bateman, 1966). The process of tissue maceration is considered to be due to destruction of protopectin. But the chemical nature of protopectin has not been elucidated, and it has been suggested that a number of enzymes may macerate or aid tissue maceration (such as polygalacturonase, pectin *trans*-eliminase, protease, and perhaps other enzymes [Naef-Roth, et al., 1961; Bateman, 1963*c*, 1966; Byrde and Fielding, 1965]), so it appears that protopectin may represent a complex of polymers that are chemically linked, and any enzyme that can attack a member of this complex and lower its molecular weight sufficiently, will macerate plant tissues. The protopectinase reported for *R. solani* most likely represents one or a complex of pectic enzymes (Deshpande, 1959*a*; Bateman, 1963*c*; Perombelon and Hadley, 1965), though it is still quite possible that other enzymes or factors may contribute to the process.

Cellulolytic enzymes.—Cellulose represents a high percentage of the dry matter of most higher plants and comprises the structural framework or the primary organized phase in the cell walls, whereas the other cell-wall constituents are often regarded as "encrusting substances" since they constitute the contin-uous matrix or amorphous fraction of the wall. The cellulose polymers within microfibrils are believed to pass through crystalline and amorphous areas (Battista, 1965). The crystalline fraction of the cellulose gives the cell wall its birefringent properties when viewed with polarized light. The portion of the cellulose polymers existing in the amorphous regions of the microfibrilar structure have been considered to be more susceptible to enzymatic attack because of spacial considerations related to enzyme diffusion and accessibility of the beta-1,4 bonds to the enzyme structure (Cowling, 1963). Many microorganisms are known to produce enzymes capable of hydrolyzing soluble or modified cellulose derivatives, but a much smaller number of organisms are known to utilize native cellulose. Only those organisms capable of utilizing native cellulose should be referred to as cellulolytic organisms (Reese, 1956). *Rhizoctonia solani* belongs to this latter group (Garrett, 1962; Bateman, 1964*b*, 1965).

The cellulolytic powers of microorganisms are often represented by a number of enzyme components. Reese (1956) considered that organisms capable of altering the crystalline structure of native cellulose must possess a cellulase that he designated as the $C_1$ enzyme, and that enzymes acting on soluble or modified cellulose derivatives possess enzymes termed Cx enzymes. The $C_1$ enzyme has not been isolated and characterized and it has been difficult to prove or disprove its existence; however, Selby, et al. (1963) have studied a cellulase in the filtrate of *Myrothecium verrucaria* that they have designated A-enzyme and that may be comparable to the $C_1$ enzyme. Many Cx enzyme components have been extensively studied from a number of organisms including *R. solani* (Reese and Gilligan, 1953; Norkrans, 1963).

The cellulolytic capabilities of *R. solani* are extensive and apparently contribute to its saprophytic ability in soil as well as to its associations with living hosts (Matsumoto, 1921; Kohlmeyer, 1956; Garrett, 1962; Daniels, 1963; Bateman, 1964*b*). Matsumoto and Hirane (1933) presented indirect evidence that cellulase is involved in pathogenesis by *H. sasakii* ( = *T. cucumeris*). Their studies revealed that less cellulose was present in camphor leaves invaded by this pathogen than in comparable healthy leaves. They also considered that cellulase was involved in cell-wall penetration by this fungus. More recent studies of *R. solani*-infected bean hypocotyl tissues have demonstrated that host cell walls lose their birefringent properties, particularly in the older portions of lesions, and that lesion extracts contain Cx as well as cellobiase activities (Bateman, 1964*b*, 1965). The alteration and destruction of cellulose within the host has been demonstrated in *Rhizoctonia*-infected tissues, but the essentiality of cellulase action to pathogenesis has not been established. Two reports are available showing that cellulase (Cx) does not macerate plant tissues and that this enzyme does not stimulate maceration by polygalacturonase (Bateman, 1963*c*; Spalding, 1963). This does not preclude in any way the possible significance of cellulase in intracellular penetration by *R. solani* or the likely involvement of this enzyme with cellular collapse in infected tissues.

Proteolytic enzymes.—Early culture studies by Matsumoto (1921) demonstrated that *R. solani* readily utilized proteins as a substrate for growth, providing ample evidence that this pathogen can produce proteolytic enzymes. In view of the mounting evidence that plant cell walls contain structural protein elements in addition to the polymeric carbohydrates (Ginzberg, 1961), it may be expected that the protein-degrading capacity of this pathogen might function in cell-wall degradation and the process of maceration of tissue in addition to a breakdown of the structural components of the host cell protoplast. Recent studies by Van Etten and Bateman (1965) revealed that the protein-degrading ability can be attributed to an inducible extra-cellular protease. The mycelium also contains an active intracellular protease. Studies of healthy and *R. solani*-infected bean hypocotyl tissues revealed that extracts of both the healthy and diseased tissue contained proteolytic activity, but in the diseased tissue, activity was much greater. The increase in protease in *R. solani*-infected tissues is limited to the tissue areas occupied by the pathogen or lesion and generally reaches a maximum level at an intermediate stage of lesion development. The role of protease in infected host tissues has not yet been established, but the potential significance of such an inducible enzyme system to pathogenesis would appear to be great.

Toxins.—A number of workers have observed discolorations and injury of host tissues in advance of penetration by certain *R. solani* isolates. The studies of Kerr (1956) and Wyllie (1962), in which host injury was demonstrated when *R. solani* was separated from the host by cellophane membranes that permitted fungal metabolites to pass through, has been considered as evidence that phytotoxic metabolic by-products of the fungus are involved in pathogenesis. Since the metabolites involved passed through cellophane, they were presumably nonenzymatic in nature. There are a number of reports demonstrating that nonenzymatic phytotoxic substances are produced in culture filtrates of *R. solani* (Newton and Mayers, 1935; Vasudeva and Sikka, 1941; Sherwood and Lindberg, 1962; Aoki, et al., 1963; Nishimura and Sabaki, 1963; Nour El Dein and Sharkas, 1964*a*), but the toxicity of a given filtrate may be dependent on the fungal isolate as well as the composition of the culture medium on which the fungus was grown (Sherwood and Lindberg, 1962; Wyllie, 1962).

Analytical studies on the nature of nonenzymatic phytotoxic products produced by *R. solani* have been limited primarily to those materials produced in culture filtrates. Sherwood and Lindberg (1962) examined two isolates of *R. solani* for phytotoxin production on two media, a defined mineral-salts–glucose medium and a cornmeal-sand medium. A phytotoxin was produced by only one of the isolates and only on the cornmeal–sand medium. The phytotoxic substance was later identified as or considered to be closely related to O-nitrophenyl-β-D-glucoside (Sherwood,

1965). This material inhibited germination of alfalfa seed, induced light brown, hydrotic, sunken lesions on pea root tissue, and prevented emergence of secondary roots from the affected area. It also induced symptoms on young cotton seedling hypocotyls resembling the cotton "soreshin" disease (Neal and Newsom, 1951).

Nishimura and Sabaki (1963) isolated several phytotoxic compounds from 40-day-old culture filtrates of *R. solani* grown on Richard's solution plus 0.5% peptone. Purification and characterization of the toxic substances in these filtrates revealed the presence of phenylacetic acid, m-hydroxyphenylacetic acid, p-hydroxyphenylacetic acid, succinic acid, lactic acid, and oxalic acid. Aoki, et al. (1963) also found that the phytotoxic metabolites of two *R. solani* isolates from potato, when grown on modified Richard's solution, were present in the acidic fraction of the filtrate. This fraction contained six phenols and 12 carboxylic acids. Five of these compounds were identified as phenylacetic acid, m-hydroxyphenylacetic acid, β-furoic acid, succinic acid, and lactic acid. Phenylacetic acid has also been identified in culture filtrates of *H. sasakii* ( = *T. cucumeris*) as well as in rice plants infected by this pathogen (Chen, 1958). It is perhaps significant that three groups of investigators have reported production of phenylacetic acid and related compounds in *R. solani*. Phenylacetic acid possesses growth-regulating properties and inhibits root development of sugar beets at a concentration of 0.05% and of rape and rice at 0.005%. This compound however, does not induce necrosis of plant tissue and thus it cannot be considered as the primary toxin in culture filtrates considered to induce a necrotic reaction (Aoki, et al., 1963).

Since only limited work has been done on the isolation of nonenzymatic phytotoxic metabolic by-products of *R. solani* from infected plant tissues, very little can be said with confidence about the nature of the materials responsible for the injurious response of the host plant which precedes hyphal penetration by certain *R. solani* isolates (Boosalis, 1950; Kerr, 1956; Flentje, 1957; Sherwood and Lindberg, 1962; Wyllie, 1962). It should be noted that Wyllie (1962) obtained the greatest phytotoxic response to soybean seedlings when the fungus was grown in a cornmeal sand medium surrounding cellophane bags containing seedlings grown in sand and that Sherwood and Lindberg (1962) obtained phytotoxin production by *R. solani* on a cornmeal-sand medium but not on a mineral-salts–glucose medium. This suggests that production of this phytotoxin is not a constitutive property of the fungal metabolism but rather may be more closely associated with the substrate on which the fungus was growing. The observation that secondary-root development may be inhibited by certain *R. solani* infections may be related to production of growth-regulating substances such as phenylacetic acid by the pathogen (Aoki, et al., 1963; Nishimura and Sabaki, 1963).

*Physiology of disease development.*—The physiological characteristics of the host-pathogen complex during disease development has received little attention. This is indeed unfortunate since the ultimate explanation of the *R. solani* diseases must reside in the elucidation of the physiological and biochemical interactions which give rise to the observed effects.

Respiratory pattern.—Studies on respiration of *R. solani*-infected tissues appear to be limited to recent studies on *Rhizoctonia*-infected bean hypocotyl tissues (Bateman, 1964a; Bateman and Daly, 1967), and the information obtained with this particular host-pathogen combination is quite limited. With the appearance of the first macroscopic symptoms associated with infection, oxygen consumption by infected bean tissues increases two- to threefold over noninfected tissues. Symptoms may appear within 14-36 hours after inoculation, depending on plant age and environmental conditions. This increased respiratory rate of infected host tissues may be maintained at least for a 9 day period, which covered all the phases of lesion development through maturation (Bateman, 1964a).

More detailed studies on the respiratory activity of *R. solani*-infected bean hypocotyl tissues exhibiting young and mature hypocotyl lesions have furnished some evidence suggesting a shift in the catabolic pathways of glucose utilization within the infected host. A comparison of the release of carbons 1 and 6 as $CO_2$ from specifically labeled $C^{14}$ glucose fed to diseased and healthy hypocotyl tissues revealed that the increased respiratory rates of the infected tissue were consistently associated with a lowering of the $C_6/C_1$ ratio of the diseased tissue as compared to healthy tissue throughout the different stages of lesion development. This response may indicate that in the diseased tissues, a greater proportion of the glucose utilized was oxidized via the hexose monophosphate shunt pathway, whereas in the healthy tissue, the Embden-Meyerhof pathway apparently played a more dominant role in glucose utilization (Bateman and Daly, 1967).

The low $C_6/C_1$ ratio was characteristic not only of the lesion areas of the infected host per se, but also of the tissues below lesions. This indicates that metabolic changes were induced in host tissues that contained little or no mycelium of the pathogen. Respiratory measurements of bean hypocotyls bearing young as well as mature lesions revealed that oxygen consumption was about two times greater than in healthy hypocotyls. Similar measurements of lesioned tissue showed that the increased oxygen consumption was two to three times greater than that in control tissue, whereas the respiratory rates for tissue immediately below lesions were 1.5-1.75 times greater than those for comparable healthy tissues. The greatest increase in respiration was associated with the lesion areas of the diseased tissues, and a considerable portion of this respiration undoubtedly was due to the pathogen. Experiments carried out with *R. solani* showed that the fungus has an extremely low $C_6/C_1$ ratio. For example, mycelium respiring at the rate of 902 $\mu l$ $O_2$/g fresh wt/hr at 30°C had a $C_6/C_1$ ratio of 0.09 after 45 minutes and 0.33 after 3.5 hours, respectively. Thus, the pathogen may be considered to play a part

in the lowering of the $C_6/C_1$ ratio of the invaded host as well as contribute to the total oxygen consumption, but this does not explain the increased res- in the lowering of the $C_6/C_1$ ratio of tissues below Rhizoctonia lesions. The best current explanation for the latter is that the fungal infection has induced changes in the metabolic activities of the host tissues (Bateman and Daly, 1967).

Changes in terminal oxidases.—Studies of the terminal oxidases present in *R. solani*-infected tissues also appear to be limited to recent studies of these enzymes in extracts of infected bean hypocotyl tissues in various stages of lesion development (Bateman and Maxwell, 1965; Maxwell and Bateman, 1967). Examination of phenol oxidase, peroxidase, catalase, cytochrome c oxidase, and ascorbic acid oxidase in extracts of *R. solani*-infected tissues revealed that there were no significant changes in the level of any of these enzymes, on a fresh weight basis, during the early phases of pathogenesis, i.e., before and during the period young lesions exhibit a water-soaked appearance and prior to brown coloration of the lesion surface and cellular distortion. The levels of peroxidase, phenol oxidase, catalase, and cytochrome c oxidase all increased markedly while the lesions acquired a brown surface coloration and the cortical cells began to collapse and exhibit marked distortions. The levels of these enzymes remained high within the lesion areas after maturation of developing lesions, i.e., after the lesions had acquired a sunken condition and exhibited the characteristic brick-red or dark-brown coloration of the dried collapsed cortical cells. Ascorbic acid oxidase did not vary appreciably through the course of disease development. Measurements of oxidative enzyme activities at three different stages of disease development within lesions, 0-2 mm and 1-2 cm from lesions revealed that where an increase in oxidative enzymes was detected, the increased activities over that of healthy tissues was always limited to the lesion area per se. There were no apparent differences between the activities of the various oxidases examined in healthy tissues and tissues 0-2 mm or 1-2 cm from lesions during any stage of lesion development, except perhaps in the case of peroxidase.

Since the increased respiratory rate of infected tissue occurs during the early stages of disease development in both the lesion area and in host tissue beyond the lesion itself (Bateman and Daly, 1967), it would appear that it is not necessary to postulate that an increase in terminal oxidases during pathogenesis is needed to account for the increased respiratory rates of the invaded host. The increased levels of phenol oxidase, peroxidase, cytochrome c oxidase, and catalase in diseased tissues appear to be associated with the collapse and destruction of the invaded host tissue and apparently play little or no role in the early phases of pathogenesis.

*Accumulation of solutes around infection loci and translocation.*—Diseased plant tissues characteristically show an increased respiratory rate, and in certain instances, a mobilization of organic metabolites as well as inorganic materials to the infection sites has been demonstrated (Yarwood and Jacobson, 1955; Shaw and Samborski, 1956; Bateman, 1964a). This phenomenon is well documented for diseases caused by obligate parasites, but fewer studies are available that demonstrate a similar phenomenon in diseases caused by facultative pathogens. Accumulation in and around infected areas has been considered to be an active process driven by the increased metabolic activity in diseased tissues (Shaw, et al., 1954). Studies with R. solani on bean have demonstrated the accumulation of calcium around developing hypocotyl lesions (Bateman, 1964a). Since the accumulation of metabolites around infection sites does not appear to be selective (Yarwood and Jacobson, 1955), there is no reason to believe that metabolites and solutes other than calcium do not accumulate around R. solani infections in plants, provided the host is not rapidly killed.

The foliage of cotton plants exhibiting severe root-rot symptoms associated with infection by R. solani and R. bataticola has a higher temperature than normal plants during daylight hours. The difference between healthy and diseased plant foliage was usually less than 1°C at 6:30 A.M. whereas by noon, the difference might be more than 2°C. These temperature differences were attributed to less water uptake by the diseased roots and thus less transpiration and dissipation of heat by the diseased plants (Vasudeva, 1944). Severe attack of potato stems by R. solani has often been associated with the formation of aerial tubers by the host. This phenomenon has been attributed to interference with translocation in plants with severe stem cankers (Güssow, 1917), but no convincing experimental data has been put forward to support this contention. In general, there appears to be a vast void in our knowledge of the physiological characteristics of R. solani-infected plants.

RESISTANCE AND INDUCTION OF RESISTANCE.—The genetics of pathogenicity and host resistance may range from a rather simple "gene to gene" relationship to an exceedingly complicated situation where variation in selected pathogenicity and in host resistance would both appear to be multigenic (Walker, 1963). At one extreme, may be the rather simple gene to gene relationship between pathogenicity and resistance described by Flor (1955) for flax rust, and at the other, the apparent multigenic and unresolved genetic complexities associated with pathogenicity and resistance as found in most Rhizoctonia diseases. Attempts to develop plants resistant to R. solani for use on an agronomic basis have generally been unsuccessful. Reported results of tests in which a large number of varieties of a given crop plant have been screened for resistance to R. solani have revealed little evidence of resistance (Luthra and Vasudeva, 1941; Kendrick, et al., 1955; Luttrell, 1962), and where resistance has been found, little or no information is available on its possible mode of inheritance, except for the inheritance of resistance to bottom rot in cabbage (Williams and Walker, 1966).

Although man has not been very successful in increasing the genetic resistance of plants to R. solani, many hosts are known to exhibit resistance to the Rhizoctonia diseases with increasing age (Shaw, 1912; Roth and Riker, 1943c; Carrera, 1951; Hassan, 1956; Bianchini, 1958; Bateman and Lumsden, 1965). Furthermore, in many instances R. solani attack of a host results in a defined and discrete lesion that fails to enlarge with time even though the pathogen is present. The nature of such resistance, which develops in infected or healthy bean hypocotyls with increasing age, has been directly related to the calcium content of the host tissues and inversely related to the methoxyl content of the pectic substances (Bateman and Lumsden, 1965). Since the polygalacturonase associated with R. solani-infected bean hypocotyls does not hydrolize calcium pectate, and older hypocotyl tissue as well as tissues adjacent to developed lesions are more difficult to macerate enzymatically with enzymes from infected tissues, it appears that the susceptibility of the host may be regulated in part by its calcium pectate content. Studies with other crops have also suggested a beneficial role of calcium in relation to resistance to R. solani (Kernkamp, et al., 1952; Castano and Kernkamp, 1956; Zyngas, 1963). Shepard and Wood (1963) revealed that the mineral nutrition of young lettuce and cauliflower seedling, illumination, and temperature had little effect on pathogenicity by R. solani, but that resistance developed with host age and was dependent on adequate nutrition and active growth of the seedlings. In the early stages of seedling development, pathogenesis depended mainly on the intrinsic properties of the host and fungal isolate, and other factors were rarely more than of secondary importance. The calcium and magnesium nutrition of the young seedlings apparently did not influence virulence or pathogenicity of R. solani in these studies.

The methoxyl content of the pectic substances of young bean seedlings is high during the period of maximum susceptibility to R. solani. The pathogen produces both polygalacturonase and pectin methylesterase, and the latter enzyme is also native to the host. Based on the knowledge of certain physiological characteristics of R. solani-infested bean tissues and the behavior of host pectin methylesterase, Bateman (1964a) proposed the following hypothesis to account for the resistance of host tissues around developed lesions to maceration by polygalacturonase. "Respiration is greatly increased in and around the point of infection by R. solani on the bean hypocotyl. Associated with this respiratory rate is the accumulation of calcium and perhaps other cations and solutes. The increased concentration of cations release and activate the pectin methylesterase associated with host cell wall material and the pectins in the affected zone are demethylated. The demethylated pectic materials form insoluble salts with calcium and perhaps other multivalent ions and are rendered resistant to hydrolysis by polygalacturonase." The conversion of pectins to calcium pectate around the developing lesions on young hypocotyls would correspond to the natural conversion of pectin to pectates in the maturing hypocotyls associated with host resistance (Bateman and Lumsden, 1965). Calcium pectate has also been considered to function in rendering plants resistant to other fungal pathogens (Wallace, et al., 1962; Thomas and Orellana, 1964). The possibility exists that if a way can be found to rapidly demethylate the pectic materials in young seedlings and, at the same time, furnish the seedling with adequate calcium, that the period of susceptibility of certain seedlings might be reduced from 2-3 weeks to a shorter period.

The natural formation or the induction of calcium pectate formation in plant tissues must be considered as a nonspecific resistance mechanism. Such a mechanism cannot explain the selective pathogenicity observed among the various strains of *R. solani* on different host species. It is thus probable that calcium pectate is only one of a complex of factors involved in host resistance to the *R. solani* diseases.

*Rhizoctonia solani* has been shown to be sensitive to alkaloids and orchinol (Greathouse and Rigler, 1940; Gäumann and Kern, 1959). Pathogenic strains of *R. solani* isolated from wheat straw, cauliflower, and tomato have been shown to exist in a symbiotic relationship with *Orchis purpurella* (Downie, 1957). The roots of orchids are often found with mycorhizal associations whereas the bulbs seldom show this phenomenon. The studies of Gäumann and Kern (1959) with *O. militaris* L. demonstrated that a chemical resistance is induced in bulb tissue on infection by certain microorganisms. Among the antimicrobial materials produced by the bulb are p-hydroxybenzyl-alcohol and orchinol. The root and stem tissues of *O. militaris* are also capable of producing orchinol on infection, but the concentration of this substance in the root may be only 1% of that reached in bulb tissues. Thus, root tissues are not completely protected.

Several plants are known to produce orchinol on infection, but some orchid species such as *O. ustulata* and *Loroglossum hircium* are unable to produce this substance. The bulb tissue from the latter plants is inhibitory to *R. repens*, however. This indicates that some chemical mechanism other than orchinol is operative in these plants (Gäumann and Kern, 1959). Neither the induction or activity spectrum of orchinol is considered to be specific toward a given microorganism. For example, a number of fungi are able to induce orchinol in a given host, but other fungi are apparently unable to do this. On the other hand, the action spectrum or orchinol may be greater than the number of organisms inducing its production; and yet still other microorganisms do not respond to it. Such a response on the part of the host plant with respect to production of antifungal metabolites and/or a differential response to such compounds by the attacking microorganism could account for the selective pathogenicity of certain pathogenic strains of a fungus.

Recently, Pierre and Bateman (1967) demonstrated the induction of two phytoalexins, phaseollin and a fungistatic substance designated Substance II, in bean in response to *R. solani* infection. These antifungal materials reached levels inhibitory to *R. solani*

soon after infection. Much more work is needed on the Rhizoctonia diseases from the point of view of the phytoalexin hypothesis of disease resistance. Further studies in this area are likely to be revealing with respect to selective pathogenicity of *R. solani* strains as well as explain the nature of certain induced resistance mechanisms operative in hosts considered to exhibit a susceptible response during the early phases of pathogenesis.

Cotton plants treated with gibberellic acid are apparently more susceptible to attack by *R. solani* (Spooner, et al., 1959). A similar response has been demonstrated with gibberellin-treated bean seedlings. (Petersen, et al., 1963). But tests with gibberelic-acid-treated beans and three *R. solani* isolates revealed that only one of the three isolates was more pathogenic on the treated seedlings and that the isolate exhibiting the greatest pathogenicity on the treated seedlings was the only one of the three that responded with increased growth when treated with gibberelic acid in culture. Although this problem has not been resolved, it appears that gibberelic acid may affect the pathogen directly and enhance its pathogenic attack.

Host exudates represent another group of physiological factors that may influence the susceptible or resistant response of higher plants to *R. solani*. There is some evidence that the virulence of *R. solani* can be influenced to a degree by culture subtrates (Sims, 1960). If this effect is widespread among *R. solani* isolates, the nature and quantity of host exudates could conceivably be an important factor in determining host susceptibility. The studies of Kerr and Flentje (1957) and Flentje, Dodman, and Kerr (1963) indicated that a specific chemical stimulus from the host was involved in infection-cushion induction by *R. solani*, whereas the work of De Silva and Wood (1964) failed to support this contention. Recent studies by Martinson (1965) conclusively demonstrated that plant exudates can induce infection-cushion formation by *R. solani*. The unresolved points of contention appear to be concerned with the specificity of the response by a fungus strain to a given host exudate. Studies with exudates from three tomato varieties showing different degrees of susceptibility to *R. solani* have been reported. Stimulation of *R. solani* growth by the exudates was correlated with the degree of disease severity incited by the pathogen on a given plant variety, i.e. the more stimulatory the exudate from a given variety of tomato, the more pathogenic the fungus on that host variety (Nour El Dein and Sharkas, 1964*b*). Preemergence damping-off of bean by *R. solani* has also been correlated with the amount of host exudate released, but in these studies the quantity of exudate appeared to be more important than the quality; the exudate was as stimulating to other fungi as it was to *R. solani* (Schroth and Cook, 1964). In still another series of studies, substances excreted from potato and tomato were more stimulatory to *R. solani* than were those from *Beta forus* and legumes or the Gramineae (Herzog, 1961). The stimulation of the fungus by plant exudates appears to be independent of its parasitic activity in many instances, i.e. nonhost exudates may be stimulatory to *R. solani*. On the other hand, the extent to which *R. solani* is stimulated by host exudates is greater with the more susceptible plant species.

RELATION OF OTHER PATHOGENIC AGENTS TO PATHOGENESIS BY RHIZOCTONIA.—*Rhizoctonia solani* is often found as a member of a pathogen complex. There are several reports of its occurrence in association with plant parasitic nematodes. *Rhizoctonia solani* and *Meloidogyne* sp. interact synergistically in relation to pathogenesis on cotton (Reynolds and Hanson, 1957; White, 1962; Brodie, 1963). Mechanical injury of the host does not materially alter fungal infection. The presence of the nematode extends the period of susceptibility of young seedling to the fungus. It seems likely that some factor other than the mechanical injury caused by the nematode is necessary to account for the synergistic response between these pathogens (Brodie, 1963). *Rhizoctonia solani* also interacts synergistically with the potato eel worm (*Heterodera rostochiensis*) in injury to both potato and tomato (Grainger and Clark, 1963; Dunn and Hughes, 1964). *Pratylenchus* sp. has been associated with *R. solani* in a root rot of wheat, but, in this instance, the destructive effects of the pathogens appear to be additive rather than synergistic (Benedict and Mountain, 1954).

*Rhizoctonia solani* commonly may be associated with other pathogenic fungi in certain root diseases (Hawn and Cormack, 1952; Bateman, 1963*b*). It is known to act synergistically with *F. solani* Snyder and Hansen in the rotting of potato tubers (Elarosi, 1958). Here, the synergistic effect was attributed to the interaction of the pectic enzyme systems of the two fungi, which resulted in a more rapid breakdown of host tissues. Virus-infected cucumber seedlings are more susceptible to Rhizoctonia damping-off, and thrips injury to cotton is known to increase the susceptibility of this crop to *R. solani*. In addition to the simultaneous interaction of *R. solani* with other pathogens in pathogenesis, *R. solani* infections may serve as portals of entry or weaken the host to such an extent that secondary organisms or pathogens may invade the host (Kharitinova, 1958*a*; Verhoeff, 1963). The physiological and biochemical explanations for the synergistic action of *R. solani* with other pathogens in pathogenesis in most instances are obscure. Almost any factor that reduces the vigor of the host or its ability to respond rapidly to *R. solani* attack—whether it be unfavorable environmental conditions, inadequate land preparation or planting methods, insect damage, removal of cotyledonary tissue, treatment with narcotics, or attack by other pathogenic agents—will increase the susceptibility of plants to *R. solani* (Neal and Newson, 1951; Flentje, 1957; Reynolds and Hanson, 1957; Bateman, 1961, Baker [R.] and Martinson, this vol.).

SUMMARY AND CONCLUSIONS.—Pathogenesis by *R. solani* is a complex phenomenon. The heterogeneity existing among strains of this pathogen as well as the diversity of its host range add further complications to this general synopsis on pathogenesis and disease. The damage caused by *R. solani* strains may appear superficially to be similar but, from the histological point of view, a broad spectrum of relationships may exist between *R. solani* and its hosts, ranging from an intercellular association to an almost purely intracellular relationship. The host-pathogen relationship in most instances involves tissue disintegration that may or may not be associated with prepenetration injury attributed to a diffusion of toxic fungal metabolites or to the production of injurious substances by the host in response to infection. All tissue systems of the host are subject to invasion by *R. solani*, but juvenile hosts are more subject to complete invasion whereas in older host plants, pathogenic attack is generally limited to cortical tissues.

The collapse of invaded host cells and tissues has been associated with an enzymatic alteration of the primary structural constituent of the host cell wall, cellulose. The pectin-degrading ability of *R. solani* is considered to function in aiding the intercellular movement of this pathogen through host tissues and, in young tissues deficient in calcium, the pectin methylesterase of the host may function synergistically with the fungal polygalacturonase in the degradation of highly methylated pectic substances. In older plant tissues or around developing lesions, this host enzyme may function in the conversion of the pectins to pectates, and, in the presence of multivalent ions such as calcium, render these materials resistant to enzymatic hydrolysis. The nature of the host response to infection by *R. solani* is poorly understood; in many instances, the pathogen is limited to a defined area of the invaded host without the appearance of morphological barriers within host tissues, whereas in other instances, phellogen layers are laid down in advance of the invading fungal hyphae.

Resistance to *R. solani* is often associated with maturation of host tissues. Resistance, or the ability of a host to limit spread of the pathogen once infection has occurred, appears to be associated with increased respiratory rates of host tissue. Factors that tend to decrease host vigor, whether they are environmental, nutritional, chemical, or biological, tend to destroy host resistance to pathogenic action by *R. solani*. Induced changes in the nature of the pectic substances, that render them more resistant to enzymatic hydrolysis in tissue surrounding *R. solani* infections have been associated with limitation of lesion size, but it is likely other biochemical factors are also involved. It appears that resistance to *R. solani* will not be explained on the basis of any one factor; it is more likely associated with a complex of factors closely linked to the metabolic activities of the host after infection.

The role of toxins (i.e. fungal metabolic by-products that are nonenzymatic and are injurious to higher plants) in pathogenesis by *R. solani* has not been firmly established. Although culture filtrates of certain pathogenic isolates may induce injury of host tissues that resembles injury associated with pathogenic invasion, the filtrates from other pathogenic isolates do not induce any marked evidence of host injury. There is no question about the ability of certain *R. solani* strains to produce phytotoxic substances other than enzymes, but this ability in certain instances appears to be dependent on the substrate on which the fungus is grown, rather than being a stable character of the pathogen's metabolism. The behavior of *R. solani* within host tissues where it is subject to the modifying influences of host metabolism is probably quite different from what it is in pure culture, where it is not subject to interaction with another living system. The need for studying the possible production of toxins as well as enzymes and their action in diseased plant tissues is obvious, and this point cannot be overstressed. Although fungal "toxins" or "protoplasmic poisons" of a nonenzymatic nature are probably involved in certain Rhizoctonia diseases, the available experimental evidence is limited and to a large degree circumstantial.

Future studies concerning pathogenesis and disease must emphasize an analysis of the host-pathogen complex if they are to contribute significantly to our understanding of disease phenomena and the mechanisms governing the susceptible and resistant host responses. Such studies may fall logically into three major areas of concern: (1) "attacking mechanisms" elaborated by the pathogen when in association with the host, (2) the possible role and significance within the host of preformed factors relating to resistance and susceptibility, and (3) induced changes within the host resulting from the host-pathogen interaction and relating to resistance and susceptibility.

The enzyme systems associated with pathogenesis by *R. solani* thus far appear to be no different from the systems elaborated by numerous saprophytic microorganisms. This does not diminish the role or significance of such enzymes in tissue destruction during pathogenesis, but it does indicate that some factor or factors beyond the ability of *R. solani* to produce tissue-degrading systems are essential to its pathogenic capabilities, even though the latter may not be permitted to function without the action of the former and vice versa. The possible significance of host systems in tissue destruction has not received adequate consideration in previous work. Most of the accumulated information regarding tissue-destroying enzymes in the *R. solani* diseases relate to pectic and cellulolytic enzyme activities within the diseased host. The potential destructive effects of proteolytic systems, as well as other enzymes capable of destroying polymeric carbohydrate fractions other than cellulose and pectic substances associated with the host cell wall, are worthy of investigation. Also, there is almost a complete lack of information regarding the effects of degradative enzymes on the membrane and protoplasmic systems of the host during pathogenesis.

The reasons for the selective pathogenicity of *R. solani* on different plant species or on different parts of a host plant are obscure, though such pathogenic variation among pathogenic isolates is well established. The answers to many of the questions raised here may not deal so much with the genetic potential of an isolate to elaborate certain enzyme systems or toxic metabolic by-products, as with the controlling influence of host metabolism or constituents on induction and possible activation (or inactivation) of "attacking mechanisms" elaborated by the pathogen when in association with the host. Research in the areas concerned with compatibility factors between unlike protoplasts or protoplasmic systems may hold considerable promise for explaining selective pathogenicity by *R. solani*.

It seems reasonable that an incompatible combination with respect to pathogenesis by *R. solani* would likely be a lack of any adverse protoplasmic interaction on the part of either member of the complex, or a detrimental interaction on the part of the plasmic system of *R. solani*. Other combinations of interactions in which the plasmic system of the fungus would not be destroyed should result in pathogenesis. In any event, the factors relating to the resistant response at this level imply a preexisting phenomenon or a rapidly evoked response. Such a resistant response at this level of interaction could be circumvented by the pathogen through the production or release of preformed toxins that kill or greatly injure host cells in advance of the attacking agent. Certain *R. solani* strains supposedly act in this manner and more information is needed on the role of "toxins" in pathogenesis by *R. solani* in order to determine whether certain *R. solani* strains behave primarily in a necrotrophic or parasitic manner during pathogenic attack.

[The fungal isolates studied by Sherwood and Lindberg (1962) that produced toxin and were previously identified as *R. solani* have been recently identified as *Ceratobasidium* spp. (Sherwood, personal communication).]

The factors governing induced mechanisms of resistance appear to relate to host vigor and the metabolic response of the host to infection. A complex of host reactions may be involved directly and indirectly linked to host metabolism, such as induced changes in the structural constituents of the host and the production of phytoalexin-like substances, etc. The induced mechanisms of tissue resistance are likely to function regardless of the manner of pathogenic attack, providing conditions are favorable for host response and the host plant possesses an adequate genetic potential.

Although data relating to pathogenesis by *R. solani* are limited, it appears unlikely that the selective pathogenicity of *R. solani* strains will be explained on the basis of tissue-degrading capability or potential of a strain to produce enzymes that degrade the structural constituents of a host. It seems more reasonable to look for an explanation of selective pathogenicity at a more fundamental level, namely, in the area of protoplasmic compatibility of unlike plasmic systems and induced changes within the host resulting from infection and the differential response of *R. solani* strains to these induced changes following ingress.

ACKNOWLEDGMENTS. — Portions of this work were supported by Public Health Service Grant No. AI 04930 from the Institute of Allergy and Infectious Diseases.

# Epidemiology of Diseases Caused by Rhizoctonia Solani

RALPH BAKER and C. A. MARTINSON—*Department of Botany and Plant Pathology, Colorado State University, Fort Collins, Colorado, and Department of Plant Pathology, Cornell University, Ithaca, New York.*

Epidemiology deals with factors affecting disease severity and disease incidence. In the discussion following the papers on soil inoculum at the 1963 Berkeley symposium (Baker [R.], 1965; Dimond and Horsfall, 1965), it was suggested that disease severity be expressed as the resultant of inoculum potential *sensu* Garrett (1956) and disease potential *sensu* Grainger (1956; see also Fig. 1). In this discussion of epidemiology of Rhizoctonia diseases, we shall attempt to follow this scheme.

Factors influencing energy available for colonization of a host at the infection court will be treated for the most part under inoculum potential. The characteristics of the inoculum itself are important and involve distribution and dissemination, survival, and the inherent ability of the organism to incite disease. Inoculum potential is thus specifically a function of inoculum density as modified by the influence of environmental or capacity factors.

As used here, disease potential involves the host and its inherent susceptibility to disease. Susceptibility varies during the life cycle of the host and markedly influences the development of diseases caused by *R. solani*. Different organs and tissues of a single plant may express at one time extremes in susceptibility and resistance. Again, capacity factors may modify disease potential by increasing susceptibility through predisposition or by decreasing susceptibility through acquired resistance.

The influence of environment on inoculum potential and disease potential will be treated in a separate section since it is difficult, in some cases, to separate those factors affecting inoculum and those affecting the host.

Plant pathologists have had to establish arbitrary values to assess disease severity. Often these values or disease measurements were assigned numerical values to permit a mathematical analysis of the information. The problem in epidemiology stems from the nonlinear relationship of the values assigned to classes of disease severity. The most profitable experiments on the epidemiology of Rhizoctonia diseases have resulted from disease-incidence measurements using the damping-off phase of the disease. However, these experiments were of a relatively short duration and fewer variables were interacting during the experiments.

A common theme throughout all papers in this symposium has been the great variability of morphology, physiology, pathogenicity, and virulence in *R. solani*. Because of this variation, we have hesitated to catalog facts from the literature. Instead, we have tried to interpret the literature in light of current concepts. Epidemiology is a dynamic study with more variables in operation than one can hope to encompass. Even the most well-controlled experiments with the best model systems are subject to criticism. For these reasons, there are few, if any, indisputable facts for *R. solani* or Rhizoctonia diseases even though there is extensive literature concerning epidemiology. It would be folly to attempt a comprehensive literature review. We only hope that in this effort, we have been able to cite the key papers.

INOCULUM.—*Sources and dissemination.*—Populations of *R. solani*, especially in soil, are significant though, until recently, difficult to detect. In many areas they are in higher density than many other soil pathogens, even to depths of 10-15 cm (Takahashi and Kawase, 1964). Ordinarily, this inoculum consists of hyphae, sclerotia, and basidiospores.

The hyphae of *R. solani* are readily observed growing through soil and have constituted a substantial portion of the hyphae isolated from soil (Warcup, 1955; Thornton, 1956). *Rhizoctonia solani* is able to grow rapidly through natural soil. Blair (1943) recorded growth velocities of more than 1 cm/day through soil. Dimock (1941) found that *R. solani* spread through nonsteamed soil in greenhouse flats at a rate exceeding 10 in./month. Brown patch of turf was observed by Shurtleff (1953*b*) to expand about 1 inch daily under favorable weather conditions despite applications of a phenyl mercury solution. These and other observations, however, do not provide satisfactory knowledge of speed and effective distance of hyphal extension because environmental influences are complex and growth is difficult to measure through soil (Hirst, 1965). It is interesting that *R. solani* in natural soil grows less vertically than horizontally (Blair, 1942). Thus, energy available for colonization of substrates is greater in a lateral direction.

Dark brown filamentous hyphae of *R. solani* were found on the surface of plant debris from soil by Boosalis and Scharen (1959), but these appeared to be devoid of protoplasm and did not resume growth. Many relatively thick-walled dark hyphae were observed within debris tissue, and 35% of these hyphae were viable. Undoubtedly these were a significant source of natural inoculum. Most of the hyphae were composed of long, monilioid aggregations of cells described many years ago by Duggar and Stewart (1901) in pure culture. Boosalis and Scharen (1959) frequently found sclerotia on the surface of plant debris. Viability of these was slightly more (44%) than that of the hyphae.

Weeds or rotation crops may be infected with *R. solani* and may function as a source of inoculum. Boosalis and Scharen (1960) presented convincing evidence that pigweed is susceptible to *R. solani* strains pathogenic on such economic hosts as bean, sugar beet, alfalfa, or potato. Strains pathogenic to these hosts were isolated from roots and stems of pigweed buried in soil for 1 year. This buried tissue also yielded several other strains. This topic is discussed further under "Survival of Inoculum."

Sclerotia are important sources of inoculum (Allison, 1951). For instance, sclerotia frequently found on the surface of seed potatoes may contaminate soil relatively free of inoculum. Dana (1925a) noted that strains carried into Washington on seed tubers were frequently more virulent than those already in the soil. However, isolates from sclerotia on tubers are often low in virulence when compared with isolates from stem cankers (Person, 1945). Also, presence of inoculum on seed pieces may not result in reduced yields. There may be an initial suppression in tuber formation, an appreciable increase in stem cankers, and even an increase in number of primary shoots killed; even so, yields may not be different between potatoes planted with either clean or infected seed (Small, 1943, 1945). The hyphae of *R. solani* penetrate into potato tubers to sufficient depths in many instances that surface chemical treatments may be ineffective in the eradication of the pathogen. Thus, these protected hyphae may be important sources of inoculum (Schaal, 1939).

*Rhizoctonia solani* is carried on and in true seeds. The early literature was comprehensively reviewed by Baker ([K.], 1947) and more recently by Neergaard (1958). The significance was discussed in some detail by Baker [K.]. Seed transmission insures the continued association of pathogenic strains of *Rhizoctonia* with the appropriate host. Dissemination of strains of the fungus to new areas may occur. Furthermore, the pathogen introduced into treated soil free from antagonistic organisms is able to spread and cause considerable damage. To this may be added the additional advantage of proximity of the pathogen to the host and the increased chance of exposure to highly susceptible plant tissues. The fungus usually enters fruits in contact with the soil. It may be carried as mycelia or sclerotia on the surface of the seed or as mycelia in the attached remnants of the funiculus. Internally, it may be found frequently in the inner layers of the seed coat, the endosperm, or the embryo. Obviously, eradicative treatments must kill the pathogen within the seed.

While Rhizoctonia inoculum is classically thought of as originating from the soil and infecting belowground parts of plants, other situations have been reported, especially from humid climates. For instance, fruits in contact with the soil were readily infected and canker infections were formed at the base of stem branches on tomato plants as a result of growth of *R. solani* over leaves touching the soil (Conover, 1949). The fungus has been observed to grow from leaf to leaf (Slooff, et al., 1947; Luttrell, 1962) or cotyledon to cotyledon (Weber and Abrego, 1958) within moisture films without contact with the soil.

Since Rolfs observed the perfect state of *R. solani* on potato in 1904, researchers have been aware that basidiospores might be important as inoculum. Observations on a wide variety of hosts, such as wheat, tomato, parsnip, cotton, beans, flax, and *Hevea* rubber (Ullstrup, 1939; Kotila, 1945a,b, 1947; Carpenter, 1949, 1951; Hawn and Vanterpool, 1952; Flentje, 1956), however, give no indication of its disease-producing potential, and actual determination of the importance of basidiospores as inoculum is difficult. Occasionally, basidiospore infection is only inferred by location of infection; for instance, Strong (1961) isolated *R. solani* from cankers on tomatoes where all cankers were associated with pruning injuries and were located at approximately the same height. This evidence suggested that the original inoculum was air-borne and that infection happened during a relatively short time. More satisfying data on the role of basidiospores in dissemination was obtained by Kotila (1945a, 1947) in studies of foliage blight of sugar beets. Plants were exposed in large chambers to mist and basidiospore inoculum produced on leaf blades and petioles of sugar beets. Air-borne basidiospores were readily transmitted from infected hosts to healthy plants. Germination of basidiospores occurred within 12 hours and symptoms usually were expressed in 4 days. Spores were disseminated 100 feet or more. These and field observations suggested that spread was principally from basidiospores.

Carpenter's (1949, 1951) intensive studies confirmed and expanded earlier indications (Kotila, 1945b; Lorenz, 1948) that basidiospores were a major factor in dissemination of the strains of *R. solani* causing target leaf spot of *Hevea* rubber. Basidium formation, basidiospore production, and basidiospore discharge occurred periodically and appeared to be diurnal phenomena regulated by light and dark periods. Maximum basidiospore discharge occurred from 6 P.M. to 6 A.M. This time, of course, correlated with dew deposition and relatively high humidities necessary for penetration. Basidiospores germinated in 2 hours, and susceptible leaf tissue was penetrated in 3 hours. Nearly 100% of the leaflets were attacked within 23 hours. Thus, under favorable conditions, the disease spread rapidly with sporulation beginning shortly after sundown and basidiospores disseminating, germinating, and establishing an infection within a relatively short period. The rapid spread of web blight of common bean in warm humid conditions

174

can also be explained by efficient production and dissemination of basidiospores (Echandi, 1965).

Basidiospores, however, may not be the only aerially disseminated propagules. Weber (1939) emphasized the ease with which small light-weight sclerotia may be detached and scattered from beans with web blight. Roth and Riker (1943a) observed that *R. solani* survived well in soil dry enough to blow as dust and attributed dissemination over a distance to passive transfer of contaminated soil. Vaartaja (1964b) isolated *R. solani* from dry soil particles and suggested dust drafts as possible sources of inoculum for reinfestation of steamed seed beds. Irrigation water may also be a very important source of inoculum. Bewley and Buddin (1921) took 41 samples of greenhouse water supplies and isolated *R. solani* from 11. In view of the great water-diversion systems used for irrigation in the arid and semiarid regions of the world, the potential for spread of plant-pathogenic microbes via irrigation water may be tremendous.

In storage, diseased market products may be sources of inoculum. Indeed, Rosenbaum (1918) reported that *R. solani* grew through paper wrappers from one fruit to another during transit. *Rhizoctonia solani* easily spread from diseased to healthy potato tubers in storage (Chamberlain, 1931).

Finally, a most interesting situation was reported by Butler [E. E.] and King (1951). Mycelium of *Rhizopus nigricans, Syncephalastrum* sp., *Mucor mucedo. Cunninghamella* sp., *Helicostylum* sp., and *Phytophthora cinnamomi* were parasitized by *Rhizoctonia* sp. pathogenic also to strawberry roots. When parasitizing these fungi, the *Rhizoctonia* sp. produced more barrel-shaped cells and sclerotia than it did in pure culture. If this phenomenon occurs with any frequency in soil, it might be an important source of inoculum.

*Inherent pathogenicity of inoculum.*—Each propagule or unit of inoculum should have a potential that would be a function of genetic capacity and stored energy (Baker [R.], 1965). Obviously, parasitism is ultimately controlled by genetic factors, and *R. solani* has been intensively investigated to determine how these factors influence pathogenicity. This subject has been reviewed by Kernkamp, et al. (1952), more recently by Flentje (1957), and by concurrent papers in this symposium and, thus, it will not be given great attention here.

It is generally conceded that *R. solani* consists of a wide variety of pathogenic strains. Some of these may be specific to only one plant while others attack a much wider range of families. Different strains may incite different symptoms in the same host (LeClerg, 1939b). While it is possible for the population of certain strains to increase and become important during repeated crop culture, this is no assurance that a strain is more active on the species from which it was isolated (Tolba and Moubasher, 1955; Sato and Soji, 1957). Daniels (1963), however, suggested that field populations of *Rhizoctonia* are continuously varying in pathogenicity. Mildly virulent strains occurring on

roots of weeds may be the unexpected source of virulent variants pathogenic to a particular crop. These variants may arise by branching and anastomosis of multinucleate hyphae. Evidence to support this scheme is not convincing, but the picture given in the literature indicates that *R. solani* is a pathogen with immense capabilities for variation and flexibility under a wide range of hosts and environmental conditions.

It is tempting to postulate that the linear-growth rate would have an influence on pathogenicity, since it appeared to be so important for occupation of non-living substrates (Lindsey, 1965). Given two strains with approximately the same genetic capacity for parasitism, would the one with the higher linear-growth rate have (apparently) higher virulence? Akai, et al. (1960) found a positive correlation between linear-growth rates of isolates from cucumber seedlings and virulence. Blair (1942), however, noted that two isolates having the slowest growth in soil were the most severe parasites, and concluded that high-linear growth rate was not necessarily associated with parasitic activity.

The inherent stored energy of units of inoculum is significant in *R. solani* as evidenced by its long survival without loss of infectivity (Pitt, 1964b, and others, see "Survival"). Inoculum grown in sterilized black loam for 180 days proved to be as virulent as some grown for only 6 days (Sanford, 1941b). Other reports also indicate that stored reserves are important. Mycelial growth was correlated with size of inoculum; fewer and shorter hyphae originated from smaller units of inoculum (Winter, 1950). Sclerotia apparently have large energy reserves. Boosalis and Scharen (1959) observed that one sclerotium produced 80 germ tubes and Shurtleff (1953b) found sclerotia viable even after 25 successive germinations. Pitt (1964b) considered this characteristic increased survival since inoculum could increase in response to stimuli.

Substrates supporting formation of inoculum may influence its virulence. Sims (1960) noted that a single-basidiospore isolate was less virulent following culture on vitamin-amended media than on the basal medium alone. There was good evidence that this was due to physiological rather than genetic changes. Shephard and Wood (1963) found that the nutritional status of the source of inoculum had a profound effect on resultant disease severity. Mycelial inoculum produced in potato-dextrose broth (PDB) incited more damping-off than did inoculum grown on a mineral-salts, medium or PDB diluted 1:10 or 1:100. They felt that nutrition probably influenced the number of hyphae reaching the host surface.

*Influence of inoculum density.*—Phytopathologists intuitively associate disease severity with the amount of inoculum present. Many important studies revolve around this principle and *R. solani* has been used widely as a tool to elaborate the basic relationships involved. Indeed, the first clear elaboration of this principle advanced by Horsfall (1932) was confirmed using this organism in damping-off experiments. Probably this pathogen has been chosen so many times for inoculum-density studies because of the distribution of economic support for research rather than its appropriateness as an experimental model. The propagule unit is difficult to define. Inoculum density in soil is not readily measured by conventional dilution-plate counts (Boosalis and Scharen, 1959); therefore inoculum densities sometimes are stated in relative rather than absolute terms. Natural soils may contain many different strains, confusing and clouding results. Even so, many of these difficulties have been surmounted and probably more has been published about inoculum-density relationships of *R. solani* than of any other soil microorganism.

Inoculum of *R. solani* is aggressive and can increase to significant levels rather rapidly. This was realized and formulated by Chester (1941) who compared seedling diseases of cotton. Those seedlings infected with *R. solani* in the same hill constituted a threat to the health of adjoining ones; plants infected with other pathogens did not. This same aggressiveness has been the source of difficulty when control of *R. solani* in field experiments was attempted by row treatments with chemicals having no residual activity. This same aggressive spread through steamed greenhouse soils has been seen repeatedly (Baker [K.], 1957).

Garrett (1956), in classifying *R. solani* as a primitive parasite, implied that there was a direct relationship between primitive parasitism and competitive saprophytic ability. Thus, the frequency of invasion of a nonliving substrate should give a relative measure of inoculum density and should be roughly correlated with disease severity (other things being equal). Experimental evidence for this assumption has been published (Martinson, 1963). Soil microbiological sampling tubes containing potato-dextrose agar (PDA) and rose bengal were used to measure competitive saprophytic ability. Disease incidence was closely correlated with frequency of invasion of the tubes by *R. solani*.

Since other methods have recently been developed for isolating *R. solani* from soil (Papavizas and Davey, 1959*b*; Menzies, 1963*b*; Ui and Ogoshi, 1964), and all of these measure competitive saprophytic ability in some way, investigators now have a variety of possibilities for determining relative inoculum densities. Sneh, et al. (1966) used plant segments colonized by the fungus, isolations from debris particles in soil, and soil microbiological sampling tubes for estimating the population of this organism. They obtained extremely good correlation between infestation density and degree of colonization of nonliving substrates, and similarly good agreement between inoculum density and disease index of infected bean seedlings. Using these data, they developed two interesting analyses based on mathematical relationships. First, the concentration of inoculum was estimated employing basic concepts treated by van der Plank (1963). The reciprocal of the probability of an exposed substrate remaining noncolonized was plotted against inoculum density. Theoretically, this should yield a straight line. In analysis, curves derived from data on percentage of dis-

eased seedlings were straightened out considerably but had little effect in other experiments involving the techniques mentioned above. Secondly, accuracy and sensitivity of methods were estimated. Slopes of curves were measured by ordinate/abscissa ratio corresponding to colonized substrate or diseased plants/infestation densities. Since the latter was known (relatively), some idea of accuracy should be inherent in the slope obtained. The other item influencing the acuracy of results was variability. Therefore, accuracy and reliability of any particular method could be determined by the ratio of mean slope of curve/range of variability of results. From this, figures were obtained indicating that colonization of bean segments was best and all methods were more precise at lower infestation levels. The greatest drawback is that Sneh, et al. (1966) used naturally infested soil as the source of inoculum. The genetic potential in the diverse strains was undoubtedly quite large and could introduce great errors into the findings.

In most experiments relating inoculum density to disease severity, multiple infections complicate interpretations of data based on symptom expression. Furthermore, various factors influencing host susceptibility after penetration may mask the initial influence of inoculum density. Sneh, et al. (1966) applied van der Plank's (1963) formula: $y = \log_e \frac{1}{1-x}$ in which $1-x$ denoting the proportion of substrate units not yet colonized or infected, is considered a correction factor for multiple infection. It would seem desirable, however, to interpret data on the basis of what is known about prepenetration stages of incubation. *Rhizoctonia*, in most instances, survives as inactive propagules. On a field basis, inoculum is randomly distributed in soil. When a substrate or infection court is introduced in the vicinity, these propagules germinate. In computing slopes of curves when inoculum density is related to disease severity, the extent in space of the influence of the substrate or infection court is important. Theoretically, a rhizosphere effect would result in a slope of 1 on a linear scale; if propagules only germinated in the rhizoplane, the relative slope should approach 0.66 on a log-log scale (Baker [R.], et al., 1967). Data are still not sufficient to confirm these models with *R. solani*.

In the studies treated above, inoculum density was determined in relative terms through the use of trapping methods. It has been possible to apply more direct methods, however. Use of seed potatoes infested with known quantities of sclerotia has yielded quantitative data. Cristinzio (1937), for instance, noted that the intensity of infection of progeny from infested seed pieces was directly proportional to inoculum density. Decreased yield of potatoes was also noted by Focke (1952) when planting tubers with numerous sclerotia (0.4 g/tuber); a "normal" load of 30 mg sclerotia/tuber had no effect. Important applications of this were developed. Tubers lifted in midsummer were each infested with only 15 mg, but 40 days later this had increased to 60 mg. Thus, for each tuber there was an increase of about 1 mg/day, a factor of importance in determining the day of harvest.

Another direct method was used by Rich and Miller (1963) who counted the number of mycelial fragments on 5-cm segments of strawberry roots. Highly significant linear regressions were obtained between number of fragments per unit area of root (whether plotted as a logarithm or not) and the number of plants surviving at the end of the growing season.

An excellent demonstration of the importance of inoculum density and its relation to disease incidence under field conditions has been reported by Boosalis and Scharen (1959). They compared the incidence of plant debris particles infested with *R. solani* in two areas of a field previously planted to sugar beets. About 8.5% of the plant residue particles from soil with a high incidence of crown rot yielded *R. solani* pathogenic to sugar beet seedlings. It was estimated that 100 grams of soil from a 1-5 inch layer in this field contained about 6,800 plant debris particles infested with the fungus. In contrast, about 2% of the particles of another portion of the field with a low incidence of crown rot yielded *R. solani*; 63% of these isolates were pathogenic to sugar beets. Only about 378 particles per 100 grams of this soil were calculated to contain the pathogenic strain in this portion of the field. Thus, the soil with a high incidence of crown rot harbored about 18 times as much inoculum as soil with a low incidence of the disease. It is of interest that these densities fall within theoretical values proposed for pathogens of this ecological type in soil (Baker [R.] and McClintock, 1965). Ui and Tochinai (1955) also noted a correlation between inoculum density and severity of sugar beet root rot in the field.

Theoretically, as inoculum density increases, disease severity should increase. In some instances, however, increased density of mycelial growth has been associated with lower incidence of disease (Sanford, 1941*a*; Blair, 1942; Das and Western, 1959). For example, if inoculum was increased in sterilized soil or 2% cornmeal-sand and mixed with natural soil, increasing the concentration above a certain point resulted in decreased disease severity. The fungus may also be much more virulent on potatoes in natural soil than in the same soil which has been steamed (Richards, 1921*b*). The reason for this is not well understood. Sanford's (1941*b*) observations suggest that the pathogen is most virulent when hyphae are very young, very thin, and still hyaline. Factors causing high densities of mycelium may depress virulence. He also suggested that substances may be released by *Rhizoctonia* that may inhibit penetration and infection.

In summary, relative densities of *Rhizoctonia* in soil may be measured by noting the frequency of invasion of nonliving substrates. This may be correlated with disease severity or disease incidence. More direct methods involving counts of sclerotia or hyphae have also been employed with success. Hopefully, studies on inoculum theory using *Rhizoctonia* will continue, as several different methods of handling the data have been suggested, all leaving important questions un-

resolved. Since disease severity usually correlates with the population of *Rhizoctonia*, control measures aimed at reducing inoculum density have been used extensively. Conventional soil treatments, while sometimes economically unfeasible, have successfully controlled the pathogen. Somewhat unconventional methods, such as redistributing inoculum beyond the infection court of deep tilling (Tupenevich, 1958), also should be explored.

*Survival of inoculum.*—The preceding sections establish and support the concept that inoculum potential is related directly to disease severity and disease incidence. *Rhizoctonia solani* is unlike the classical foliar pathogens (except where it's a foliar pathogen) where rapid production of secondary inoculum and dissemination of inoculum are important for epiphytotics. The inoculum potential of *R. solani* must be maintained in the soil in the absence of a suscept through some mechanism of survival. This generally connotes survival of propagules but it also involves the infective capacity of these propagules. The mere existence of a propagule, which can be readily demonstrated by passive isolation techniques, does not mean that the propagule has the energy to establish infection. Certain active techniques for isolation depend on the competitive saprophytic ability of the organism for isolation, and these techniques may be more useful for determining the inoculum potential and survival of *R. solani* in soil.

Park (1965) clearly outlined the modes of survival of pathogens in soil. Survival may occur by either inactive or active processes. Inactive survival involves "either a passive inactivity imposed by the environment, or a positive dormancy governed by the physiology of the resting structure." Active survival includes parasitism of other hosts, commensal reproduction within a rhizosphere, and saprophytic growth on the organic matter in soil. Papavizas (this vol.) discusses certain aspects of the survival of *R. solani*.

Inactive survival.—Many have reported the survival of *R. solani* in pure culture. Chowdhury (1944) and Pitt (1964*b*) noted that dry sclerotia survived for 2½ years in the laboratory, and Gadd and Bertus (1928) reported survival for 6 years. Martinson has stock cultures in dry soil that have remained viable when stored at room temperatures for 6 years. Longevity of sclerotia is decreased by high temperatures or high moisture conditions during storage (Newton, 1931; Kernkamp, et al., 1952). Basidiospores lose viability quite rapidly in dry storage; Muller (1924) found that viability was lost after 6 weeks.

It is commonly thought that *R. solani* exists in soil as sclerotia associated with plant debris or as thick-walled hyphae within the debris particles (Boosalis and Scharen, 1959). Several investigators studied the survival of sclerotia buried in nonsterile soil. Pitt (1964*b*) observed 79% survival after 6 months and Chowdhury (1944) found that all sclerotia lost viability when buried in wet soil for 4-5 months and in dry soil for 6-7 months. Sclerotia and hyphae associated with plant debris appear to be relatively short-lived. Townsend (1934) buried Rhizoctonia-infected samples of lettuce refuse in muck soil over winter and found that the organism germinated slowly and poorly and produced "feeble" cultures the next spring. *Rhizoctonia solani* in lettuce refuse on the soil surface survived the New York winter weather and germinated readily in the spring. Pitt (1964*b*) buried naturally infected wheat straws in soil. He was able to isolate *R. solani* from only 14.7% of the straws after 110 days burial. Originally, 95% of the straws carried the fungus. The pathogen persisted in the straws for a short time, but soon was suppressed or killed by the other colonizers. Papavizas (1964) allowed single-basidiospore isolates of *R. solani* and *R. praticola* to thoroughly colonize buckwheat stem segments. He then studied the "decolonization" of these segments in soil. He was unable to isolate the test fungi from 21-90% of the segments after 6 weeks soil exposure. Incorporation of PCNB into the soil increased the rate of decolonization. The heterokaryotic parent cultures survived better in colonized substrate than the single-basidiospore progeny did, except for two isolates that were better than the parent culture. Heterokaryosis obviously aided the parent culture in survival, but the experiment also pointed out the benefits of genetic recombination.

The survival of *R. solani* on the "seed" tubers of potato is well known and accounts for a major portion of the literature on the fungus prior to 1950. The organism also persists on or within the true seed of many plants (Baker [K.], 1947; Neergaard, 1958). As we previously stressed, this is a major means for dissemination of *R. solani*.

*Active survival.*—*Rhizoctonia solani* is considered a primitive parasite with simple food requirements and a high mycelial growth rate (Garrett, 1956). With this arsenal of activity and a quite broad host range, it is common to find the organism associated with other crop and weed plants (LeClerg, 1934; Sanford, 1952; Herzog and Wartenburg, 1958; Herzog, 1961; Daniels, 1963; Oshima, et al., 1963; Pitt, 1964*b*). Pitt (1964*b*) found that parasitic survival of the sharp eyespot fungus on other susceptible crops was very important. He observed that the organism formed sclerotia on potato tubers and stolons and that a high incidence of sharp-eyespot disease occurred on wheat following potato crops. Sanford (1952) determined that *R. solani* strains attacking potato depended essentially on parasitic nutrition for survival. The hyphae of *R. solani* commonly inhabited the living roots of host and nonhost plants but survival was better in the former. Herzog (1961), Daniels (1963), and Oshima, et al. (1963) found strains on weed hosts that were pathogenic to crop plants.

Numerous cross inoculation tests were performed with various isolates of *R. solani* and various crop plants. In many experiments of this type, the test plants were subjected to supernormal inoculum potentials in steamed soils and the plants succumbed. Under more nearly normal conditions in naturally infested soils, an isolate may have failed to establish even a parasitic relationship with a test plant. Yet some of the host-range studies have been well controlled and the information is quite useful for determining crop rotation sequences.

Reproduction in the rhizosphere of higher plants as a type of commensal survival has hardly been investigated for *R. solani*. Reports by Herzog and Wartenburg (1958) and Herzog (1961) suggest that this means for growth and survival may warrant further investigation. Materials exuded from living plant roots supported the growth of *R. solani* and the formation of sclerotia by the organism (Herzog and Wartenburg, 1958).

The saprophytic phase of *R. solani* in soil has been recognized for a long time (Rolfs, 1904) and the vast literature pertaining to this phase is covered by Papavizas (this vol.). The various isolates and single-basidiospore cultures are quite variable in saprophytic abilities, yet saprophytic growth through soil and reproduction on plant debris is a very important means for survival.

The active and inactive survival mechanisms are both involved in the survival of *R. solani*. The relative significance of each survival mechanism will vary in every instance, yet combined they generally maintain the inoculum potential of *R. solani* for extended periods in soil.

Capacity factors.—A given quantity of inoculum has a definite capacity for penetration and infection of a host only under certain specified conditions. Under a different set of conditions, this capacity normally will not be the same. The capacity is changed by various factors of nutrition and environment (Dimond and Horsfall, 1960; Martinson, 1963), termed capacity factors. In the same sense, disease potential (as used by Baker [R.], 1965) also is influenced by these capacity factors. To appreciate fully the importance of capacity factors in inoculum potential and disease development, the effects should be studied in terms of the potential pathogen, the potential suscept, and, after ingress, the host-parasite and suscept-pathogen interactions. These distinctions rarely are evident in the literature. Much has been written concerning environment and the Rhizoctonia diseases. Much of this information is observational, where experience has shown disease severity or disease incidence to be increased by certain factors of the environment. There are also a number of reports on the effects of environment on growth, morphology, and physiology of *R. solani*. Much of this information on the fungus per se may be indirectly related to inoculum potential, and the information is covered in separate papers by Sherwood, Papavizas, and Bateman in this volume, where they discuss the physiology of *R. solani* in vitro, as a soil saprophyte, and a pathogen, respectively. In this section, we will discuss pertinent papers on effects of capacity factors on the development of Rhizoctonia diseases and we will attempt to separate the specific effects as they may relate to inoculum potential and disease potential.

Experimental systems have been devised for studying some of the environmental factors singularly and factorially. The reliability of the accumulated data may be questioned, since many of the experimental models were poorly designed. Most of the authors failed to test the validity of their methods either by analyzing the model system or by checking with dissimilar models. Garrett (1955) warned that the common use of sterilized soil in many of the model systems has resulted in little information beyond that involving the etiology of disease. Few thorough studies have been made of the environment and disease development by *R. solani*.

*Nutrition.*—Nutrition has three broad facets: (1) the supply of energy-yielding organic compounds, (2) the supply of organic compounds an organism requires but cannot synthesize, and (3) the supply of mineral nutrients essential for growth. Nutrition of both host and pathogen are involved in the epidemiology of Rhizoctonia diseases and, in a broader sense, the nutrition of all the microorganisms in the soil environment must be included.

Higher plants are autotrophic and normally all the organic compounds needed for plant growth and development are synthesized by the plant. The nutrimental status of the *R. solani* inoculum per se and its influence on inoculum potential was discussed earlier in this paper.

Numerous experiments have been performed to determine the effects and the role of externally supplied nutrients in Rhizoctonia-disease development. Fertilizers and organic materials have been added to the soil system to determine their effects on disease control, Rhizoctonia activity, microbial activity, or available inorganic nutrients, but no reports have been found where all of these effects were studied simultaneously.

Fertilization experiments on agricultural soils with inorganic chemical fertilizers revealed that potassium, nitrogen, or calcium deficiencies increased the disease potential. Disease potential may also increase with excessive nitrogen fertilization. Fertilization with potash greatly decreased Rhizoctonia disease of potato in three separate trials (Janssen, 1929; Van Beekom, 1945; Zaleski and Blaszczak, 1954). Hynes (1937) observed that the readily available calcium was much lower in wheat and oat fields affected with Rhizoctonia root rot. An application of calcium controlled the disease. Although they do not mention the pathogen involved, Albrecht and Jenny (1931) found that damping-off of soybean seedlings occurred when the calcium supply was low. The availability of calcium ions were more effective in disease control than adsorbed, exchangeable calcium ions. Calcium found within the crystal lattice of minerals was unable to prevent disease. Calcium was superior to magnesium which was better than potassium for disease control.

In addition to the control of Rhizoctonia root rot of wheat with calcium, Hynes (1937) also controlled the disease by fertilizing with ammonium sulfate. The ammonium sulfate treatment appreciably con-

trolled purple patch of cereals (Anon., 1938; Anon., 1950), Rhizoctonia disease of potato (Janssen, 1929) and Rhizoctonia foot rot of tomato (Small, 1927). In contrast, excessive nitrogen increased the brown-patch disease of grasses (Hearn, 1943) and the damping-off of broad-leaf tree seedlings by *R. solani* and *Pythium* spp. (Wright, 1941). Wright found that if the excess nitrates were tied up by the microflora decomposing organic amendments, then damping-off decreased considerably.

The above results from agricultural soils agree generally with data from sand-culture experiments where mineral imbalance in "standard" nutrient solutions also influenced disease severity. Zyngas (1963) observed that Rhizoctonia damping-off of cotton seedlings was less prevalent when additional calcium, potassium and boron were present in the nutrient solution. Brown patch of Seaside bent grass (*Agrostis palustris*) was more severe when the concentration of nitrogen in the nutrient solution was decreased. (Bloom and Couch, 1958). Damping-off of pea and cucumber was more severe when the amount of nitrogen and phosphorus in the nutrient solution was decreased, and tomatoes were damped-off more readily when additional nitrogen and phosphorus were present (Sayed, 1961). *Rhizoctonia solani* and *P. ultimum* were the pathogens in Sayed's experiments. Castano and Kernkamp (1956) found that root rot of soybeans was increased in sand culture when calcium, iron, nitrogen, phosphorus, and magnesium were omitted from the nutrient solution. An absence of potassium did not affect the disease readings. Plants grown in the absence of calcium had poor root systems and very soft cortical tissues with large intercellular spaces. The middle lamella was difficult to distinguish. The role of calcium, and some other multivalent cations, in disease resistance is probably to form insoluble pectates in the plant cell wall which are resistant to hydrolysis by Rhizoctonia polygalacturonases (Bateman, 1964a, and this vol.). The use of sand-culture techniques removes much of the effects from the realm of inoculum potential and places them on disease potential. This deduction is based on the fact that the rhizosphere is diluted with "sand drip culture" and "sand 'slop' culture" techniques and most of the exudates are lost from the rhizosphere. *Rhizoctonia solani* is virtually unable to make any extended growth through sand without some source of energy. (Blair, 1943; Winter, 1951; Garrett, 1962; Shephard and Wood, 1963). Furthermore, any factors stimulatory to the infection process (Dodman and Flentje, this vol.) would be diluted and washed away from the site of exudation. Based on present experimental evidence, inorganic mineral nutrients appear to influence disease potential more than inoculum potential.

High salt concentrations predispose plants to disease by *R. solani*. Beach (1949) found that damping-off of tomatoes and cucumbers was very severe when the plants were grown in 8× or 16× Knop's solutions. Again, the effects were probably on disease potential because damping-off even occurred at the 8× and 16× concentrations of the nutrient solutions in the absence of pathogens.

Organic manures have been evaluated for disease control. Singh and Singh (1957) found that well-decomposed farmyard manure used in either greenhouse or field soils, resulted in less root-rot damage and less killing of *Cyamopsis psoraloides* by *R. solani*. They attributed the success to increased plant vigor and tissue maturity resulting from better nutrition. Blodgett (1940) reduced black scurf on potatoes somewhat with barnyard manures and rye cover crops. Dorst (1923) found that fresh organic manures increased the incidence of *R. solani* on potatoes. Stable manures also increased Rhizoctonia foot rot of tomato (Small, 1927). Quanjer (1940) attributed conflicting results to varying degrees of freshness of the stable manure applied. Fresh manures tended to increase disease and well-decomposed manures tended to suppress disease. Quanjer (1940) felt that these responses were linked to the rate of $CO_2$ evolution from the different manures and the relative tolerance of *R. solani* and potato shoots to high $CO_2$ concentrations. Unfortunately, data were not presented to support his ideas.

Crop residues and processed organic materials have been added to soils where the investigators were attempting to control Rhizoctonia diseases. Biological control was the general objective. However, these amendments obviously affected the nutritional balance of the soil. The organic matter provided the energy for growth and the carbon for the formation of new cells by *R. solani*. The mineral ions essential to both microbial and plant growth are involved in organic matter decomposition. During the process of organic matter decomposition and utilization, mineral ions essential to microbial growth are assimilated from the soil solution and certain mineral ions may become deficient. Alexander (1961) noted that immobilization of nitrogen, phosphorus, and sulfur may occur with organic matter decomposition and that the amount of potassium and manganese immobilization is probably not important in soils. As decomposition progresses through an ecological succession of microbes, mineral ions are released to the soil solution, a process termed mineralization (Alexander, 1961). Thus, the incorporation of any organic matter, manure, or crop residue into the soil affects the nutrient status in the soil for host and pathogen and for the other microbes in the soil.

Kommedahl and Young (1956) observed a decrease in *R. solani* infection on wheat seedlings following incorporation of corn stalks into soil in either the presence or absence of ammonium sulfate at 20 lb/ton of stalks. Snyder, et al. (1959) found that the addition of barley straw, wheat straw, corn stover, and pine shavings at 1%(w/w) reduced the infection of bean hypocotyls by *R. solani*. Alfalfa and soybean residues similarly used increased the amount of infection. These data would support the idea that residues low in nitrogen suppress disease, but there is ample evidence to the contrary. Sanford (1947, 1952) found that corn meal, nitrogenous substrates, and nitrogenous salts tended to reduce disease and the

persistence of *R. solani* in soil. Davey and Papavizas (1960) used various organic amendments in experiments to control *R. solani* on beans. They observed that soybean, corn, and oat amendments reduced the disease and sawdust had no effect. The disease reduction was generally greater when additional nitrogen was added with the residue and was correlated with lower competitive saprophytic activity of *R. solani*. The reduction in disease development seemed to be greatest when the amendment was undergoing rapid decomposition. However, when the carbon/nitrogen ratio of the residue was reduced to 5 or 10, the disease severity was increased again. Blair (1943) added grass, lucerne, and straw meals and various organic fertilizers to soils, and observed a striking decrease in the velocity of mycelial growth through the soil. This decrease in growth rate by *R. solani* was attributed to immobilization of available nitrogen from the soil solution and to accumulation of $CO_2$ in the soil environment. Nitrate and ammonium nitrogen sources added with organic amendments reduced the depressing effects of the organic materials on *R. solani* growth. Papavizas (1963) reported that the addition of ammonium nitrate with organic amendments decreased the number of antagonists toward *R. solani* in the soil. Thus, the overall role of organic amendments in disease control seems to be quite complex. Present evidence supports the thesis that the major effects of fresh organic amendments in disease control are directed toward decreasing the inoculum potential of the pathogen. Occasionally, the organic amendments increased disease even though the inoculum potential was decreased (Martinson, 1959). Disease potential was possibly increased through predisposition by: (1) the immobilization of essential plant nutrients resulting in mineral deficiencies in the host plant (Alexander, 1961), or (2) the formation of phytotoxic compounds during decomposition (Cochrane, 1948; Patrick and Koch, 1958) affecting plant metabolism directly or increasing the exudation of substances stimulatory to the pathogen (Toussoun and Patrick, 1963).

The status of the mineral ions during decomposition of organic amendments in soil has not been followed critically in relation to *R. solani* activity and disease development. The vast literature assembled by soil microbiologists (Alexander, 1961) points out the dynamic status of minerals; however, a preoccupation with biological control by the plant pathologists has allowed the finer points of mineral nutrition to escape study.

Blair (1943) demonstrated vividly the ability of *R. solani* to grow through soil. A food base helped to initiate growth, but for extended growth, energy-yielding nutrients were needed from the soil. The fungus showed a greater nutrient requirement for sclerotium formation than for growth through soil. The energy requirement has not been determined for penetration of a host by *R. solani* and the establishment of a parasitic relationship with the host. Plant exudates are obviously important for the formation of infection structures (Kerr and Flentje, 1957; Flentje,

Dodman, and Kerr, 1963; Martinson, 1965) and the penetration processes. The roles in the infection process have not been determined for the individual components of plant exudates. Host exudates will stimulate growth of *R. solani* in soil (Martinson and Baker, 1962), and provide nutrients for growth prior to penetration (Schroth and Cook, 1964). The amount of exudate regulates the number of infection cushions (Flentje, et al., 1964) and the amount of disease (Toussoun and Patrick, 1963; Schroth and Cook, 1964). Exudation may be increased by restricted aeration (Woodcock, 1962) or by the presence of phytotoxic decomposition products from organic matter (Toussoun and Patrick, 1963). Infection of foliar parts of plants may be aided by guttation fluids from injured leaves as Rowell (1951) has shown for *R. solani* on bent grasses. The guttation fluids also supported the leaf to leaf spread of the pathogen. Irrigation water would not substitute for the guttation fluid.

Exogenous energy-yielding compounds in plant exudates aid *R. solani* during the processes leading to host penetration and thereby they increase the inoculum potential. Evidence has not been presented to show a reliance on exudates for penetration. The endogenous food reserves within *R. solani* inoculum are also important and this subject was covered under inoculum density.

*Temperatures.*—The influence of temperature on the various physiological and morphological processes by *R. solani* and on the development of Rhizoctonia diseases has been studied by many throughout the world. Yet, the investigations have generally been inadequate because it has been common to include temperature experiments along with other investigations.

The effects of temperature on the growth of *R. solani* in vitro have been covered thoroughly by Sherwood (this vol.). A number of investigators have successfully correlated growth rates from in vitro studies with the development of disease. Deductions from these correlations often leave the impression that disease severity is related primarily to fungal growth and that host response is not important. Although some data may support these conclusions, the early work of Balls (1906) and Richards (1921b, 1923a,b) established that disease severity was not a mere function of fungal growth. Richards (1921b, 1923a) found that the optimum soil temperature for disease development on potatoes, peas, and beans was near 18°C and little disease developed above 21-24°C. The optimum temperature for fungal growth was about 26°C. Balls (1906) and Jones (Jones [L. R], et al., 1926) found the same response with cotton and the *R. solani* causing sore shin. Jones repeated an often-quoted statement about *R. solani*, "that neither the nature of the host nor its normal temperature relations materially influence the temperature range for the parasitic action of this fungus. This evidence rather favors the idea that this relation of temperature to parasitism with *Rhizoctonia* is a fixed character of the fungus." The evidence referred to is the work of

Balls, F. R. Jones, and Richards, whose conclusions are not in accord with the findings of Peltier (1916). Peltier had found that higher temperatures (near 30°C) favored disease development by *R. solani* on carnations. Richards (1921b) recognized the temperature effect on host vigor and the velocity of potato-shoot emergence. Potato shoots emerged at 24°C in about one-third the time required at 15°C, and he indicated that disease "escape" was possibly involved at higher temperatures. Jones, et al. (1926) also recognized the possibility of disease "escape," the ability of the host to "grow away" from *R. solani*. Richards (1923b) advanced the idea that the enzyme or group of enzymes responsible for tissue destruction are secreted more abundantly at 18°C or they react more with the host tissue at this particular temperature. He had no evidence to support this hypothesis and we were unable to find any work since then that would support it. Richards also noted a tendency for *R. solani* to produce more aerial growth on artificial substrates at higher temperatures, while closely adhering to the substrate at the lower temperatures. This change in morphology of fungous growth with different temperatures may have some significance in inoculum potential.

Richards (1923a) also observed a difference in the prevalence of symptom types at different temperatures. Growing-point destruction on potato was much more severe at 18° than at 23°C, and lesions on the side of the stem were more prevalent at 23° than at 18°C, though not as severe. Sanford (1938b) ran numerous tests with 12 different isolates and could not corroborate Richards' observations. He was unable to see any difference in the prevalence of either type of injury at either 17° or 23°C. One type of injury occurred with near absence of the other type at both 23° and 16°C. Other capacity factors influenced the results. Severe sprout-tip injury merely indicated that the capacity factors (temperature included) favored either high inoculum potential or high disease potential prior to and during emergence. When stem lesions were very prevalent and growing-tip injury was slight, inoculum potential or disease potential was low initially, then after emergence, the capacity factors favored high inoculum potential and disease potential. Sanford found that recovery by means of secondary sprouts was important if the primary sprout was severely lesioned. These secondary sprouts appeared to be very resistant and equally common at the two temperatures even though emergence of the secondary sprouts was more rapid at 23°C.

Probably the most significant work relating to temperature and disease incidence was performed by Leach (1947) with four damping-off pathogens, *R. solani*, *Phoma betae*, *Pythium ultimum*, and *Aphanomyces cochlioides*. He found, just as had many of the above authors, that disease incidence did not correspond closely with the growth rate of either the host or the pathogen at different temperatures. Instead, disease incidence was inversely related to the ratio between the coefficient of velocity of seedling emergence (CVE) and the growth rate of the fungus.

The CVE, calculated by Kotowski's (1927) formula, is an indication of host vigor and possibly was related to the chances of the host to escape infection as mentioned by Richards (1921b). The growth rate of R. solani was determined by the velocity of mycelial extension on synthetic media. When disease incidence was compared with the host/pathogen ratio at various soil temperatures, disease incidence increased as the ratio decreased. The host/pathogen growth ratio is a complex term and is not expressed in units. The definition of the ratio will vary because the procedures for measuring host growth and pathogen growth are not always the same. Thus, the ratio expresses only relative values.

The strain of R. solani used by Leach (1947) grew on solid media from 8-40°C. Rapid growth was observed between 20° and 30°C with a maximum between 25° and 30°C. With cool temperature crops such as spinach and sugar beets, there was a low incidence of preemergence damping-off at the lower temperatures (4-12°C), where host growth was favored over pathogen growth. At higher soil temperatures, the relative growth rates favored the pathogen, especially at 20-30°C, and preemergence damping-off was severe. A warm temperature crop, watermelons, showed opposite results; R. solani caused extensive seed decay and preemergence damping-off of this crop at soil temperatures below 25°C. but disease incidence was low at 30° and 35°C, where the CVE was much higher and the host/pathogen growth ratio was highest. Leach found, with all of the many host-pathogen combinations he tested, that the host/pathogen growth ratio accurately predicted the incidence of damping-off. Beach (1949) confirmed Leach's results and conclusions in several experiments with R. solani as a damping-off pathogen of tomatoes, cucumbers, spinach, and peas. Graham, et al. (1957) were unable to consistently correlate the host/pathogen growth ratio with either preemergence or postemergence damping-off of either lespedeza or Ladino clover. Emergence in the presence of pathogens was more closely correlated with the effects of temperature on the host than on growth of R. solani, P. debaryanum, or F. roseum. The preemergence damping-off data for lespedeza did follow the host/pathogen growth ratio fairly well when R. solani was tested. Rhizoctonia solani caused little preemergence damping-off of Ladino clover at any temperature. Graham, et al. tried unsuccessfully to use CVE in the host/pathogen ratio to predict postemergence damping-off. The use of CVE as a measure of host resistance after emergence is probably invalid. Yu (1940) observed a temperature-disease development response on broad beans similar to the response Leach (1947) observed with R. solani and watermelons. Yu felt that the decreased disease at higher temperatures could not be explained solely on the increased velocity of hypocotyl growth, but that the increased physiological activities of the host plant obviously imparted some additional resistance to infection.

Leach (1947) found that damping-off was more severe at a given temperature (and host/pathogen ratio) when heavily infested soil rather than lightly infested soil was used in the tests. Martinson (1959, 1963) expanded this to show that inoculum density modifies the effects of temperature on preemergence damping-off of radishes and watermelons. At high inoculum densities, R. solani caused severe damping-off even if the host/pathogen growth ratio was quite high. However, at one inoculum density in natural soil, disease incidence was inversely proportional to the host/pathogen growth ratio, which confirmed Leach's conclusions. Martinson (1963) felt that the seedling emergence in the presence of R. solani correlated more closely with a host CVE/inoculum-potential ratio than with a host/pathogen ratio.

Various physiological and pathogenic strains are reported for R. solani in the literature, and many authors used temperature criteria for differentiation of strains. Richter and Schneider (1953) separated 176 diverse strains into six groups according to morphological characters and to their ability to form hyphal anastomoses. In four of the groups, optimal hyphal growth occurred from 25-29°C, and these strains showed little evidence of physiological specialization. Strains in the other two groups grew optimally at 21-25°C, would grow at lower temperatures than the other strains, and attacked primarily potato and crucifer plants (Parmeter and Whitney, this vol.). Ui, et al. (1963) found two distinct strains attacking flax, which they called "spring strain" and "summer strain." The spring strain had a lower growth range than the summer strain and a lower optimal temperature (25°C) than the summer strain (28°C). The spring strain readily attacked seedlings and the summer strain was able to invade mature stems much easier than seedlings. At soil temperatures of 13°C, all strains trapped from the soil were the spring strains. At 20°C, they trapped both strains. Barker (1961) observed that bean isolates from Wisconsin caused the greatest disease development on bean from 16-24°C while North Carolina isolates incited disease best at 28°C. Hunter, et al. (1960) isolated three strains of R. solani from cotton and found that disease development with the different strains varied with temperature. The mildly virulent strain caused most disease at 24°C; the moderately virulent strain caused most disease at 32°C; and the highly virulent strain caused severe disease at 24°, 27.5°, and 32°C. The presence of different "temperature strains" of R. solani, as Hunter, et al. (1960) have shown, may resolve the differences in some of the conflicting reports (Balls, 1906; Jones, et al., 1926; Massey, 1928; Hansford, 1928; Walker, 1928; Fahmy, 1931; Forteneichner, 1931; Vasudeva and Ashraf, 1939) where sore-shin was listed as a "low temperature" or a "high temperature" disease. Flentje and Saksena (1957) reported an interesting case of a temperature interaction with physiological specialization. Rhizoctonia solani of the "praticola" type in contrast to the "filamentosa" type, appeared to be unspecialized and it attacked a wide range of hosts above 18°C. However, below 18°C, the praticola type was pathogenic only to beets.

Certain strains of R. *solani* incite diseases at very low temperatures, where the fungus growth rate is quite slow; then disease severity decreases at higher temperatures. Lettuce damping-off strains are most active near 8°C (Abdel-Salam, 1933; Jacks, 1951; Shephard and Wood, 1963), whereas strains causing bottom rot of lettuce are most active above 24°C (Townsend, 1934). The strains that incite either the sharp eyespot disease or the root rot of cereals caused the most disease at the lowest temperatures tested, 9.3° and 12°C, respectively, (Samuel and Garrett, 1932; Pitt, 1964*a,b*) and disease severity appeared to be quite low at 20°C. Potato tubers may be attacked at low temperatures in storage. Henriksen (1961) studied the development of pit rot at different temperatures. Optimum temperature for pathogenesis was 8°C and the fungus incited some pit rot at 4°C. Above 8°C, disease severity decreased. The "punky" stem-end tuber rot described by Thatcher (1942) developed quite well at 5°C and not at all at 14°C or above. Tubers held in very wet sand at 15-20°C were also infected by the same strain. Both the very wet sand and 5°C conditions inhibited wound periderm formation by the tuber and the fungus advanced uninhibited. Disease potential increased due to a decrease in resistance. In most of these instances where disease was most severe at very low temperatures, the increase in disease was probably a result of higher disease potential and the effects were not mediated through the pathogen.

The Rhizoctonia root canker of alfalfa is seasonal in development as lesions develop mainly during the summer months (Smith [O. F.], 1943). Smith [O. F.] (1946) observed in controlled experiments that cankers developed at temperatures of 20-35°C, with optimal development near 30°C. He correlated this with field observations where cankers did not develop much before the soil warmed to 20°C and they ceased to develop further in the fall after the soil cooled below 20°C. The fungus grew below 20°C, but it appeared that disease resistance was sufficient.

Several attempts have been made to correlate weather data with epiphytotics of Rhizoctonia diseases. Two of the best examples have involved bottom rot of lettuce and brown patch of lawn grasses.

Townsend (1934) observed that bottom rot was a problem in New York during periods of warm, humid summer weather and the disease problem seldom disappeared until the first cool nights of September. Lettuce plants were grown to maturity in greenhouses, a sclerotium of R. *solani* was placed on the soil near the stem, and then the plants were covered with bell jars to maintain the humidity. At air temperatures of 24-32°C, 23 of 27 inoculated plants were destroyed with bottom rot; at 16-24°C, only 10 of 27 plants showed any symptoms and none of these were severe. At the higher temperature, the incubation period averaged 46 hours, while at the lower temperature, the incubation period was 84 hours. This work was repeated five times with essentially the same results each time. Townsend studied the weather data and the development of bottom rot over 4 years, 1928-

1931. During these years, there were 11 periods when bottom rot reached epiphytotic levels. The mean daily temperature was from 20-24°C during these periods (the mean summer temperature was 20°C) and some precipitation was associated with each epiphytotic. Showers occurred on half the days in the 11 periods. Very warm night temperatures coupled with rain over several days resulted in the most severe epiphytotics. Minimum daily temperatures as low as 10°C stopped bottom-rot development.

Dickinson (1930) studied the effects of air temperature on R. *solani* pathogenic to grasses on putting-green turf. Sclerotia of R. *solani* germinated optimally between 17.8-20°C and the quickness of germination was determined by the prior temperature exposure. If sclerotia were held at a higher temperature then cooled to the critical 17.8-20°C range, the sclerotia germinated within several hours by producing short hyphae. When sclerotia held at a lower temperature were warmed to 17.8-20°C, the temperature had to remain at 17.8-20°C for 8-10 hours before germination occurred. The short germination hyphae remained relatively inactive unless the temperature was increased to 22.8°C above. Parasitism began at 22.8°C and ceased at 32.2°C, and it depended also on high humidity. If the air temperature decreased below 16.7°C, the germination hyphae on the sclerotia were destroyed. Dickinson used this laboratory data as a guide and was able to forecast with 100% accuracy the development of brown patch on 7 days that were conducive for disease out of a 123-day period in 1928, and on 5 days that were conducive for disease out of 123 days in 1929. The amazing thing is that he never missed. Although this work has not been confirmed entirely, there are some published works that support Dickinson's results in part. Hearn (1943) reported that brown patch development was favored by air temperatures between 22.2-34.5°C. Rowell (1951) found that parasitic and saprophytic growth of R. *solani* was stopped when minimum temperatures fell below 21.1°C. Shurtleff (1953*b*) observed germination of sclerotia from 15.6-37°C and the sclerotia would regerminate at least 25 times.

Dahl (1933) analyzed the temperature data for five consecutive summers and compared these data with the occurrence of brown patch. He found that the disease varied directly with the temperature and occurred on 82% of days with minimum temperature above 21°C. The sclerotia would germinate at 8-12°C and optimum rate of germination occurred at 28-32°C. Dahl observed that sclerotia did not require a drop in temperature or a low temperature for germination to be initiated, as Dickinson (1930) reported, and in most cases these treatments decreased the rate at which germination proceeded.

Isolates of R. *solani* from grass have exhibited quite different temperature optima and ranges (Monteith and Dahl, 1928; Shurtleff, 1935*b*; Endo, 1963) in pure culture. One cannot question the validity of the findings of Dickinson (1930) or Dahl (1933) because of the great variability of R. *solani*.

Temperature also affects basidiospore production

and germination by *Thanatephorus cucumeris*. Carpenter (1949) observed production and discharge of basidiospores at temperatures of 21.5-28°C with strains of *Hevea* rubber. Kotila (1929) observed an optimum of 20.7°C for basidiospore formation by sugar beet strains, but temperatures of 21-25°C with 100% relative humidity (RH) were required for epiphytotics where basidiospores functioned as aerial inoculum (Kotila, 1947). At temperatures below 21°C, growth of the fungus was confined to the crown region of the sugar beets (Kotila, 1947). Müller (1924) found that basidiospores germinated optimally at 21-25°C and failed to germinate above 30°C.

*Moisture.*—Water in the soil and in the air is an important factor of inoculum potential and of disease potential. The greatest confusion in the literature relating to epidemiology concerns the effects of soil moisture. Confusion was generated from the frequent use of poorly defined terminology and the failure of authors to separate the effects of moisture tension and aeration. The moisture condition has generally been expressed as a percentage of some moisture characteristic of the particular soil, such as water-holding capacity, moisture-holding capacity, moisture equivalent, saturation, or field capacity. The values for these terms may be measured, but the measurements will not be constant values. The moisture concentration for a certain characteristic will be different for every soil, and even for one soil it is not constant. Furthermore, it is difficult or often imposible to relate the relative values for these different terms. Griffin (1963) recently reviewed the literature on soil moisture as it relates to plant pathology. He concluded that "there is remarkably little unambiguous evidence on the effect of the soil moisture regime on the activities of soil fungi or on their influence on root diseases caused by fungi." Most of the literature on soil moisture and Rhizoctonia diseases may be classified as ambiguous information, but there are certain apparent generalities. Where necessary, we have reinterpreted the published data in light of the basic concepts of soil moisture.

Soil moisture may be expressed under two basic concepts: the amount of water in a given mass or volume of soil, and the energy of retention and thermodynamic concept of water activity. The former concept is useful in biology, when measurements can be compared to a moisture-tension curve and known physical characteristics such as field capacity, permanent-wilting point, and saturation point. If aeration is of primary importance, then additional soil characteristics such as porosity, bulk density, and noncapillary porosity must be known. The thermodynamic concept of water activity is most important when the water is held with such force that it will move only very slowly under the influence of gravity (Taylor, 1958). These soils are at "field capacity" or drier and the hydraulic flow of water is negligible. Problems with aeration are not so great below field capacity if the noncapillary porosity or air-capacity is not too low. Moisture stress is generally more important below field capacity; above field capacity, aeration is probably the most important factor. The relation between moisture content and water activity is not linear, nor is it a single valued function (Taylor, 1958). Temperature influences water activity appreciably but has only a small influence on moisture content. Thus, one can demonstrate water flow against a concentration gradient (Taylor, 1958). However, in future work, if the amount of water in the soil is relative to the problem and needs to be known accurately, then it should be measured directly. Likewise, if the free energy of soil moisture is pertinent to the studies, then it should be measured directly. With the aid of previously plotted moisture-tension curves, it is possible to estimate one from the other. Taylor, et al. (1961) have thoroughly evaluated and discussed the methods for measuring soil water.

In damping-off experiments (Abdel-Salam, 1933; Roth and Riker, 1943*b*; Beach, 1949; Wright, 1957), optimum damping-off has been obtained at moisture concentrations of 20-80% of saturation. Disease was less consistent at saturation. A lettuce strain that primarily attacked the stem near the soil line or above the soil caused maximum damping-off at 80% of saturation (Abdel-Salam, 1933). The higher soil moisture was obviously necessary to keep the relative humidity high near the soil surface. Flentje and Hagedorn (1964) observed a greater incidence of tip blight on emerging pea seedlings if the relative humidity near the soil line was very high. Since the pathogen was most active near the soil surface, aeration was not a limiting factor. Roth and Riker (1943*b*) found that damping-off of red pine seedlings was severe from 13-70% of saturation; however, disease was very severe at any soil moisture when the plants were covered with nonmoisture-proof cellophane. Obviously, the pathogen was sensitive to restricted aeration and high moisture stresses. Bateman (1961) observed an almost linear inverse relationship between severity of poinsettia root rot and soil moisture content between 30-80% saturation. Little root rot developed at 80% saturation. Root growth of poinsetia in the absence of pathogens was not affected between 30-87% saturation. Thus, inoculum potential was probably decreased by the restricted aeration at the higher moisture levels. Moisture stress probably did not affect either the pathogen or the host due to the nature of the experimental setup.

Pitt (1964*a,b*) found that the sharp eyespot disease of cereals was favored by low soil moisture. In this instance, low temperatures (near 9°C) were most favorable also, so the effects of these factors were probably manifested through an increase in the disease potential. Peterson, et al. (1965) studied the association of *R. solani* with wheat roots in two different soils and noticed that all *R. solani* activity occurred above the moisture equivalent (tensions less than 0.3 atmosphere). Bloom and Couch (1960) varied soil moisture between field capacity and permanent-wilting point and found no differences in the severity of brown patch on Seaside bent grass. They gave no data but they implied that moisture concentrations approaching saturation may have resulted in a higher disease severity.

Rolfs (1904) found that heavy, poorly drained soils favored the development of *R. solani* on potatoes. Good soil aeration and aeration around the foliage suppressed the development of the fungus on the stems. Heavy irrigation increased infection of young potato plants (Schaal, 1935). The development of scurf on potatoes also was more severe in poorly drained soils (Weber, 1923) and when heavy irrigation was practiced the month before harvesting (Schaal, 1935). Lenticels were probably the infection court (Schaal, 1935) so the enlarged lenticels found on potatoes under high soil-moisture conditions probably increased the disease potential. Water-logging was a predisposing factor in the root rot of coconut palms incited by *R. solani* (Menon, et al., 1952). Coconut palms along river banks had a high incidence of the disease.

The growth of *R. solani* through soil was restricted by poor aeration (Blair, 1943; Das and Western, 1959) and, likewise, its competitive saprophytic activity was decreased by poor aeration at high soil-moisture contents (Papavizas and Davey, 1961). Sanford (1938b) demonstrated the effects of soil-moisture stress on the growth of *R. solani* throughout the available moisture range. With greater moisture stresses, linear growth decreased. There was some evidence that lower temperatures increased the moisture stress, which would agree with the thermodynamics of the situation (Taylor, 1958). *Rhizoctonia solani* grows optimally at 100% RH and growth is definitely retarded at 99.5% RH (Schneider [R.] 1953). Schmiederknecht (1960) designated *R. solani* as a hygrophile, a fungus with a straight line or a concave growth curve with an optimum at 100% RH.

Moisture stresses in soil are related to the water activity and the relative humidity of the soil air. Thus, if aeration could be excluded as a variable, the optimum growth should occur at the saturation point where there is no moisture stress. A relative humidity of 98.8% exists in soil at the permanent-wilting point (Taylor, et al., 1961); thus, with Schneider's (1953) and Schmiederknecht's data, we may assume that the growth of *R. solani* is affected over the available moisture range for plant growth. The relative effects will depend on the species of higher plant involved.

The development of diseases on aerial portions of the plant has depended on free moisture or near 100% RH conditions. Sclerotia of the strains of *R. solani* inciting brown patch of turf germinated at 98% RH or above (Shurtleff, 1953b) but the fungus needed free water or nearly 100% RH for growth and infection (Dickinson, 1930; Rowell, 1951; Shurtleff, 1953b; Kerr, 1956). Successful infection of soybean leaves (Ikeno, 1933) and rice plants (Hemmi and Endo, 1933) by the "sasakii" type of *R. solani* required free moisture for at least 18-24 hours. Precipitation and high humidities were required for the development of bottom rot epiphytotics on lettuce (Townsend, 1934) and cabbage (Wellman, 1932). In fact, the only resistance found in lettuce to bottom rot was

functional in nature where the bottom leaves were more erect and there was better aeration beneath the leaves (Townsend, 1934). Fruit rot of tomatoes occurs when the fruit touches moist soil (Ramsey and Bailey, 1929). This disease has increased in importance with overhead sprinkler irrigation (Crossan and Lloyd, 1956). *Rhizoctonia solani* required free moisture or a very high relative humidity for infection-formation cushion on tomato fruits, but once the fungus was established on the fruit, external moisture did not affect disease development (Gonzalez and Owen, 1963). Free moisture or essentially 100% RH was required for germination and growth of the basidiospores (Schenk, 1924; Kotila, 1943, 1947) and the development of foliar epiphytotics.

*Aeration.* — In the previous section, the primary effect of high soil moisture was restricted aeration. Respiration in the soil system results in a depletion of the oxygen supply and an accumulation of $CO_2$ or bicarbonates. The large noncapillary pores in the soil provide much of the "air capacity" of soils and the avenues for renewal of the soil air (Baver, 1959). The effective air capacity is a function of pore size, air permeability of the soil, and air diffusion. In loam soils, diffusion was restricted with moisture tensions less than 30 cm of water (Baver, 1959). Air-diffusion rates and yields of crops have correlated positively. The air capacity (volume air/volume soil) required for optimum growth of plants was a function of soil texture, soil structure, and the crop plant (Baver, 1959). For optimum plant growth, Kopecky (1927) proposed air-capacity requirements of: 6-10% for sudan grass, 10-15% for wheat and oats, and 15-20% for barley and sugar beets. A 12-16% stand loss was experienced with sugar beets during the period following blocking to harvest, when the air capacity was 8% (Baver and Farnsworth, 1940). With a 2% air capacity, about 50% of the beets died. Baver (1959) later attributed this loss to "blackroot."

A $CO_2$ concentration of 2.5% in the atmosphere retarded the linear growth of *R. solani* on PDA (Blair, 1943) and concentrations of 20-25% $CO_2$ greatly inhibited growth of *R. solani* (Vasudeva, 1936; Blair, 1943; Durbin, 1959a) and the competitive saprophytic activity of the organism (Papavizas and Davey, 1962a). Blair (1943) studied the depressing effect of organic amendments to soil on growth of *R. solani* through soil and found that $CO_2$ accumulation was the primary factor. He found that alkaline soils acted as "carbon dioxide acceptors" so the depressing effect was less in these soils. Equilibria between $CO_2$ and bicarbonate and between bicarbonate and carbonate are dependent on the pH of the soil solution. Griffin (1963) discussed this further. Durbin (1959a) found that strains of *R. solani* showed a differential tolerance to high concentrations of $CO_2$ depending on their ecological niche. Strains that primarily cause root rots grew much better than foliar- or stem-attacking isolates in the presence of high concentrations of $CO_2$. Papavizas (1964) found that the single-basidiospore progeny from a culture expressed quite different tolerances toward high concentrations of $CO_2$. Some of the single-basidiospore isolates were more tolerant to $CO_2$ than the heterokaryotic parent culture was. Storage of fresh vegetables in atmospheres containing about 25% $CO_2$ will inhibit the growth of *R. solani* on the produce (Brooks, et al., 1936).

Bateman (personal communication) and Papavizas and Davey (1962a) found that the oxygen supply does not become limiting to *R. solani* growth in vitro or in soil until the concentration in the atmosphere is 1%. Thus, from the available literature on oxygen and $CO_2$ concentrations in soil (Baver, 1959), it appears that with restricted aeration, $CO_2$ toxicity was a more probable explanation than oxygen deficiency for reduced growth of *R. solani,* especially in the neutral or acidic soils. When disease severity increased with restricted aeration, restricted aeration increased the disease potential relatively more than it decreased the inoculum potential.

Both soil texture and soil structure have a profound effect on soil aeration (Baver, 1959). These two characteristics influence both air capacity and the noncapillary porosity. Rhizoctonia diseases of the roots that are most severe under moist soil conditions are also most severe on poorly structured, heavy clay soils during moist seasons (Rolfs, 1904; Small, 1927; Fahmy, 1931; Abdel-Salam, 1933; Afanasiev and Morris, 1942). Improvement of the soil structure (Afanasiev and Morris, 1942; Baver, 1959), and drainage (Rolfs, 1904), and the addition of sand, cinders, manure, or plant debris (Abdel-Salam, 1933; Baver, 1959) increased aeration and decreased the disease potential. Roth and Riker (1943a) and Harter and Whitney (1927) reported that *R. solani* damage was more severe on light sandy soils. Here, poor aeration probably decreased inoculum potential more than it increased disease potential, or the hosts (red pine seedlings and sweet potatoes, respectively) were sensitive to high moisture stresses.

*Hydrogen ion concentration.*—Sherwood (this vol.) reviewed the literature pertinent to pH and growth of *R. solani*. In general, most strains of *R. solani* grow quite well over a broad pH range and certainly within the pH limits of most agricultural soils and host tissues. Plants in general grow poorly at the extremes of pH, both acidity and alkalinity, that *R. solani* tolerates. Over the pH ranges for good plant growth, most authors (Schaffnit and Meyer-Herman, 1930; Frederikson, et al., 1938; Vasudeva and Ashraf, 1939; Sanford, 1947; Bloom and Couch, 1958; Bateman, 1962, 1963; Pitt, 1964a) failed to find any effect of pH on severity of diseases. Roth and Riker (1943b, c) found that damping-off of red pine seedlings by *R. solani* increased in soils more acid than pH 5.5. In contrast, Jackson (1940) decreased damping-off of ponderosa pine and Douglas fir by *R. solani* by decreasing pH from pH 6.5 to values as low as pH 3.5. Jackson's experiments were performed in liquid and sand cultures. Weindling and Fawcett (1936) lowered soil pH to 4 and successfully controlled damping-off of citrus seedlings by *R. solani*. Lowering the

pH of steamed soil did not control the disease when the soil was infested later with *R. solani*. Thus, they suggested that a change in the soil microflora to the detriment of *R. solani* had occurred after the pH was lowered. The inconsistency of results obtained with steamed soil by other investigators (see section on inoculum density) may negate their conclusion. Weindling and Fawcett used aluminum sulfate and acid peat moss to lower the soil pH. Jackson (1940) presented some evidence that aluminum ions at low pH decreased the pathogenic activity of *R. solani* and Kahn and Silber (1959) reported a great retardation in linear growth of *R. solani* by sphagnum peat moss.

At present, there is no evidence that pH affects the inoculum potential of *R. solani* to any extent. However, if the processes of host penetration are primarily enzymatic (Bateman, this vol.), then the pH of the rhizosphere soil solution may affect enzyme production and enzymatic activity. Soil reaction influences the availability of certain mineral ions that may function as cofactors for enzymes. (Tisdale and Nelson, 1960). Calcium, magnesium, potassium, nitrogen, iron, phosphorus, and sulfur have been linked to disease development (see section on nutrition) and the availability of all is directly or indirectly related to pH. Thus, pH could indirectly affect the disease potential through mineral nutrition.

*Fungitoxic compounds.*—Fungitoxic materials have been added to soil and seeds to eradicate, exclude, or hinder *R. solani*. These compounds definitely influence inoculum potential and they may be phytotoxic and increase the disease potential. The addition of large quantities of crop residue to soil may result in the production of phytotoxic compounds during residue decomposition (Cochrane, 1948; Patrick and Koch, 1958). These definitely increase the disease potential toward *R. solani* (Martinson, 1959; Toussoun and Patrick, 1963).

The effectiveness of fungitoxic compounds in decreasing inoculum potential has been modified by many factors. Isolates of *R. solani* have differed in their tolerance to a toxic chemical (Kernkamp, et al., 1952; Sinclair, 1960) and the continued use of an effective toxicant such as PCNB has favored a shift in the *R. solani* population in soil toward biotypes with greater tolerance to PCNB (Shatla and Sinclair, 1965). Inoculum density and the capacity factors function together as regulators of inoculum potential, so a change in one factor influences the inoculum potential. Fungitoxic compounds act by decreasing the inoculum potential so the effectiveness of a chemical treatment must be judged in light of the residual inoculum potential. Martinson (1963) demonstrated that the effectiveness of PCNB in soil for control of damping-off was determined by the concentration of PCNB and the inoculum density of *R. solani*. At low inoculum densities, damping-off was controlled with 10 ppm PCNB. As the inoculum density was increased, additional PCNB was required for disease control. Richardson and Munnecke (1964) expanded this work when they used thiram [bis(dimethylthiocarbamoyl) disulfide] and MMDD (methylmercuric dicyandiamide), the active ingredient of Panogen, to control *R. solani* and *Pythium* irregulare as preemergence damping-off pathogens. They obtained very precise log/log correlations between inoculum density and effective fungicide dosage. Reavill (1954) demonstrated that the capacity factors of temperature, fungous nutrition, and pH can affect the fungitoxic activity of TCNB (tetrachloronitrobenzene) on fungi. Rushdi and Jeffers (1956) observed the effects of these same factors and of soil texture and soil amendments on fungicide activity toward *R. solani* in soil. Thus, all the interactions among factors demonstrate and certainly increase one's appreciation of the dynamics of inoculum potential.

DISEASE POTENTIAL.—Obviously, host susceptibility has much to do with disease severity and incidence of disease. The ability to contract disease is a function of the inherent resistance of the host, determined by its genetic capacity, its susceptibility during the period of its life cycle (which is also controlled to a significant extent genetically), and the environmental effect or predisposition (Grainger, 1956; Yarwood, 1959; Baker [R.], 1965). Other papers in this symposium have reviewed some of these areas, especially host range and basic mechanisms of resistance.

*Rhizoctonia solani* has been considered a primitive parasite (Garrett, 1956) and both intercellular and intracellular invasion of parenchymatous tissue is common (Kernkamp, et al., 1952; Christou, 1962; Chi and Childers, 1964). Thus, it may be difficult to obtain significant genetic resistance in a particular host species (McNew, 1960). Some reports in the literature confirm this belief, but there are exceptions. This is discussed by other authors in this volume.

Most of the hosts of *R. solani* are particularly susceptible to penetration and tissue invasion during the seedling and juvenile stages of growth and the plants become more resistant with age (Person, 1944a; Luttrell and Garren, 1952; Fulton and Hanson, 1960; Christou, 1962; Holly and Baker [R.], 1963; Shephard and Wood, 1963; Pitt, 1964a). Figure 1 illustrates an example of this phenomenon and depicts a fine example of the interaction of disease potential and inoculum potential in the severity of disease. Disease potential of carnation cuttings was measured at different times after transplanting. Transplanted cuttings became more resistant with age to Rhizoctonia stem rot; 15 days after transplanting, they were very resistant to *R. solani*, even at high inoculum densities. The increased resistance associated with plant age has been attributed to thickening host cell walls (Flentje, 1957), lignification (McClure and Robbins, 1942), wound periderm formation (Thatcher, 1942), and calcium content of host tissue (Bateman and Lumsden, 1965). (See Bateman [this vol.] for a more thorough discussion.) Apparently, maturing tissue is not uniformly resistant. Discrete areas of young radish tissue may be susceptible, but, with age, these decrease and eventually become entirely resistant (Flentje, Dodman, and Kerr, 1963).

Resistance of mature tissue does not imply freedom from disease in older plants, however. Neal (1942) reported a situation in which infection by *R. solani* ordinarily confined to early planted cotton seedlings persisted and the fungus caused considerable damage in older plants. Again, infected seedlings, growing under conditions unfavorable for symptom development or seedlings growing from seeds treated with chemicals not entirely eradicating the pathogen, have succumbed after being transplanted (Weber, 1931; Baker [R.] and Sciaroni, 1952).

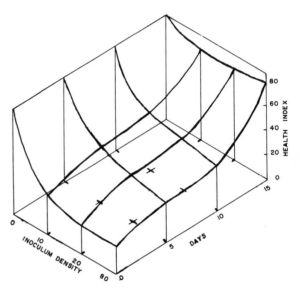

*Fig. 1.* Health indices of carnations inoculated with *R. solani* at transplanting time and at successive 5-day intervals thereafter with 10, 20, or 80 cc of cornmeal-sand cultures/1,000 cm² surface area of soil. Health index on 0-64 scale: 0 = plant dead, 64 = plant symptomless. This is a representative experiment based on averages from 36 plants per treatment divided into four replications.

In some cases, relatively mature tissue may still be susceptible. For instance, Dimock (1940) observed stocks to be susceptible after 2 months or longer. Determination of susceptibility of sugar beets at the seedling stage did not reflect the mature plant reaction to the crown-rot phase (Schuster, et al., 1958). Bottom rot of lettuce does not occur until plants are headed and approach maturity (Townsend, 1934). Ellis (1951) also reported a fruit rot of cucumber developing after the fruits were more than 3 inches long. On smaller fruits, a yellowish-brown superficial discoloration was observed but *R. solani* mycelium, while present on the surface, had not penetrated deeply into the flesh. Finally, strain differences have determined the age at which a host was attacked. Ui, et al. (1963) reported an interesting situation in flax where one strain of *R. solani* attacked seedlings in the spring, but not mature plants, while another strain infected mature plants in the summer, and was only weakly pathogenic to seedlings.

MICROBIAL INTERACTIONS AFFECTING INOCULUM POTENTIAL.—Beginning with the oft-cited work of Weindling (1932) and Weindling and Fawcett (1936), interactions of *R. solani* with other microorganisms in soil have been a favorite subject of research and speculation by a number of experimenters. Sanford (1946) reviewed the early papers on this subject and Papavizas reviewed them in this symposium. In summary, some investigators have noted synergism with supposedly nonpathogenic organisms in soil (e.g. Wright, 1945); more frequently however, evidence for antagonism has been demonstrated. Dunleavy (1952) added an antagonistic strain of *Bacillus subtilis* to steamed soil; this soil and soil not infested with the bacterium were infested 7 days later with *R. solani*. Damping-off of sugar beets was markedly less severe in the presence of the antagonistic bacterium. *Fusarium* sp. frequently isolated from roots of apparently healthy conifer seedlings in Saskatchewan forest nurseries were added to soils with *R. solani* (Vaartaja and Cram, 1956). Less seedling mortality occurred when goth genera were present than when *R. solani* was alone.

In demonstrating suppressive effects of microflora on *R. solani*, most investigators have first shown these to be antagonistic (usually antibiotic) in culture. This model was then adopted to the greenhouse situation where steam treated soils and suitable hosts were used to demonstrate the antagonistic microbial interaction and its effect on inoculum potential. Similar results, predictably, have been difficult to obtain in raw soils. Rich and Miller (1962), however, were able to reduce the incidence of root disease of strawberry by adding wheat kernels infested with antagonistic *Trichoderma viride* to field soil 10 days before transplanting. Even so, the effect of the *Trichoderma* was not isolated, since an organic amendment (wheat) was added also and the experimental design shares the same disadvantage of all others so far reported in the literature: failure to use a suitable control fungus that was as nearly identical to the original antagonist as possible except that it was not antagonistic in culture.

MICROBIAL INTERACTIONS OR ASSOCIATIONS AFFECTING DISEASE POTENTIAL.—The association of *R. solani* with other pathogens has often resulted in greater injury to a host. Thus, infection of olive rootlets by *R. solani* has increased symptoms of Verticillium wilt (Wilhelm, et al., 1962). An interaction, as measured by reduced yield in wheat, was noted between *R. solani* and *F. solani* (Price and Stubbs, 1963). Damage from *R. solani* inducing crown-bud rot of alfalfa was significantly increased when cultivars of *F. roseum* were present (Hawn and Cormack, 1952). The entry of *Aspergillus niger* and *A. flavus* into peanut kernels was associated with breakdown in pod structure incited primarily by *R. solani*, sometimes in combination with the lesser cornstalk borer (Ashworth and Langley, 1964). Finally, other pathogens have been instrumental in predisposing hosts to

attack by *R. solani*. Maier (1961a) suggested that *Thielaviopsis basicola* functioned in this way on beans.

The association of *F. solani* and *R. solani* on potatoes was studied in some detail by Elarosi (1956, 1957a,b). Sites originally occupied by *R. solani* were inoculated 30 days later with *F. solani*. The resultant rotting, greater than that produced by either alone, was not a case of primary and secondary infection since *F. solani* was the first to advance intercellularly and it also positively influenced pathogenesis by *R. solani*. Pimple-like outgrowths resulting from elongated parenchyma were observed to contain conidia of *F. solani* (Elarosi, 1956). From studies of the fungi in pure cultures, Elarosi (1957b) suggested a number of mechanisms for the synergism: (1) *R. solani* altered the pH of the substratum to a pH suitable for growth of *F. solani*, (2) vitamins produced by *F. solani* had a stimulatory effect on *R. solani*, and (3) *F. solani* partially degraded the pectic substances to forms that *R. solani* could readily utilize.

Synergism with virus diseases also has been demonstrated. For example, *Fragaria vesca* plants inoculated with chlorosis virus were considerably more susceptible to Rhizoctonia root rot than were plants free from the virus (Skiles, 1953). Cucumber mosaic virus infection in cucumber seedlings increased damping-off due to *R. solani* from 10-15% to 60-87% (Bateman, 1961). This was probably due to increased susceptibility to fungal attack following transport of food materials from the roots to the virus-infected cotyledons. Observations were made that "Ratoon" sugar cane was particularly susceptible to root rots caused by *R. solani* (Anon., 1922).

The relation of nematodes to pathogenesis by *R. solani* has already been adequately discussed by Bateman (this vol.).

Finally, Neal and Newsom (1951) observed an apparent interaction between soreshin disease of cotton and thrips damage. Plants treated with appropriate insecticides had 70% more leaf area and much less soreshin than control plants, which had severe symptoms of soreshin. This increased susceptibility was linked with decreased photosynthetic ability and translocation of elaborated food in the weakened insect-infested plants.

Occasionally, *R. solani* was found to have a protective effect on other diseases. Klein (1959) noted a significant reduction in number of dead Harosoy soybeans when PDA slant cultures of *R. solani* were mixed with pathogenic *Phytophthora* sp. *Rhizoctonia solani* caused the soybean plants to emerge slowly, creating abnormalities especially in stem and embryo root tips. *Rhizoctonia solani* apparently incited disorganization and necrosis of epidermal cells of potato tubers, which retarded tumor formation by *Synchytrium endobioticum* (Björling, 1948). Again, Bald (1947) found a high incidence of rhizoctonia disease when scab of potatoes (*Streptomyces scabies*) was low, and vice versa. This was confirmed statistically. Runner hyphae of *R. solani* usually reached the edge of a scab lesion but rarely into it, and those that did resembled exhausted hyphae found in culture.

CONCLUSIONS.—The essential components of epiphytology of Rhizoctonia diseases have been reviewed, but the relative importance of each component has not been advanced. It may be inane and possibly misleading to generalize on a subject where much of the available data are inconsistent. The great variability of the pathogen, the taxonomic surmises, and the carelessness of investigators have led undoubtedly to much of the disagreement in the literature. The significant factor is that disease is the product of a host-pathogen interaction or, as we have stated: disease severity is the product of inoculum potential and disease potential. Survival and the dispersal of inoculum influence inoculum potential and the capacity factors influence both inoculum and disease potential. Each capacity factor, be it temperature, soil moisture, potassium availability, or one of many others, independently and dependently influences disease severity. Thus, each additional factor compounds the complexity of a single host-pathogen interaction. Since there appears to be a central trend of host-pathogen responses with respect to type of disease, it may be safe to generalize on these bases. For the most part, we must resort to philosophy, and we expect our colleagues to suggest that conclusions may be unnecessary.

For damping-off diseases and many of the canker diseases, the period when infection courts are susceptible is usually brief. The energy for infection must be quickly available. This requirement is realized most successfully with large amounts of inoculum. Environmental influences, while operative on disease potential and the metabolic activities of *R. solani*, are of less importance unless they fluctuate to the extremes. At the environmental extremes for good growth of the host, capacity factors become important, and then almost anything could attack the host in its weakened condition. So, the influence of inoculum density predominates. Experimental evidence for this reasoning is available (Martinson, 1963) and the futile attempts by growers to decrease losses from damping-off by manipulation of the environment supports the concept. The best control measures are those that decrease the inoculum density or survival of inoculum; these embrace the principles of exclusion and eradication.

With root rots and storage rots, infection courts are exposed to the pathogen for extended periods of time and the environment mediates in disease development. Although disease severity is still influenced by inoculum density, pathogen activity and host susceptibility will determine the disease severity. These are greatly influenced by capacity factors, and disease control may be accomplished by manipulating the environment. Also, disease potential is an important factor in the pathogen-suscept interaction. The persistent irritation of the host by *R. solani* may lead to infection, but the disease severity will be limited by host reaction or by limiting the duration of exposure to the pathogen. From the viewpoint of control, the principles of protection and immunization are probably the most important.

With the foliar diseases such as web blight and brown patch, the period for disease development is often short. The pathogen is operating in a widely fluctuating environment, and its activity is dependent on a certain combination of environmental conditions. Thus, sufficient inoculum must be available when the environmental conditions are perfect. The duration and frequency of the periods for optimum activity by *R. solani* is obviously important in disease development. All four principles of disease control—exclusion, eradication, protection, and immunization—may be utilized with foliar diseases, but the principle of protection appears to be most important.

# Control of Rhizoctonia

LYSLE D. LEACH and R. H. GARBER—*Department of Plant Pathology, University of California, Davis, California, and Crops Research Division, Agricultural Research Service, United States Department of Agriculture, Cotton Research Station, Shafter, California.*

With *Rhizoctonia* as with other organisms, successful control measures are determined by the characteristics of the pathogen, the host crops, and the environment. All of these factors are discussed in detail in previous papers, so discussion here is limited to a summary and classification of the consequent control measures applicable to diseases caused by *R. solani.* We will also identify areas of research that appear to offer opportunity for improving control measures in the future.

The basic considerations have been discussed already: *R. solani* is worldwide in its distribution; it is pathogenic to a wide range of crop plants; and the strains are extremely variable in specialized responses to host plants and in symptom production, temperature relations, physiological response, and resistance to fungicides.

Recent reviews containing important information on the control of *R. solani* cover chemical seed-bed treatments (Vaartaja, 1964*a*) and soil fungicides (Domsch, 1964); reviews covering the survival of microorganisms in soil and the relation of ecological factors include those by Garrett (1950), Hawker (1957), Menzies (1963*a*), and Parks (1965). Biological control was admirably covered in the Berkeley Symposium of 1963, published as the *Ecology of Soil-Borne Plant Pathogens* (Baker [K.] and Snyder, 1965).

Information on the control of diseases caused by *R. solani* is most extensive on the modification of cultural practices, including biological control, but there is an impressive body of information and experience on the use of heat and fungicides in special cases. Selection for resistance or tolerance to infection by *R. solani* plays less of a part than with some other types of diseases, but there have been numerous attempts to develop such control.

The discussion here is divided into four topics: modification of cultural practices; resistance; eradication or suppression; and interactions of organisms and chemicals. The examples included are only to illustrate principles; no attempt is made at a complete literature review of the various topics.

MODIFICATION OF CULTURAL PRACTICES.—The ideal control measure would be to exclude a pathogen from areas or fields where it does not now exist. Unfortunately, as stated, *R. solani* is already worldwide in distribution, being found in most agricultural soils at various levels of infestation. Strain differences nevertheless make it desirable to avoid introducing the pathogen from other sources through transport of infested soil, infested transplants, or contaminated or infected seeds.

*Avoiding transmission with propagating material.*— Seeds of vegetables, ornamentals, or field crops may be contaminated or infected by *R. solani.* Reports of infected bean seeds as early as 1904 (Hedgecock) have since been confirmed (Leach and Pierpont, 1956). Bean pods in contact with moist soil are invaded and the mycelium grows through the pod into the seed coat or cotyledons. Seeds that are severely damaged are readily removed in cleaning operations, but seeds that are lightly infected or only contaminated may be used in planting. These seeds may decay or may produce infected seedlings from which the fungus spreads into the soil or to adjacent seedlings.

Baker [K.] (1947) described the process of infection of pepper seeds by *R. solani* and suggested hot-water treatment of seed and modified cultural practices to eliminate this source of inoculum. Neergaard (1958) found infection on seeds of lettuce, spinach, radish, and several other crucifers, but not on grass seeds examined. Transmission of *R. solani* with seed of *Agrostis tenuis* was reported by Leach and Pierpont (1958). Ordinary seed protectants do not control seed-borne *R. solani* on most of these crops, so improved methods of culture or special heat or chemical treatments are needed to combat seed infection.

Potato tubers commonly carry sclerotia of *R. solani* on their surface, so the use of infected tubers for seed may result in infection of stems and new tubers.

Tuber-borne sclerotia are considered the most important source of inoculum in some areas, and plant pathologists have attempted to influence the survival of the fungus by destroying inoculum sources on the tuber. Many reports indicate both successes and failures from chemical treatments. Van Emden (1958) reviewed the literature on seed-tuber disinfection and stated that treatment is desirable even when no sclerotia are present, since inoculum occurs as hyphae in cavities around tuber eyes. Schaal (1939) reported that, even then, some hyphae may penetrate tuber periderm and lenticels, escaping chemical treatment and thus remaining viable. The fungus continues to grow in stored tubers and, as storage time increases, the hyphae penetrate deep enough into the tubers to escape the toxicant effect (Blaszczak, 1954). The time required to kill sclerotia is a function of chemical concentration and sclerotia size (Sanford and Marritt, 1933). Some toxicants, such as mercuric chloride, are fungistatic, and their effects are negated if tubers are washed after treatment (Elmer, 1942). Chemicals used to kill sclerotia may be more toxic to tuber tissue than to fungus tissue, causing yield losses even when the pathogen is controlled (Rolfs, 1904). Hot-water treatments (Newton, 1931) failed because the tubers could not survive the high temperatures required to kill the fungus tissue.

In other areas, because of the high inoculum potentials of the pathogen in soil (Rolfs, 1904; Gilman and Melhus, 1923; Hurst, 1926; Schlumberger, 1927; Goss and Werner, 1929; Sanford, 1937), chemical treatments are no longer recommended. Infections are minimized by extended rotations, cover crops, and fertilizers.

Vegetables or ornamentals grown in greenhouses or plant beds for transplanting may become infected by *R. solani*, which is then introduced into commercial or garden plantings. Healthy transplants are assured by disease-free seed, pasteurized planting medium, and precautions to avoid contamination (Baker [K.], 1957).

*Time of seeding.*—Seeds have characteristic optimum and minimum temperatures for germination and seedling maturation. Pathogens also have characteristic growth responses to changes in temperature. As demonstrated repeatedly, seeds of high-temperature crops germinated near their minimum temperatures are frequently infected by soil-borne fungi, resulting in seed decay or damping-off. Fulton, et al. (1956), working with cotton seedlings, found that *R. solani* played a very important role as a pathogen when soil temperatures were low early in the season, and a less prominent role as the temperature increased later in the season. In contrast, low-temperature crops planted at soil temperatures below the minimum for favorable growth of *R. solani* usually escape early-season infection (Leach, 1947). Spinach or sugar beets planted in California in November, December, or January usually escape seedling infection because low temperatures inhibit the pathogen while permitting germination of the crop seeds. Consideration of the temperature response of both host and pathogen often permits selection of the most favorable seeding dates. This topic is discussed in greater detail by Baker ([K.], this vol.). These relationships can be altered appreciably by low-temperature or high-temperature strains of the fungus.

*Seed bed preparation.*—Many investigators have observed that deep planting of seed exposes more seedling tissue to infection and prolongs the period of susceptibility. Injury is usually minimized by planting the seed as shallow as soil moisture permits. In rainfall areas, seeding can usually be quite shallow without excessive risk from moisture deficiency. In irrigated areas, it is customary to preirrigate the fields and then prepare the seed bed for crops such as beans or cotton. Because of loss of surface moisture, bean seeds are planted 2-4 inches deep in rows spaced 30 inches apart, and, after emergence, irrigation furrows are prepared and weeds are controlled by cultivating the soil against the hypocotyls, thus covering an additional 3-5 inches of stem tissue. Rhizoctonia infection commonly occurs on hypocotyls during and soon after emergence, and also on the buried stem tissue following the first irrigation. The first type of infection can be reduced by incorporating fungicides in the seed furrow (by methods discussed later), but it would be impractical to treat all of the soil crowded against the upper stems. In repeated field experiments conducted by ourselves and by Kendrick, Paulus, and Davidson (1957), the incidence of early-season infection was reduced by localized placement of fungicides in the seed furrow, but yield was not increased, partly because of later infection by *R. solani* and partly because other root-rot organisms, such as *Fusarium solani* f. *phaseoli*, were not controlled by any of the fungicides used. Another limitation on fungicidal control of *R. solani* on beans is the fact that bean varieties differ considerably in tolerance to concentrations of pentachloronitrobenzene (PCNB), the fungicide most commonly used for control. With some bean varieties, especially certain processing types, the concentration of PCNB required to prevent infection affects root development seriously and causes stunting of seedlings (Tolmsoff and Leach, unpub.).

Under such circumstances, Rhizoctonia infection can be minimized by modifying the system of culture. An example is found in the San Joaquin Valley of California. Extension personnel developed a practical system of shallow planting that effectively reduced infection without the use of fungicides. Figures 1 and 2 contrast the conventional planting method with the modified procedure.

In the modified method, weeds are controlled by a preplant application of herbicide; the soil is formed into beds on 42-inch centers; the beans are seeded ½-1 inch deep in two rows 12 inches apart on each bed; and germination is promoted by subirrigation of the beds from irrigation water in the furrows. As a result, less than 1 inch of hypocotyl tissue is exposed to infection, and this only for short periods.

Accurate experimental comparisons are not available and would be difficult to secure. Observations by growers and farm advisors, however, indicate that the modified shallow planting reduces infection and increases yields. Other benefits are a reduction in cultivation and hand weeding and an increase in total length of bean rows of 40% per acre (which may be responsible for part of the yield increase). They also observed a reduction of pod infection from contact with the soil. This last would, of course, reduce seed transmission.

The modified shallow seeding appeared so attractive that within 3 years 35% of all bean acreage in San

# CONVENTIONAL SEEDING

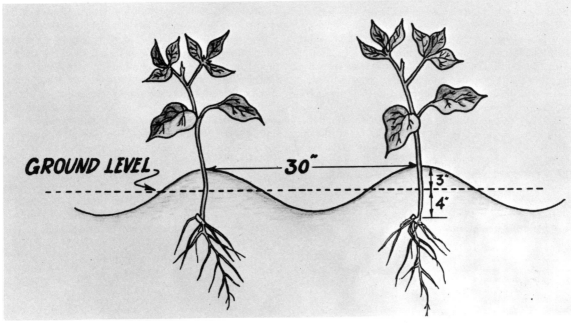

*Fig. 1.* Conventional bean seedling, with 3-4 in. of hypocotyl covered with soil and exposed to infection by *R. solani.*

# SHALLOW SEEDING

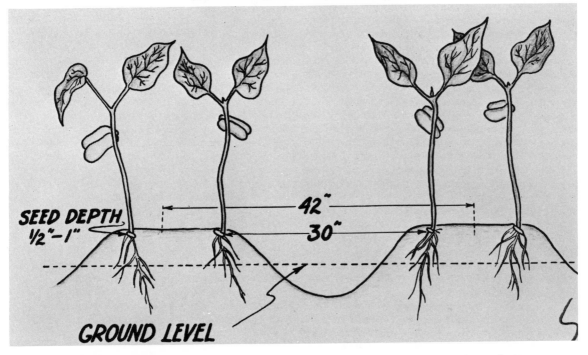

*Fig. 2.* Shallow seedling of bean, with only ½-1 in. of hypocotyl exposed to infection by *R. solani.*

Joaquin County has been converted to this system. This seeding procedure, similar to the usual bed plantings of lettuce and sugar beets in irrigated areas, is followed only where weeds can be controlled by preplant herbicide application and where beds can be formed that will subirrigate uniformly.

It is obvious that the system would have to be modified for use in other areas or with other crops, but an attractive control measure is offered by the principle of providing minimum contact of susceptible plant tissue with infested moist soil to reduce the probability of Rhizoctonia infection.

Another example where excessive soil contact contributes to infection is the crown phase of dry-rot canker of sugar beets, a disease studied extensively by Richards (1921a). When soil is cultivated against the petioles of flat-planted sugar beets in forming irrigation furrows, soil-borne R. solani infects the petioles and grows into the crown, causing partial or complete destruction of the crown and leaves. Bed planting and care in cultivation minimize soil contact with the petioles, reducing this type of infection.

Another case of Rhizoctonia infection induced by excessive contact with moist contaminated soil is crater spot of celery petioles (Houston and Kendrick, 1949).

*Crop sequence.*—In soils already infested by R. solani, infection on susceptible crops can be minimized by rotation with crops that do not support the growth of the fungus. Crop sequence is undoubtedly the best known, most widely adopted, and perhaps oldest method of controlling plant diseases.

There are many reasons for rotating crops or for planting them in a certain sequence. This discussion centers primarily on selection of the crop sequence that will minimize Rhizoctonia infection on succeeding susceptible crops. In controlled greenhouse experiments, Coons and Kotila (1935) showed that growing alfalfa or sweet clover greatly increased damping-off in sugar beet plantings that followed. Corn as a preceding crop reduced damping-off of sugar beets, whereas beans neither reduced nor increased the incidence of infection. Field observations confirmed these results. Presumably, R. solani was one of the pathogens involved, though this was not stated. Maxon (1938) reported that continuous growing of sugar beets increased Rhizoctonia rot, the most serious disease in the Great Plains area. Rotation with grain or other nonhost crops for 1-4 years reduced the severity of this disease.

In the San Joaquin Valley, Rhizoctonia seedling infection of cotton is usually more severe following plantings of alfalfa, legume pastures, or sugar beet than even with continuous cotton plantings. Rhizoctonia infection of cotton is less following cereal crops, and is highest following a cowpea green-manure crop.

Although tomato plants are usually damaged by R. solani only in the seedling stage, contact with infested moist soil during and after harvest may result in abundant infection of fruit and vines. This infected tissue supports the pathogen during the winter, and in the following spring may contribute to extremely destructive infection on sugar beet seedlings growing in the contaminated soil. If a crop of grain sorghum intervenes, however, the tomato tissue decays and the inoculum potential drops to a low level, minimizing infection on the following crop of sugar beets.

The general objective of high profits in crop production occasionally leads to monoculture of susceptible crops or to crop sequences that allow pathogens such as R. solani to survive parasitically on alternate crops or weed hosts. Some crops increase the inoculum level of R. solani because the fungus multiplies as a parasite on the plant tissue and survives as a saprophyte on plant residue, whereas other crops do not support the fungus, so that the inoculum is reduced either because the fungus dies out passively or because it is actively reduced by biological activity.

*Biological control.*—The term biological control is used by some writers in a very broad sense to include many cultural relations, whereas others use a more restricted meaning.

With specific reference to plant disease, Garrett (1965) used the following definition: "Biological control is any condition under which, or practice whereby, survival or activity of a pathogen is reduced through the agency of any other living organism with the result that there is a reduction in the incidence of the disease caused by the pathogen."

Garrett's definition is fulfilled by the processes by which plant residues, soil amendments, or nutrients stimulate the multiplication of organisms that are competitive, antagonistic, or parasitic to the pathogen.

In an outstanding series of papers, Weindling (1932) showed that *Trichoderma lignorum* could reduce the pathogenicity of R. solani by the production of a toxin. Acidification of the soil favored *Trichoderma* and led to increased control of damping-off on citrus seedlings. Moje, et al. (1957) reported that acetylenedicarboxlic acid treatments stimulated the production of an almost pure culture of *T. viride* in soil. Those workers suggested that this might provide an indirect method of controlling the disease. Similarly, Richardson (1954) found that thiram protected pea seedlings for a longer time than the chemical persisted in soil. He suggested that thiram-resistant species such as *T. viride* became dominant, and thus suppressed the pathogens through competition or direct antagonism. He pointed out that *T. viride* is known to produce two antibiotics, gliotoxin and viridin, and can protect seedlings of several species of plants from damping-off organisms. Other researchers (Daines, 1937; Wood [R. K. S.], 1951; Wood [R. K. S.] and Tveit, 1955) have also described the use of antagonistic organisms for controlling R. solani.

Garrett (1965) pointed out, however, that only a transient change in the soil population is likely to result from the inoculation of unsterilized soil with a selected antagonistic microorganism.

An alternative to the introduction of antagonistic organisms is the amendment of soils by adding organic or nutrient materials or incorporating selected plant residues to stimulate the activities of competitive or antagonistic organisms. Wright (1941) found, in both field and greenhouse tests, that the damping-off of broad-leaf tree seedlings was less severe following the incorporation of corn, wheat, or oat residue than following legumes. He also showed that damping-off increased directly with the nitrate content of the soil. Applications of glucose reduced nitrate content and strikingly reduced damping-off.

According to Sanford (1947), isolates of *R. solani* soon disappeared from the soil in the presence of wheat, oats, barley, or corn. Similarly, Kommedahl and Young (1956) found that infection of wheat seedlings was strikingly reduced after two crops of corn or oats. The effect on parasitism by *R. solani* of amending unsterilized soil with green manure was described by Boosalis (1956).

Snyder, et al. (1959) demonstrated repeatedly that the infection of beans by *Rhizoctonia, Fusarium,* and *Thielaviopsis* was reduced by incorporating barley or wheat straw, corn stover, or even pine shavings into the soil. In contrast, alfalfa and soybean residues increased the incidence of Rhizoctonia infection. They suggested that reduction of the disease was associated with high carbon/nitrogen (C/N) ratios.

Residues of lettuce or tomato increased Rhizoctonia root rot of beans, whereas residues of alfalfa, barley, soybean, sorghum, or sesame reduced infection (Maier, 1959a).

The effects of inorganic fertilizers were studied by Das and Western (1959), who reported that the growth of *R. solani* in sterilized soils was increased by balanced fertilizers or by moderate doses of phosphorus or potassium. Growth was decreased by high doses of nitrogen or potassium.

In a series of papers, Davey and Papavizas (1959, 1960) and Papavizas and Davey (1959c, 1960) reported studies on the effects of crop residues and soil amendments on the growth of *R. solani* and its pathogenicity on beans. They concluded that inhibition of *R. solani* was maximum in soils adjusted to C/N ratios of 40-100; and was less at ratios in higher (200-400) or lower (5-20) ranges.

Reduction in fungus activity has been attributed to: (1) increase in $CO_2$ concentration in the soil atmosphere; (2) scarcity of available N in soil solution; and (3) antagonistic effects of other organisms.

Although most reports indicate that *R. solani* activity is favored by high nitrogen, Hills and Axtell (1950), Walker [A. C.], et al. (1950), and Williams [W. A.] and Ririe (1957) reported that dry-rot canker of sugar beets was reduced significantly by high doses of nitrogenous fertilizer. The high nitrogen in these cases may have been related to increased resistance of the host to infection rather than to suppression of the pathogen.

Several investigators have stated that practical biological control of root diseases, though a possibility in the future, is not promising at present. Actually we are probably using forms of biological control in selection of crop sequence, incorporation of green manure or crop residue, and even in our fertilization practices.

Because our information is incomplete, however, it is often difficult to explain the results, and more precise information will be needed before the full potential of biological control can be realized.

RESISTANCE TO RHIZOCTONIA INFECTION.—Although crop plants differ considerably in susceptibility to infection, it apparently has been extremely difficult for plant breeders to develop, among susceptible varieties or species, strains resistant to *R. solani*. This is understandable when we consider the nature of the parasitism of the fungus (Bateman, this vol.); its extremely wide host range, extending over many genera and species; and the lack of sharp differentiation among specialized strains of the pathogen. This relationship was summarized by Hanna (1956) as follows: "On theoretical grounds, the odds would seem to be against finding in a single variety, resistance to a fungus such as *Rhizoctonia solani*, which is not selective in its parasitism. In this particular area of plant protection, chemical control seems to offer greater promise of success than control by resistant varieties."

Improvement of the resistance of sugar beets to several root-rot pathogens was reported by Downie, et al. (1952). Nebraska 525, a selection, was definitely more tolerant to Rhizoctonia infection than its parent. Afanasiev and Morris (1952, 1954, 1956) reported testing several hundred varieties and inbred lines of sugar beets in greenhouse and field trials. They found that 19 varieties showed a high degree of resistance to root rot.

A selection of large-seeded lima bean with outstanding tolerance to root rot was reported by Kendrick and Allard (1952, 1955). This favorable character is being included in breeding programs but has not yet appeared in any commercial varieties released.

Luthra and Vasudeva (1941) were unable to find any resistant varieties of cotton, and Richter and Schneider (1953) encountered the same situation with wild species and varieties of *Solanum*. Some potato varieties, however, have some resistance (Hofferbert and Orth, 1951; Komlossy, 1954; Kulmatycka, et al., 1955). Hofferbert, et al. (1953) correlated this resistance with germination vigor and rapid growth, vigorous sprouts, extensive root and stolon production, regenerative power, and an undefined resistance of shoots to invasion by *R. solani*. Other examples in which varietal resistance has been reported are beans (Machacek and Brown, 1948; Schroth and Cook, 1964), gladiolus (Creager, 1945), lettuce (Poole, 1952), and rice (Hashioka, 1951). Steinwat, et al. (personal communication) reported in a line of lima bean a resistance to stem rot that appeared to be inherited as a single dominant factor.

An unusual case of resistance to bottom rot of cabbage was recently reported by Williams and Walker (1966). They identified a resistant variety, 3564R, and found that the resistance was inherited as a monogenic dominant character.

In general, while it has been possible to identify differences among varieties or selections in susceptibility to Rhizoctonia infection, it is extremely rare that a high degree of resistance has been found or produced by selection or breeding within a susceptible host species.

ERADICATION OR SUPPRESSION.—An organism that is as widespread and omnipresent as *R. solani* cannot yet be considered in terms of eradication on anything other than a very localized basis. It can be eliminated from greenhouse soils, plant beds, or even portions of field soils by heating soil above the thermal death point of the organism or by applying chemicals at lethal rates.

*Control by heat.*—Heating soil to kill *R. solani* and other pathogenic organisms is used widely with container-grown plants in greenhouses and nurseries, and less extensively with plant beds and small-scale field plots. This topic is discussed in detail in a service manual by Baker ([K.], 1957). Dry or moist heat (usually steam or hot water) may be used.

Complete or near-complete sterilization of soil has dangerous consequences if a pathogen is reintroduced, for in the absence of competition it can grow luxuriantly and cause serious disease losses. Ferguson (1953) showed that pepper plants sown in steam-sterilized flats could be protected from accidental contamination by *R. solani* if selected antagonistic but nonpathogenic fungi were introduced into the soil immediately after sterilization. This method has not come into general use, however.

Since most plant pathogens in soil can be killed at a pasteurizing temperature of 60°C for 30 minutes Baker [K.] and Olsen (1960) and Baker ([K.], 1962) proposed the use of aerated steam at moderate temperature instead of complete sterilization, with the attendant dangers mentioned above.

Steaming the soil to about 60°C will destroy plant pathogens such as *R. solani* without eliminating saprophytic organisms that may be antagonistic to any introduced pathogens. Treatment at 100°C, on the other hand, produces overkill—a biological vacuum—and the first organism to return will luxuriate; if this is a pathogen, serious disease loss may result (Baker [K.] and Roistacher, 1957; Baker [K.], 1962).

Steaming at temperatures below 100°C is possible with aerated steam (Baker [K.] and Olsen, 1960; Baker [K.], 1967). Such treatments reduce the danger of recontamination (Baker [K.], et al., 1967; Olsen, et al., 1967; Olsen and Baker [K.], 1967), post-treatment phytotoxicity (Dawson, et al., 1965), and also the cost of treatment (Morris, 1954; Johnson [D.] and Asa, 1960; Baker [K.] 1967).

Mixing air with steam lowers its temperature to a level determined by the ratio of air to steam. At 100°C, the ratio is 0:1, at 71.1°, it is 3.4:1 by weight and 2.1:1 by volume; at 60°, it is 6.5:1 and 4.1:1, respectively. Valve-regulated steam is usually injected into an air stream from a straight-radial-blade centrifugal blower whose flow is regulated by an input damper (Baker [K.], 1967). The mixture is usually injected into the soil mass from below through a plenum chamber. Aerated steam moves through the soil in

the same way as regular steam, and at the same relative speed, and may be used in the same types of equipment (Baker [K] and Roistacher, 1957). It is as easy to use as regular steam, and somewhat less expensive. At 60°C for 30 minutes, pathogenic fungi, bacteria, and nematodes are killed, nearly all viruses are destroyed, and most weed seeds are killed if the soil is kept moist for 3 days prior to treatment (Baker [K.], 1967).

*Control by fungicides.*—Chemical control of disease caused by *R. solani* is practical with many crops and, in some instances, has become the principal method of control. The method of fungicide application varies with the nature of the host-pathogen relationship, the cultural practices employed, the activity of the fungicide, and the value of the crop (Burchfield, 1960; Kreutzer, 1960).

As discussed earlier, an extremely versatile species such as *R. solani* may incite foliar diseases as well as many soil-borne diseases, such as seed rot, preemergence and postemergence seedling disease, crown and root diseases of older plants, and tuber diseases such as black scurf of Irish potatoes. Fungicide control of such a diverse array of diseases ranges from general and repeated applications to localized application of volatile or nonvolatile chemicals or even simple seed treatment.

The host-pathogen relationship determines in part the area to be chemically treated, but the cultural practices used for the host plant influence the method of application and/or even the formulation of fungicide used. Plants may be grown in pots, flats, or greenhouse benches, or transplanted to beds as seedlings. Plants grown in rows in the field may be seeded directly or transplanted. They may be seeded continuously in rows or planted in regularly spaced hills. Vegetables and field crops may be planted into flat fields, with the soil cultivated into raised beds around them at a later time, or may be planted into preformed beds. Propagating units may be planted shallow or deep. Practices in rainbelt areas differ from those in the dry western United States, where crops are planted either in preirrigated soil or in plant beds irrigated following planting. Each of these variables may influence timing, rate, placement, and formulation of fungicide application.

The area to be treated and the method of fungicide application are important, but fungicide effectiveness is also dependent on the activity of the fungicide. Fungicidal activity is related to chemical qualities such as volatility, solubility, or, in some instances, systemic activity in the host tissue. Selective fungicides may control *R. solani* and not other pathogens. Broad-spectrum materials may eradicate or inactivate the entire soil flora within their effective range. All chemicals must be used at rates that are fungistatic or fungicidal to *R. solani* but not phytotoxic to the host. Fungicides must be compatible with other pesticides or chemical fertilizers applied to the plant, and must not leave toxic residues in host tissue harvested.

Finally, control measures are limited to fungicide rates not in excess of the minimum economic limit imposed by the monetary or esthetic value of the host.

The above questions must be answered by the individual exigencies of each situation. No general rules can be established; these questions must be answered before the most desirable material and method for chemical control can be chosen.

Foliar diseases incited by *R. solani* have been controlled by spraying or dusting leaves or other aboveground parts with fungicides. Brown [W.] (1935), in one of the first published reports on the use of PCNB as a fungicide, reported controlling Rhizoctonia disease of lettuce by dusting the plants and soil.

Fumigation of soil with general biocides is expensive, and therefore limited to greenhouse soils, plant beds, nurseries, and a few field-grown crops that offer high financial returns. Usually, in addition to *R. solani,* such a treatment is expected to kill other fungi, nematodes, soil insects, and even weed seeds. Improved materials and automated methods of fumigation have revolutionized the production of forest nursery plants (Vaartaja, 1964*a*) and field-grown strawberries (Wilhelm, et al., 1961). Because most fumigants eliminate almost all competitors and leave little or no residue, great care must be exercised to prevent the reintroduction of pathogens such as *R. solani* into fumigated soils.

The general incorporation of solid fungicides into large soil areas can be illustrated by Livingston's experiments (1962) on the control of Rhizoctonia infection of potatoes. In this case it was necessary to treat the soil in the area where the tubers were formed. PCNB was broadcast over the entire area and disked into the surface soil. In surface-irrigated areas, the treated surface soil is cultivated into raised beds prior to seeding. Livingston and associates found that disease control and yield increases were related directly to dose rates, though phytotoxicity also increased at the higher rates.

Volatile chemicals have also been used for similar purposes, either incorporated into the soil or injected at intervals over the area to be treated (Newhall, 1955). To obtain control in the latter case, the volatile phase of the chemical must move for some distance in the soil.

Thaxter (1891) made perhaps the first attempt to control a plant disease by localized placement of fungicide in a seed furrow. He placed sulfur in the furrow to control soil-borne spores of the onion smut fungus, *Urocystis cepulae* Frost. Later, Selby (1900, 1902) developed the formaldehyde drip method for the same purpose. Interest in direct soil treatment with protective fungicides was stimulated by the report of Leach and Snyder (1947) that a seed row application of nabam was effective in controlling the root-rot complex, including *R. solani,* on seedling beans and peas. At about the same time, Hildebrand, et al. (1949) and McKeen (1950) reported that thiram in the seed furrow controlled seedling diseases of sugar beet. By 1952, nabam was being field-tested on cotton seedlings. When the effectiveness of PCNB

against Rhizoctonia infection of cotton seedlings was confirmed, field experiments were soon conducted across the cotton belt (Brinkerhoff, et al., 1954; Ranney and Bird, 1956; Bird, et al., 1957).

In Rhizoctonia seedling disease of row crops, the main plant part to be protected is the underground root or stem from the depth of seeding to the soil surface. The soil volume treated is small, varying from 1-4 square inches times the number of linear feet of crop per acre. The fungicide rate, determined in greenhouse trials, is calculated on the basis of the parts per million of active ingredient per unit weight of air-dried soil that are required to control heavy infestations of *R. solani.* To obtain protection, it is necessary to treat the maximum area permitted by the limitations of cultural practices and crop economics. Either volatile or nonvolatile fungicides may be applied at nonphytotoxic rates during the planting operation.

Table 1 shows a simple method of converting the effective concentration of fungicides determined in greenhouse experiments to the dose required to produce equivalent concentrations in localized field applications.

Localized placement of fungicides proved to be practical and is used each year on millions of acres of cotton to control *R. solani* and other fungus-caused seedling diseases. The soil area treated varies but, for economy of operation, is usually about 2 inches wide and as deep as the seed is planted. Whether fungicides are applied as sprays, dusts, or granules, the intent is to apply a portion of the protectant to the seed and the rest to the covering or surrounding soil. Specially designed equipment attached to the planter can apply fungicides at the time of seeding.

In recent years, large areas of cotton have been treated during planting by adding relatively large volumes of dilute dusts to seed—the so-called planterbox or hopper-box method. The dust falls by gravity flow, along with the seed, into the bottom of the seed furrow. Very little dust is mixed into the covering soil.

Studies have been conducted to determine the method of application of chemicals that will give the best distribution throughout the soil profile. Garber and Leach (1957) and Luick, et al. (1959) traced the distribution with chemical indicators, including radioactive rubidium 86, an oil-soluble safranin dye, and a fluorescent dye; Ranney and Hillis (1958) and Johnson [T. R.] and Hillis (1958) used a fluorescent dye. Figure 3 shows distribution patterns produced by a single spray nozzle directed either into the bottom of the seed furrow or onto the covering soil, contrasting them with the pattern produced by a double nozzle arrangement.

Ideally, a fungicide is applied best in the area where the host is most susceptible or liable to fungus attack. Proper placement of fungicides in the seed furrow has been stressed by several workers and has been assumed to be from the seed level to the soil surface. In cotton-growing areas of the Southwest, however, where little or no rain falls after seeding, fungicides applied to the immediate seed areas and not in the

TABLE 1. Relationship of concentration of fungicide (ppm of dry soil) to rate of field application (active ingredient lb per acre) with row treatments. Conversion of field doses to ppm of fungicide[1]

| | Lb/acre | Cross section of treated area, sq. in. | | | | |
|---|---|---|---|---|---|---|
| | | 1 x 1"<br>ppm | 2 x 1"<br>ppm | 2 x 2"<br>ppm | 2 x 3"<br>ppm | 2 x 4"[2]<br>ppm |
| 20" row[3] | 0.25 | 17 | 8 | 4 | 3 | 2 |
| | 0.50 | 33 | 17 | 8 | 6 | 4 |
| | 1 | 67 | 33 | 17 | 11 | 8 |
| | 2 | 133 | 67 | 33 | 22 | 17 |
| | 4 | 266 | 133 | 67 | 44 | 33 |
| 30" row[3] | 0.25 | 25 | 12 | 6 | 4 | 3 |
| | 0.50 | 50 | 25 | 12 | 8 | 6 |
| | 1 | 100 | 50 | 25 | 17 | 12 |
| | 2 | 200 | 100 | 50 | 33 | 25 |
| | 4 | 400 | 200 | 100 | 67 | 50 |
| 40" row[3] | 0.25 | 33 | 17 | 8 | 6 | 4 |
| | 0.50 | 67 | 33 | 17 | 11 | 8 |
| | 1 | 133 | 67 | 33 | 22 | 17 |
| | 2 | 266 | 133 | 67 | 44 | 33 |
| | 4 | 533 | 266 | 133 | 88 | 67 |

[1]Assumes apparent specific gravity of 1.3 equivalent to 81 lb dry soil/cubic foot. Modify proportionately for higher or lower apparent specific gravity.

[2]Cross section of treated area is based on depth of planting and width of furrow opener. For example, 2-in. deep and 2-in. wide gives a cross section of 4 sq. in. to be treated.

[3]With a 20-in. row, there are 26,136 ft of row per acre; with a 30-in. row, there are 17,424 ft of row per acre; and with a 40-in. row, there are 13,068 ft of row per acre.

covering soil have many times given excellent seedling disease protection from organisms such as *R. solani*.

In greenhouse trials, PCNB at 50 ppm localized in the seed area gave complete protection from *R. solani* even though the soil was heavily infested from the seed level to the soil surface. We can only speculate about the nature of the fungicidal activity involved in these cases, where disease protection is apparently achieved more than 1½ inches above the area of application. In any case, earlier concepts of proper fungicide placement have been revised, regardless of whether the fungicide is giving protection because of volatility or solubility in soil solution, or is systemic in host tissue. Bell and Owen (1963) found the most effective placement of the fungicide in the rainbelt area to be uniform distribution from the bottom of the furrow to the surface of the soil.

Fungicides, such as PCNB, are commonly used that are highly specific against *R. solani*. Although such narrow selectivity may give very effective control of Rhizoctonia seedling disease, it is nevertheless a questionable practice unless some provision is made to protect the seedling against other seedling pathogens. Numerous workers have found that controlling *R. solani* with PCNB may accentuate problems with other organisms, such as *Pythium* spp. (Gibson, et al., 1961): In one of our greenhouse trials, PCNB was applied at 50 ppm by soil weight to control an anticipated *R. solani* problem in a soil mix. The intent was to avoid this organism while studying a Fusarium-wilt root-knot-nematode complex of cotton. A high percentage of seedlings died in the PCNB-treated soil but none in untreated soil. The dying seedlings

yielded *Pythium aphanidermatum* and *P. ultimum*. Problems such as these will be avoided by using less-selective fungicides or a combination of fungicides effective against different organisms.

A common practice for many years has been seed treatment of row crops for protection against seed decay and preemergence damping-off caused primarily by *Pythium* sp. or *R. solani*. This often improves emergence but provides little protection against postemergence damping-off or root rot caused by *R. solani*.

In many experiments (Leach, et al., 1959, 1960), postemergence damping-off has been reduced by combining PCNB or other fungicides effective against *R. solani* with the usual seed treatments, such as mercury compounds, thiram, captan, or Dexon. With relatively nontoxic fungicides, higher doses to increase protection may be applied to the seed as an "overcoating" by the use of moisture or adhesives.

It is obvious that several methods are available for controlling Rhizoctonia infection of row crops with fungicides, including various forms of seed treatments, planter-box mixtures of fungicidal dust with seed, spraying, dusting, or granular application into the seed furrow and incorporation of fungicide into the soil. The choice of material and method will depend on the degree of protection required and the cost that can be reasonably justified.

*Interaction of Pesticides.*—Although the most widely used controls for some diseases caused by *R. solani* are perhaps single and combination fungicide treatments, certain important aspects of chemical control remain unclear. For example, Nash (1967) reported

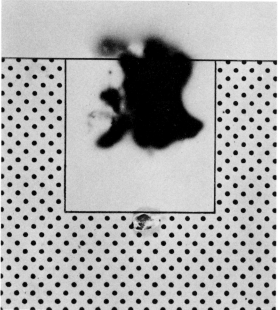

*Fig. 3.* Autoradiograms showing the distribution of radioactive rubidium applied in seed furrows through: (*a*) single nozzle directed onto seeds in bottom of furrow, (*b*) single nozzle directed into covering soil behind seed drop, and (*c*) two nozzles, one directed into furrow and one into covering soil.

that cotton production currently uses a great number of chemicals: at least 12 fungicides, 10 herbicides, and 28 insecticides. Other crops are treated similarly. It is important to recognize that most of the research work in the development of chemical controls has been conducted by pathologists without particular regard to other pesticides used in plant culture. It has been known for a number of years that certain insecticides used as seed treatments are harmful if not used in conjunction with a fungicide (Leach, et al., 1954; Richardson, 1960; etc.). Motsinger (1967) summarized research projects on pesticide interactions affecting cotton seedling diseases, citing a number of synergistic or additive effects, mostly deleterious.

Pinckard and Standifer (1966) reported an apparent interaction between cotton herbicidal injury and seedling blight due to *R. solani*. Infection was increased by deep incorporation of the herbicide trifluralin, but was almost eliminated if PCNB was used in combination with it.

As modern plant culture trends are away from hand labor and toward complete mechanization, involving the use of a multiplicity of chemicals, future research efforts will need to involve cooperative studies between pathologists and scientists from other disciplines.

*Interaction of Nematodes and Rhizoctonia.*—It is well known that nematodes and pathogenic fungi interact to enhance disease incidence and severity. Among numerous examples that can be cited is the classic relation between root-knot nematode and *F. oxysporum vasinfectum*. Similar relations can be shown for other diseases, including *R. solani* (Reynolds and Hansen, 1957). It is very difficult to establish the role of the nematode in these situations. Most

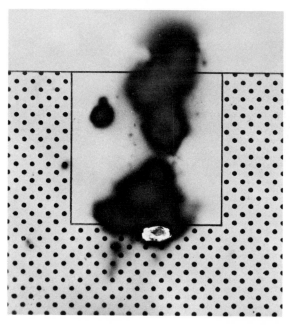

of the pathogenic fungi involved, including *R. solani*, are quite capable of invading or damaging plant cells in the absence of nematodes. As of now, we must still speculate as to whether the nematode enhances plant diseases by providing invasion sites simply by injuring root tissue, or whether some more sophisticated mechanism in some way predisposes plants to fungus attack. This subject has been discussed in recent years by Sadasivan (1961).

NEW FUNGICIDES.—Fungicides used to control *Rhizoctonia* have included both selective and broadspectrum types. Several chemicals used at fungistatic rates can effectively control the fungus on seedlings under most circumstances. PCNB is a good example of a narrow-spectrum chemical that has been used at low rates to control *R. solani* on leaves and in the soil. Zineb, maneb, or captan, in contrast, are examples of broad-spectrum chemicals that will control *R. solani,* though at comparatively higher rates. Several compounds reported to have systemic action have been introduced recently. Demosan, registered for use as a seed overcoating, is effective against *R. solani* but when phycomycetes are also involved, is used in combination with mercury or other seed treatments. Experimental materials such as Vitavax and Du Pont 1991 are effective against *R. solani* as seed or soil fungicides, but are not yet registered for commercial use.

It can be anticipated that new and more effective fungicides will become available for specific purposes in the future.

# Rhizoctonia Solani: Special Methods of Study

J. B. SINCLAIR—*Department of Plant Pathology, Louisiana State University, Baton Rouge.*

ISOLATION AND CULTURE OF RHIZOCTONIA SOLANI. *Media used for isolation.*—*Rhizoctonia solani* is comparatively easy to isolate from infected plant tissues, soil, and rhizospheres. The fungus grows on a wide range of solid and liquid media ranging from water agar to highly specialized media (see Table 1). Recipes for most of the common media were presented by: Riker and Riker (1936), *Difco Manual* (Anon., 1953), Johnson [L. F.] et al. (1959). Many workers prefer to use simple or modified potato-dextrose agar (PDA) or other decoction agars. Maier and Staffeldt (1963). for example, found *R. solani* isolates from cotton varied in their ability to grow on carrot-decoction agar, PDA, and nitrate-dextrose agar. All isolates, however, grew well on PDA. Ashour and El-Kadi (1959) compared the growth of *R. solani* on five different media. They found that the fungus grew most rapidly on PDA and Richard's media.

For routine isolation of *R. solani*, these agars are often modified with acids, bacteriostatic or fungistatic compounds. Acidified PDA helps control bacterial contamination and may favor growth of *R. solani*, since the fungus prefers an acid medium (Samuel and Garrett, 1932; Sanford and Marritt, 1933; Allington, 1936; Kotila, 1945a; Houston and Kendrick, 1949; Warcup, 1950).

Three acids are commonly used for acidification of agar: acetic, lactic, and phosphoric. A few drops of a 10% solution of one of these may be added to a culture plate prior to pouring of agar or directly into the medium before dispensing. Acidification usually gives a pH range of 3.5-5.0. The response of *R. solani* to pH is highly variable. In general, the optimum for growth of most isolates of *R. solani* is between 5.5-7.0 (Ragheb and Fabian, 1955; Elarosi, 1957b; Bateman, 1962). There is probably some reduction in growth on media with a pH lower than that for optimum growth. Latham and Linn (1961) found media at pH 4.0 using lactic acid suppressed growth of the fungus. Slower growth isolates may be lost in the presence of more rapidly growing ones.

Some workers (LeClerg, 1939; Papavizas and Davey, 1959a; Latham and Linn, 1961; Schmitthenner, 1964) incorporated antibiotics into media to decrease bacterial contamination and to increase frequency of isolating *R. solani*. Papavizas and Davey (1959a) used a mixture of fresh solution containing 50 ppm each of aureomycin hydrochloride, neomycin sulfate, and streptomycin sulfate in water agar to increase the frequency of isolating *Rhizoctonia* spp. from soil. Latham and Linn (1961) found that PDA containing either streptomycin sulfate at 100 μg/ml or vancomycin at 50 μg/ml was superior in all instances to media containing rose bengal. They reported also that novobiocin did not suppress the growth of *R. solani*. Schmitthenner (1964) reported 50 mg/ml of streptomycin sulfate added to media helped reduce bacterial contaminants. Rose bengal (Schmitthenner, 1964) and methylene blue (LeClerg, 1939a) also were used to reduce contamination.

Isolates of *R. solani* should be routinely grown on a liquid medium, such as potato-dextrose or potato-sucrose broth, to detect contamination by either bacteria or yeasts that may not be apparent on agar.

*Isolation of R. solani from infected plant tissue.*—Standard procedures for isolation of plant pathogenic fungi are generally used to obtain *R. solani* from infected plant tissues. Small sections of infected roots stems, leaves, or reproductive organs are aseptically removed from the plant after either surface sterilization or washing in tap or distilled water. Rao (personal communication) found that *R. solani* was unable to tolerate highly competitive conditions on agar plates and classed the fungus as having limited "competitive saprophytic ability." Diseased tissue therefore should be either thoroughly washed or surface sterilized before plating.

Parmeter (personal communication) found that plating isolation pieces on water agar minimizes competition, allows *R. solani* to grow away from most competitors, and facilitates hyphal-tip isolation.

Solutions of 1.0% sodium hypochlorite (Kernkamp, et al., 1949; Schmitthenner, 1964) or 0.1% mercuric chloride (Mohamed, 1962) have been used to disinfest plant tissues. Luttrell (1962), however, stated that since *R. solani* is frequently killed by usual procedures of surface sterilization, isolations from infected leaves, stems, and roots should be made by washing bits of tissue in sterile water before plating out fragments on agar media. Bourn and Jenkins (1928a) washed infected portions of various aquatic plants before plating out. Hansen and Curl (1964) prepared clover stolons for isolation by cutting them

199

TABLE 1.  Media used for isolating and culturing *R. solani* and literature cited.

| Medium | Citation |
| --- | --- |
| **Decoction and extract agars** | |
| Bran | Ray (1943) |
| Bean hypocotyl | Bateman (1963a, 1964b) |
| Carrot | Maier and Staffeldt (1963) |
| Coen's | Person (1944a) |
| Cornmeal | Ullstrup (1936); Saksena (1960); Schmitthenner (1964) |
| Dextrose-yeast extract + V-8 juice | LeClerg (1939a) |
| Leomin | Person (1944a) |
| Lima bean | Ullstrup (1936); Schmitthenner (1964) |
| Malt | Ullstrup (1936); Gibson, et al. (1961) |
| Nutrient | Chesters (1948); Warcup (1950) |
| Nitrate-dextrose | Maier and Staffeldt (1963) |
| Potato | Shephard and Wood (1963) |
| Potato-cerelase | Kilpatrick, et al. (1954) |
| Potato-carrot-dextrose | Maier (1962) |
| Potato-dextrose | Samuel and Garrett (1932); Allington (1936); Ullstrup (1936); Weber (1939); Neal (1944); Kotila (1945a); Chesters (1948); Houston and Kendrick (1949); Allison (1952); Papavizas and Davey (1959a, 1962b); Saksena (1960); Martinson and Baker (1962); Bateman (1963a); Shephard and Wood (1963) |
| Potato-dextrose + germinating radish-seed exudate | Martinson and Baker (1962) |
| Potato-marmite | Whitney (1964a) |
| Potato-sucrose | Shatla and Sinclair (1963); Elsaid and Sinclair (1964) |
| Richard's | Ullstrup (1936); Ashour and El-Kadi (1959) |
| Soil-decoction | Carpenter (1949) |
| Tromer's malt | Neal (1944) |
| Water | Boosalis and Scharen (1959); Shephard and Wood (1963); Whitney (1964) |
| Water + germinating radish-seed-exudate | Martinson and Baker (1962) |
| Water + cold-sterilized aster straw | Flentje and Stretton (1964) |
| **Synthetic** | |
| Czapek's | Allington (1936); Maier (1961b); Ruppel, et al. (1965) |
| Czapek-Dox | Kaufman and Williams (1965) |
| Czapek-Dox + yeast-extract | Warcup (1950); Gibson, et al. (1961) |
| Edward's synthetic | Edwards and Newton (1937) |
| **Liquid** | |
| Fries' nutrient solution | Millikan and Field (1964) |
| Potato-broth | LeClerg (1939a); Shephard and Wood (1963) |
| Potato-dextrose-broth | Bateman (1963a); Ruppel, et al. (1965) |
| Potato-sucrose-broth | Shatla and Sinclair (1963); Elsaid and Sinclair (1964) |
| **Other** | |
| Barley-grain | Creager (1945) |
| Cornmeal-vermiculite | Kernkamp, et al. (1949) |
| Cornmeal-sand | Gibson, et al. (1961) |
| Cornmeal-water | Dickinson (1930) |
| Corn, sand, or soil | Tervet (1937); LeClerg (1941b); Ray (1943); Demaree (1945); Kotila (1945b); Boosalis (1956); Flentje and Stretton (1964) |
| Oat | Ullstrup (1936); Person (1944a) |
| Oatmeal-Sand | Shephard and Wood (1963); Ruppel, et al. (1965) |
| Oat-wheat and water | Person (1944a) |
| Sorghum | Brodie and Cooper (1964) |
| Sterile moist black soil | Bruehl (1951) |
| Vermiculite + nutrients | Varney (1961) |

into segments 3-6 inches long and washing in tap water for 20 minutes; then through two changes of sterile, demineralized water on a mechanical shaker for 10 minutes each and finally by hand for 2 minutes in 50 ml of sterile water. The excess water was blotted aseptically and the stolons were then cut into three segments. These segments were plated on PDA at pH 4.2 and incubated at 26°C for 6 days.

Schmitthenner (1964) washed diseased alfalfa seedlings for 10 minutes under running tap water in a stainless steel 5-inch sieve inserted in a beaker. Approximately 1.0 ml Tween 20 (Hercules Powder Company) was added at the beginning and halfway through the washing cycle. Excess liquid was blotted with sterile filter paper. The seedlings were placed on agar media in petri plates and the agar disk was inverted over the tissues.

The fungus could be either transferred to new agar plates after it grew out from the tissue and hyphal-tip isolates recovered, or, if the fungus appeared to be in pure culture, hyphal-tip isolates could be made directly from the original plate.

*Isolation of R. solani from soil.*—There were four books and six references published recently on the biology and ecology of soil-borne microorganisms and plant pathogens, portions of which are pertinent to this subject (Garrett, 1956; Burgess, 1958; Parkinson and Waid, 1960; Park, 1963; Maciejowska, 1964; Baker and Snyder, 1965). There were several review articles (Montegu, 1960; Warcup, 1960; Durbin, 1961; Menzies, 1963a,b) and a handbook (Johnson, [L. F.] et al., 1959) published on techniques for isolating and assaying microorganisms from soil. Those techniques that have had some degree of success for the isolation of *R. solani* shall be discussed.

Dilution plate (Johnson [L. F.], et al., 1959).—A water suspension using a 25 gram sample of dry soil was shaken for 30 minutes. A 10 ml sample was added to 90 ml sterile-water blank and followed through a successive dilution series until the final test dilution was reached. After shaking, 1.0 ml of suspension was aseptically transferred to agar medium just above solidifying temperature. After incubation, the average number of colonies per dish was multiplied by the dilution factor to obtain the number of microorganisms per gram in the original soil.

Maier (1961b) reported fairly reliable recovery of *Rhizoctonia* spp. from soil by making initial dilutions of 1:10 in pasteurized sand, with subsequent dilution in distilled water to $1:10^3$ or $1:10^4$, pipetted into sterile plates and swirled into potato-carrot-dextrose agar.

Microscopic detection on plant debris (Bourn and Jenkins, 1928a; Boosalis and Scharen, 1959).—This method was based on the distinct morphological characters of *R. solani*, which make it easy to identify. Plant debris fragments screened from field soil were plated directly onto modified water agar, and the fungus was then detected by direct microscopic observation. The principal disadvantage of the technique was that it dislodged all of the sclerotia from some debris

tissues and thereby introduced error in estimating the number of infested substrate sites. Another disadvantage was the difficulty in distinguishing the mycelium of *R. solani* from certain other fungi. These authors found that the standard soil-dilution plate method was unsatisfactory, since it did not yield *R. solani* from the same soil samples.

In the technique of Bourn and Jenkins (1928a), soil to be tested was suspended in tap water and then passed through a 600-mesh screen. The settled soil was resuspended and repeatedly passed through the screen until the supernatant was clear. Pieces of wood, root, straws, lumps of soil, and other foreign objects were removed with forceps, then a strong jet of tap water was run through the screen to remove any soil adhering to the organic debris particles. Washed debris particles were blotted dry between sterile paper towels. From the towels, 100 particles were picked at random and transferred to 25 plates of water agar at pH 4.8. Plates were marked at four points on the bottom and one drop of 2.0% solution of streptomycin sulfate was placed above each mark. A debris particle was placed on top of the streptomycin sulfate drop and incubated at 24°C for 48 hours. *Rhizoctonia solani* grew from the particles, and hyphal-tip isolates were transferred to PDA.

Hyphal isolation method (Warcup, 1955).—Soil was suspended in beakers of water and allowed to settle. The supernatant was poured off and the soil resuspended and allowed to settle until the supernatant was clear. The soil was spread in a small amount of water in a culture dish. The individual hyphae or groups of hyphae were detected under a binocular dissecting microscope and 20-50 were removed with a fine forceps to sterile culture plates with a small amount of sterile water. Cooled, but still liquid, agar was poured into the plate and the hyphae were dispersed by rotating the agar before it solidified. Hyphae showing growth were transferred to fresh media.

Improved immersion plate.—This technique was described by Wood [F. A.] and Wilcoxson (1960) as a variation of Chesters (1940) technique. After 25 ml of water agar had solidified in sterile culture plates, a sterile disk was placed on the agar surface and the cover was closed (Fig. 1). To isolate fungi from soil, the covers were removed and the plates were pressed against the soil profile so that soil and disk were in contact. The plates were covered with aluminum foil and held in place with soil. After 2 days or more, the plates were taken to the laboratory, the disks were removed immediately, and the fungi isolated were transferred to fresh agar plates. This technique favored the isolation of nonsporulating fungi, such as *Rhizoctonia* spp.

Plate-profile.—*Rhizoctonia* spp. were isolated 15-18% of the time by the plate-profile method and 0-1% by the dilution plate method in studies by Anderson and Huber (1962). The plate-profile method used a 20 × 20 × 1.5 cm autoclavable plastic plate with 77 holes 1.0 cm deep and 0.5 cm in diameter spaced at

*Fig. 1.* Improved immersion plate used for isolating fungi from soil showing placement of sterile, perforated disk in agar plate (Wood and Wilcoxon, 1960).

2.5 cm intervals (Fig. 2). There were 11 horizontal rows of seven holes each. The plate was autoclaved and, after cooling, each hole was filled aseptically with sterile agar. The holes were covered with autoclaved plastic electricians' tape. The plate was exposed for about 5 days to a soil profile after small holes were punched in the tape directly above the agar-filled holes. The tape was removed in the laboratory and the agar plugs were transferred to agar plates. The frequency of isolation of various soil fungi and their location at specific levels in the soil profile were determined with this method. The method lends itself to a study of microorganisms active in the vicinity of plant roots.

Glass-cloth and nylon-mesh traps.—Maier (personal communication), employing 1 × 1 cm square of multifilament glass cloth embedded in soil, reported that *R. solani* readily colonized the cloth and formed abundant sclerotia (Fig. 3). He found that percentage colonization was a more reliable method for estimating populations than the immersion-tube method described below. Monofilament nylon mesh with 400 squares per square centimeter was used by Old and Nicolson (1962) to study the spread of *R. solani* in soil. These authors suggested that the method might be used to isolate the fungus from the soil. It might also be used to make population estimations.

Immersion tubes.—This method was devised by Chesters (1940), who used a large glass tube, closed at one end with holes blown out of the lower half of the tube. The tube was filled with nutrient agar and placed in soil. Pieces of agar from the immersion tube were plated on PDA after 7 days. *Rhizoctonia* spp. were among the fungi collected. Mueller and Durrell (1957) used agar-filled plastic centrifuge tubes with holes arranged in a spiral. The holes were covered with electrician's tape, which was pierced with a sterile needle just before immersion into the

soil. Martinson and Baker (1962) increased the frequency of isolation of *R. solani* by adding exudate of radish seed to agar placed in the tubes.

Thornton (1958) designed another modification of the immersion-tube technique, using perforated rods and tubes. Sterile-water agar was dispensed aseptically with a hypodermic syringe into holes in a sterile rod. The rod was inserted into a perforated tube with the agar-filled holes turned away from the tube apertures. A hole was dug in the soil, the tube inserted horizontally into the soil profile, and the rod was then turned so that the agar-filled holes were opposite the tube apertures. Just before removal, after 4-7 days, the rod was rotated to turn the agar-filled holes away from the apertures. The agar plugs were punched out of the rod in the laboratory and plated on nutrient agar.

A modification of Chesters' immersion tube was designed by Gochenaur (1964). This method reduced selectivity for fungi that grow rapidly and tolerate low-oxygen tension. Pyrex immersion tubes were invaginated with 2-4 capillaries and then filled with 1.0 g of dry soil in place of nutrient agar. The tube was moistened, plugged with cotton, capped with aluminum foil, and autoclaved. Tubes were placed in soil for 7 days, then withdrawn. After the exterior was cleaned with 70% ethanol, the contents were removed and dispensed on three agar plates. The technique provided a well-aerated medium in which moisture content rapidly came into equilibrium with surrounding soil. This technique has not been tested specifically for the isolation of *R. solani* from soil, but it should be tried since it offered some advantages over other immersion-tube methods.

Varney (1961) showed that *R. solani* grew satisfactorily in vermiculite saturated with nutrients. Substituting nutrient vermiculite for agar or soil in immersion tubes might improve recovery of *R. solani* and should be explored. The advantages would be a standardized medium that would be loose and well aerated and in which moisture levels could be controlled. The mixture could easily be broken apart and uniformly distributed on agar plates for assay.

Trapping and baiting.—Plant segments of bean, cotton, wheat, barley, oat, carrot, and potato have been used successfully as plant baits for *R. solani* (Sneh, et al., 1966). They found that autoclaved green segments buried in nonautoclaved soil both increased and decreased the acceleration in percentage of colonized plant fragments, whereas ripe cereal segments were less influenced by autoclaving.

Sterile lima bean segments (Chu, 1966) and split-stem segments of rush (Downes and Longhnane, 1965) were successfully used to isolate *Rhizoctonia* spp. from various soils.

The use of buckwheat stems for trapping *R. solani* was worked out by Papavizas and Davey, 1959b, 1961, 1962b). Segments of mature buckwheat stems about 5.0 mm long were mixed with soil and incu-

*Fig. 2.* Plate-profile technique for isolating soil microorganisms. (A) Preparation of profile. (B) Pin holes punched through tape over agar-filled holes. (C) Plate placed against soil profile in vicinity of roots. (D) Electricians' tape stripped from plate for transferring invaded agar samples to petri plate (Anderson and Huber, 1965.)

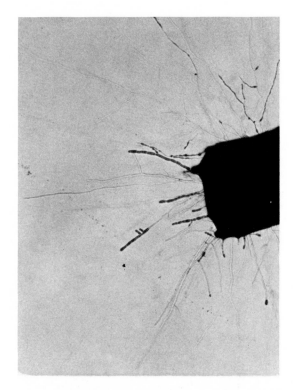 

*Fig. 3.* Low-power photomicrograph of fungus hyphae growing from colonized glass-cloth fiber 24 hours after planting. Several different fungi were recovered in this isolation (× 40), Maier (1967).

bated for 4 days at 25°C. The segments were recovered from soil, placed on 10-mesh screens, and washed thoroughly under tap water. Excess water was removed by blotting between sterile paper towels. Segments were either soaked in an antibiotic solution before plating to reduce contaminants or plated on 2% water agar containing 50 ppm of streptomycin sulfate, aureomycin hydrochloride, and neomycin sulfate. Hyphal-tip isolations were made after 20-24 hours incubation.

Sneh, et al. (1966) reported stem segments of cotton and bean gave higher percentage of colonization than buckwheat.

El-Zarka (1963) developed a modification of the buckwheat method using *Cochorus olitorius* L. as a substrate. Stainless steel insect pins 4 cm long were inserted perpendicularly through pieces of the substrate. The stem pieces were placed in the soil and recovered after 24 hours. These were washed thoroughly in tap water and surface sterilized in 1.0% sodium hypochlorite solution for 1-2 minutes, rinsed in several changes of sterile water, and soaked for 2 hours in a fresh solution of 100 ppm terramycin hydrochloride at room temperature. The pieces were removed, dried in paper towels, and plated on agar.

Messiaen (1957) compared five different methods for the isolation and culturing of fungi from soil in France. He found that *R. solani* was detected best by incubating corn grains in soil and then isolating on prune agar. A similar technique was found satisfactory by Kendrick and Johnson (1958) when they recovered *R. solani* from corn seed buried in soil and later plated on water agar.

Bourn and Jenkins (1928a) planted disinfected seed of the host plant in *R.solani*-infested soil. The soil and seed were incubated in total darkness for 7-10 days. Seedlings were then lifted from the soil and examined for lesions. Segments with lesions were cut from the stems and roots and plated on agar. A variation of this technique was used by Wilhelm (1956) to recover *R. solani* from soil in strawberry fields in California. Nonblemished roots were immersed in 0.1% mercuric chloride for 1.5-2.0 minutes, rinsed, and incubated in coarse, sterile, moist sand in petri dishes covered with lids containing a layer of 2.0% water agar. Smith [L. F.] and Ashworth (1965) used the number of lesions on seedlings of small white bean to measure inoculum potential of *R. solani* in greenhouse experiments.

*Evaluation of soil isolation techniques.*—Thornton (1956) compared the screened immersion-plate technique and Warcup's plating method for isolating *R. solani* from grassland soils. He reported that typical mycelia of *R. solani* were predominant in all sites and levels tested by the immersion-plate technique. Warcup's technique failed to yield comparable results. Since results with the Rossi-Cholodny slide, agar-soil film, and direct examination methods demonstrated the presence of the fungus, he concluded that Warcup's technique did not give a true picture of the fungus flora in soil.

Davey and Papavizas (1962*b*), using two naturally infested soils, compared four methods: debris-particle, immersion-tube, buckwheat-colonization, and infected-host. The infected-host method was found best for studying the range of fungus isolates pathogenic to certain hosts. The immersion-tube method recovered only a limited number of fungus isolates. The debris-particle method in conjunction with the buckwheat-colonization method gave the most complete information. The debris-particle method measured the quiescent isolates in soil and the buckwheat-colonization method measured the active. Isolates of the fungus obtained by the four methods varied from highly virulent to avirulent.

Further studies by Papavizas and Davey (1962*b*) used saprophytic-colonization methods to determine frequency of isolation of *Rhizoctonia* spp. from 15 naturally infested soils. They found, in general, that stem segments of buckwheat, cotton, and lima bean were satisfactory, while those of corn, oats, and soybean were not. Livingston, et al. (1962) reported the plant-debris method and the buckwheat-stem method were comparable in the frequency of isolation of *R. solani*.

Ui and Ogoshi (1964) compared several methods for frequency of isolation of *R. solani* from both artificially infested and naturally infested soils. They reported a higher frequency was consistently obtained with the flax-trap technique when compared to the soil-dilution plate, Warcup's soil-plate, LaTouche's slide-trap, Mueller and Durrell's immersion-tube, and Blair's contact-slide techniques. Mueller and Durrell (1957) stated that the soil-dilution plate method was not suitable for the isolation of *R. solani*.

Agnihothrudu (1962) compared the following methods for the isolation of various fungi, including *R. solani*, from tea soils in India: direct-isolation, Rossi-Cholodny slide, washed-root, Warcup's soil-plate, and dilution-plate. He was able to recover the fungus with the first three methods, but not with the last two.

A high degree of correlation was found between infestation level and colonization potential of *Rhizoctonia* spp. in soil with the plant-colonization, plant-debris particle, or immersion-tube methods (Sneh, et al., 1966).

One should not therefore rely on any single method for studying *R. solani* populations in soils. Warcup (1951) pointed out that "different methods of sampling soil fungi are complementary rather than exclusive."

*Media used for culturing.*—*Rhizoctonia solani*, as shown above, can be recovered on a wide range of substrates. Shaw (1912) cultured the fungus on raw plant material, such as carrots and potatoes, and on bread, meal, and filter paper. Balls (1906) grew the fungus on filter paper containing 2% of either asparagin or sucrose. Varney (1961) successfully grew the fungus in vermiculite saturated with nutrients. *Rhiz-octonia solani* is commonly cultured on agar or liquid media ranging from water agar to highly specialized media (Table 1). The basic nutrient requirements of the fungus were discussed by Sherwood (This vol.).

Many workers prefer to use a nonmodified PDA or slightly modified PDA for culturing the fungus. Successful culturing of certain isolates may require the use of an infusion broth or agar using host tissue similar to that from which the original isolates were obtained. Other specialized media may be required for specific studies concerned with nutrient requirements or physiologic studies and are reviewed in other sections of this book.

Czapek's agar or modifications of it and Edward's medium were the chief synthetic media used for culturing *R. solani*.

*Rhizoctonia solani* can be cultured successfully on various liquid media. It usually forms a tough, crusty mycelial mat on the surface of potato-sucrose broth in nonshake cultures. The mat is easily removed and washed. Most staling products can be separated from the fungus mat by decanting off liquid medium before the mycelial mat is broken up. *Rhizoctonia solani* also grows well in liquid shake cultures.

Steamed-sterilized soil was used successfully by a number of workers to increase inoculum of *R. solani* (Sanford, 1952). This substrate is available and easy to prepare. The constituents of soils are variable, however, and may not be repeatable from one area to another. Cornmeal-sand medium (Gibson, et al., 1961) could be used as a standard for this method.

Recent descriptions of the use of foam rubber (Domsch and Schicke, 1960) and synthetic sponge (Sulba Rao, 1964) for the culturing of fungi should be noted. These synthetic materials were moistened with liquid-nutrient media, autoclaved, and inoculated with various fungi including *R. solani*. The advantages of using these techniques were that the fungus penetrated readily and the substrate afforded abundant aeration and increased surface area for mycelial growth. The use of these synthetic media for culturing *R. solani* should be further explored.

*Rhizoctonia solani* is therefore a relatively easy organism to isolate and culture on a wide variety of media. Some of these are easy to prepare and the growth obtained is suitable for most studies. It would be of material advantage to future studies of *R. solani* not requiring specialized media if either one or two agar media could be universally used as standards for in vitro studies. Variation in *R. solani* attributed to the variety of media used to study this fungus might be better understood if routine culture methods could be standardized.

*Incubation and storage.*—Incubation conditions for isolating and culturing *R. solani* are not as critical as for some other pathogenic fungi. Most isolates will grow well in vitro over a relatively wide range of temperatures, relative humidity, pH, $CO_2$ concentrations, light, and other environmental factors. The literature concerning the effects of environment was reviewed by Sherwood (this vol.).

Much variability in optimum temperatures for in vitro growth of *R. solani* is reported. Incubation temperature might play an important role in recovering and maintaining the fungus only if high- or low-temperature isolates were studied. Cochrane (1958) stated that temperature optima and ranges reported were valid under the specific conditions of the experiment, with regard to time, medium, other environmental factors, and the method of measurement. He concluded that there was no single temperature optimum for either growth or any enzymatic type reaction in *R. solani*.

Most in vitro experiments in closed culture dishes would have close to 100% relative humidity, the optimum for the growth of the fungus. The fungus is known to survive in air-dried soil and may be considered tolerant to low relative humidities in soil. The author observed the fungus surviving on desiccated agar in culture tubes for many weeks, which would indicate the fungus could survive low relative humidities in vitro (unpub.).

Most isolates of *R. solani* prefer an acid medium, with optimum pH for growth being less variable than that reported for temperature. The optimum pH for in vitro growth of most isolates of *R. solani* is between 5.5-7.0 (LeClerg, 1934; Jackson, 1940; Blair, 1943; Khan, 1950; Ragheb and Fabian, 1955; Elarosi, 1957*b*; Bateman, 1962). Matsumoto (1963) found that the effect of pH varies according to the nature of the media used and thought it was almost impossible to designate a definite optimum pH value for growth of the fungus.

Blair (1943) reported that aeration of soil was necessary for optimum growth of *R. solani* as $CO_2$ accumulation was very toxic. The summary statement of Garrett (1956): "the toxic effect of carbon dioxide upon a fungus is greatest when the energy growth is least" is appropriate.

Light had an important effect on in vitro growth (Durbin, 1959*b*; Weinhold and Hendrix, 1962), sclerotia production (Valdez, 1955; Durbin, 1959*b*, and basidial production (Whitney, 1964*a*) of *R. solani*.

Storage and maintenance of stock-culture collections of *R. solani* and other phytopathogenic fungi involves certain problems. Stock cultures of *R. solani* stored in a refrigerator or at room temperature require frequent subculturing mainly because of desiccation. This may involve monthly transfers depending on environmental conditions. Viability and pathogenicity may then be altered. Shaw (1912) reported that *R. solani* remains viable for 5-6 months on agar media in culture plates and tubes. This was the experience of the author (unpub.), who recovered the fungus from 4–6-month-old agar slants by plating on fresh agar. It was not determined, however, if the fungus survived as mycelium or as sclerotia. Parmeter (personal communication) believed that most isolates will remain viable on PDA for at least 1 year. Sims (1960) found certain isolates of the fungus lost virulence following culture on amended media.

Loss of pathogenicity after prolonged storage in vitro may be a problem. Schroeder and Provvidenti (1961) found that tomato-fruit-rot isolates of *R. solani* lost their infectivity after 2 months in culture. They reported that 4-day-old cultures gave five times the infectivity on tomato that 2-month-old cultures did. Sherwood (personal communication) stated that certain cultures of the fungus stored in a refrigerator could not be recovered after 3 months. He maintained his stock culture collection of the fungus on PDA fortified with 0.5 gram each of malt extract, yeast extract, and casein hydrolysate. These cultures stored in a cabinet at room temperature and transferred at least once every 3 months have maintained their morphological and pathogenic characters for 6-12 years. Parmeter (personal communication) reports having similar results.

There are a number of techniques for the storage of phytopathogenic microorganisms. A review of these techniques was made by Iyenger, et al. (1959). One of the best methods for the preservation of agar cultures of *Rhizoctonia* spp. is with sterile mineral oil. They found that the fungus remains viable under oil after 5 but not after 7 years. Schneider [C. L.] (1957) reported that 70 of 76 cultures of *Rhizoctonia* spp. survived 6-7.5 years on PDA amended with vitamins $B_1$ and C, covered with sterile mineral oil, and sealed with cotton plugs and Parafilm. In a similar experiment, he reported that 31 of 40 cultures survived 5.5-9.0 years. Pathogenicity studies on the surviving cultures were not made. Pumpyanskaya (1964) showed that a mineral-oil depth of 0.2-0.5 cm did not prevent oxygen penetration, while 1.0 cm gave relatively anaerobic conditions, and 6.0 cm gave complete anaerobiosis.

The fungus could also be preserved either on grain or in soil. Kreitlow (1950) showed that *R. solani* could be stored for at least 16 months by growing isolates on a steamed mixture of two parts wheat and one part oats in a 250 ml flask for 3 weeks at 15°C and allowing them to dry at room temperature. The material was ground to no less than 40 mesh and stored at 5°C.

Soil has been used as a storage medium by Tolmsoff (personal communication). He reported that sugar beet isolates were stored for several months in soil by infesting autoclaved soil with cultures grown on sterile sugar beet seed and incubated until a high population built up. The soil was then air-dried and stored at cool temperatures. If the soil was maintained at room temperature in a moist condition, cultures lost infectivity to sugar beet within 2-3 months. This is opposite to results found using agar. Vaartaja (1964*b*) isolated *R. solani* from a sand-soil mixture stored for 2 years with only 0.2% by weight of water.

Sulba Rao (1964) recommended the use of a washed, synthetic sponge cut into the shape of agar slopes, inserted into test tubes, moistened with 3-5 ml of Richard's medium, and autoclaved, as an alternative to agar culture of fungi. It afforded abundant aeration and increased surface areas of mycelial ramification.

The best methods of incubation of *R. solani* in vitro are to grow the fungus: (1) only on fresh standardized, buffered medium (pH 6.0); (2) at temperatures optimum for growth of the particular isolates being studied; and (3) in a closed cabinet to prevent due exposure to sunlight or artificial light.

The best conditions for long-term storage of the fungus are fresh, standardized, buffered medium, immersion of the culture under sterile mineral oil, and storage at temperatures below 0°C. Containers should be sealed before storage. Comprehensive studies are needed, however, to determine the effects of prolonged storage under oil on pathogenicity of *R. solani* isolates.

The use of foam rubber (Domsch and Schicke, 1960) and synthetic sponge (Sulba Rao, 1964) moistened with liquid nutrients for storage of *R. solani* should be explored.

METHODS USED IN DETERMINING INOCULUM POTENTIAL AND POPULATIONS OF *R. solani* IN SOIL.—The concept of inoculum potential was reviewed by several authors (Garrett, 1960; Horsfall and Dimond, 1963; Dimond and Horsfall, 1960, Baker [R.], 1965) and is discussed in other sections of this volume. The term inoculum potential was first used by Horsfall (1932) to define "variations in the content of pathogenic fungi or inoculum in the soil." The meaning of the term has changed and become broader, but it still carries the concept that is a measure of the ability of a soil-borne organism to cause disease in a susceptible host under certain conditions, i.e. to act as inoculum. It connotes measurement of this ability in a carrier, such as soil, sand, or an artificial medium infested with the organism. Inoculum potential should not replace "pathogenicity" or "virulence," terms used to describe characteristics of organisms.

*Rhizoctonia solani* exists in soil both as a potential parasite and a saprophyte and is probably distributed throughout the soil (Garrett, 1956; Livingston, et al., 1964). Some *Rhizoctonia* spp. are active saprophytes in soil and can survive in the absence of hosts by competitive colonization of dead organic matter (Garrett, 1956; Boosalis and Scharen, 1959; Papavizas and Davey, 1961). Papavizas and Davey (1962b) emphasized that pathogenic and nonpathogenic strains in soil were morphologically indistinguishable from one another. *Rhizoctonia solani* has been classed as a primitive parasite (Garrett, 1956) and as a weakly competitive saprophyte (Rao, personal communication). Papavizas (1964) pointed out that the fungus may lose its pathogenicity after passage through soil or following changes in nutritional levels.

Susceptibility of hosts may be influenced by environmental conditions. These facts create a problem for developing an accurate method of measuring the inoculum potential of *R. solani* in soil. This is discussed below under testing for pathogenicity.

Since *R. solani* exists as mycelium in the soil, most immersion methods are useful in estimating populations of this pathogen. Martinson (1963) found that the frequency for isolation of *R. solani* with soil microbiological sampling tubes correlated with preemergence damping-off of radish. The correlation held when inoculum density, temperature, and concentration of pentachloronitrobenzene (PCNB) were varied. These experiments were conducted under controlled conditions; an isolate of the fungus known to be pathogenic to radish and watermelon was used. Under these conditions, particularly where soil was infested with a pathogenic isolate of the test fungus, the inoculum potential was measured adequately. Under field conditions, the sampling tube method would measure only population density and thus not give a true picture of the inoculum potential of the soil.

Some studies and techniques measuring the inoculum potential of *R. solani* in field soil have in reality estimated only the population of the fungus in soil, including saprophytic and pathogenic isolates. In order to measure the true inoculum potential of *R. solani* in a given soil sample, it must be shown that all isolates of the fungus in the sample are pathogenic. If an isolate is not pathogenic, it cannot be considered part of the inoculum. The problem of selective pathogenicity among isolates of *R. solani* may also become involved.

Some of the methods for isolation of *R. solani* from soil were used to compare the amount of disease with estimated populations of the fungus. Papavizas and Davey (1959b), using buckwheat stem pieces and naturally or artificially infested soil, found some correlation between infection index and frequency of isolation of *R. solani*. Their infection index, however, took into account only invisible symptoms and not preemergence damping-off. They found that 60% of 50 *R. solani* cultures isolated by the buckwheat-stem method were pathogenic. They concluded, however, that the infection index was not always reliable for detecting differences in soil infestation by the pathogen.

Livingston, et al. (1964) found general agreement between the plant-debris fragment and the buckwheat-stem-segment methods for indexing *Rhizoctonia* spp. inoculum levels. The buckwheat technique was particularly sensitive. Comparisons made by them in 1960 and 1961 showed that disease severity was greatly influenced by seasonal differences, soil types, and general field conditions.

The most successful method for predicting the severity of either Rhizoctonia root or crown rot is by estimating the inoculum potential of soil from the number of infested plant debris particles or mycelial fragments on the roots of host plants. Rich and Miller (1963) showed that the severity of Rhizoctonia root rot of strawberry was related to the number of *Rhizoctonia* spp. fragments adhering to a 5.0 cm length of strawberry root.

The debris-particle method was used by Boosalis and Scharen (1959) to compare the incidence of sugar beet crown rot in two soil samples from the same field. The first sample came from an area of the field that had a high incidence of crown rot the previous season, the second from an area with low incidence. The soil with the highest infection index also had the highest percentage of disease plant debris particles supporting strains of *R. solani* pathogenic to sugar beet. Both soil samples were infested with the same pathogenic strain of the fungus. The second sample, however, had a mixture of at least two strains, the second being pathogenic only to potato stems. This technique gave both a population estimate and an estimate of the inoculum potential of the soil, i.e. indicating the ratio of pathogenic to nonpathogenic strains.

Smith [L. F.] and Ashworth (1965) estimated inoculum potential under field conditions by counting the number of lesions that developed within a 96-hour period on small, green tomato fruit buried in soil. Tomato fruit were found to be immune to *Pythium* spp. which obscured *R. solani* damage to tomato and bean seedlings. Sanford (1952) found the disease rating of host plants to be a satisfactory measure of the persistence of *R. solani* in soils, whereas data from buried glass slides and root colonization of indicator plants were unreliable.

The application of the theory of inoculum potential to *R. solani* may best be studied under controlled conditions using artificially infested soil. Sanford (1952), Papavizas and Davey, 1959b), and Martinson (1963) infested soil with pathogenic strains of *R. solani*. Measured inoculum mixed with a known quantity of soil was varied and correlated with a disease index. The relationship among the pathogen, host, and environment then could be expressed in mathematical terms (Dimond and Horsfall, 1965). Study of inoculum potential and variables affecting it under controlled conditions might lead to methods for predicting disease severity.

STUDYING PATHOGENICITY.—Pathogenicity tests are necessary in fulfilling Koch's postulates. It is particularly important that pathogenicity be established for isolates of *R. solani* being studied.

Tolba and Moubasher (1964) stated that *Rhizoctonia* isolates are nonspecific in their pathogenicity. Most standard techniques used to establish pathogenicity of other phytopathogens must therefore be modified for *R. solani*. The fungus does not produce conidia. Basidiospore production is either rare, difficult to obtain in culture, or lacking. Basidiospore infection in nature appears to be rare, but this may be due to ignorance rather than fact. Strong (1961) attributed stem canker of greenhouse tomatoes in Michigan to basidiospore infection. Pinckard and Luke (1967) thought basidiospore infection of cotton bolls played an important role in boll rot initiation. Basidiospore infection appears to be common in the tropics (Echandi, personal communication). More studies are needed to determine the role of basidiospores in spread of disease.

Mycelia and/or sclerotia are widely used in place of spores. Some isolates produce few, if any, sclerotia in culture and mycelia must be used. If mycelia are used for testing of pathogenicity, hyphal-tip cultures should be obtained. These can be obtained by plating mass transfers of the fungus on water agar and, after several days' incubation, examining under a dissecting microscope. Single hyphal-tip cells can be cut from hyphae with a sharp dissecting knife or razor-blade chip. Cut cells become almost transparent, while intact cells remain translucent and are easily selected (Whitney and Parmeter, 1963).

*Use of basidiospores.*—The perfect state of *R. solani* forms more or less freely in nature on a wide range of hosts (Kotila, 1929, 1947; Exner, 1953; Pinckard, 1964; Whitney, 1964a; Pinckard and Luke, 1967). Successfully induced fruiting of isolates in vitro would aid materially in taxonomic and cytological, as well as pathogenicity studies. A number of methods were derived to induce basidiospore formation.

The common method is to transfer the fungus from a rich medium to a poor one (Fig. 4). The transfer of the fungus directly from host tissue or enriched agar cultures to 2% water agar was reported to be successful by a number of workers (Kotila 1929, 1947; Ullstrup, 1939; Exner and Chilton, 1944; Carpenter, 1949; Exner, 1953; Sims, 1956). Sims (1956) recorded basidiospore formation after transfer of the fungus from amended potato broth to alligator weed (*Alternanthera philoxerioides*) stems and roots. Kilpatrick (1966) induced sporulation of *R. solani* by transferring agar plugs of the fungus to a small "moist chamber" consisting of filter paper, moistened with distilled water, placed on the top and bottom of a culture plate.

A second method is to place fungus hymenium and/or mycelium on the inside of a culture dish cover and place it over either sterile water, water agar, or amended agar (Exner, 1953; Hawn and Vanterpool, 1953; Sims, 1956; Whitney, 1963).

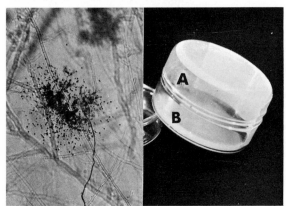

Fig. 4. Cluster of *Rhizoctonia* basidia produced on water agar (*left*) (Whitney and Parmeter, 1963) and experimental set-up for collecting basidiospores (*right*). Upper plate (*A*) contains perfect state of the fungus and lower plate (*B*) contains water agar (Whitney, 1963).

Basidiospore formation was successful when the fungus was grown directly on either Coons' medium (Kotila, 1929) or alligator weed decoction agar (Sims, 1956). Stretton, et al. (1964) and Whitney and Parmeter (1964) found that either covering or mixing an agar culture of the fungus with soil or sand would induce sporulation (Fig. 5.).

*Fig. 5.* R. *solani* fruiting in culture plates after being covered with soil to induce sporulation (Whitney and Parmeter, 1964).

Warcup and Talbot (1962) successfully used modifications of three methods to induce fruiting of basidiomycetous fungi isolated from soil. These were: Warcup's chaff method (Warcup, 1959); Tamblyn and Da Costa's wood-block method (Tamblyn and Da Costa, 1958); and Flentje's soil method (Flentje, 1956).

The perfect state of R. *solani* usually occurs in nature at the base of plants and on the underside of soil clods during periods of high rainfall and humidity (Exner and Chilton, 1943; Warcup and Talbot, 1962; Pinckard, 1964, 1967). This suggested that humidity and light were important factors in inducing fruiting. The requirement of high RH for spore production was shown by Kotila (1929), Carpenter (1949), Exner (1953), Hawn and Vanterpool (1953), and Sims (1956). Heaviest sporulation and discharge by the fungus in nature was shown to occur during the first 6 hours of darkness at night with little or none occurring during the day (Carpenter, 1949). Several investigators suggested that light inhibited the initiation of the perfect state (Kotila, 1929; Carpenter, 1949; Hawn and Vanterpool, 1953; Flentje, Kerr, Dodman, McKenzie, and Stretton, 1963; Whitney, 1963, 1964a). Whitney (1964a) suggested that alternating light and dark periods would produce a correspondingly rhythmatic cycle of sporulation.

Temperature, pH, $CO_2$, and oxygen requirements for sporulation are not critical (Kotila, 1929; Carpenter, 1949; Hawn and Vanterpool, 1953; Whitney, 1963).

A technique to induce sporulation at will may have to be developed for particular isolates. Pathogenicity studies using cultures from single spores could then be made. Once an isolate is stimulated to fruit, single basidiospores can be harvested by several methods. Kotila (1929) described a method of collecting single basidiospores using a micromanipulator. Even with modern equipment, this is tedious and is recom-

mended only where the basidiospores from a single basidium are required. Olsen et al. (1967) removed single basidiospores from a tetrad by touching them with a small water-agar block on the tip of a fine glass needle. The block with spore was then placed in the center of a culture plate containing agar.

Allowing the discharge of basidiospores from the lid of a culture dish to an agar medium below was shown to be a successful means of obtaining single basidiospores (Exner, 1953; Hawn and Vanterpool, 1953). Isolated germinating spores were located, marked, and removed by using a "biscuit cutter" technique as described by Keitt, et al. (1915). Each spore was then transferred to suitable media in a culture tube.

Other methods used for single-spore isolations of fungi may be applicable to *Rhizoctonia*. These are the dry-needle technique (Hildebrand, 1938); or dog hair attached to a metal probe or needle (Guthrie, 1963). These workers picked single spore from either agar or glass slides. Khair, et al. (1966) found it difficult to recover spores from agar, because some types of spores became embedded in the water film on the agar surface, and from glass slides because of their smoothness. They found that single spores, deposited on dialysis tubing cellophane taped over a rectangle of 4% water agar, were more easily harvested.

*Use of mycelium for inoculating aerial plant parts.—* Young plant tissue is generally considered more susceptible than mature tissue to R. *solani*. Testing for pathogenicity on aerial plant parts necessitates, in most cases, the use of either seedlings or newly formed leaves and stems. Luttrell (1962), using fescue and soybeans, found that spraying plants with a suspension of macerated agar plate cultures of R. *solani* was effective on tender seedlings and ineffective on mature leaves. Alfalfa seedlings in the unfoliate-leaf stage were found by Schmitthenner (1964) to be less susceptible to R. *solani* than were seedlings in the cotyledonary-leaf stage. Parmeter (personal communication) found that R. *solani* readily attacked older leaves, but that *Ceratobasidium* sp. attacked only young leaves.

Schroeder and Provvidenti (1961) and Ruppel, et al. (1965) worked with *Tephrosia vogelii* and were successful in establishing infection on 6-week-old seedlings by atomizing a mycelial suspension from 1-week-old PDA cultures onto plants held for 24 hours in a constant mist. They both recorded variation in virulence among isolates of the fungus. Atkins and Lewis (1954), however, obtained infection of all aerial plant parts of soybeans, regardless of age, with R. *solani* in a moist chamber at about 30°C, 24 hours after spray-inoculation. Tandon (1961) reported systemic infection on guava seedlings in a greenhouse after inoculation with R. *solani*. After inoculation, small, irregular spots occurred on upper leaves. These spots coalesced and spread downward until the entire plant succumbed.

Luttrell (1962) compared methods of foliar inoculation with R. *solani* on fescue grass. Good infection was obtained by spraying plants with a suspension of macerated agar-plate cultures. Better infection resulted, however, when alfalfa meal was added to the suspension. The best and most convenient method was dusting moistened plants with dried oat-grain cultures (Kreitlow, 1950). Allison and Hanson (1951) sprayed plants with a blended suspension of sclerotial fragments and mycelium and placed them in a moist chamber.

In order to successfully establish infection on tomato fruit, Schroeder and Provvidenti (1961) found that a PDA biscuit with mycelium from a 4-day-old culture had to be placed over an epidermal puncture made with a fine-pointed, sterile needle. Their inoculation studies showed that the type of symptoms produced depended on fruit maturity at the time of infection.

Two additional techniques described by Luttrell (1962) were the use of single oat grains from whole oat-grain cultures and single sclerotia on leaves. The use of sclerotia permitted quantitative comparisons in terms of size of lesion produced from a single point of inoculation during a specific period in an incubation chamber.

Schuster, et al. (1958) described a toothpick method for determining the pathogenicity of *R. solani* isolates on sugar beet seedlings in the field. They found, however, that the reaction of the seedling was not correlated with mature plant reaction to crown rot.

Regardless of the method used, infection seems to be influenced by available nutrients in either sclerotia or in nonliving substrates (Luttrell, 1962), and this should be taken into consideration in making inoculations, especially on aerial plant parts.

*Inoculating plant parts at or below the soil line.*—There are a number of standard techniques used to test pathogenicity of *R. solani* isolates on plant parts either at or below the soil line. Most involve using infested soil or other media sown with seed or seedlings of host plants either before or after soil infestation. Methods have been designed for laboratory, greenhouse, and field experiments. Media can be infested with fungus hyphae, either in suspension or growing on a suitable medium, or sclerotia.

Infection of stem and hypocotyls of jute and avocado was accomplished by techniques worked out by Shaw (1912) and Mircetich and Zentmyer (1964), respectively. Shaw tested for pathogenicity by placing small pieces of an agar culture on stems of susceptible jute plants, while Mircetich and Zentmyer placed 2.0 mm disks of agar with fungus in cuts in avocado stems and covered with adhesive tape. Portions of the host were surface sterilized and placed on glass slides with the test organism in a moist chamber, using a technique described by Flentje (1965).

Root inoculation with *R. solani* has received consideration, particularly on fleshy-root crops such as potato and sugar beet. Schwinn (1961), to establish the causality of *R. solani* for dry core of potato, inoculated both mechanically wounded and natural lenti-

cel excrescences. Cormack and Moffatt (1961) studied the pathogenicity of *R. solani* on sugar beet roots by removing a core of tissue, placing the inoculum in the hole, and replacing the core. The amount of decay at different incubation temperatures was recorded.

Mircetich and Zentmyer (1964) studied the pathogenicity of *R. solani* on avocado roots by using seedlings grown in aerated, complete nutrient solution in 5-gallon, ceramic crocks at pH 6.5. One-half petri dish cultures of the test fungus were blended with 100 ml of de-ionized water for 1.0 minute and poured into the crocks with the seedlings. The other half of the culture was placed in cheesecloth and suspended in the nutrient solution. At 35 days after inoculation, 60-95% of the roots were infected (Fig. 6).

*Infesting soil and other media.*—There are a variety of methods using infested soil for testing pathogenicity of *R. solani*. Some techniques have been developed in conjunction with the evaluation of soil fungicides for control of Rhizoctonia seedling diseases. A few of these were selected for review.

Infested oat and wheat grain were used by LeClerg (1941) to infect the underground stems of potato and sugar beet. The grain inoculum was placed in direct contact with the stems. He also mixed the grain inoculum with steamed soil and allowed it to stand for 2 weeks. The resulting infested soil was added to tops of greenhouse pots containing potato seed pieces. Infested oat, wheat, and/or barley grains were used to infest soil or to infest underground stems of crop plants (Newton and Mayers, 1935; Le Clerg, 1941a, Ashworth, et al., 1964).

Staten and Cole (1948) prepared infested soil by mixing diseased plant tissue with the soil. To study *R. solani* on cotton seedlings, Lehman (1940) infested soil by stirring a culture of *R. solani* in half of the soil contained in greenhouse flats.

Sinclair (1958, 1960) used 2-week-old macerated broth cultures of *R. solani* to infest greenhouse flats for the study of cotton seedling damping-off by the fungus. After 2 weeks incubation in the greenhouse, the soil was sown with cotton seed.

Bruehl (1951) prepared inoculum by incubating the fungus in flasks containing 150 grams of black soil at 20-23°C for 8 days. A portion of this inoculum was mixed in the top half of sterile soil in greenhouse pots and used to test for pathogenicity on oat and wheat seedlings.

Root canker of alfalfa was studied by Smith [O. F.] (1943) by placing transplants directly in infested soil or by allowing plants to grow about a month in noninfested soil and then placing inoculum in holes in the soil.

The pathogenicity of *R.solani* on celery was tested by pouring a suspension of macerated agar cultures around the crown and between the petioles before soil was pushed up around the base of the plants (Houston and Kendrick, 1949).

The pathogenicity of various fungi to turfgrass seedlings was tested by Endo (1961) by growing seedlings in steam-sterilized soil in clay pots. The soil was infested after germination with small plugs of washed mycelium placed in the soil. Various isolates of *R. solani* were tested for their potential to incite damping-off of *Tephrosia vogelii* (Ruppel, et al., 1965), by growing fungus isolates on sand-oatmeal medium for 3 weeks, then mixing with 300 grams of sterile sand in partially filled greenhouse pots, and covering them with a thin layer of inoculum.

Schmitthenner (1964) used three different methods to study the virulence of *R. solani* isolates on alfalfa seedlings: (1) The vermiculite-culture technique employing a 1 liter flask plus 500 ml of vermiculite moistened with autoclaved yeast-extract–glucose solution; (2) the injection-infestation technique using shake cultures of the fungus filtered, resuspended in distilled water, blended, and injected into the soil around roots of 10-day-old seedlings; and (3) the inoculum-layer technique, using fungus cultures placed in a layer below, but separated from, the seed at sowing. Using the inoculum-layer and injection-infestation methods, differences in virulence were detected.

Owen and Gay (1964) used greenhouse ground beds artificially infested with a mixture of seven isolates of *R. solani* from cotton seedlings to study control of the fungus with soil fungicides (Fig. 7). Temperature and moisture were controlled to assure rea-

sonable kill of plants in nontreated soil. They obtained good results in field trials by mixing *R. solani*-infested cottonseed and seedlings into the soil three weeks before planting and by using overhead irrigation.

Whether inoculations should be made with sterilized soil or with soil containing a normal soil microflora is open to question.

The advantages of using pure culture inoculations are that they provide: (1) effective means of controlling inoculum levels; (2) precise measurement for fungus effects; (3) severest disease conditions possible; (4) reproducible results; and (5) elimination of stimulatory and antagonistic effects of other ecosystems. The disadvantages are that, because of the alteration of soil ecosystems, conditions are not comparable to field situations because: (1) disease will likely be more severe than might be expected under field conditions; and (2) plant vigor may be increased or decreased.

The advantage of using field soil with or without infestation of the test organism is that it approximates the disease situation encountered in the field. The disadvantages may be: (1) difficulty in obtaining reproducible results because of variability in the ecosystems from experiment to experiment; (2) data may be appropriate only for the area from which the soil

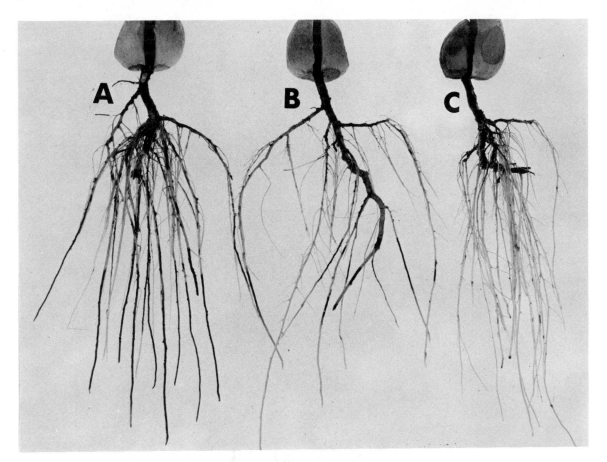

*Fig. 6.* *A, B.* Typical root lesions produced by *R. solani* on avocado seedlings in inoculated nutrient solution. (*C*) Noninoculated control. (Afetr Mircetich and Zentmyer, 1964.)

sample was indigenous; and (3) the test may not represent the severest disease situation possible.

*Use of infested sand and agar media.*—Endo (1961) was able to detect nonpathogenic and pathogenic isolates of *R. solani* from turfgrass by sowing seed on sterile quartz sand in culture plates watered with sterile water. Newton and Mayers (1935) tested the susceptibility of plants to *R. solani* by growing surface-sterilized seed on nutrient agar previously inoculated with the fungus. A similar technique was used by Bruehl (1951), who placed germinated wheat and oat seed in culture plates of the fungus. Hansen and Curl (1964) found that differences in pathogenicity

among isolates of *R. solani* could be detected on white clover seedlings established on low-nutrient substrates. Another method, used by Vaartaja et al. (1961), detected differences in pathogenicity by placing aseptically germinated tree seed in tubes containing half-strength cornmeal agar and the fungus.

Sims (1960) showed that various amended agar media significantly reduced virulence and dry weight below that of nonamended media among three isolates of *R. solani*.

Flentje (1965), in summarizing the various methods for testing pathogenicity of *Rhizoctonia* isolates, outlined three complementary approaches: (1) The direct addition of the test organism to natural or

*Fig. 7.* Soil in floor of greenhouse beds was artificially infested with 7 isolates of *R. solani* to evaluate soil fungicides. (Owen and Gay, 1964.)

amended soil in which the host plants are grown and the natural soil microflora is present. This method is often used to study the ability of the fungus to grow through soil. It does not indicate how the organism enters the host or how pathogenesis is effected. (2) The direct opposition of the test organism to the host in sterile sand or soil or on surfaces such as glass slides in a moist chamber, allowing for direct study on host penetration, type of tissue damage, and rate of damage in the absence of other organisms. (3) The indirect opposition of the test organism to the host by testing culture filtrates or extracts of the organism on host tissues, providing for the study of possible mechanisms involved in pathogenesis.

*Effects of environment on pathogenicity studies.*—Classically, the study of any plant disease and the pathogenicity of the pathogen involved must take into account the interaction of host and pathogen and the effect of the environment on this interaction. Much of this information on factors affecting pathogenicity is summarized elsewhere in this volume.

There seems to be no limit to the host range of the fungus. The whole concept of host range for *R. solani* may be confused because the host range can be altered by time, kind of plants used, varietal resistance and susceptibility, and by a number of soil microecological factors (Papavizas and Ayers, 1965).

*Rhizoctonia solani* can survive, grow, and remain pathogenic in soil under a wide range of environmental conditions because of its variability. Shephard and Wood (1963) thought pathogenicity of *R. solani* depended mainly on intrinsic properties of the host and isolate, and environmental factors were rarely of more than secondary importance in determining whether an isolate would attack a particular host plant. But the environment does play more than a secondary role in the infection of *R. solani*. The effect may not be directly on the fungus, but an indirect effect on the host plant, as with soil temperature and soil reaction.

Khan (1950) stated that the generalization that *R. solani* is either a high- or low-temperature pathogen or is destructive in an acid or alkaline soil, is not justified.

Results from studies concerning the effect of soil moisture on *R. solani* and disease development are not in agreement. This may be due to lack of standardization in technique. In a study by Bateman (1963*b*), in which he made a factorial analysis of soil temperature, moisture, and pH, using *R. solani* and other soilborne organisms, he found no combination of the three factors prevented infection of poinsettia by the fungus. Shephard and Wood (1963) showed that high humidity appreciably increased pathogenicity of *R. solani*.

The effect of soil moisture on the susceptibility of host plants needs study. Standardization experiments using field capacity (field-moisture capacity) as defined by the Soil Science Society (Anon., 1962) should be used in future studies. Couch, et al. (1967) reviewed the application of soil-moisture principles to the study of plant disease.

The effect of nutrient balance on saprophytic activity of *R. solani* was considered by Papavizas and Davey (1961), Papavizas, et al. (1962), and Papavizas (1964). Gibson (1956) demonstrated that the activity of *R. solani* as a damping-off pathogen is largely dependent on the nutrients available in the soil.

*Predisposition to infection.*—Testing for pathogenicity of *R. solani* is not always successful, as was pointed out by Huber and Finley (1959) and Brigham (1961). Such variable results may be due, in part, to the fact that the fungus was pathogenic on weakened seedlings under field conditions. Several agents are reported to predispose plants to infection by *R. solani*.

Certain plant viruses predisposed hosts to infection by *R. solani*, such as cucumber-mosaic virus (Bateman 1961); barley yellow-dwarf virus (Smith, 1962); and beet curly-top virus (Erwin and Flack, 1963).

Neal and Newsom (1951) showed a relationship between thrip (*Frankliniella fusca*) injury and cotton soreshin caused by *R. solani*.

Smith [A. M.] and Prentice (1929) observed that an association existed between *R. solani* and *Heterodera schachtii*. Grainger and Clark (1963) reported a synergistic effect between the fungus and *H. rostochiensis*. Mountain and Benedict (1956) in greenhouse and field experiments found that the combination of *R. solani* and *Pratylenchus minyus* reduced growth of wheat plants by half. In greenhouse tests, Taylor and Wyllie (1959) showed a combination of either *Meloidogyne hapla* plus *R. solani* or *M. javanica* plus *R. solani* reduced emergence of soybean to 2% and 17% of the check, respectively. Sayed (1961) showed a significant increase in damping-off of pea caused by *R. solani* in association with *M. hapla*. Reynolds and Hanson (1957) and White (1962) showed an increase in Rhizoctonia disease of cotton seedlings associated with an increase in the incidence of both *M. incognita* and *M. incognita acrita*. Similar results were obtained by Brodie and Cooper (1964), who showed that cotton seedlings grown in soil infested with either of three *Meloidogyne* spp. *Rotylenchus reniformis*, or *Hoplolaimus tylenchiformis* were susceptible longer to *R. solani* than were seedlings in nematode-free soil.

Fulton and Hanson (1950) clipped foliage of red clover plants at frequent intervals and increased the susceptibility of the host to *R. solani*. Reynolds and Hanson (1957) and Brodie and Cooper (1964) showed that excising either all or portions of cotyledons of cotton seedlings increased their susceptibility to *R. solani*. Webster and Roth (1965) increased susceptibility of lima bean seedlings to *R. solani* by damaging the seed coat before planting.

Plant damage due to agricultural chemicals was shown to increase the incidence of *R. solani* on cotton seedlings (Erwin and Reynolds, 1958; Miller and Ahrens, 1964; Ranney, 1964; Pinckard and Standifer, 1966).

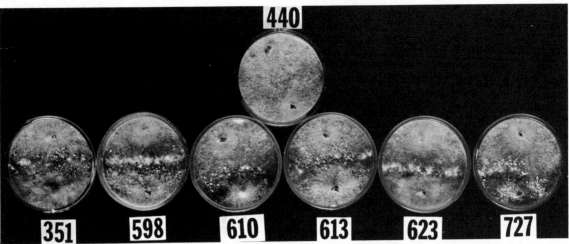

*Fig. 8.* Example of paired homokaryons forming tufts at their common line of convergence. Upper petri dish shows reaction of homokaryon 440 when paired with itself: colonies completely merge but no tufts are produced. Lower 6 petri dishes show results of pairing homokaryon 440 with 6 other homokaryons in which tuft formation is induced. Tufts can be seen as erumpent cottony wisps of mycelium at line of convergence between paired colonies. Sclerotium initials can also be seen in some plates; these later harden and darken. Colonies photographed after growth on Frie's agar for 10 days (Parmeter, et al., University of California, Berkeley).

The physiologic responses of host plants to these predisposing factors should be studied to determine the exact nature of increased susceptibility to *R. solani.*

*Heterokaryosis and pathogenicity.*—The effect(s) of heterokaryosis on pathogenicity and other characters of *R. solani* isolates also needs study.

The common technique for developing heterokaryotic lines is to make mycelial transfers of two homokaryotic lines on opposite sides of culture plates containing agar. When marked tufts of aerial mycelium form at the lines of contact, bits of mycelia are transferred to agar tubes or plates (Fig. 8). The characters from hyphal-tip isolates are then compared with that of the two orginal strains (Fig. 9).

Whitney and Parmeter (1963) used a semiliquid medium in which equal amounts of two strains of *R. solani* were blended. Droplets of the mixture were spotted on water agar and hyphal tips from subsequent growth were transferred to agar slants and studied.

Cellophane was used by Whitney and Parmeter (1963) as a substrate for anastomosis studies and the development of heterokaryons in *R. solani.* Sterile pieces of cellophane cut to fit culture plates were placed on the surface of water agar. Single-spore strains that had given tuft reactions were plated at opposite sides of the cellophane. When anastomosis had occurred, a small piece of cellophane bearing the anastomosis figure was cut from the cellophane and transferred to agar slants.

METHODS OF STUDYING TOLERANCE TO CHEMICALS. —*Rhizoctonia solani* is highly variable. This fact has been emphasized throughout all sections of this book. This variability allows for rapid adaptation to changes

in the environment including tolerance to chemicals. The development of tolerance in microorganisms to drugs is well documented in human medicine beginning with the work of Kossiakoff (1887). Only within the past 25 years have plant pathologists examined

*Fig. 9.* Comparison of characteristics of heterokaryon (XI) with 2 original strains (8 and 52). (Whitney and Parmeter, 1963.)

the development of tolerance to fungicides used for the control of plant-disease-causing organisms. The literature of fungicide adaptation was reviewed recently in theses written by Shatla (1959), Elsaid (1962), Raffray (1965), and Ricaud (1965). Kernkamp, et al. (1952) noted three papers dealing with tolerance to toxic substances in their review.

*Rhizoctonia solani*, like other members of the Mycelia Sterilia, offers some limitations in methods that can be used to study tolerance to fungicides and other chemicals, because of the lack of sufficient spore production. This fungus does not produce asexual spores. Basidiospore production is either lacking in most isolates or is difficult to obtain in sufficient amounts for use in techniques requiring spore suspensions. This limitation may be overcome in the future with better methods of inducing basidiospore production. Fungicide tolerance was studied using both in vitro and in vivo methods.

It is a prerequisite for any *R. solani* isolate used for chemical-tolerance studies that it be in pure culture, stable in cultural characteristics, and able to withstand continued subculturing. The author has found that all isolates of *R. solani* are not capable of repeated subculturing, particularly where weekly serial transfers may be involved. Much of the work has been done using cultures from hyphal-tip isolations. Studies on chemical tolerance in *R. solani* were conducted by growing the fungus on a suitable agar medium containing various concentrations of the test chemical.

Test fungicides can be incorporated into nutrient media by several means. Most workers add chemicals in known concentration to the agar medium. Ricaud (1965) recrystallized PCNB and TCNB (tetrachloronitrobenzene) from acetone from commercial preparations and used them in either the vapor phase or dispersed them as a fine suspension in the medium. These fungicides were added to the medium before autoclaving. Concentration was based on ppm of actual ingredient. For studies in our laboratory, commercial fungicides were added to media after autoclaving just before the agar became solid (Sinclair, 1960; Shatla and Sinclair, 1963; Elsaid and Sinclair, 1964; Raffray and Sinclair, 1965). Concentration also was based on ppm of active ingredient. These different approaches may have an effect on results.

The use of radial growth as a measure of adaptation requires that either a standardized "seeding" or "plant inoculation" method is used. A sterile corkborer should be used to cut plugs, biscuits, or cores of agar with mycelium from agar plates. For *R. solani*, cores should be cut as uniformly as possible from the outer growth of the colony in an old culture (more than 7-10 days old at optimum incubation). Various studies have shown that the older mycelium loses its vigor and therefore might not be suitable for culture plate studies. Domsch (1958) stated, however, that age of mycelium does not substantially influence its sensitivity. If very young cultures (2-5 days old at optimum temperature) are used, most of the colony is suitable for inoculation. Use of a standardized my-celial suspension for seeding plates is not satisfactory because the inoculum does not spread uniformly over the agar. Agar plugs should be placed in the center of each culture plate. Radial growth is measured either at the radius or diameter. If growth is irregular, the widest part of the colony should be measured.

Isolates should be incubated at a constant, controlled temperature optimum for growth. Incubation should be in a closed cabinet, since light has been shown to have an adverse effect on growth of *R. solani* in vitro. The length of incubation will be determined, in part, by the rate of growth of fungus isolates in nontreated plates. This may range from 3-6 days.

An obvious limitation of this technique is that radial-growth measurements may not give a true picture of differences in mycelial growth between treatments (Brown, 1923). Some chemicals may limit horizontal growth, but not vertical growth; others may prevent the fungus from penetrating into the agar. Check plates may have fungus mycelium growing in quite a different manner and may thus influence final results. Brown (1923) compared dry-weight measurements and radial growth for studying growth of fungi and found the use of radial growth: (1) made it possible to carry out experiments on a large scale; (2) allowed for abundant repetition; and (3) provided for effective controls. In radial-growth measurement, a culture can be followed throughout all its stages while dry weight methods involve destruction of a culture for each measurement taken. Brancato and Golding (1963) concluded that the use of diameter growth of fungi was sufficiently reliable for determining growth rates and for comparing the effect of environmental factors.

The use of liquid shake cultures was explored by Ricaud (1965). He inoculated 100 ml of medium with five agar plugs taken 2 mm from the margin of a 2-day-old culture of *R. solani*. PCNB was added to the medium at either 2.5, 5.0, or 10.0 ppm before autoclaving. Only a single isolate was studied because another isolate proved unsuitable for growth in shake culture. He found that the percentage of inhibition was almost constant for the three concentrations. Only a slight increase in inhibition with increased fungicide concentration was noted. This confirmed his results on agar. From these results, it appeared that all isolates of *R. solani* would not be suitable for shake-culture studies. The possibility of using broth media for this type of study and comparing results on agar should be explored further. Nonshake-culture growth should be compared with shake-culture and agar cultures grown simultaneously under identical conditions.

It has been the experience of this author that most isolates of *R. solani* will form a thick mat on the surface of nonshaken potato-dextrose broth. This mat is easily removed for either the wet and/or dry weight.

*Factors affecting studies on tolerance.*—Shatla (1959), Elsaid (1962), Maier (1962), Shatla and Sinclair (1963), and Elsaid and Sinclair (1964) found as concentrations of PCNB in agar were increased, there was a corresponding decrease in growth among isolates of *R. solani*.

Shatla and Sinclair (1963) compared the growth of 6 of 36 isolates that had differed in their growth rates on agar with 1,370 ppm PCNB. The amount of PCNB required to inhibit completely the growth of these isolates ranged from 6,000-10,000 ppm. Ricaud (1965) found different concentrations of PCNB and TCNB incorporated in agar produced only a slight response with increased concentration beyond the saturation point (10 ppm). The solubility of chlorinated nitrobenzenes is less than 2.5 ppm (Reavill, 1950). The actual amount needed to obtain saturation of assay medium may exceed this value through gradual loss by diffusion (Ricaud, 1965). There seems to be no question that *R. solani* isolates vary in their ability to grow on agar containing different concentrations of PCNB and TCNB. It appears that this effect may involve more than a response to fungicide concentration.

Many studies of chemical tolerance assume that either reduction or absence of growth indicates toxicity and that increased growth in agar with the toxicant indicates tolerance to it. In cases where the toxicant has limited solubility, increased concentrations may have little effect on increase in the quantity of toxic material. Increased concentration of a fungicide in powder form, for example, may increase the osmotic value, change the pH, or affect available nutrients of the agar medium (Ricaud, 1965).

Ricaud (1965) found when PCNB and TCNB were suspended in agar at different concentrations, there was only a very slight response with increasing concentration beyond the saturation point. The fungicides also were only slightly more effective when impregnated in the medium than when applied in the vapor phase. These results suggested that the effect on mycelial growth of contact toxicity associated with undissolved solids in an assay medium was slight.

Many of the techniques described for fungicide evaluation under greenhouse conditions could be modified to study fungicide tolerance among isolates of *R. solani*. Greenhouse techniques for evaluating fungicides to control *R. solani* on cotton have been described by Ranney and Bird (1956), Sinclair (1958, 1960, 1965a), Elsaid and Sinclair (1962), Maier and Staffeldt (1963), Owen and Gay (1964), Gayed and Temple (1965), and Raffray and Sinclair (1965). A clay-pot technique was used to screen soil fungicides for control of root-rotting organisms isolated from infected lily, poinsettia, and other ornamentals by Schaffer and Haney (1956) and also to screen soil fungicides for control of bean-root-rot organisms by Davison and Vaughn (1957).

A simple and rapid technique for the evaluation of chemicals for control of *R. solani* was reported by Gayed and Temple (1965). Dried flax stems were placed in infested soil treated with a chemical. After 24 hours, the "traps" were removed and laid on 2% water agar containing bacterial-growth inhibitors.

The extent of fungal growth after 24 hours was used to measure the efficacy of the chemical.

In general, soil in either greenhouse pots or flats is mixed with known quantities of either fungicide, fungus, or both and used to study the tolerance in the fungus to the fungicides. Greenhouse environmental conditions are not usually as controlled as in laboratory experiments. Experiments of this nature, conducted in controlled environmental chambers will give valuable information about fungicide tolerance in *R. solani* and other phytopathogens. For instance, Raffray and Sinclair (1965) described a modification of a technique of Elsaid and Sinclair (1962) using greenhouse flats to determine if three isolates of *R. solani* became more tolerant to 1,370 ppm PCNB captan, and maneb after three serial transfers through treated soil.

Flats of steam-sterilized soil were either nontreated or treated with each of the three fungicides. A strip of soil was removed from one side of the flat and replaced with soil infested with one of the isolates of *R. solani*. The preparation of inoculum to infest steam-sterilized soil was essentially the same as described by Sinclair (1960). Fifty seeds of the susceptible cotton variety Deltapine 15 were sown in each of five rows in a greenhouse flat. Rows were approximately equidistant from one another. The progress of the fungus through the soil was measured by the progressive kill of the cotton seedlings. The fungus was reisolated from infected seedlings on the opposite side of the flat and plated on potato-sucrose agar, without and with the respective fungicides. A strip of infested soil from the first flat then was taken from the side opposite the original strip of infested soil, placed in a second flat, and the procedure repeated.

In a technique described by Domsch (1958), a soil-compost mixture was placed in shallow trays, the fungicide worked in or applied by drenching, and *R. solani* introduced by watering with a mycelial suspension on the same day. Alderman peas were sown and covered, and trays kept for 6 days at 8°C, and then for 6 days in the dark at 20°C. Captan, PCNB, and an organic mercurial gave results reproducible to an exceptional degree.

Shatla (1959) for 2 years collected cotton seedlings showing symptoms of infection by *R. solani* from various locations in cotton-growing areas of Louisiana. Seedlings came from either nontreated soil or soil treated with PCNB alone or in combination with other soil fungicides. The isolates were plated on agar with 1,370 ppm PCNB. The mean diameter of the isolates from PCNB-treated soil was significantly greater than of isolates from nontreated soil after 7-day incubation at room temperature.

The advantages and disadvantages of using either pure culture inoculation or inoculation in the presence of normal soil microflora were discussed earlier in this volume.

CONCLUSIONS.—*Rhizoctonia solani* appears to be a good organism for exploring chemical tolerance in fungi. Tolerance to fungicides has been established in this fungus, but better techniques and further studies are needed. Tolerance to a larger number of chemicals needs to be determined. A chemical to which strains of *R. solani* are tolerant, and that would overcome some of the disadvantages of PCNB and TCNB, needs to be found.

Additional in vitro and in vivo studies need to be made to determine the extent, potential, and nature of this resistance. Effects of prolonged exposure of *R. solani* to these fungicides should be studied. The mechanism of variation in *R. solani* has been reported earlier in this volume (Flentje, et al., this vol). If fungicide resistance is due to mutation, as suggested by Flentje and Stretton (1964) and Ricaud (1965), then definite proof could be obtained by genetical analysis.

# BIBLIOGRAPHY

ABDEL-SALAM, M. M. 1933. Damping-off and other allied diseases of lettuce. J. Pomol. Hort. Sci. 11: 259-275.

AFANASIEV, M. M. 1956. Resistance of inbred varieties of sugar beets to Aphanomyces, Rhizoctonia, and Fusarium root rots. J. Am. Soc. Sugar Beet Tech. 9: 178-179.

———, and H. E. MORRIS. 1942. Control of seedling diseases of sugar beets in Montana. Phytopathology 32: 477-486.

———, and ———. 1952. Resistance and soil treatments for control of Rhizoctonia of sugar beets. Proc. Am. Soc. Sugar Beet Tech. 7:562-567.

———, and ———. 1954. Testing sugar beet varieties for their resistance to Aphanomyces, Rhizoctonia, and Fusarium root rots. Proc. Am. Soc. Sugar Beet Tech. 8: 90-93.

AGNIHOTHRUDU, V. 1962. A comparison of some techniques for the isolation of fungi from tea soils. Mycopathologia 16: 235-242.

AKAI, S., H. OGURA, and T. SATO. 1960. Studies on Pellicularia filamentosa (Pat.) Rogers. I. On the relation between pathogenicity and some characters on culture media. (Japanese, with English summary.) Ann. Phytopathol. Soc. Japan 25: 125-130.

ALBRECHT, W. A., and H. JENNY. 1931. Available soil calcium in relation to "damping-off" of soybean seedlings. Botan. Gaz. 92: 263-278.

ALEXANDER, M. 1961. Introduction to soil microbiology. John Wiley & Sons, Inc., New York. 472 p.

———. 1964. Biochemical ecology of soil microorganisms. Ann. Rev. Microbiol. 18: 217-252.

ALLINGTON, W. B. 1936. Sclerotial formation in Rhizoctonia solani as affected by nutritional and other factors. Phytopathology 26: 831-844.

ALLISON, J. L. 1951. Rhizoctonia blight of forage legumes and grasses. Plant Disease Reptr. 35: 372-373.

———, 1952. Rhizoctonia solani, a foliage pathogen of forage legumes and grasses in southeastern United States. (Abstr.) Phytopathology 42: 281.

———, and C. H. HANSON. 1951. Methods of determining pathogenicity of Sclerotinia trifolium on alfalfa and Rhizoctonia solani on Lotus. (Abstr.) Phytopathology 41: 1.

———, and J. C. WELLS. 1951. Forage crop diseases. N. Carolina Agr. Ext. Circ. 361: 1-11.

ANDERSEN, A. L., and D. M. HUBER. 1962. Bacterial necrosis and isolation frequency of Rhizoctonia from rhizospheres under several cropping sequences. (Abstr.) Phytopathology 52: 721.

———, and ———. 1965. The plate-profile technique for isolating soil fungi and studying their activity in the vicinity of roots. Phytopathology 55: 592-594.

ANDERSON, E. J. 1939. Effect of nutrient variations on

host and parasite in the Rhizoctonia stem-rot disease of bean. (Abstr.) Phytopathology 29: 1.

ANDERSON, R. L., and E. J. ORDAL. 1961. CO$_2$-dependent fermentation of glucose by Cytophaga succinicans. J. Bacteriol. 81: 139-146.

ANDREUCCI, E. 1964. A damping-off of olive in the seedbed caused by Corticium solani. Riv. Pat. Veg. Pavia 4: 249-268.

ANONYMOUS. 1922. Report on the Department of Agriculture, Barbados, for the financial year, 1921-22. (Abstr.) Rev. Appl. Mycol. 2: 260.

———. 1938. Plant diseases. Notes contributed by the Biological Branch. Agr. Gaz. N.S. Wales 49: 388-390.

———. 1950. Purple patch of cereals. Agr. Gaz. N.S. Wales 61: 295-296.

———. 1953. Difco manual of dehydrated culture media and reagents for microbiological and clinical laboratory procedures. 9th ed. Difco Laboratories, Inc., Detroit, Mich. 350 p.

———. 1962. Glossary of terms approved by the Soil Science Society of America. Soil Sci. Proc. 26: 305-317.

———. 1965a. Estimates of crop losses and disease-control costs in California, 1963. California Agr. Expt. Sta., Berkeley. 102 p.

———. 1965b. Losses in agriculture. U. S. Dept. Agr. Agr. Handbook 291: 1-120.

AOKI, H., T. SASSA, and T. TAMURA. 1963. Phytotoxic metabolites of Rhizoctonia solani. Nature (London) 200: 575.

ASHOUR, W. E., and M. M. EL-KADI. 1959. Cultural studies on Fusarium semitectum, Alternaria tenuis and Rhizoctonia solani which cause damping-off of tomato seedlings. A'in Shams Sci. Bull. 1958: 57-68.

ASHWORTH, L. J., JR., and J. V. AMIN. 1964. A mechanism for mercury tolerance in fungi. Phytopathology 54: 1459-1463.

———, and B. C. LANGLEY. 1964. The relationship of pod damage to kernel damage by molds in Spanish peanut. Plant Disease Reptr. 48: 875-878.

ATKINS, J. G., JR., and W. D. LEWIS. 1954. Rhizoctonia aerial blight of soybeans in Louisiana. Phytopathology 44: 215-218.

ATKINSON, G. F. 1892. Some diseases of cotton. Alabama Agr. Expt. Sta. Bull. 41: 1-65.

AUBE, C., and W. E. SACKSTON. 1965. Distribution and prevalence of Verticillium species producing substances with gibberellin-like biological properties. Can. J. Botany 43: 1335-1342.

AYERS, W. A., and G. C. PAPAVIZAS. 1963. Violet-pigmented pseudomonads with antifungal activity from the rhizosphere of beans. Appl. Microbiol. 11: 533-538.

BAKER, K. F. 1947. Seed transmission of Rhizoctonia solani in relation to control of seedling damping-off.

Phytopathology 37: 912-924.

———. 1957. Damping-off and related diseases. California Agr. Expt. Sta. Manual 23: 34-51.

———. 1962. Principles of heat treatment of soil and planting material. J. Australian Inst. Agr. Sci. 28: 118-126.

———. 1968. Control of soil-borne plant pathogens with aerated steam. Proceedings Greenhouse Growers Inst., Pullman, Wash.

———, N. T. FLENTJE, C. M. OLSEN, and H. M. STRETTON. 1967. Effect of antagonists on growth and survival of *Rhizoctonia solani* in soil. Phytopathology 57: 591-597.

———, and C. M. OLSEN. 1960. Aerated steam for soil treatment. Phytopathology 50: 82.

———, and C. N. ROISTACHER. 1957. Heat treatment of soil. Principles of heat treatment of soil. Equipment for heat treatment of soil. California Agr. Expt. Sta. Manual 23: 123-196, 290-293.

———, and R. H. SCIARONI. 1952. Diseases of major floricultural crops in California. California State Florist's Assoc., Los Angeles. 57 p.

———, and W. C. SNYDER [eds.]. 1965. Ecology of soil-borne plant pathogens. Univ. California Press, Berkeley, Los Angeles. 571 p.

BAKER, R. 1965. The dynamics of inoculum. Pp. 395-403. *In* K. F. Baker and W. C. Snyder [eds.], Ecology of soil-borne plant pathogens. Univ. California Press, Berkeley, Los Angeles.

———, and D. L. McCLINTOCK. 1965. Populations of pathogens in soil. Phytopathology 55: 495.

BALD, J. G. 1947. The treatment of cut potato setts with zinc oxide. 2. Infection of stems and tubers with *Rhizoctonia* and scab. J. Coun. Sci. Industr. Res. Australia 20: 190-206.

BALFE, I. G. 1935. An account of sclerote-forming fungi causing diseases in *Matthiola*, *Primula*, and *Delphinium* in Victoria. Proc. Roy. Soc. Victoria (N.S.) 47: 369-386.

BALLS, W. L. 1906. Physiology of a simple parasite. Khediv. Agr. Soc. Yearbook (Cairo) 1905-1906: 173-195.

———. 1908. Temperature and growth. Ann. Botany 22: 557-591.

BARKER, K. R. 1961. Factors affecting the pathogenicity of *Pellicularia filamentosa*. Ph.D. Thesis, Univ. of Wisconsin, Madison. 92 p.

———, and J. C. WALKER. 1962. Relationship of pectolytic and cellulytic enzyme production by strains of *Pellicularia filamentosa* to their pathogenicity. Phytopathology 52: 1119-1125.

BATEMAN, D. F. 1961. Synergism between cucumber mosaic virus and *Rhizoctonia* in relation to Rhizoctonia damping-off of cucumber. (Abstr.) Phytopathology 51: 574-575.

———. 1962. Relation of soil pH to development of Poinsettia root rots. Phytopathology 52: 559-566.

———. 1963a. Pectolytic activities of culture filtrates of *Rhizoctonia solani* and extracts of Rhizoctonia-infected tissues of bean. Phytopathology 53: 197-204.

———. 1963b. Factorial analysis of environment and pathogens in relation to development of the poinsettia root rot complex. Phytopathology 53: 509-516.

———. 1963c. The "macerating enzyme" of *Rhizoctonia solani*. Phytopathology 53: 1178-1186.

———. 1964a. An induced mechanism of tissue resistance to polygalacturonase in Rhizoctonia-infected hypocotyls of bean. Phytopathology 54: 438-445.

———. 1964b. Cellulase and Rhizoctonia disease of bean. Phytopathology 54: 1372-1377.

———. 1965. Loss of birefringence of cell walls in Rhizoctonia-infected tissue. Phytopathology 55: 494.

———. 1966. Hydrolytic and *trans*-eliminative degradation of pectic substances by extracellular enzymes of *Fusarium solani* f. *phaseoli*. Phytopathology 56: 238-244.

———. 1967. Alteration of cell wall components during pathogenesis by *Rhizoctonia solani*. Pp 58-79. *In* C. J. Mirocha and I. Uritani, The dynamic role of molecular constituents in plant-parasite interaction. Bruce Publishing Co., St. Paul, Minn. 372 p.

———, and J. M. DALY. 1967. The respiratory pattern of Rhizoctonia-infected bean hypocotyls in relation to lesion development. Pytopathology 57: 127-131.

———, and A. W. DIMOCK. 1959. The influence of temperature on root rots of poinsettia caused by *Thielaviopsis basicola*, *Rhizoctonia solani*, and *Pythium ultimum*. Phytopathology 49: 641-647.

———, and R. D. LUMSDEN. 1965. Relation of calcium content and nature of the pectic substances in bean hypocotyls of different ages to susceptibility to an isolate of *Rhizoctonia solani*. Phytopathology 55: 734-738.

———, and D. P. MAXWELL. 1965. Phenoloxidase activity in extracts of *Rhizoctonia solani*-infected hypocotyls of bean and its distribution in relation to lesion areas. (Abstr.) Phytopathology 55: 127.

———, and J. C. ROGOWICZ. 1965. Host pectins and pathogenesis by *Rhizoctonia*. Cornell Plantations 20: 54-56.

BATTISTA, O. A. 1965. Colloidal macromolecular phenomena. Amer. Scientist 53: 151-173.

BAUGH, C. L., J. W. LANHAM, and M. J. SURGALLA. 1964a. Effects of bicarbonate on growth of *Pasteurella pestis* II. Carbon dioxide fixation into oxaloacetate by cell-free extracts. J. Bacteriol. 88: 553-558.

———, A. W. ANDREWS, and M. J. SURGALLA. 1964b. Effects of bicarbonate on growth of *Pasturella pestis* III. Replacement of bicarbonate by pyrimidines. J. Bacteriol. 88: 1394-1398.

BAVER, L. D. 1959. Soil physics. 3d ed. John Wiley & Sons, Inc., New York. 489 p.

———, and R. B. FARNSWORTH. 1940. Soil structure effects in the growth of sugar beets. Soil Sci. Soc. Am. Proc. 5: 45-48.

BEACH, W. S. 1949. The effects of excess solutes, temperature and moisture upon damping-off. Pennsylvania Agr. Expt. Sta. Bull. 509: 1-29.

BEEVERS, H. 1961. Respiratory metabolism in plants. Row, Peterson & Co., Evanston, Ill. 232 p.

BEGIN, N., and P. G. SCHOLEFIELD. 1965. The uptake of amino acids by mouse pancreas in vitro II. The specificity of the carriers. J. Biol. Chem. 240: 332-337.

BELL, D. K., and J. H. OWEN. 1963. Effect of soil temperature and fungicide placement ·on cotton seedling damping-off caused by *Rhizoctonia solani*. Plant Disease Reptr. 47: 1016-1021.

BENEDICT, W. G. 1954. Stunt of clovers, caused by *Rhizoctonia solani*. Can. J. Botany 32: 215-220.

———, and W. B. MOUNTAIN. 1954. Studies on the association of *Rhizoctonia solani* and nematodes in a root-rot disease complex of winter wheat in southwestern Ontario. Proc. Can. Phytopathol. Soc. 22: 12.

BENEKE, E. S. 1963. *Calvatia*, calvacin, and cancer. Mycologia 55: 257-270.

BERKELEY, M. J., and M. A. CURTIS. 1873. Notices of American fungi. Grevillea 1: 179.

BERLINER, M. D., and H. DUFF. 1965. Ultrastructure of *Armillaria mellea* hyphae. Can. J. Botany 43: 171-172.

BEWLEY, W. F., and W. BUDDIN. 1921. On the fungus flora of glasshouse water supplies in relation to plant disease. Ann. Appl. Biol. 8: 10-19.

BIANCHINI, P. 1958. Las llagas del cafe en Costa Rica. Bol. Tec. Min. Agr. Industr. Costa Rica 21. 27 p.

———, and F. L. WELLMAN. 1958. Experimentos en el control de *Pellicularia* del Cafe y ciertas diferencias en Pellicularias de cinco huespedes. Turrialba 8: 73-92.

BIGGS, R. 1937. The species concept in *Corticium coronilla*. Mycologia 29: 686-706.

BIRD, L. S., C. D. RANNEY, and G. M. WATKINS. 1957. Evaluation of fungicides mixed with the covering soil at planting as a control measure for the cotton seedling disease complex. Plant Disease Reptr. 41: 165-173.

BJÖRLING, K. 1948. Bidrag till kännedomen om Potatiskräftsvampens (*Synchytrium endobioticum* [Schilb.] Perc.) biologi. Medd. Växtskyddsanst. 52: 21p.

BLAIR, I. D. 1942. Studies on the growth in soil and the parasitic action of certain *Rhizoctonia solani* isolates from wheat. Can. J. Res. 20: 174-185.

———. 1943. Behaviour of the fungus *Rhizoctonia solani* Kühn in the soil. Ann. Appl. Biol. 30: 118-127.

BLASZCZAK, W. 1954. [The effect of the date of disinfecting potatoes infected by black scurf (*Rhizoctonia solani* Kuehn) on the rapidity of germination and health of the sprouts.] Acta Agrobot. 2: 21-38.

BLODGETT, F. M. 1940. A second report on the effect of agronomic practices on the incidence of *Rhizoctonia* and scab of potatoes. Am. Potato J. 17: 290-295.

BLOOM, J. R., and H. B. COUCH. 1958. Influence of pH, nutrition, and soil moisture on the development of large brown patch. (Abstr.) Phytopathology 48: 260.

———, and ———. 1960. Influence of environment on diseases of turf-grasses. I. Effect of nutrition, pH, and soil moisture. Phytopathology 50: 523-535.

BOIDIN, J. 1958. Essai taxonomique sur les Hydnés résupines et les Corticiés. Théses No. 202, Univ. de Lyon: 99-102.

BOOSALIS, M. G. 1950. Studies on the parasitism of *Rhizoctonia solani* Kuhn on soybeans. Phytopathology 40: 820-831.

———. 1956. Effect of soil temperature and green-manure amendment of unsterilized soil on parasitism of *Rhizoctonia solani* by *Penicillium vermiculatum* and *Trichoderma* sp. Phytopathology 46: 473-478.

———. 1964. Hyperparasitism. Ann. Rev. Phytopathol. 2: 363-376.

———, and A. L. SCHAREN. 1959. Methods for microscopic detection of *Aphanomyces euteiches* and *Rhizoctonia solani* associated with plant debris. Phytopathology 49: 192-198.

———, and ———. 1960. The susceptibility of pigweed to *Rhizoctonia solani* in irrigated fields of western Nebraska. Plant Disease Reptr. 44: 815-818.

BOSCH, E. 1948. Untersuchungen uber die Biologie und Bekampfung der Vermehrungspilze *Moniliopsis aderholdii* und *Rhizoctonia solani*. Ann. Agr. Suisse 62: 791-825.

BOULTER, D., and E. DERBYSHIRE. 1957. Cytochromes of fungi. J. Exptl. Botany 8: 313-318.

BOURDOT, H., and A. GALZIN. 1911. Hyménomycétes de France III. Corticiés. Bull. Soc. Mycol. France 27: 223-266.

———, and ———. 1928. Hyménomycétes de France. Lechevalier, Sceaux. 761 p.

BOURN, W. S., and B. JENKINS. 1928a. Rhizoctonia disease on certain aquatic plants. Botan. Gaz. 85: 413-425.

———, and ———. 1928b. *Rhizoctonia* on certain aquatic plants. Contrib. Boyce Thompson Inst. 1: 383-396.

BOYLE, L. W. 1956a. Fundamental concepts in the development of control measures for southern blight and root rot on peanuts. Plant Disease Reptr. 40: 661-665.

———. 1956b. The role of saprophytic media in the development of southern blight and root rot on peanuts. (Abstr.) Phytopathology 46: 7-8.

BOXER, G. E., and T. M. DELVIN. 1961. Pathways of intracellular hydrogen transport. Absence of enzymatic hydrogen-carrying systems is a factor in aerobic glycolysis of malignant tissue. Science 132: 1495-1501.

BRACKER, C. E., and E. E. BUTLER. 1963. The ultrastructure and development of septa in hyphae of *Rhizoctonia solani*. Mycologia 55: 35-58.

———, and ———. 1964. Function of the septal pore apparatus in *Rhizoctonia solani* during protoplasmic streaming. J. Cell Biol. 21: 152-157.

BRANCATO, F. P., and N. S. GOLDING. 1963. The diameter of the mold colony as a reliable measure of growth. Mycologia 45: 848-864.

BRAUN, H. 1930. Der Wurzeltöter der Kartoffel. Monograph. Pflanzenschutz 5: 1-136.

BRESADOLA, G. 1903. Fungi polonici. Ann. Mycol. 1: 65-131.

———. 1925. New species of fungi. Mycologia 17: 68-77.

BRESSLER, R., and S. J. WAKIL. 1961. Studies on the mechanism of fatty acid synthesis IX. The conversion of malonyl coenzyme A to long chain fatty acids. J. Biol. Chem. 236: 1643.

BRIGHAM, R. D. 1961. Castor bean diseases in Texas in 1960. Plant Disease Reptr. 45: 124-125.

BRINKERHOFF, L. A., E. S. OSWALT, and J. F. TOMLINSON. 1954. Field tests with chemicals for the control of *Rhizoctonia* and other pathogens of cotton seedlings. Plant Disease Reptr. 38: 467-475.

BRITON-JONES, H. R. 1924. Strains of *Rhizoctonia solani* Kühn (*Corticium vagum* Berk. and Curt). Trans. Brit. Mycol. Soc. 9: 200-210.

———. 1925. Mycological work in Egypt during the period 1920-1922. Min. Agr. Egypt, Tech. & Sci. Service Bull. 49 (Bot. Sect.). 129 p.

BRODIE, B. B. 1963. Pathogenicity of certain parasitic nematodes on cotton seedlings and their relationship to post-emergence damping-off caused by *Rhizoctonia solani* Kühn and *Pythium debaryanum* Hesse. Diss. Abstr. 23: 4491-4492.

———, and W. E. COOPER. 1964. Relation of parasitic nematodes to post-emergence damping-off of cotton. Phytopathology 54: 1023-1027.

BROOKS, C., C. O. BRATLEY, and L. P. McCOLLOCH. 1936. Transit and storage diseases of fruits and vegetables as affected by initial carbon dioxide treatments. U. S. Dept. Agr. Tech. Bull. 519. 24 p.

BROWN, W. 1923. Experiments on the growth of fungi on culture media. Ann. Botany 37: 105-129.

———. 1935. On the Botrytis disease of lettuce with reference to its control. J. Pomol. 13: 247-259.

———, F. T. BROOKS, and F. C. BAWDEN. 1948. A discussion on the physiology of resistance to disease in plants. Proc. Roy. Soc. London (B) 135: 171-195.

———, and C. C. HARVEY. 1927. Studies in the physiology of parasitism. X. On the entrance of parasitic fungi into the host. Ann. Botany 41: 643-662.

BRUEHL, G. W. 1951. *Rhizoctonia solani* in relation to cereal crown and root rots. Phytopathology 41: 375.

BULLER, A. H. R. 1933. Researches on fungi. Vol. V. Longmans, Green & Co., Ltd., London. 416 p.

BURCHFIELD, H. P. 1960. Performance of fungicides on plants and in soil—physical, chemical, and biological considerations. P. 477-520. *In* J. G. Horsfall and A. E. Dimond [eds.], Plant Pathology, an advanced treatise. III. Academic Press, Inc., New York, London.

BURGES, A. 1939. Soil fungi and root infection, a review. Brotéria 8: 64-81.

————. 1958. Micro-organisms in the soil. Hutchinson Co., Ltd., London. 187 p.

BURT, E. A. 1916. The Thelephoraceae of North America. VI. *Hypochnus.* Ann. Mo. Botan. Garden 3: 203-241.

————. 1918. Corticiums causing Pellicularia disease of the coffee plant, Hypochnose of pomaceous fruits, and Rhizoctonia disease. Ann. Mo. Botan. Garden 5: 119-132.

————. 1926. The Thelephoraceae of North America. XV. *Corticium.* Ann. Mo. Botan. Garden 13: 173-354.

BUTLER, E. E. 1957. *Rhizoctonia solani* as a parasite of fungi. Mycologia 49: 354-373.

————, and T. H. KING. 1951. Fungus parasitism by *Rhizoctonia.* (Abstr.) Phytopathology 41: 5.

BUTLER, E. J., and S. G. JONES. 1949. Plant Pathology. Macmillan & Co., Ltd., London. 979 p.

BUTLER, F. C. 1953a. Saprophytic behaviour of some cereal root-rot fungi. I. Saprophytic colonization of wheat straw. Ann. Appl. Biol. 40: 284-297.

————. 1953b. Saprophytic behaviour of some cereal root-rot fungi. II. Factors influencing saprophytic colonization of wheat straw. Ann. Appl. Biol. 40: 298-304.

————. 1953c. Saprophytic behaviour of some cereal root-rot fungi. III. Saprophytic survival in wheat straw buried in soil. Ann. Appl. Biol. 40: 305-311.

BUXTON, E. W. 1960. Heterokaryosis, saltation, and adaptation. P. 359-405. *In* J. G. Horsfall and A. E. Dimond [eds.], Plant Pathology, Vol. II. Academic Press Inc., New York. 715 p.

BYRDE, R. J. W., and A. H. FIELDING. 1965. An extracellular — $\alpha$ — Arabinofuranosidase secreted by *Sclerotinia fructigena.* Nature (London) 205: 390-391.

CAMERINO, P. W., and L. SMITH. 1964. The mechanism of stimulation of the reduced diphosphopyridine nucleotide oxidase of heart muscle particles and mitochondria by soluble cytochrome c. J. Biol. Chem. 239: 2345-2350.

CARPENTER, J. B. 1949. Production and discharge of basidiospores by *Pellicularia filamentosa* (Pat.) Rogers on *Hevea* rubber. Phytopathology 39: 980-985.

————. 1951. Target leaf spot of the *Hevea* rubber tree in relation to host development, infection, defoliation, and control. U. S. Dept. Agr. Tech. Bull. 1028. 34 p.

CARRERA, C. J. M. 1951. La enfermedad de las almacigas de coniferas en la Republica Argentina. Bol. Fac. Agron. Vet. Buenos Aires. 30. 51 p.

CASTANO, J. J., and M. F. KERNKAMP. 1956. The influence of certain plant nutrients on infection of soybeans by *Rhizoctonia solani.* Phytopathology 46: 326-328.

CASTELLANI, E. 1935. Ricerche morphologico-systemiche su alcune Rizottonie. Ann Ist. Super. Agr. For., Firenza (2) 5: 63-78.

————. 1936. Una nuova specie di Rhizoctonia: "*Rh. muneratii*" n. sp. Nuov. Giorn. Botan. Ital. 43: 563-567.

CHAMBERLAIN, E. E. 1931. Corticium disease of potatoes. I. Propagation and spread of the disease. New Zealand J. Agr. 43: 204-209.

CHANCE, B., and B. HESS. 1959. Spectroscopic evidence of metabolic control. Science 129: 700-708.

————, B. SCHOENER, and S. ELSAESSER. 1965. Metabolic control phenomena involved in damped sinusoidal oscillations of reduced diphosphopyridine nucleotide in a cell-free extract of *Saccharomyces carlsbergensis.* J. Biol. Chem. 240: 3170-3181.

CHANDRA, P., and W. B. BOLLEN. 1960. Effect of wheat straw, nitrogenous fertilizers, and carbon-to-nitrogen ratio on organic decomposition in a subhumid soil. J. Agr. Food Chem. 8: 19-24.

CHEN, SHAN-MING. 1943. Studies on *Rhizoctonia solani* Kühn. Ph.D. Thesis, Univ. of Minnesota, Minn. 96 p.

CHEN, Y. 1958. Studies on metabolic products of *Hypochnus sasakii* Shirai. Isolation of p-hydroxyphenylacetic acid and its physiological activity. Bull. Agr. Chem. Soc. Japan 22: 136-142.

CHEREWICK, W. J. 1948. Two important diseases of alfalfa in Manitoba. (Abstr.) Phytopathology 38: 5.

CHESTER, K. S. 1941. The probability law in cotton seedling disease. Phytopathology 31: 1078-1088.

CHESTERS, C. G. C. 1940. A method of isolating soil fungi. Trans. Brit. Mycol. Soc. 24: 352-354.

————. 1948. A contribution to the study of fungi in the soil. Trans. Brit. Mycol. Soc. 30: 100-117.

CHI, C. C., and W. R. CHILDERS. 1964. Penetration and infection in alfalfa and red clover by *Pellicularia filamentosa.* Phytopathology 54: 750-754.

CHOWDHURY, S. 1944. Diseases of pan (*Piper betle*) in Sylhet, Assam. IV. Rhizoctonia root-rot. Proc. Indian Acad. Sci., Sect. B. 20: 229-244.

————. 1946. A Rhizoctonia leaf blight of *Dioscorea.* Current Sci. (India). 15: 81-82.

CHRISTIANSEN, M. P. 1959. Danish resupinate fungi. Part 1. Ascomycetes and Heterobasidiomycetes. Dansk Botan. Arkiv 19: 1-55.

————. 1960. Danish resupinate fungi. Part II. Homobasidiomycetes. Dansk Botan. Arkiv 19: 57-388.

CHRISTOU, T. 1962. Penetration and host-parasite relationships of *Rhizoctonia solani* in the bean plant. Phytopathology 52: 381-389.

CHU, M. 1966. Incidence of *Rhizoctonia* in a cultivated and a fallow soil in Hong Kong. Nature (London) 211: 862-863.

CHUPP, C., and A. F. SHERF. 1960. Vegetable diseases and their control. The Ronald Press Company. New York. 693 p.

CIFERRI, R. 1938. Mycoflora domingensis exsiccata. Ann. Mycol. 36: 198-245.

COCHRANE, V. W. 1948. The role of plant residues in the etiology of root rot. Phytopathology 38: 185-196.

————. 1958. Physiology of fungi. John Wiley & Sons, Inc. New York. 524 p.

CONOVER, R. A. 1949. Rhizoctonia canker of tomato. Phytopathology 39: 950-951.

CONWENTZ, H. 1890. Monographie der baltischen Bernsteinbäume. Wilhelm Englemann, Danzig. 151 p.

COOK, R. J., and W. C. SNYDER. 1965. Influence of host exudates on growth and survival of germlings of *Fusarium solani* f. *phaseoli* in soil. Phytopathology 55: 1021-1025.

COOKE, M. C. 1876a. Report on diseased leaves of coffee and other plants. Indian Mus. Rept.: 1-7.

————. 1876b. Some Indian fungi and affinities of *Pellicularia.* Grevillea 4: 116, 134-135.

————. 1876c. Two coffee diseases. Popular Sci. Rev. 15: 161-168.

————. 1881. The coffee disease in South America. J. Linn. Soc. (Bot.) 18: 461-467.

COOLEY, J. S. 1942. Defoliation of American holly cuttings by *Rhizoctonia.* Phytopathology 32: 905-909.

COONS, G. H., and J. E. KOTILA. 1935. Influence of

preceding crops on damping off of sugar beets. (Abstr.) Phytopathology 25: 13.

CORMACK, M. W., and J. E. MOFFATT. 1961. Factors influencing storage decay of sugar beets by *Phoma betae* and other fungi. Phytopathology 51: 3-5.

COSTANTIN, J., and L. DUFOUR. 1895. Nouvelle flore des champignens pour la détermination facile de toutes les espéces de France et de la plupart des espéces européonnes. 2nd ed. Dupont, Paris. 295 p.

COUCH, H. B. 1962. Diseases of turfgrasses. Reinhold Publishing Corporation, New York. 289 p.

———, and J. R. BLOOM. 1958. Influence of soil moisture, pH, and nutrition on the alteration of disease proneness in plants. Trans. New York Acad. Sci. 20: 432-437.

———, L. H. PURDY, and D. H. HENDERSON. 1967. Application of soil moisture principles to the study of plant disease. Virginia Polytech. Instit. Res. Div. Bull. 4. 23 p.

COWLING, E. B. 1963. Structural features of cellulose that influence its susceptibility to enzymatic hydrolysis. P. 1-32. *In* E. T. Reese [ed.], Advances in enzymatic hydrolysis of cellulose and related materials. Pergamon Press, New York. 290 p.

CRANDALL, B. S., and J. G. ARILLAGA. 1955. A new *Rhizoctonia* from El Salvador associated with root rot of coffee. Mycologia 47: 403-407.

CREAGER, D. B. 1945. Rhizoctonia neck rot of gladiolus. Phytopathology 35: 230-232.

CRISTINZIO, M. 1937. Experienze intorno alla capacita infettiva della *Rhizoctonia solani* Kühn a mezzo di tuberi di Patata infetti. Ric. Osserv. Reg. Fitopatol. Portici (Sez. Patol. Veg.) 6: 71-94.

CROSIER, W. F. 1936. Prevalence and significance of fungous associates of pea seeds. Proc. Assoc. Offic. Seed Analysts N. Am. 1936: 101-107.

CROSSAN, D. F., and P. J. LLOYD. 1956. The influence of overhead irrigation on the incidence and control of certain tomato diseases. Plant Disease Reptr. 40: 314-317.

CUNNINGHAM, G. H. 1953. Thelephoraceae of New Zealand. Part II. The genus *Pellicularia*. Trans. Roy. Soc. New Zealand 81: 321-328.

———. 1963. The Thelephoraceae of Australia and New Zealand. Bull. 145. D.S.I.R., Wellington. 359 p.

DAHL, A. S. 1933. Effect of temperature on brown patch of turf. (Abstr.) Phytopathology 23: 8.

DAINES, R. H. 1937. Antagonistic action of *Trichoderma* on *Actinomyces scabies* and *Rhizoctonia solani*. Amer. Potato J. 14: 85-93.

DANA, B. F. 1925a. The Rhizoctonia disease of potato. Washington Agr. Expt. Sta. Popular Bull. 131. 30 p.

———. 1925b. The Rhizoctonia disease of potato. Washington Agr. Expt. Sta. Bull. 191: 1-78.

DANIELS, J. 1963. Saprophytic and parasitic activities of some isolates of *Corticium solani*. Trans. Brit. Mycol. Soc. 46: 485-502.

DAS, A. C., and J. H. WESTERN. 1959. The effect of inorganic manures, moisture and inoculum on the incidence of root disease caused by *Rhizoctonia solani* Kühn in cultivated soil. Ann. Appl. Biol. 47: 37-48.

DAVEY, C. B., and G. C. PAPAVIZAS. 1959. Effect of organic soil amendments on the Rhizoctonia disease of snap beans. Agron. J. 51: 493-496.

———, and ———. 1960. Effect of dry mature plant materials and nitrogen on *Rhizoctonia solani* in soil. Phytopathology 50: 522-525.

———, and ———. 1962. Comparison of methods for isolating *Rhizoctonia* from soil. Can. J. Microbiol. 8: 847-853.

———, and ———. 1963. Saprophytic activity of *Rhizoctonia* as affected by the carbon-nitrogen balance of certain organic soil amendments. Soil Sci. Soc. Am. Proc. 27: 164-167.

DAVIS, P. H., and V. H. HEYWOOD. 1963. Principles of Angiosperm taxonomy. Oliver & Boyd Ltd., Edinburgh and London. 556 p.

DAVIS, W. C. 1941. Damping-off of longleaf pine. Phytopathology 31: 1011-1016.

DAVISON, A. D., and J. R. VAUGHN. 1957. Effect of several antibiotics and other organic chemicals on isolates of fungi which cause bean root rot. Plant Disease Reptr. 41: 432-433.

DAWSON, J. R., R. A. H. JOHNSON, P. ADAMS, and F. T. LAST. 1965. Influence of steam/air mixtures, when used for heating soil, on biological and chemical properties that affect seedling growth. Ann Appl. Biol. 56: 243-251.

DE BEER, J. F. 1965. Studies on the ecology of *Rhizoctonia solani* Kühn. Ph.D. Thesis, Univ. of Adelaide, South Australia.

DE CANDOLLE, A. P. 1815. Mémoire sur les rhizoctones, nouveau genre de champignons qui attaque les racines, des plantes et en particulier celle de la luzerne cultivée. Mém. Mus. d'Hist. Nat. 2: 209-216.

DEMAREE, J. B. 1945. Rhizoctonia bud rot of strawberry plants. Phytopathology 35: 710-713.

DEMEL, R. A., S. C. KINSKY, and L. L. M. VAN DEENEN. 1965. Penetration of lipid monolayers by polyene antibiotics. Correlation with selective toxicity and mode of action. J. Biol. Chem. 240: 2749-2753.

DEMENTYEVA, M. E. 1962. [Diseases of cultivated fruits.] Moscow. 240 p.

DESHPANDE, K. B. 1959a. Pectolytic enzyme system of *Rhizoctonia solani*: properties of protopectinase. Biol. Plantarum 2: 139-151.

———. 1959b. Studies on nitrogen metabolism of *Rhizoctonia solani* Kühn. J. Biol. Sci. 2: 1-5.

———. 1960a. Studies on the pectolytic enzyme system of *Rhizoctonia solani* Kühn. I. Production of protopectinase. Biol. Sci. 3: 1-8.

———. 1960b. Studies on the pectolytic enzyme system of *Rhizoctonia solani* Kühn. IV. Viscosity reducing enzymes. Enzymologia 22: 295-306.

———. 1961. Studies on the pectolytic enzyme system of *Rhizoctonia solani* Kühn. III. Pectinesterase and polygalacturonase. J. Indian Botan. Soc. 40: 456-464.

DE SILVA, R. L., and R. K. S. WOOD. 1964. Infection of plants by *Corticium solani* and *C. praticola*—effect of plant exudates. Trans. Brit. Mycol. Soc. 47: 15-24.

DICKINSON, L. S. 1930. The effect of air temperature on the pathogenicity of *Rhizoctonia solani* parasitizing grasses on putting-green turf. Phytopathology 20: 597-608.

DIMOCK, A. W. 1940. Rhizoctonia foot rot of stocks can be controlled. Florists' Rev. 87: 13-14.

———. 1941. The Rhizoctonia foot-rot of annual stocks (*Matthiola incana*). Phytopathology 31. 87-91.

DIMOND, A. E., and J. G. HORSFALL. 1960. Prologue-inoculum and the disease population. P. 1-22. *In* J. G. Horsfall and A. E. Dimond [eds.], Plant pathology, an advanced treatise. III. Academic Press Inc., New York.

———, and ———. 1965. The theory of inoculum. P. 404-415. *In* K. F. Baker and W. C. Snyder [eds.], Ecology of soil-borne plant pathogens. Univ. California Press, Berkeley, Los Angeles.

DOBBS, C. G., and W. H. HINSON. 1953. A widespread fungistasis in soil. Nature (London) 172: 197-199.

DODGE, B. O., and N. E. STEVENS. 1924. The Rhizoctonia brown rot and other fruit rots of strawberry. J. Agr. Res. 28: 643-648.

DODMAN, R. L. 1965. Studies on plant exudates and the mode of penetration by *Thanatephorus cucumeris.* Ph.D. Thesis, Univ. of Adelaide, S. Australia.

———, K. R. BARKER, and J. C. WALKER. 1966. Auxin production by *Rhizoctonia solani.* (Abstr.) Phytopathology 56: 875.

DOMSCH, K. H. 1955. Die Kultivierung von Bodenpilzen auf bodenahnlichen Substraten. Arch. Mikrobiol. 23: 79-87.

———. 1958. Die Prufung van Bodenfungiciden. I. Pilz-Substrat-Fungicid-Kombinationen. II. Pilz-Wirt-Fungicid-Kombinationen. Plant Soil 10: 114-146.

———. 1960a. Das Pilzspektrum einer Bodenprobe. I. Nachweis der Homogenität. Arch. Mikrobiol. 35: 181-195.

———. 1960b. Das Pilzspektrum einer Bodenprobe. III. Nachweis der Einzelpilze. Arch. Mikrobiol. 35: 311-339.

———. 1964. Soil fungicides. Ann. Rev. Phytopathol. 2: 293-320.

———, and P. SCHICKE. 1960. Erfolgkontralle fungizides Bodenbehandlungen. Nachrbl. Deut. Pflantzenschutzdienst 12: 121-124.

DONK, M. A. 1931. Revisie van de Nederlandse Heterobasidiomycetae en Homobasidiomycetae Aphyllophoraceae. Deel I. Mededel. Ned. Mycol. Ver. 18-20: 116-118.

———. 1941. Nomina generica conservanda and confusa for Basidiomycetes (Fungi). Bull. Botan. Garden Buitenzorg (3) 17: 155-197.

———. 1954a. Notes on resupinate Hymenomycetes - 1. On *Pellicularia* Cooke. Reinwardtia 2: 425-434.

———. 1954b. A note on sterigmata in general. Bothalia 6: 301-302.

———. 1956a. Notes on resupinate Hymenomycetes - 2. The tulasnelloid fungi. Reinwardtia 3: 363-379.

———. 1956b. Notes on resupinate Hymenomycetes - 3. Fungus 26: 3-24.

———. 1957a. The generic names proposed for Hymenomycetes - VII. "Thelephoraceae." Taxon 6: 106-123.

———. 1957b. Notes on resupinate Hymenomycetes - IV. Fungus 27: 1-29.

———. 1958a. Notes on resupinate Hymenomycetes - V. Fungus 28: 16-36.

———. 1958b. Notes on the basidium. Blumea Suppl. 4: 96-105.

———. 1964. A conspectus of the families of Aphyllophorales. Persoonia 3: 199-324.

DORST, J. C. 1923. Aantasting van de Aardappelplant door *Rhizoctonia solani* enhaar bestrijding door sublimaat. Tijdschr. Plantenziekten 29: 97-106.

DOWLER, W. M., P. D. SHAW, and D. GOTTLIEB. 1963. Terminal oxidation in cell-free extracts of fungi. J. Bact. 86: 9-17.

DOWNES, M. J., and J. B. LOUGHNANE. 1965. A baiting material for the isolation of *Rhizoctonia* from soil. Euopean Potato J. 8: 179-190.

DOWNIE, D. G. 1957. *Corticium solani*—an orchid endophyte. Nature (London) 179: 160.

———. 1959a. *Rhizoctonia solani* and orchid seed. Trans. Botan. Soc. Edinburgh 37: 279-285.

———. 1959b. The mycorrhiza of *Orchis purpurella.* Trans. Botan. Soc. Edinburgh 38: 16-29.

DOWNY, A. R., M. L. SCHUSTER, and R. K. OLDEMEYER. 1952. Cooperative field testing of strains of sugar beets for resistance to several root rots. Proc. Am. Soc. Sugar Beet Tech. 6: 557-561.

DRAYTON, F. L. 1915. The Rhizoctonia lesions on potato stems. Phytopathology 5: 59-63.

DUGGAR, B. M. 1915. *Rhizoctonia crocorum* (Pers.) DC. and *R. solani* Kühn (*Corticium vagum* B. & C.) with notes on other species. Ann. Missouri Botan. Garden 2: 403-458.

———. 1916. *Rhizoctonia solani* in relation to the "Mopopilz" and the "Vermehrungspilz." Ann. Missouri Botan. Garden 3: 1-10.

———, and F. C. STEWART. 1901. The sterile fungus *Rhizoctonia* as a cause of plant diseases in America. New York (Cornell) Agr. Expt. Sta. Bull. 186: 51-76.

DUNCAN, D. P. 1954. A study of some of the factors affecting the natural regeneration of tamarack (*Larix laricina*) in Minnesota. Ecology 35: 498-521.

DUNLEAVY, J. M. 1952. Control of damping-off of sugar beets by *Bacillus subtilis.* (Abstr.) Phytopathology 42: 465.

DUNN, E., and W. A. HUGHES. 1964. Interrelationship of potato root eelworm *Heterodera rostochiensis* Woll., *Rhizoctonia solani* Kühn and *Colletotrichum atramentarium* (B. and Br.) Toub. on the growth of the tomato plant. Nature (London) 201: 413-414.

DURBIN, R. D. 1955. The effect of $CO_2$ on the vertical distribution of various strains of *Rhizoctonia solani.* (Abstr.) Phytopathology 45: 693.

———. 1957. Importance of variation and quantity of pathogens. California Agr. Expt. Sta. Manual 23: 255-262.

———. 1958. Factors influencing the ecology of clones of *Rhizoctonia solani.* Ph.D. Thesis, Univ. California, Berkeley. 77 p.

———. 1959a. Factors affecting the vertical distribution of *Rhizoctonia solani* with reference to $CO_2$ concentration. Am. J. Botany 46: 22-25.

———. 1959b. Some effects of light on the growth and morphology of *Rhizoctonia solani.* Phytopathology 49: 59-60.

———. 1961. Techniques for the observation and isolation of soil microorganisms. Botan. Rev. 27: 527-560.

ECHANDI, E. 1965. Basidiospore infection by *Pellicularia filamentosa* (= *Corticium microsclerotia*), the incitant of web blight of common bean. Phytopathology 55: 698-699.

ECKERT, J. W. 1957. Fungistatic and phytotoxic activity of certain aromatic nitro compounds. Ph.D. Thesis, Univ. California, Davis. 80 p.

ECKSTEIN, Z., and E. ZUKOWSKI. 1958. O wtasnosciach grzybolojczych pewnych pochodnych benzoksazolonu. [The fungicidal properties of some benzoxazolone derivatives.] Przem. Chem. 37: 418-420. [Abstr. *in* Rev. Appl. Mycol. 38: 350-351, 1959.]

EDSON, H. A. 1915. Seedling diseases of sugar beets and their relation to root-rot and crown-rot. J. Agr. Res. 4: 135-168.

———, and M. SHAPOVALOV. 1918. Potato stem lesions. J. Agr. Res. 14: 213-219.

EDWARDS, H. I., and W. NEWTON. 1937. The physiology of *Rhizoctonia solani* Kühn. V. The activity of certain enzymes of *Rhizoctonia solani* Kühn. Sci. Agr. 17: 544-549.

EIDAM, E. 1887. Untersuchungen zweier Krankheifserscheinungen, die an den Wurzeln der Zuckerrübe in Schlesien seit letzten sommer ziemlich laufig vorgekommen sind. Jahresber. Schles. Gesells. Vaterl. Kultur, Breslau 1887: 261-262.

ELAROSI, H. M. 1956. Sporulation of fungi inside the plant host cell. Nature (London) 177: 665-666.

———. 1957a. Fungal associations. I. Synergistic relation between *Rhizoctonia solani* Kühn and *Fusarium solani* Snyder and Hansen in causing a potato tuber rot. Ann. Botany (NS) 21: 555-568.

———. 1957b. Fungal associations. II. Cultural studies on

*Rhizoctonia solani* Kühn, *Fusarium solani* Snyder and Hansen, and other fungi, and their interactions. Ann. Botany (NS) 21: 569-585.

———. 1958. Fungal associations. III. The role of pectic enzymes on the synergistic relation between *Rhizoctonia solani* Kühn and *Fusarium solani* Snyder and Hansen in the rotting of potato tubers. Ann. Botany (NS) 22: 399-416.

EL-HELALY, A. F., I. A. IBRAHIM, M. W. ASSAWAH, H. M. ELAROSI, and S. H. MICHAIL. 1962. Seasonal prevalence of the main pathogens causing damping-off in tomato in Egypt. Phytopath. Medit. 1: 152-156.

ELLIS, D. E. 1951. Noteworthy diseases of cucurbits in North Carolina in 1949 and 1950. Plant Disease Reptr. 35: 91-93.

———, and R. S. COX. 1951. The etiology and control of lettuce damping-off. N. Carolina Agr. Expt. Sta. Tech. Bull. 94: 1-33.

ELMER, O. H. 1942. Effect of environment on prevalence of soil-borne *Rhizoctonia*. Phytopathology 32: 972-977.

ELSAID, H. M. 1962. Screening and adaptation of *Rhizoctonia solani* Kühn to various soil fungicides. M.S. Thesis, Louisiana State Univ., Baton Rouge. 73 p.

———, and J. B. SINCLAIR. 1962. A new greenhouse technique for evaluating fungicides for control of cotton sore-shin. Plant Disease Reptr. 46: 852-856.

———, and ———. 1964. Adapted tolerance to organic fungicides by isolates of *Rhizoctonia solani* from seedling cotton. Phytopathology 54: 518-522.

EL ZARKA, A. M. 1963. A rapid method for the isolation and detection of *Rhizoctonia solani* Kühn from naturally infested and artificially inoculated soils. Mededel. Landbouwhogeschoolen Opzoekingssta. Staat Gent: 28: 877-885.

———. 1965. Studies on *Rhizoctonia solani* Kühn, the cause of the black scurf disease of potato. Mededel. Landbouwhogeschoolen, Wageningen 217. 73 p.

ENDO, R. M. 1961. Turf grass diseases in southern California. Plant Disease Reptr. 45: 869-873.

———. 1963. Influence of temperature on rate of growth of five fungus pathogens of turf grass and on rate of disease spread. Phytopathology 53: 857-861.

ENDO, S. 1930. On the influence of the temperature upon the development of *Hypochnus*. Ann. Phytopathol. Soc. Japan 2: 280-283. [Abstr. *in* Japan. J. Botany 5: 89-90.]

———. 1931. Studies on the Sclerotium diseases of the rice plant. V. Ability of overwintering of certain important fungi causing Sclerotium diseases of the rice plant and their resistance to dry conditions. Forsch. Geb. Pflanzenkrankh. (Kyoto) 1: 149-167. [Abstr. *in* Japan. J. Botany 6: 28, 1932.]

———. 1933. Influence of salt on the pathogenicity of *Hypochnus sasakii* Shirai. Trans. Tottori Soc. Agr. Sci. 4: 362-367. [Abstr. *in* Biol. Abstr. 8: 17424.]

———. 1935. Effect of sunlight on the infection of the rice plant by *Hypochnus sasakii* Shirai. Bull. Miyazaki Coll. Agr. Forestry 8: 75-78.

———. 1940. Physiological studies on the causal fungi of Sclerotium diseases of rice plant with special reference to some factors controlling the occurrence of the disease. Bull. Miyazaki Coll. Agr. Forestry 11: 55-218.

ERIKSSON, J. 1958a. Studies in Corticiaceae (*Botryohypochnus* Donk, *Botryobasidium* Donk, and *Gloeocystidiellum* Donk). Svensk Botan. Tidskr. 52: 1-17.

———. 1958b. Studies in the Heterobasidiomycetes and Homobasidiomycetes-Aphyllophorales of Muddus National Park in North Sweden. Symbolae Botan. Upsaliensis 16: 1-172.

———. 1958c. Studies of the Swedish Heterobasidiomycetes and Aphyllophorales with special regard to the family Corticiaceae. Almqvist and Wiksells, Uppsala. 26 p.

ERWIN, D. C. 1954. Relation of *Stagonospora*, *Rhizoctonia*, and associated fungi to crown rot of alfalfa. Phytopathology 44: 137-144.

———. 1956. Important diseases of alfalfa in southern California. Plant Disease Reptr. 40: 380-383.

———, and R. A. FLACK. 1963. Association of curly top virus and *Rhizoctonia solani* with a late season decline of flax. Plant Disease Reptr. 47: 36-40.

———, and H. T. REYNOLDS. 1958. The effect of seed treatment of cotton with Thimet, a systemic insecticide, on Rhizoctonia and Pythium seedling diseases. Plant Disease Reptr. 42: 174-176.

EXNER, B. 1953. Comparative studies of four Rhizoctonias occurring in Louisiana. Mycologia 45: 698-719.

———, and S. J. B. CHILTON. 1943. Cultural differences among single basidiospore isolates of *Rhizoctonia solani*. Phytopathology 33: 171-174.

———, and ———. 1944. Comparative studies of basidiospore cultures of *Rhizoctonia solani*. (Abstr.) Phytopathology 34: 999.

FAHMY, T. 1931. The sore-shin disease and its control. Min. Agr. Egypt, Tech. and Sci. Service (Plant Protect. Sect.) Bull. 108. 24 p.

FAWCETT, D. E. 1966. An atlas of fine structure: The cell, its organelles and inclusions. W. B. Saunders Company, Philadelphia. 448 p.

FELIX, E. L. 1955. Notes on some plant diseases in Tennessee. Plant Disease Reptr. 39: 275-276.

FERGUSON, J. 1953. Controlled reinfestation of treated soil. (Abstr.) Phytopathology 43: 586-587.

FINCHAM, J. R. S., and P. R. DAY. 1963. Fungal genetics. F. A. Davis Company, Philadelphia. 300 p.

FLENTJE, N. T. 1952. *Corticium praticola* Kotila: an interesting Basidiomycete occurring in England. Nature (London) 170: 892.

———. 1956. Studies on *Pellicularia filamentosa* (Pat.) Rogers. I. Formation of the perfect stage. Trans. Brit. Mycol. Soc. 39: 343-356.

———. 1957. Studies on *Pellicularia filamentosa* (Pat.) Rogers. III. Host penetration and resistance and strain specialization. Trans. Brit. Mycol. Soc. 40: 322-336.

———. 1959. The physiology of penetration and infection. P. 76-87. *In* C. S. Holton, et al. [eds.], Plant pathology problems and progress 1908-1958, Univ. Wisconsin Press, Madison.

———. 1965. Pathogenesis by soil fungi. P. 255-268. *In* K. F. Baker and W. C. Snyder [eds.], Ecology of soilborne plant pathogens, Univ. California Press, Berkeley, Los Angeles.

———, R. L. DODMAN, and A. KERR. 1963. The mechanism of host penetration by *Thanatephorus cucumeris*. Australian J. Biol. Sci. 16: 784-799.

———, and D. J. HAGEDORN. 1964. Rhizoctonia tip blight and stem rot of pea. Phytopathology 54: 788-791.

———, A. KERR, and R. L. DODMAN. 1961. Rep. Waite Agr. Res. Inst. (S. Australia) 1958-1959. 58 p.

———, ———, ———, A. R. McKENZIE, and H. M. STRETTON. 1963. *Thanatephorus investigations. In* Rep. Waite Agr. Res. Inst., (S. Australia) 1962-1963: 54-55.

———, ———, ———, ———, and ———. 1964. *In* Rep. Waite Agr. Res. Insti., (S. Australia) 1962-63: 53-61.

———, and H. K. SAKSENA. 1957. Studies on *Pellicularia filamentosa* (Pat.) Rogers. II. Occurrence and distribution of pathogenic strains. Trans. Brit. Mycol. Soc. 40: 95-108.

——, and H. M. STRETTON. 1964. Mechanisms of variation in *Thanatephorus cucumeris* and *T. praticolus*. Australian J. Biol. Sci. 17: 686-704.

——, ——, and E. J. HAWN. 1963. Nuclear distribution and behaviour throughout the life cycles of *Thanatephorus, Waitea* and *Ceratobasidium* species. Australian J. Biol. Sci. 16: 450-467.

——, ——, and A. R. McKENZIE. 1967. Mutation in *Thanatephorus* cucumeris. Australian J. Biol. Sci. 20: 1173-1180.

FLOR, H. H. 1955. Host-parasite interaction in flax rust —its genetics and other implications. Phytopathology 45: 680-685.

FOCKE, R. 1952. Der Einfluss von Auspflanzzeit und Vorkeimung der Kartoffel auf die Höhe der durch *Rhizoctonia solani* hervorgerufenen Schäden. Wiss. Z. Univ. Rostock 1: 47-54.

FORSBERG, J. L. 1946. Diseases of ornamental plants. Colorado Agr. Mech. Coll., Fort Collins. 172 p.

FORSTENEICHNER, F. 1931. Die Jugendkrankheiten der Baumwolle in der Turkei. Phytopathol. Z. 3: 367-419.

FOSTER, D. W., and J. B. TAYLOR. 1966. The metabolism of reduced pyridine nucleotides in tumor tissue. J. Biol. Chem. 241: 38-44.

FRANK, B. 1883. Uber einige neue u. weniger bekannte Pflanzenkrankheiten. Bericht Deutsch. Botan. Gesells. 1: 62-63.

FREDERIKSEN, T., C. A. JØRGENSEN, and O. NEILSEN. 1938. Undersøgelser over Kartoflens Rodfiltsvamp og dens Bekaempelse. Tidsskr. Planteavl 43: 1-64.

FREY-WYSSLING, A., E. GRIESHABER, and K. MULETHALER. 1963. Origin of sphaerosomes in plant cells. J. Ultrastruct. Res. 8: 506-516.

FRIES, N. 1938. Uber die Bedeutung von Wuchsstoffen fur das Wachstum verschiedener Pilze. Symbolae Botan. Upsaliensis. 3(2) 188 p.

FRUTON, J. S., and S. SIMMONDS. 1959. General Biochemistry. John Wiley & Sons, Inc., New York. 1077 p.

FUKANO, H. 1932. Cytological studies in *Hypochnus sasakii* Shirai, causing a sclerotial disease of the rice plant. Bull. Sci. Kyushu Imp. Univ. 5: 117-136.

FULTON, H. R. 1908. Diseases of pepper and beans. Louisiana Agr. Expt. Sta. Bull. 101: 1-21.

FULTON, N. D., and E. W. HANSON. 1960. Studies on root rots of red clover in Wisconsin. Phytopathology 50: 541-550.

——, B. A. WADDLE, and R. O. THOMAS. 1956. Influence of planting dates on fungi isolated from diseased cotton seedlings. Plant Disease Reptr. 40: 556-558.

GADD, C. H., and L. S. BERTUS. 1928. *Corticium vagum* B. & C.—The cause of a disease of *Vigna oligosperma* and other plants in Ceylon. Ann. Roy. Botan. Garden (Peradeniya) 11: 27-49.

GARBER, R. H., and L. D. LEACH. 1957. The use of chemical indicators in the study of distribution of row-treatment fungicides. (Abstr.) Phytopathology 47: 521.

GARNJOBST, L. 1953. Genetic control of heterocaryosis in *Neurospora crassa*. Am. J. Botany 40: 607-613.

——. 1955. Further analysis of genetic control of heterocaryosis in *Neurospora crassa*. Amer. J. Botany 42: 444-448.

GARREN, K. H. 1955. Disease development and seasonal succession of pathogens of white clover. Part II. Stolon diseases and the damage-growth cycle. Plant Disease Reptr. 39: 339-341.

GARRETT, S. D. 1938. Soil conditions and the root-infecting fungi. Biol. Rev. 13: 159-185.

——. 1950. Ecology of the root inhabiting fungi. Biol. Rev. 25: 220-254.

——. 1951. Ecological groups of soil fungi: A survey of substrate relationships. New Phytologist 50: 149-166.

——. 1955. A century of root-disease investigation. Ann. Appl. Biol. 42: 211-219.

——. 1956. Biology of root-infecting fungi. Cambridge Univ. Press, London, New York. 293 p.

——. 1960. Inoculum potential. P. 23-56. *In* J. G. Horsfall and A. E. Dimond [eds.], Plant pathology, an advanced treatise, III, Academic Press, Inc., New York.

——. 1962. Decomposition of cellulose in soil by *Rhizoctonia solani* Kühn. Trans. Brit. Mycol. Soc. 45: 115-120.

——. 1965. Toward biological control of soil-borne plant pathogens. P. 4-17. *In* K. F. Baker and W. C. Snyder [eds.], Ecology of soil-borne pathogens, Univ. California Press, Berkeley, Los Angeles.

GARZA-CHAPA, R., and N. A. ANDERSON. 1966. Behavior of single-basidiospore isolates and heterokaryons of *Rhizoctonia solani* from flax. Phytopathology 56: 1260-1268.

GATTANI, M. L. 1957. Studies on the control of damping-off of safflower with antibiotics. Plant Disease Reptr. 41: 160-164.

GAUMANN, E. 1957. Uber die toxine der *Endothia parasitica* (Marr) And. Pflanzenschuzber 19: 9-16.

——, and H. KERN. 1959. Uber chemische Abwehrreaktionen bei Orchideen. Phytopathol. Z. 36: 1-26.

GAYED, S. K., and A. TEMPEL. 1965. A simple and rapid method for screening chemicals against damping-off of cotton caused by *Rhizoctonia solani* Kühn. Mededel. Landbouwhogeschoolen Opzoekingssta. Staat Gent 30: 1722-1727.

GEORGOPOULOS, S. G., and S. WILHELM. 1962. Effect of nonsterile soil on *Rhizoctonia solani* mycelium in the presence of PCNB. (Abstr.) Phytopathology 52: 361.

GERM, H. 1960. Methodology of the vigour test for wheat, rye and barley in rolled filter paper. Proc. Intern. Seed Testing Assoc. 25: 515-518.

GIBSON, I. A. S. 1956. Sowing density and damping-off of pine seedlings. E. African Agr. J. 21: 183-188.

——, M. LEDGER, and E. BOEHM. 1961. An anomalous effect of pentachloronitrobenzene on the incidence of damping off caused by a *Pythium* sp. Phytopathology 51: 531-533.

GIESY, R. M., and P. R. DAY. 1965. The septal pores of *Coprinus lagopus* in relation to nuclear migration. Am. J. Botany 52: 287-293.

GILMAN, J. C., and I. E. MELHUS. 1923. Further studies on potato seed treatment. Phytopathology 13: 341-358.

GILMOUR, J. S. L. 1961. Taxonomy. P. 27-45. *In* A. M. MacLeod and L. S. Cobley [eds.], Contemporary botanical thought. Oliver & Boyd Ltd., Edinburgh, London. 197 p.

GINZBURG, B. Z. 1961. Evidence for a protein gel structure cross-linked by metal cations in the intercellular cement of plant tissue. J. Exptl. Botany 12: 85-107.

GIRBARDT, M. 1961. Licht—und elektronenmikroskopische Untersuchungen an *Polystictus versicolor*. II. Die Feinstrucktur von Grundplasma und Mitochondrien. Arch. Mikrobiol. 39: 351-359.

GOCHENAUR, S. E. 1964. A modification of the immersion tube method for isolating soil fungi. Mycologia 56: 921-923.

GONZALEZ, L. C., and J. H. OWEN. 1963. Soil rot of

tomato caused by *Rhizoctonia solani.* Phytopathology 53: 82-85.

GOSS, R. W., and H. O. WERNER. 1929. Seed potato treatment tests for control of scab and *Rhizoctonia.* Nebraska Agr. Expt. Sta. Res. Bull. 44. 42 p.

GOTTLIEB, D. 1946. The utilization of amino acids as a source of carbon by fungi. Arch. Biochem. 9: 341-351.

———, H. E. CARTER, J. H. SLONEKER, LUNG CHI WU, and E. GAUDY. 1961. Mechanisms of inhibition of fungi by filipin. Phytopathology 51: 321-330.

GRAHAM, J. H. 1958. Effect of gibberellic acid on damping-off of ladino white clover. Plant Disease Reptr. 42: 963-964.

———, V. G. SPRAGUE, and R. R. ROBINSON. 1957. Damping-off of ladino clover and lespedeza as affected by soil moisture and temperature. Phytopathology 47: 182-185.

GRAHAM, K. M., M. P. HARRISON, and C. E. SMITH. 1962. Control of storage rot of strawberry plants. Can. Plant Disease Survey 42: 205-207.

GRAINGER, J. 1956. Host nutrition and attack by fungal parasites. Phytopathology 46: 445-456.

———, and M. R. M. CLARK. 1963. Interactions of *Rhizoctonia* and potato root eelworm. European Potato J. 6: 131-132.

GRATZ, L. O. 1925. Wire stem of cabbage. New York (Cornell) Agr. Expt. Sta. Mem. 85: 1-60.

GREATHOUSE, G. A., and N. E. RIGLER. 1940. The chemistry of resistance of plants to Phymatotrichum root rot. V. Influence of alkaloids on growth of fungi. Phytopathology 30: 475-485.

GREEN, D. E. 1958. Studies in organized enzyme systems. P. 177-227. *In* The Harvey lectures 1956-57, Academic Press Inc., New York.

GREENSPAN, G. 1960. Oxygenation of steroids by strains of the genus *Rhizoctonia.* Diss. Abstr. 21: 1041-1042.

GREGOR, M. J. F. 1935. A disease of bracken and other ferns caused by *Corticium anceps* (Bres. et Syd.) Gregor. Phytopathol. Z. 8: 401-418.

GRIFFIN, D. M. 1963. Soil moisture and the ecology of soil fungi. Biol. Rev. 38: 141-166.

GROOSHEVOY, S. E., P. M. LEVYKE, and E. I. MALBIEVA. 1940. [Methods of disinfecting seed-bed soil by natural sources of heat.] A. I. Mikoyan Pansoviet Sci. Res. Inst. Tob. Indian Tob. Publ. 141: 49-61. [*Abstr. in* Rev. Appl. Mycol. 20: 87, 1941.]

GROSSBARD, E. 1958. Autoradiography of fungi through a layer of soil and in agar culture. Nature (London) 182: 854-856.

———, and D. R. STRANKS. 1959. Translocation of cobalt-60 and caesium-137 by fungi in agar and soil cultures. Nature (London) 184: 310-314.

GROVE, S. N., D. J. MORRE, and C. E. BRACKER. 1966. Dictyosomes in vegetative hyphae of *Pythium ultimum.* Indiana Acad. Sci. Proc. [In press.]

GUPTA, M. N. 1963. Studies in the pathogenicity of certain parasitic fungi with special reference to their pectic enzyme system. Agrar. Univ. J. Res. Sci. 12: 161-162.

GUSSOW, H. T. 1917. The pathogenic action of *Rhizoctonia* on potato. Phytopathology 7: 209-213.

GUTHRIE, E. J. 1963. Two useful techniques for rust work. Robigo 14: 3-4.

GUTZEVITCH, S. A. 1934. ["Black leg" disease of the cabbage, *Moniliopsis aderholdi* Ruhl.]. [Russian, with English summary.] Trav. Soc. Nat. St.-Petersb. (Leningr.) 63: 69-82.

HACSKAYLO, J., and R. B. STEWART. 1962. Efficacy of phorate as a fungicide. Phytopathology 52: 371-372.

HADLEY, G., and M. PEROMBELON. 1963. Production of pectic enzymes by *Rhizoctonia solani* and Orchid endophytes. Nature (London) 200: 1337.

HAFIZ, A., and D. J. NIEDERPRUEM. 1963. Studies on basidiospore germination in *Schizophyllum commune.* Am. J. Botany 50: 614-615.

HAIG, I. T. 1936. Factors controlling initial establishment of western pine and associated species. Yale Univ. School Forestry Bull. 41: 1-149.

HALLIWELL, G. 1961. The action of cellulolytic enzymes from *Myrothecium verrucaria.* Biochem. J. 79: 185-192.

HANSEN, J. D., and E. A. CURL. 1964. Interactions of *Sclerotium rolfsii* and other microorganisms isolated from stolon tissue of *Trifolium repens.* Phytopathology 54: 1127-1132.

HANSEN, L. R. 1963. Skarp øyeflekk på korn paråsaket av *Rhizoctonia solani* Kühn. Meld. Norges Landbrttøjsk 42: 1-12.

HANSFORD, C. G. 1928. Annual Report of the Government Mycologist for the period October 11th, 1926, to December 31st, 1927. Ann. Rept. Uganda Dept. Agr. 1927: 37-42.

HARLEY, J. L. 1965. Mycorrhiza. P. 218-230. *In* K. F. Baker and W. C. Snyder [eds.], Ecology of soil-borne plant pathogens, Univ. California Press, Berkeley, Los Angeles.

HARTER, L. L., and W. A. WHITNEY. 1927. Mottle necrosis of sweet potatoes. J. Agr. Res. 34: 893-914.

HARTLEY, C. 1921. Damping-off in forest nurseries. U. S. Dept. Agr. Bull. 934: 1-99.

———, T. C. MERRILL, and A. S. RHOADS. 1918. Seedling diseases of conifers. J. Agr. Res. 15: 521-558.

HASEGAWA, T., and T. TAKAHASHI. 1958. Studies on the oxidation of steroids by fungi Part I. The steroid oxidizing property of *Corticium sasakii.* Bull. Agr. Chem. Soc. Japan 22: 212-217.

———, and ———. 1959. Microbiological oxidations of Reichstein's Compound S by *Corticium solani.* Bull. Agr. Chem. Soc. Japan 23: 137-138.

HASEGAWA, Y., T. NAKAI, Y. FUJIMURA, Y. KANEKO, and S. DOI. 1964a. Studies on nucleases produced by micro-organisms I. Production of phosphodiesterase (t'-nucleotide forming) by *Pellicularia* sp. J. Agr. Chem. Soc. Japan 38: 461-466.

———, ———, ———, ———, and ———. 1964b. Studies on nucleases produced by micro-organisms II. Properties of phosphodiesterase (5'-nucleotide forming) produced by *Pellicularia* sp. J. Agr. Chem. Soc. Japan 38: 467-471.

HASHIOKA, Y. 1951. Studies on pathological breeding of rice, IV. Varietal resistance of rice to the sclerotial diseases. Japan J. Breeding 1: 21-26.

HASSAN, S. F. 1956. Pathogenicity of root rotting fungi of oats. Plant Disease Reptr. 40: 890-897.

HAWKER, L. E. 1957. Ecological factors and the survival of fungi. P. 238-258. *In* R. E. O. Williams and C. C. Spicer [eds.], Microbial ecology, 7th Symp. Soc. Gen. Microbiol., Cambridge Univ. Press, London.

———, M. A. GOODAY, and C. E. BRACKER. 1966. Plasmodesmata in fungal cell walls. Nature (London) 212: 635.

HAWN, E. J. 1959. Histological study on crown and bud rot of alfalfa. Can. J. Botany 37: 1247-1249.

———, and M. W. CORMACK. 1952. Crown bud rot of alfalfa. Phytopathology 42: 510-511.

———, and T. C. VANTERPOOL. 1952. The perfect stage of *Rhizoctonia solani* from flax. (Abstr.) Proc.

Can. Phytopathol. Soc. 19: 16.

———, and ———. 1953. Preliminary studies on the sexual stage of *Rhizoctonia solani* Kühn. Can. J. Botany 31: 699-710.

HEALD, F.D. 1933. Manual of plant diseases. McGraw-Hill Book Company, New York. 953 p.

HEARN, J. L., JR. 1943. *Rhizoctonia solani* Kühn and the brownpatch disease of grass. Texas Acad. Sci. Proc. 26: 41-42.

HEDGCOCK, G. G. 1904. A note on *Rhizoctonia*. Science 19: 268.

HEMMI, T., and S. ENDO. 1928. On a staining method for testing the viability of sclerotia of fungi. Mem. Coll. Agr. Kyoto Imper. Univ. 7: 39-49.

———, and ———. 1931. Studies on Sclerotium diseases of the rice plant. III. Some experiments on the sclerotial formation and the pathogenicity of certain fungi causing Sclerotium diseases of the rice plant. Forsch. Geb. Pflanzenkrankh. (Kyoto) 1: 111-125.

———, and ———. 1933. Studies on Sclerotium diseases of the rice plant. VI. On the relation of temperature and period of continuous wetting to the infection of the rice plant by *Hypochnus sasakii* Shirai. [Japanese, with English summary.] Forsch. Geb. Pflanzenkrankh. Kyoto 2: 202-218.

———, and K. YOKOGI. 1927. Studies on Sclerotium diseases of the rice plant. I. Agr. and Hort. 2: 955-1094.

HENRIKSEN, J. B. 1961. *Corticium solani* as the cause of pit-rot. European Potato J. 4: 243-252.

HERZOG, W. 1961. Das Überdauern und der Saprophytismus des Wurzeltöters *Rhizoctonia solani* Kühn im Boden. Phytopathol. Z. 40: 379-415.

———, and H. WARTENBERG. 1958. Untersuchugen über die Lebensdauer der Sklerotien von *Rhizoctonia solani* Kühn im Boden. Phytopathol. Z. 33: 291-315.

HESLOP-HARRISON, J. 1953. New concepts in flowering-plant taxonomy. William Heinemann, Ltd. London. 135 p.

HILDEBRAND, A. A., W. E. McKEEN, and L. W. KOCH. 1949. Row treatment of soil with tetramethylthiuram disulphide for control of blackroot of sugarbeet seedlings. I. Greenhouse tests. Can. J. Res. (C) 27: 23-43.

HILDEBRAND, E. M. 1938. Techniques for the isolation of single microorganisms. Botan. Rev. 4: 627-664.

HILLS, F. J., and J. D. AXTELL. 1950. The effect of several nitrogen sources on beet sugar yields in Kern County, California. Proc. Am. Soc. Sugar Beet Tech. 1950: 356-361.

———, and L. D. LEACH. 1962. Photochemical decomposition and biological activity of P-dimethylamino benzenediazo sodium sulfonate (Dexon). Phytopathology 52: 51-56.

HINE, R. B. 1963. Effect of streptomycin and pimaricin on growth and respiration of *Pythium* species. Mycologia 54: 640-646.

HIRST, J. M. 1965. Dispersal of soil microorganisms. P. 69-81. *In* K. F. Baker and W. C. Snyder [eds.], Ecology of soil-borne plant pathogens, Univ. California Press, Berkeley, Los Angeles.

HOCHSTER, R. M. 1961. 5-Bromouracil: A new inhibitor of glucose-6-phosphate dehydrogenase. Biochem. Biophys. Res. Communications 6: 289-292.

HOFFERBERT, W., and H. ORTH. 1951. Unsere Arbeiten zur *Rhizoctonia*-Frage bei der Kartoffel. I. Gibt es für den Züchter Möglichkeiten der Rhizoctonia methodisch zu begegnen? Z. Pflanzenkrankh. 58: 245-256.

———, ———, and G. ZU PUTLITZ. 1953. Unsere Arbeiten zur *Rhizoctonia*-Frage der Kartoffel. II. Teil. Z. Pflan-

zenkrankh. 60: 385-397.

HOLLEY, W. D., and R. BAKER. 1963. Carnation production, including the history, breeding, culture and marketing of carnations. Wm. C. Brown Company, Dubuque, Iowa. 142 p.

HOLLOWAY, B. W. 1955. Genetic control of heterocaryosis in *Neurospora crassa*. Genetics 40: 117-129.

HORSFALL, J. G. 1932. Dusting tomato seed with copper sulfate monohydrate for combating damping-off. New York Agr. Expt. Sta. (Geneva) Tech. Bull. 198. 34 p.

———, and A. E. DIMOND. 1963. A perspective on inoculum potential. Maheshwari Comm. Vol. J. Indian Botan. Soc. 42A: 46-57.

HOUSTON, B. R. 1945. Culture types and pathogenicity of isolates of *Corticium solani*. Phytopathology 35: 371-393.

———. 1946. The physiology and pathogenicity of strains of *Corticium solani*. (Abstr.) Phytopathology 36: 401.

———, and J. B. KENDRICK. 1949. A crater spot of celery petioles caused by *Rhizoctonia solani*. Phytopathology 39: 470-474.

HOWARD, F. L., J. B. ROWELL, and H. L. KEIL. 1951. Fungus diseases of turf grasses. Rhode Island Agr. Expt. Sta. Bull. 308: 1-56.

HUBER, D. M., and A. M. FINLEY. 1959. *Gliocladium*, a causal agent in the bean root rot complex in Idaho. Plant Disease Reptr. 43: 626-628.

HUNTER, R. E., E. E. STAFFELDT, and C. R. MAIER. 1960. Effects of soil temperature on the pathogenicity of *Rhizoctonia solani* isolates. Plant Disease Reptr. 44: 793-795.

HURST, R. R. 1926. Report of the Dominion Field Laboratory of Plant Pathology, Charlottetown, P.E.I. P. 20-29. *In* Rept. Dominion Botanist for Year 1925, Div. Botany, Can. Dept. Agr.

HUSSAIN, S. S., and W. E. McKEEN. 1963. *Rhizoctonia fragariae* sp. nov. in relation to strawberry degeneration in southwestern Ontario. Phytopathology 53: 532-540.

HYNES, H. J. 1937. Studies on Rhizoctonia root-rot of wheat and oats. Dept. Agr. N.S. Wales Sci. Bull. 58. 42 p.

IKENO, S. 1933. Studies on Sclerotium diseases of the rice plant. VIII. On the relation of temperature and period of continuous wetting to the infection of soybean by the sclerotia of *Hypochnus sasakii* Shirai and on autolysis of the same fungus. [Japanese, with English summary.] Forsch. Geb. Pflanzenfrankh. Kyoto 2: 238-256.

ISAAC, P. K. 1964a. Metabolic specialization in the mycelium of *Rhizoctonia solani* Kühn. Can. J. Microbiol. 10: 621-622.

———. 1964b. Cytoplasmic streaming in filamentous fungi. Can. J. Botany 42: 787-792.

ISRAEL, O. P., and M. S. ALI. 1964. Effect of carbohydrates on the growth of *Rhizoctonia solani* Kühn. Biol. Plantarum 6: 84-87.

ITO, K., and S. KONTANI. 1952. The causal fungus of the web-blight of leguminous woody plants. Bull. Forestry Expt. Sta. Meguro 54: 45-78.

———, ———, and H. KONDO. 1955. Web-blight fungus of Japanese larch seedlings. Bull. Forestry Expt. Sta., Meguro 79: 43-70.

ITO, T. 1958. *Pellicularia koleroga* Cooke causing the thread blight of *Ginko biloba*. Bull. Forestry Expt. Sta., Meguro 105: 11-18.

IYENGER, M. R. S., D. N. SRIVASTAVA, and M. L. N. SASTRY. 1959. A study on the preservation of fungal cultures by the mineral oil method. Indian

Phytopathol. 12: 90-98.

JAARSVELD, A. 1940. De invloed van verschillende bodemschimmels op de virulentie van *Rhizoctonia solani* Kühn. De Acadamische Boekwinkel N.V., P. H. Vermeulen, Amsterdam. 101 p.

———. 1944. Der Einfluss verschiedener Bodenpilze auf die Virulenz von *Rhizoctonia solani* Kühn. Phytopathol. Z. 14: 1-75.

JACKS, H. 1951. The efficiency of chemical treatments of vegetable seeds against seed-borne and soil-borne organisms. Ann. Appl. Biol. 38: 135-168.

JACKSON, H. S. 1949. Studies of Canadian Thelephoraceae IV. *Corticium anceps* in North America. Can. J. Res. (C) 27: 241-252.

JACKSON, L. W. R. 1940. Effects of H-ion and Al-ion concentrations on damping-off of conifers and certain causative fungi. Phytopathology 30: 563-579.

JAFFE, L. F. 1966. An autotropism in *Botrytis*: Measurement technique and control by $CO_2$. Plant Physiol. 41: 303-306.

JANSSEN, J. J. 1929. Invloed der Bemesting op de Gezondheid van de Aardappel. Tijdschr. Plantenziekten 35: 119-151.

JAUCH, C. 1947. Observations on natural and artificial infections with "*Pellicularia filamentosa*" (*Corticium solani*). Min. Agr. Buenos Aires, Misc. Publ. (A) 3. 7 p.

JOHNSON, D., and K. AAS. 1960. Investigations of the technique of soil steaming. Acta Agr. Scand. Suppl. 9: 1-69.

JOHNSON, L. F., E. A. CURL, J. H. BOND, and H. A. FRIBOURG. 1959. Methods for studying soil microflora-plant disease relationships. Burgess Publishing Company, Minneapolis. 178 p.

JOHNSON, T. R., and A. M. HILLIS. 1958. A fluorescent mineral tracer technique to determine fungicide placement in the soil profile. Plant Disease Reptr. 42: 287.

JONES, L. R., J. JOHNSON, and J. G. DICKSON. 1926. Wisconsin studies upon the relation of soil temperature to plant disease. Wisconsin Agr. Expt. Sta. Bull. 71. 144 p.

JONES, W. 1952. Infection of basal shoots of hop plants by *Rhizoctonia solani*. Sci. Agr. 32: 114-115.

JOSHI, S. D. 1924. The wilt disease of safflower. Mem. Dept. Agr. India (Botan. Ser.) 13: 39-46.

KAGAWA, T., D. R. WILKEN, and H. R. LARDY. 1965. Control of choline oxidation in liver mitochondria by adenine nucleotides. J. Biol. Chem. 240: 1836-1842.

KAHN, R. P., and G. SILBER. 1959. Movement of four soil-borne pathogens in sterile soil and sphagnum moss. (Abstr.) Phytopathology 49: 542.

KAMIYA, N. 1959. Protoplasmic streaming. Protoplasmatologia 8: 1-199.

KAUFMAN, D. D., and L. E. WILLIAMS. 1965. Influence of soil reaction and mineral fertilization on numbers and types of fungi antagonistic to four soil-borne plant pathogens. Phytopathology 55: 570-574.

KEITT, G. W. 1915. Simple technique for isolating single strains of certain types of fungi. Phytopathology 5: 266-269.

KENDRICK, J. B., JR. 1951. The influence of temperature upon the incidence of Rhizoctonia root rot of lima beans. Phytopathology 41: 20.

———. 1956, 1957. *In* Proc. Pacific Coast Res. Conf. on Control of Soil Fungi 3: 18; 4: 6.

———, and R. W. ALLARD. 1952. A root rot tolerant lima bean. (Abstr.) Phytopathology 42: 515.

———, and ———. 1955. Lima bean tolerant to stem rot.

California Agr. 9: 8, 15.

———, and A. R. JACKSON. 1958. Factors influencing the isolation of certain soil-borne plant pathogens from soil (Abstr.) Phytopathology 48: 394.

———, A. O. PAULUS, and J. DAVIDSON. 1957. Control of Rhizoctonia stem canker of lima bean. Phytopathology 47: 19-20.

———, and G. A. ZENTMYER. 1957. Recent advances in control of soil fungi. P. 219-275. *In* R. L. Metcalf [ed.], Advances in pest control research. Vol. 1, Interscience Publishers, Inc., New York.

KERNKAMP, M. F., D. J. DE ZEEUW, S. M. CHEN, B. C. ORTEGA, C. T. TSIANG, and A. M. KHAN. 1952. Investigations on physiologic specialization and parasitism of *Rhizoctonia solani*. Minnesota Agr. Expt. Sta. Tech. Bull. 200. 36 p.

———, J. W. GIBLER, and L. J. ELLING. 1949. Damping-off of alfalfa cuttings caused by *Rhizoctonia solani*. Phytopathology 39: 928-935.

KERR, A. 1956. Some interactions between plant roots and pathogenic soil fungi. Australian J. Biol. Sci. 9: 45-52.

———, and N. T. FLENTJE. 1957. Host infection of *Pellicularia filamentosa* controlled by chemical stimuli. Nature (London) 179: 204-205.

KHADGA, B. B., J. B. SINCLAIR, and B. B. EXNER. 1963. Infection of seedling cotton hypocotyl by an isolate of *Rhizoctonia solani*. Phytopathology 53: 1331-1336.

KHAIR, J., G. FLEISCHMANN, and A. DINOOR. 1966. Rapid isolation of single spores of fungi from dialysis tubing cellophane. Phytopathology 56: 346.

KHAN, A. M. 1950. Factors affecting the pathogenicity of *Rhizoctonia solani* on legumes. Ph.D. Thesis, Univ. Minnesota, Minneapolis. 53 p.

KHARITINOVA, Z. M. 1958a. [The biology and harmful activity of the fungus *Rhizoctonia solani* Kühn.] Sb. Inst. Prickl. Zool. Fitopat. 1958: 206-221. [Abstr. *in* Referat. Zhur. Biol. 1959: 211.]

———. 1958b. [Comparative study of the fungi *Rhizoctonia solani* Kühn and *R. aderholdii* (Ruhl.) Kolosh.] Trav. Soc. Nat. (Moscow-Leningrad) 43: 75-81.

KICKX, J. 1867. Flore cryptogamique des Flandres. II. Libraire de J. B. Bailliere et Fils, Paris. 490 p.

KILLERMANN, S. 1928. Eubasidii, Reihe Hymenomycetae. *In* Engler and Prantl, Natürliche Pflanzenfamilien 2. Ausg. 8: 136.

KILPATRICK, R. A. 1966. Induced sporulation of fungi on filter paper. Plant Disease Reptr. 50: 789-790.

———, E. W. HANSEN, and J. G. DICKSON. 1954. Root and crown rots of red clover in Wisconsin and the relative prevalence of associated fungi. Phytopathology 44: 252-259.

KING, M. K., and P. K. ISAAC. 1964. The uptake of glucose-6-T and glycine-2-T by *Rhizoctonia solani* Kühn. Can. J. Botany 42: 815-821.

KIRK, B. T., and J. B. SINCLAIR. 1966. Plasmodesmata between hyphal cells of *Geotrichum candidum*. Science 153: 1646.

KLEIN, H. H. 1959. Etiology of *Phytophthora* disease of soybeans. Phytopathology 49: 380-383.

KOHLMEYER, J. 1956. Über den Cellulose-Abbau durch einige phytopathogene Pilze. Phytopathol. Z. 27: 147-182.

KOMLÖSSY, G. 1954. [Significant results of resistance examinations in variety tests.] Mezögazd. Kiadv. 1953: 94-132. [Abstr. *in* Hungarian Agr. Rev. 4: 1, 1955.]

KOMMEDAHL, T., and H. C. YOUNG. 1956. Effect of host and soil substrate on the persistence of *Fusarium* and *Rhizoctonia* in soil. Plant Disease Reptr. 40: 28-29.

KONTANI, S., and K. MINEO. 1962. Studies on web-blight of *Alnus* seedlings. Bull. Forestry Expt. Sta. Meguro 134: 1-19.

KOPECKY, J. 1927. Investigations of the relations of water to soil. Proc. 1st Intern. Congr. Soil Sci. 1: 495-503.

KOSSIOKOFF, M. G. 1887. De la propriete que possedent les microbes de s'accomoder aux milieux antiseptiques. Ann. Inst. Pasteur 1: 465.

KOTILA, J. E. 1929. A study of the biology of a new spore-forming *Rhizoctonia, Corticium praticola*. Phytopathology 19: 1059-1099.

———. 1943. A new sugar beet leaf blight caused by a strain of *Corticium solani*. (Abstr.) Phytopathology 33: 6-7.

———. 1945a. Rhizoctonia foliage disease of *Hevea brasiliensis*. Phytopathology 35: 739-741.

———. 1945b. Cotton-leaf-spot *Rhizoctonia* and its perfect stage on sugar beets. Phytopathology 35: 741-743.

———. 1947. Rhizoctonia foliage blight of sugar beets. J. Agr. Res. 74: 289-314.

KOTOWSKI, F. 1927. Temperature relations to germination of vegetable seed. Am. Soc. Hort. Sci. Proc. 23: 176-184.

KOVOOR, A. T. A. 1954. Some factors affecting the growth of *Rhizoctonia bataticola* in the soil. J. Madras Univ. (B) 24: 47-52.

KREITLOW, K. W. 1950. Longevity of inoculum of *Sclerotinia trifoliorum* prepared from cultures grown on grain. Phytopathology 40: 16.

KREUTZER, W. A. 1960. Soil treatment. P. 431-476. *In* J. G. Horsfall and A. E. Dimond [eds.], Plant pathology, an advanced treatise, Academic Press Inc., New York, London.

KÜHN, J. G. 1858. Die Krankheiten der Kulturegewächse, ihre ursachen und ihre Verhütung. Gustav Bosselmann, Berlin. 312 p.

KULMATYCHKA, I., P. LESZCZENKO, and K. MALEC. 1955. Rizoktonioza Ziemniaków. Acta Agrobot. 3: 27-43.

KURODANI, K., K. YOKOGI, and M. YAMAMOTO. 1959. On the effect of 2,4-dichlolophenoxy acetic acid to the mycelial growth of *Hypochnus sasakii* Shirai. [Japanese with English summary.] Forsch. Geb. Pflahzenkrankh, Kyoto 6: 132-135.

LABROUSSE, F., and J. SAREJANNI. 1930. Recherches physiologiques sur quelques champignons parasites. Phytopathol. Z. 2: 1-38.

LAMPEN, J., J. W. GILL, P. M. ARNOW, and I. MAGANA-PLAZA. 1963. Inhibition of the pleuropneumonia-like organism *Mycoplasma gallisepticum* by certain polyene anti-fungal antibiotics. J. Bact. 945-949.

LARGE, J. R. 1949. Parasitic diseases of tung. Plant Disease Reptr. 33: 22-30.

LARPENT, J. P. 1962. La notion de dominance apicale chez *Rhizoctonia solani* Kühn. C. R. Acad. Sci. (Paris) 254: 1137-1139.

———. 1965. Relation entre vitesse de croissance et ramification du mycélium jeune de quelqces champignons. C. R. Acad. Sci. (Paris) 260: 265-267.

———. 1966. Caractères et déterminisme des corrélations d'inhibition dans le mycélium jeune de quelques champignons. Ann. Sci. Nat. (Ser. 12) 7: 1-130.

LATHAM, A. J., and M. B. LINN. 1961. A non-fungistatic bacteriostatic medium containing streptomycin and vancomycin. Plant Disease Reptr. 45: 866-867.

LAURITZEN, J. I. 1929. Rhizoctonia rot of turnips in storage. J. Agr. Res. 38: 93-108.

LAVATE, W. V., and R. BENTLEY. 1964. Distribution of normal isoprenologs of coenzyme Q and dihydro

coenzyme $Q_{10}$ in various molds. Arch. Biochem. Biophys. 108: 287-291.

LEACH, C. M., and M. PIERPOINT. 1958. *Rhizoctonia solani* may be transmitted with seed of *Agrostis tenuis*. Plant Disease Reptr. 42: 240.

LEACH, L. D. 1947. Growth rates of host and pathogen as factors determining the severity of pre-emergence damping-off. J. Agr. Res. 75: 161-179.

———, R. H. GARBER, and W. H. LANGE. 1959. Cotton seed treatment trials in California, 1954-1958, with special reference to specific fungicides. Plant Disease Reptr. Suppl. 259: 213-221.

———, ———, and W. J. TOLMSOFF. 1960. Selective protection afforded by certain seed and soil fungicides. (Abstr.) Phytopathology 50: 643-644.

———, W. H. LANGE, F. J. HILLS, and J. B. KENDRICK, JR. 1954. Lima bean seed treatment trial in California, 1950-52. Plant Disease Reptr. 38: 193-199.

———, and W. C. SNYDER. 1947. Localized chemical applications to the soil and their effects upon root rots of beans and peas. (Abstr.) Phytopathology 37: 363.

LeCLERG, E. L. 1934. Parasitism of *Rhizoctonia solani* on sugar beets. J. Agr. Res. 49: 407-431.

———. 1939a. Studies on a culture variant of *Rhizoctonia solani*. Phytopathology 29: 267-274.

———. 1939b. Methods of determination of physiologic races of *Rhizoctonia solani* on the basis of parasitism on several crop plants. Phytopathology 29: 609-616.

———. 1939c. Studies on dry-rot canker of sugar beets. Phytopathology 29: 793-800.

———. 1941a. Pathogenicity studies with isolates of *Rhizoctonia solani* obtained from potato and sugar beet. Phytopathology 31: 49-60.

———. 1941b. Comparative studies of sugar beet and potato isolates of *Rhizoctonia solani*. Phytopathology 31: 274-278.

———, L. H. PERSON, and S. B. MEADOWS. 1942. Further studies on the temperature relations of sclerotial isolates of *Rhizoctonia solani* from potatoes. Phytopathology 32: 731-732.

LEGATOR, M., and D. RACUSEN. 1959. Mechanism of allyl alcohol toxicity. Bact. 77: 120-121.

LEHMAN, S. G. 1940. Cotton seed dusting in relation to control of seedling infection by *Rhizoctonia* in the soil. Phytopathology 30: 847-853.

LINDENFELSER, L. A., T. G. PRIDHAM, O. L. SHOTWELL, and F. H. STODOLA. 1958. Antibiotics against plant diseases. IV. Activity of duramycin against selected microorganisms. Antibiot. Ann. 1957/1958: 241-247.

LINDER, D. H. 1942. A contribution towards a monograph of the genus *Oidium* (Fungi Imperfecti). Lloydia 5: 165-207.

LINDSEY, D. 1965. Ecology of plant pathogens in soil. III. Competition between soil fungi. Phytopathology 55: 104-110.

LINSKENS, H. F., and P. HAAGE. 1963. Cutinase-Nachweis in phytopathogenen Pilzen. Phytopathol. Z. 48: 306-311.

LITTLEFIELD, L. J., R. D. WILCOXSON, and T. W. SUDIA. 1964. Effects of temperature on translocation of phosphorus 32 by several isolates of *Rhizoctonia solani*. (Abstr.) Phytopathology 54: 899.

———, ———, and ———. 1965. Translocation of phosphorus-32 in *Rhizoctonia solani*. Phytopathology 55: 536-542.

LIVINGSTON, C. H., N. OSHIMA, and M. D. HARRISON. 1964. Terraclor as a control for various levels of Rhizoctonia inoculum in potato soil. Am. Potato J. 41: 239-243.

———, ———, and C. MORRILL. 1962. Evaluation of ter-raclor as a control measure for Rhizoctonia disease of potatoes. (Abstr.) Phytopathology 52: 18.

LOCKWOOD, J. L. 1959. *Streptomyces* spp. as a cause of natural fungi-toxicity in soils. Phytopathology 49: 327-331.

———. 1960. Lysis of mycelium of plant-pathogenic fungi by natural soil. Phytopathology 50: 787-789.

LORENZ, R. C. 1948. A new leaf disease of *Hevea* in Peru. J. Forestry 46: 27-30.

LUCAS, R. L. 1955. A comparative study of *Ophiobolus graminis* and *Fusarium culmorum* in saprophytic col-onization of wheat straw. Ann. Appl. Biol. 43: 134-143.

LUICK, J. R., R. H. GARBER, N. B. AKESSON, and L. D. LEACH. 1959. A method for the study of fungi-cide distribution by farm machinery using Rb⁸⁶. J. Applied Radiation and Isotopes. 5: 147-148.

LUTHRA, J. C., and R. S. VASUDEVA. 1941. Studies on the root rot disease of cotton in the Punjab. IX. Varietal susceptibility to the disease. Indian J. Agr. Sci. 11: 410-421.

———, ———, and M. ASHRAF. 1940. Studies on the root-rot disease of cotton in the Punjab. VIII. Further studies on the physiology of the causal fungi. Indian J. Agr. Sci. 10: 653-662.

LUTTRELL, E. S. 1962. Rhizoctonia blight of tall fescue grass. Plant Disease Reptr. 46: 661-664.

———, and K. H. GARREN. 1952. Blights of snap bean in Georgia. Phytopathology 42: 607-613.

LYDA, S. D. 1963. The physiological effect of sodium pentachlorophenate on *Sclerotinia fructicola* (Wint) Rehm. Ph.D. Thesis, Univ. of California, Davis, 48 p.

MACHACEK, J. E., and A. M. BROWN. 1948. Experi-ments on vegetable seed disinfection and observations on varietal resistance of beans, peas, and sweet corn to some diseases in Manitoba. Sci. Agr. 28: 145-153.

MACIEJOWSKA, Z. 1964. Choroby korzeni roslin na tle ekologii grzybow glebowych. Biul. Inst. Ochr. Rosl. Poznan 26: 177-203.

MacLEAN, N. A. 1948. Rhizoctonia rot of tulips in the Pacific Northwest. Phytopathology 38: 156-157.

MADARANG, S. A. 1941. Rhizoctonia damping-off of Cinchona seedlings. Philippine J. Forestry 4: 105-121.

MAIER, C. R. 1959a. Effect of certain crop residues on bean root pathogens. Plant Disease Reptr. 43: 1027-1030.

———. 1959b. Cultural and pathogenic variability of *Rhizoctonia solani* isolates from cotton-growing areas of New Mexico. Plant Disease Reptr. 43: 1063-1066.

———. 1961a. Black root-rot development on pinto beans, incited by selected *Thielaviopsis basicola* isolates, as influenced by different soil temperatures. Plant Dis-ease Reptr. 45: 804-807.

———. 1961b. Selective effects of barley residue on fungi of the pinto bean root-rot complex. Plant Disease Reptr. 45: 808-811.

———. 1962. Response of selected *Rhizoctonia solani* isolates to different soil chemicals in cultural tests. (Abstr.) Phytopathology 52: 19.

———, and E. E. STAFFELDT. 1960. Cultural variabili-ty of selected isolates of *Rhizoctonia solani* and *Thiela-viopsis basicola*, and the variability in their pathogeni-city to acala and pima cotton, respectively. Plant Dis-ease Reptr. 44: 956-961.

———, and ———. 1963. Cotton seedling disease complex in New Mexico. New Mexico State Univ. Agr. Expt. Sta. Bull. 574. 41 p.

MALAGUTI, G. 1951. Mancha de la hoja envainadora del arroz causada por *Rhizoctonia solani*. Agron. Trop. (Maracay) 1: 71-75.

MANDELS, M., F. W. PARRISH, and E. T. REESE. 1962. Sophorose as an inducer of cellulase in *Tri-choderma viride*. J. Bact. 83: 400-408.

MARCHIONATTO, J. B. 1946. Nota crítica sobre 'Moni-liopsis aderholdi.' Rev. Fac. Agron. La Plata 26: 1-4.

MARTIN, G. W. 1945. The classification of the Tre-mellales. Mycologia 37: 527-542.

———. 1948. New or noteworthy tropical fungi. IV. Lloydia 11: 111-122.

———. 1957. The tulasnelloid fungi and their bearing on basidial terminology. Brittonia 9: 25-30.

MARTINSON, C. A. 1959. Inoculum potential studies of *Rhizoctonia solani*. M.S. Thesis, Colorado State Univ., Fort Collins. 132 p.

———. 1963. Inoculum potential relationships of *Rhizoc-tonia solani* measured with soil microbiological sam-pling tubes. Phytopathology 53: 634-638.

———. 1965. Formation of infection cushions by *Rhizoc-tonia solani* on synthetic films in soils. (Abstr.) Phyto-pathology 55: 129.

———, and R. R. BAKER. 1962. Increasing relative fre-quency of specific fungus isolations with soil micro-biological sampling tubes. Phytopathology 52: 619-621.

MASSEY, R. E. 1928. Work of the section of plant physiology and pathology. Sudan Agr. Res. Rept. 1926-1927: 120-142.

MATHEW, K. T. 1952. Growth-factor requirements of *Pellicularia koleroga* Cooke in pure culture. Nature (London) 170: 889-890.

———. 1953. Thiamine, its intermediates and growth of *Corticium microsclerotia* (Matz) Weber. Proc. Indian Acad. Sci. (B) 38: 1-10.

———. 1954a. Studies on the black rot of coffee. I. The disease in South India and some general considerations. Proc. Indian Acad. Sci. (B) 39: 133-170.

———. 1954b. Studies on the black rot of coffee. II. Nutritional requirements of *Pellicularia koleroga* Cooke with special reference to growth substances. Proc. In-dian Acad. Sci. (B) 39: 179-211.

MATSUMOTO, T. 1921. Studies in the physiology of the fungi. XII. Physiological specialization in *Rhizoctonia solani* Kühn. Ann. Missouri Botan. Garden 8: 1-62.

———. 1923. Further studies on physiology of *Rhizoctonia solani* Kühn. Imp. Coll. Agr. Forestry, Morioka (Japan) Bull. 5. 63 p.

———. 1934. Some remarks on the taxonomy of the fungus *Hypochnus sasakii* Shirai. Trans. Sapporo Nat. Hist. Soc. 13: 115-120.

———, and S. HIRANE. 1933. Physiology and parasitism of the fungi generally referred to as *Hypochnus sasakii* Shirai. III. Histological studies in the infection by the fungus. J. Soc. Trop. Agr. (Formosa) 5: 367-373.

———, and W. YAMAMOTO. 1935. *Hypochnus sasakii* Shirai in comparison with *Corticium stevensii* Burt and *Corticium koleroga* (Cooke) v. Hohn. Trans. Nat. Hist. Soc. (Formosa) 25: 161-175.

———, ———, and S. HIRANE. 1932. Physiology and para-sitism of the fungi generally referred to as *Hypoch-nus sasakii* Shirai. I. Differentiation of strains by means of hyphal fusion and culture in differential media. J. Soc. Trop. Agr. (Formosa) 4: 370-388.

———, ———, and ———. 1933. Physiology and parasitism of the fungi generally referred to as *Hypochnus sa-sakii* Shirai. II. Temperature and humidity relations. J. Soc. Trop. Agr. (Formosa) 5: 332-345.

MATZ, J. 1917. A *Rhizoctonia* of the fig. Phytopathology 7: 110-118.

———. 1921. The Rhizoctonias of Porto Rico. Porto Rico

Dept. Agr. J. 5: 1-31.

MAXSON, A. C. 1938. Root-rots of the sugar beet. Proc. Am. Soc. Sugar Beet Tech. 1938: 60-64.

MAXWELL, D. P., and D. F. BATEMAN. 1967. Changes in the activities of some oxidases in extracts of Rhizoctonia-infected bean hypocotyls in relation to lesion maturation. Phytopathology 57: 132-136.

McCLURE, T. T., and W. R. ROBBINS. 1942. Resistance of cucumber seedlings to damping-off as related to age, season of year, and level of nitrogen nutrition. Botan. Gaz. 103: 684-697.

McKEEN, C. D. 1950. Arasan as a seed and soil treatment for the control of damping-off in certain vegetables. Sci. Agr. 30: 261-270.

McKENZIE, A. R. 1966. Studies on genetically controlled variation in *Thanatephorus cucumeris*. Ph.D. Thesis, Univ. Adelaide, S. Australia. 183 p.

McNEW, G. L. 1960. The nature, origin, and evolution of parasitism. P. 19-69. *In* J. G. Horsfall and A. E. Dimond [eds.], Plant pathology, an advanced treatise, II, Academic Press Inc., New York, London.

McRAE, W. 1934. Foot-rot diseases of *Piper betle* L. in Bengal. Indian J. Agr. Sci. 4: 585-617.

MELOCHE, H. P., JR., 1962. Enzymatic utilization of glucose by a basidiomycete. J. Bact. 83: 766-774.

MENON, K. P. V., U. K. NAIR, and K. M. PANDALAI. 1952. Influence of waterlogged soil conditions on some fungi parasitic on the roots of the coconut palm. Indian Coconut J. 5: 71-79.

MENZIES, J. D. 1963a. Survival of microbial plant pathogens in soil. Botan. Rev. 29: 79-122.

———. 1963b. The direct assay of plant pathogen populations in soil. Ann. Rev. Phytopathol. 1: 127-142.

MESSIAEN, C. M. 1957. L'influence des méthodes d'isolement et des milieux de culture dans l'échantillonage de la flora fongique des sols. Compt. Rend. Acad. Agr. France 43: 384-386.

MEYER, R. W. 1965. Heterokaryosis and nuclear phenomena in *Rhizoctonia*. Ph.D. Thesis, Univ. California, Berkeley. 118 p.

MILLER, P. M., and J. F. AHRENS. 1964. Effect of a herbicide, a nemotocide and a fungicide on Rhizoctonia infestation of *Taxus*. (Abstr.) Phytopathology 54: 901.

———, and E. M. STODDARD. 1956. Hot water treatment of fungi infecting strawberry roots. Phytopathology 46: 694-696.

MILLIKAN, D. F., and M. L. FIELDS. 1964. Influence of some representative herbicidal chemicals upon the growth of some soil fungi. (Abstr.) Phytopathology 54: 901.

MIRCETICH, S. M., and G. A. ZENTMYER. 1964. Rhizoctonia seed and root rot of avocado. Phytopathology 54: 211-213.

MOHAMED, H. A. 1962. Effect of date of planting on fungi and other microorganisms isolated from cotton seedlings. Plant Disease Reptr. 46: 801-803.

MOJE, W. J., J. P. MARTIN, and R. C. BAINES. 1957. Structural effect of some organic compounds on soil organisms and citrus seedlings grown in an old citrus soil. J. Agr. Food Chem. 5: 32-36.

MOLITORIS, H. P., and J. VAN ETTEN. 1965. Aging and enzyme activity of *Rhizoctonia solani* and *Sclerotium bataticola*. (Abstr.) Phytopathology 55: 1069.

MOLLENHAUER, H. H., and D. J. MORRÉ. 1966. The golgi apparatus and plant secretion. Ann. Rev. Plant Physiol. 17: 27-46.

MONSON, A. M. 1960. Movement of radioisotopes in *Rhizoctonia solani*. (Abstr.) Phytopathology 50: 646.

———, and T. W. SUDIA. 1963. Translocation in *Rhizoc-*

*tonia solani*. Botan. Gaz. 124: 440-443.

MONTEGUT, J. 1960. Value of the dilution method. P. 43-52. *In* D. Parkinson and J. S. Waid [eds.], The ecology of soil fungi. Liverpool Univ. Press, Liverpool, Eng.

MONTEITH, J. 1926. The brown-patch disease of turf: its nature and control. U. S. Golf Assoc. Green Comm. Bull. 6: 127-142.

———, and A. S. DAHL. 1928. A comparison of some strains of *Rhizoctonia solani* in culture. J. Agr. Res. 36: 897-903.

———, and ———. 1932. Turf diseases and their control. U. S. Golf Assoc. Green Sect. Bull. 12: 85-188.

MOORE, R. T. 1965a. Distribution and characterization of the golgi complex in the Phycomycetes. J. Cell Biol. 27: 69A.

———. 1965b. The ultrastructure of fungal cells. P. 95-118. *In* G. C. Ainsworth and A. S. Sussman [eds.], The fungi, I, Academic Press Inc., New York.

———, and J. H. McALEAR. 1961. Fine structure of mycota. 5. Lomasomes — previously uncharacterized hyphal structures. Mycologia 53: 194-200.

———, and ———. 1962. Fine structure of Mycota. 7. Observations on septa of Ascomycetes and Basidiomycetes. Am. J. Botany 49: 86-94.

———, and ———. 1963. Fine structure of mycota. 9. Fungal mitochondria. J. Ultrastructure Res. 8: 144-153.

MOORE, W. C. 1959. British parasitic fungi. Cambridge Univ. Press, London. 430 p.

MOREAU, C., and M. MOREAU. 1956. Examen comparatif du mycélium et des sclérotes chez diverses souches du *Rhizoctonia solani* Kühn et du *Morchella hortensis* Boud. Bull. Soc. Botan. France 103: 117-120.

MORRIS, L. G. 1954. The steam sterilising of soil. Experiments on fine soil. Natl. Inst. Agr. Eng. Rept. 14: 1-32.

MORSE, W. J., and M. SHAPOVALOV. 1914. The Rhizoctonia disease of potatoes. Maine Agr. Expt. Sta. Bull. 230: 193-216.

MOTSINGER, R. E. 1967. Fungicide and herbicide interactions as they affect stands. Beltwide Cotton Production Mechanization Conf. Dallas, 1967: 26-28.

MOUNTAIN, W. B., and W. G. BENEDICT. 1956. The association of *Rhizoctonia solani* and nematodes in root rot of winter wheat. (Abstr.) Phytopathology 46: 241-242.

MUELLER, K. E., and L. W. DURRELL. 1957. Sampling tubes for soil fungi. Phytopathology 47: 243.

MULLER, H. 1955. Untersuchungen uber die Wirkung des Cyanamide im Kalkstickstoff auf pathogene und nichtpathogene Mikroorganismen des Bodens. Arch. Mikrobiol. 22: 285-306.

MULLER, K. O. 1923a. Über die Beziehungen von *Moniliopsis aderholdii* zu *Rhizoctonia solani*. Arbeit. Biol. Reichsanst. Land-u Forstw. 11: 321-325.

———. 1923b. Über die Beziehungen zwischen *Rhizoctonia solani* Kühn und *Hypochnus solani* Prill. et Del. Arbeit. Biol. Reichsanst. Land-u Forstw. 11: 326-330.

———. 1924. Untersuchungen zur Entwickelungsgeschichte und Biologie von *Hypochnus solani* P.u.D. (*Rhizoctonia solani* K.). Arbeit. Biol. Reichsanst. Land- u. Forstw. 13: 198-262.

NAEF-ROTH, ST., E. GAUMANN, and P. ALBERSHEIM. 1961. Zur Bildung eines mazerierenden Fermentes durch *Dothidea ribesia* Fr. Phytopathol. Z. 40: 283-302.

NAGATA, Y. 1960. Biochemical studies on the *Corticium centrifugum*. Part 13. On the terminal oxidase systems II. Gifu Univ. (Japan) Res. Bull. Faculty Agr. 12: 137-144.

NAITO, N., and T. TANI. 1952-3. The effect of 2,4-D in culture media on the mycelial growth, sporulation, and sclerotial formation of various pathogenic fungi. Tech. Bull. Kagawaken Agr. Coll. 3: 119-125; 4: 50-55. [Abstr. *in* Rev. Appl. Mycol. 32: 325, 1953.]

NAKAYAMA, T. 1940. [A study on the infection of cotton seedlings by *Rhizoctonia solani*.] Ann. Phytopathol. Soc. Japan 10: 93-103.

NASH, R. G. 1967. Phytotoxic pesticide interactions in soil. Agron. J. 59: 227-230.

NEAL, D. C. 1942. Rhizoctonia infection of cotton and symptoms accompanying the disease in plants beyond the seedling stage. Phytopathology 32: 641-642.

———. 1944. Rhizoctonia leaf spot of cotton. Phytopathology 34: 599-602.

———, and L. D. NEWSOM. 1951. Soreshin of cotton and its relationship to thrips damage. Phytopathology 41: 854.

NEERGAARD, P. 1958. Infection of Danish seeds by *Rhizoctonia solani* Kuehn. Plant Disease Reptr. 42: 1276-1278.

NESS, H. 1927. The distribution limits of the long-leaf pine and their possible extension. J. Forestry 25: 852-857.

NEU, H. C., and L. A. HEPPEL. 1965. The release of enzymes from *Escherichia coli* by osmotic shock and during the formation of spheroplasts. J. Biol. Chem. 240: 3685-3692.

NEUBERT, D., and A. L. LEHNINGER. 1962. The effect of oligomycin, gramicidin and other antibiotics on reversal of mitochondrial swelling by adenosine triphosphate. Biochem. Biophys. Acta. 62: 556-565.

NEUHOFF, W. 1924. Zytologie und systematische Stellung der Auriculariaceen und Tremellaceen. Botan. Arkiv. 8: 250-297.

NEWHALL, A. G. 1955. Disinfestation of soil by heat, flooding and fumigation. Botan. Rev. 21: 189-250.

NEWTON, W. 1931. The physiology of *Rhizoctonia*. Sci. Agr. 12: 178-182.

———, and N. MAYERS. 1935. The psysiology of *Rhizoctonia solani* Kühn. III. The susceptibility of different plants as determined by seedling infection. IV. The effect of a toxic substance produced by *Rhizoctonia solani* Kühn when grown in liquid culture, on the growth of wheat, carrots and turnips. Sci. Agr. 15: 393-401.

NIEDERPRUEM, D. J. 1963. Role of carbon dioxide in the control of fruiting of *Schizophyllum commune*. J. Bact. 85: 1300-1308.

———. 1964. Respiration of basidiospores of *Schizophyllum commune*. J. Bact. 88: 210-215.

———, and D. P. HACKETT. 1961. Cytochrome system in *Schizophyllum commune*. Plant Physiol. 36: 79-84.

NIGHTINGALE, A. A., and G. B. RAMSEY. 1936. Temperature studies of some tomato pathogens. U. S. Dept. Agr. Tech. Bull. 520. 36 p.

NISHIMURA, S., and M. SABAKI. 1963. [Isolation of the phytotoxic metabolites of *Pellicularia filamentosa*.] Ann. Phytopathol. Soc. Japan 28: 228-234.

NISIKADO, Y., and K. HIRATA. 1937. Studies on the longevity of sclerotia of certain fungi, under controlled environmental factors. Bericht. Ohara Inst. Landwirts. Forsch. Kurashiki 7: 535-547.

———, ———, and T. HIGUTI. 1938. Studies on the temperature relations to the longevity of pure culture of various fungi, pathogenic to plants. Bericht. Ohara Inst. Landwirts. Forsch. Kurashiki 8: 107-124.

NISIZAWA, K., I. MORIMOTO, N. HANDA, and Y. HASHIMOTO. 1962. Cellulose-splitting enzymes VII. Starch zone electrophoresis of cellulase and other carbohydrases from *Irpex lacteus*. Arch. Biochem. Biophys. 96: 152-157.

NOBLE, M., J. De TEMPE, and P. NEERGAARD. 1958. An annotated list of seedborne diseases. Commonwealth Mycol. Inst., Kew. 159 p.

NORKRANS, B. 1963. Degradation of cellulose. Ann. Rev. Phytopathol. 1: 325-350.

NOUR EL DEIN, M. S., and M. S. SHARKAS. 1963. Pectolytic enzyme activity of *Rhizoctonia solani*. Phytopathol. Z. 48: 439-444.

———, and ———. 1964a. Isolation of a toxic substance from the culture filtrate of *Rhizoctonia solani*. Phytopathol. Z. 52: 53-58.

———, and ———. 1964b. The pathogenicity of *Rhizoctonia solani* in relation to different tomato root exudates. Phytopathol. Z. 51: 285-290.

OBRIG, T. G., and D. GOTTLIEB. 1966. The effect of age on substrate permeability of *Rhizoctonia solani* and *Sclerotium bataticola*. (Abstr.) Phytopathology 56: 893.

ODA, H., H. SUMI, and Y. TANAKA. 1961. Fungicide for sheath blight control I. Synthesis of organic arsine xanthates and their in vitro fungicidal activities II. Effect of organic arsine xanthates against sheath blight on rice. Kenkyusho Nempo 11: 193-201. [Abstr. *in* Chem. Abstr. 55: 6768f, 1961.]

OGURA, H., S. AKAI, and T. SATO. 1961. Studies on *Pellicularia filamentosa* (Pat.) Rogers: II. On the formation of the free amino acids and sugars in *P. filamentosa*. Ann. Phytopath. Soc. Japan 26: 31-36. [Abstr. *in* Rev. Appl. Mycol. 41: 15, 1962.]

OLD, K. M., and T. H. NICOLSON. 1962. Use of nylon mesh in studies of soil fungi. Plant Disease Reptr. 46: 616.

OLIVE, L. S. 1957. Two new genera of the Ceratobasidiaceae and their phylogenetic significance. Am. J. Botany 44: 429-435.

OLSEN, C. M., and K. F. BAKER. 1968. Selective heat treatment of soil, and its effect on the inhibition of *Rhizoctonia solani* by *Bacillus subtilis*. Phytopathology 58: 79-87.

OLSEN, C. M., N. T. FLENTJE, and K. F. BAKER. 1967. Comparative survival of monobasidial cultures of *Thanatephorus cucumeris* in soil. Phytopathology 57: 598-601.

ONTKO, J. A., and D. JACKSON. 1964. Factors affecting the rate of oxidation of fatty acids in animal tissues. Effect of substrate concentration, pH and coenzyme A in rat liver preparations. J. Biol. Chem. 239: 3674-3682.

O'REILLY, H. J., W. A. KREUTZER, C. E. HORNER, G. W. BRUEHL, L. CAMPBELL, and J. D. MENZIES. 1958. The prevalence and importance of fungal root diseases on the Pacific Coast. Proc. Pacific Coast Res. Conf. on Control of Soil Fungi 5. 7 p.

OSHIMA, N., C. H. LIVINGSTON, and M. D. HARRISON. 1963. Weeds as carriers of two potato pathogens in Colorado. Plant Disease Reptr. 47: 466-469.

OWEN, J. H., and J. D. GAY. 1964. Tests of soil fungicides under uniform conditions for control of cotton damping-off caused by *Rhizoctonia solani*. Plant Disease Reptr. 48: 480-483.

PACKER, L. 1960. Metabolic and structural states of mitochondria I. Regulation by adenosine diphosphate. J. Biol. Chem. 235: 242-249.

PALO, M. A. 1926. Rhizoctonia disease of rice. I. A study of the disease and of the influence of certain conditions upon the viability of the sclerotial bodies of the causal fungus. Philippine Agr. 15: 361-376.

PAPAVIZAS, G. C. 1963. Microbial antagonism in bean

rhizosphere as affected by oat straw and supplemental nitrogen. Phytopathology 53: 1430-1435.

———. 1964. Survival of single-basidiospore isolates of *Rhizoctonia praticola* and *Rhizoctonia solani*. Can. J. Microbiol. 10: 739-746.

———. 1965. Comparative studies of single-basidiospore isolates of *Pellicularia filamentosa* and *Pellicularia praticola*. Mycologia 57: 91-103.

———. 1966. Biological methods for the control of plant diseases and nematodes. P. 82-94. *In* E. F. Knipling [ch.], Pest control by chemical, biological, genetic, and physical means, U. S. Dept. Agr. Publ. Agr. Res. Service 33-110. 214 p.

———, and W. A. AYERS. 1965. Virulence, host range, and pectolytic enzymes of single-basidiospore isolates of *Rhizoctonia praticola* and *Rhizoctonia solani*. Phytopathology 55: 111-116.

———, and C. B. DAVEY. 1959a. Evaluation of various media and antimicrobial agents for isolation of soil fungi. Soil Sci. 88: 112-117.

———, and ———. 1959b. Isolation of *Rhizoctonia solani* Kuehn from naturally infested and artificially inoculated soils. Plant Disease Reptr. 43: 404-410.

———, and ———. 1959c. Investigations on the control of the Rhizoctonia disease of snap beans by green organic soil amendments. (Abstr.) Phytopathology 49: 525.

———, and ———. 1960. Rhizoctonia disease of bean as affected by decomposing green plant materials and associated microfloras. Phytopathology 50: 516-522.

———, and ———. 1961. Saprophytic behavior of *Rhizoctonia* in soil. Phytopathology 51: 693-699.

———, and ———. 1962a. Activity of *Rhizoctonia* in soil as affected by carbon dioxide. Phytopathology 52: 759-766.

———, and ———. 1962b. Isolation and pathogenicity of *Rhizoctonia* saprophytically existing in soil. Phytopathology 52: 834-840.

———, ———, and R. S. WOODARD. 1962. Comparative effectiveness of some organic amendments and fungicides in reducing activity and survival of *Rhizoctonia solani* in soil. Can. J. Microbiol. 8: 915-922.

PARK, D. 1956. Effect of substrate on a microbial antagonism, with reference to soil conditions. Trans. Brit. Mycol. Soc. 39: 239-259.

———. 1963. The ecology of soil-borne fungal diseases. Ann. Rev. Phytopathol. 1: 241-258.

———. 1965. Survival of microorganisms in soil. P. 82-97. *In* K. F. Baker and W. C. Snyder [eds.], Ecology of soil-borne plant pathogens, Univ. California Press, Berkeley.

———, and L. S. BERTUS. 1932. Sclerotial diseases of rice in Ceylon. I. *Rhizoctonia solani* Kühn. Ann. Roy. Botan. Garden (Peradeniya) 11: 319-331.

———, and ———. 1934a. Sclerotial diseases of rice in Ceylon. 3. A new Rhizoctonia disease. Ceylon J. Sci. (A) 12: 1-10.

———, and ———. 1934b. Sclerotial diseases of rice in Ceylon. 5. *Rhizoctonia solani* B strain. Ceylon J. Sci. (A) 12: 25-36.

PARKINSON, D., and J. S. WAID. 1960. The ecology of soil fungi, Liverpool Univ. Press, Liverpool, England. 324 p.

PARMETER, J. R., JR. 1965. The taxonomy of sterile fungi. Phytopathology 55: 826-828.

———, H. S. WHITNEY, and W. D. PLATT. 1967. Affinities of some *Rhizoctonia* species that resemble mycelium of *Thanatephorus cucumeris*. Phytopathology 57: 218-223.

PATOUILLARD, N. 1900. Essai taxonomique sur les familles et les genres des Hymenomycetes. Declume, Lons-le-Saunier. 184 p.

———, and G. DE LAGERHEIM. 1891. Champignons de l'Equateur. Bull. Soc. Mycol. France 7: 163.

PATRICK, Z. A., and L. W. KOCH. 1958. Inhibition of respiration, germination, and growth by substances arising during decomposition of certain plant residues in the soil. Can. J. Botany 36: 621-647.

PEACE, T. R. 1962. Pathology of trees and shrubs with special reference to Britain. Oxford Univ. Press, London. 723 p.

PELTIER, G. L. 1916. Parasitic Rhizoctonias in America. Illinois Agr. Expt. Sta. Bull. 189: 281-390.

PEROMBELON, M., and G. HADLEY. 1965. Production of pectic enzymes by pathogenic and symbiotic Rhizoctonia strains. New Phytol. 64: 144-151.

PERSON, L. H. 1944a. The occurrence of a variant in *Rhizoctonia solani*. Phytopathology 34: 715-717.

———. 1944b. Parasitism of *Rhizoctonia solani* on beans. Phytopathology 34: 1056-1068.

———. 1945. Pathogenicity of isolates of *Rhizoctonia solani* from potatoes. Phytopathology 35: 132-134.

PETERSEN, L. J., J. E. DeVAY, and B. R. HOUSTON. 1961. The effect of gibberellic acid on the parasitism of *Rhizoctonia solani* on beans and cotton. (Abstr.) Phytopathology 51: 67.

———, ———, and ———. 1963. Effect of gibberellic acid on development of hypocotyl lesions caused by *Rhizoctonia solani* on Red Kidney bean. Phytopathology 53: 630-633.

PETERSON, E. A., J. W. ROUATT, and H. KATZNELSON. 1965. Microorganisms in the root zone in relation to soil moisture. Can. J. Microbiol. 11: 483-489.

PIA, J. 1927. Thallophyta. *In* M. Hirmer [ed.], Handbuch der Paläo-botanik. I, 31-136. R. Oldenbourg, Berlin.

PIERRE, R. E., and D. F. BATEMAN. 1967. Induction and distribution of phytoalexins in Rhizoctonia-infected bean hypocotyls. Phytopathology 57: 1154-1160.

PILAT, A. 1957. Prehled evropskych druhu radu prakyjankotvarych-Protclavariaceae Heim. Ceska Mykol. 11: 66-95.

PINCKARD, J. A. 1964. *Pellicularia filamentosa* a common saprophyte on mature cotton stems in Louisiana. (Abstr.) Phytopathology 54: 626.

———, and W. J. LUKE. 1967. *Pellicularia filamentosa*, a primary cause of cotton boll rot in Louisiana. Plant Disease Reptr. 51: 67-70.

———, and L. C. STANDIFER. 1966. An apparent interaction between cotton herbicidal injury and seedling blight. Plant Disease Reptr. 50: 172-174.

PITT, D. 1964a. Studies on sharp eyespot disease of cereals. I. Disease symptoms and pathogenicity of isolates of *Rhizoctonia solani* Kühn and the influence of soil factors and temperature on disease development. Ann. Appl. Biol. 54: 77-89.

———. 1964b. Studies on sharp eyespot disease of cereals. II. Viability of sclerotia: persistence of the causal fungus, *Rhizoctonia solani* Kühn. Ann. Appl. Biol. 54: 231-240.

PITTMAN, H. A. 1937. The Rhizoctonia and common scab diseases of potatoes. J. Dept. Agr. W. Australia (Ser. 2) 14: 288-301.

POOLE, C. F. 1952. Lettuce improvement in Hawaii. Proc. Am. Soc. Hort. Sci. 60: 397-400.

POTGIETER, H. J. 1965. Lysis and the chemical composition of *Fusarium solani, Neurospora crassa* and *Rhizoctonia solani* cell walls. Ph.D. Thesis, Cornell University, Ithaca, 89 p.

PRICE, R. D., and L. L. STUBBS. 1963. An investigation of the barley yellow dwarf virus as a primary or associated cause of premature ripening or "deadheads" of wheat. Australian J. Agr. Res. 14: 154-164.

PRILLIEUX, E. 1891. Sur la penetration de la Rhizoctone violette dans les racines de betterase et de luzerne. Bull. Soc. Botan. France (3) 43: 9-11.

———, and G. DELCROIX. 1891. *Hypochnus solani* Nov. Sp. Bull. Soc. Mycol. France 7: 220-221.

PUMPYANSKAYA, L. V. 1964. Khranenie nidroorganizmov pod mineral'nym moslom. [Storage of microorganisms under mineral oil.] Mikobiologiya 33: 1065-1070.

QUANJER, H. M. 1940. Rhizoctonia-ziekte in Aardappelen en bemesting. Tijdschr. Plantenziekten 46: 175-176.

RABINOVITZ-SERENI, D. 1932. Ricerche biologiche sulla *Rhizoctonia* di semenzai di citrus. Boll. R. Staz. Pat. Veg. (N.S.) 12: 187-209.

RADHA, K., and K. P. V. MENON. 1957. The genus *Rhizoctonia* in relation to soil moisture. I. Studies on *Rhizoctonia solani* and *Rhizoctonia bataticola*. Indian Coconut J. 10: 29-36.

RAFFRAY, J. B. 1965. Factors influencing tolerance of *Rhizoctonia solani* Kühn to three organic fungicides. M.S. Thesis, Louisiana State Univ., Baton Rouge. 61 p.

———, and J. B. SINCLAIR. 1965. Fungicide tolerance in *Rhizoctonia solani* influenced by temperature and serial transfer in fungicide-treated soil. Plant Disease Reptr. 49: 500-503.

RAGHEB, H. S., and F. W. FABIAN. 1955. Growth and pectolytic activity of some tomato molds at different pH levels. Food Res. 20: 614-625.

RAMAKRISHNAN, K., and T. S. RAMAKRISHNAN. 1948. Banded lead blight of arrowroot, *Maranta arundinacea*. Indian Phytopathol. 1: 129-136.

RAMAKRISHNAN, T. S. 1960. Notes on some fungi from South India. VII. Proc. Indian Acad. Sc. (B) 51: 164-168.

RAMSEY, G. B. 1917. A form of potato disease produced by *Rhizoctonia*. J. Agr. Res. 9: 421-426.

———, and A. A. BAILEY. 1929. The development of soil rot of tomatoes during transit and marketing. Phytopathology 19: 383-390.

———, and G. K. K. LINK. 1932. Market diseases of fruits and vegetables: tomatoes, peppers, eggplants. U. S. Dept. Agr. Misc. Publ. 121: 1-44.

———, and M. A. SMITH. 1961. Market diseases of cabbage, cauliflower, turnips, cucumbers, melons, and related crops. U. S. Dept. Agr., Agr. Handbook 184: 1-49.

———, and J. S. WIANT. 1941. Market diseases of fruits and vegetables: asparagus, onions, beans, peas, carrots, celery, and related vegetables. U. S. Dept. Agr. Misc. Publ. 440: 1-70.

———, ———, and M. A. SMITH. 1949. Market diseases of fruits and vegetables: potatoes. U. S. Dept. Agr. Misc. Publ. 98: 1-60.

RANNEY, C. D. 1964. A deleterious interaction between a fungicide and systemic insecticides on cotton. Plant Disease Reptr. 48: 241-245.

———, and L. S. BIRD. 1956. Greenhouse evaluation of in-the-furrow fungicides at two temperatures as a control of cotton seedling necrosis. Plant Disease Reptr. 40: 1032-1040.

———, and A. M. HILLIS. 1958. A study of the distribution of in-the-furrow applied fungicides. Phytopathology 48: 345.

RAO, A. S. 1959. A comparative study of competitive saprophytic ability in twelve root-infecting fungi by an agar plate method. Trans. Brit. Mycol. Soc. 42: 97-111.

———, and G. V. M. RAYUDU. 1964. Nitrogen nutrition of *Rhizoctonia solani* Kühn in relation to its occurrence on root surfaces. Current Sci. 33: 186-187.

RAPER, J. R. 1960. The control of sex in fungi. Am. J. Botany 47: 794-808.

RASULEV, U. U. 1959. [On variability and the taxonomic position of some forms of fungi of the genus *Rhizoctonia*.] Uzbek Biol. Z. 5: 21-30.

RAY, W. W. 1943. The effect of cotton seed dusting on emergence of seedlings in soil infested with *Rhizoctonia*. Phytopathology 33: 51-55.

REAVILL, M. J. 1950. The effect of certain chloronitrobenzenes on plant growth. Ph.D. Thesis, Univ. London.

———. 1954. Effect of certain chloronitrobenzenes on germination, growth, and sporulation of some fungi. Ann. Appl. Biol. 41: 448-460.

REESE, E. T. 1956. A microbiological progress report. Appl. Microbiol. 4: 39-45.

———, and W. GILLIGAN. 1953. Separation of components of cellulolytic enzymes systems by paper chromatography. Arch. Biochem. Biophys. 45: 74-82.

REYNOLDS, H. W., and R. G. HANSON. 1957. Rhizoctonia disease of cotton in presence or absence of the cotton root-knot nematode in Arizona. Phytopathology 47: 256-261.

RICAUD, B. I. 1965. Strains of *Rhizoctonia solani* resistant to chlorinated nitrobenzenes. Ph.D. Thesis, Univ. London. 107 p.

RICH, S., and P. M. MILLER. 1962. Effect of soil fungicides and antifungal fungi on the incidence of strawberry root disease. (Abstr.) Phytopathology 52: 926.

———, and ———. 1963. Efficiency of infective propagules of *Rhizoctonia solani* in reducing vigour of strawberry plants. Nature (London) 197: 719-720.

RICHARDS, B. L. 1921a. A dryrot canker of sugar beets. J. Agr. Res. 22: 47-52.

———. 1921b. Pathogenicity of *Corticium vagum* on the potato as affected by soil temperature. J. Agr. Res. 21: 459-482.

———. 1923a. Further studies on the pathogenicity of *Corticium vagum* on the potato as affected by soil temperature. J. Agr. Res. 23: 761-770.

———. 1923b. Soil temperature as a factor affecting the pathogenicity of *Corticium vagum* on the pea and the bean. J. Agr. Res. 25: 431-451.

RICHARDSON, L. T. 1954. The persistence of thiram in soil and its relationship to the microbiological balance and damping-off control. Can. J. Botany 32: 335-346.

———. 1960. Effect of insecticide-fungicide combinations on emergence of peas and growth of damping-off fungi. Plant Disease Reptr. 44: 104-108.

———, and D. E. MUNNECKE. 1964. A bioassay for volatile toxicants from fungicides in soil. Phytopathology 54: 836-839.

RICHTER, H. 1936. Fusskrankheit und Wurzelfäule de Lupine. (Erreger: *Rhizoctonia solani* K.). Zentr. Bakteriol. (2) 94: 127-133.

———, and R. SCHNEIDER. 1953. Untersuchungen zur morphologischen und biologischen Differenzierung von *Rhizoctonia solani* K. Phytopathol. Z. 20: 167-226.

———, and ———. 1954. Untersuchungen zur *Rhizoctonia*-Anfälligkeit der Kartoffelsorten II. Zuchter 24: 264: 271.

RIKER, A. J., and R. S. RIKER. 1936. Introduction to research on plant diseases. John S. Swift Company, Inc., St. Louis. 117 p.

ROGER, L. 1942. Les champignons a sclerotes parasites du riz. Bull. Econ. Indochine 6. 302 p.

ROGERS, D. P. 1935. Notes on the lower Basidiomycetes. Univ. Iowa Studies Nat. Hist. 17: 1-43.

———. 1939. The genus *Hypochnus* and Fries's observations. Mycologia 31: 297-307.

———. 1943. The genus *Pellicularia* (Thelephoraceae). Farlowia 1: 95-118.

———. 1949. Nomina conservanda proposita and nomina confusa—fungi. Farlowia 3: 425-493.

———. 1951. *Trechispora* and *Pellicularia*. Mycologia 43: 111.

ROLFS, F. M. 1902. Potato failures. Colorado Agr. Expt. Sta. Bull. 70. 20 p.

———. 1903. *Corticium vagum* B. and C. var. *solani* Burt. A fruiting stage of *Rhizoctonia solani*. Science (NS) 18: 729.

———. 1904. Potato failures. A second report. Colorado Agr. Expt. Sta. Bull. 91. 33 p.

ROSENBAUM, J. 1918. The origin and spread of tomato fruit rots in transit. Phytopathology 8: 572-580.

———, and M. SHAPOVALOV. 1917. A new strain of *Rhizoctonia solani* on the potato. J. Agr. Res. 9: 413-419.

ROSS, D. J. 1960. Physiological studies of some common fungi from grassland soils. New Zealand J. Sci. 3: 219-257.

ROTH, C. 1935. Untersuchungen uber den Wurzelbrand der Fichte (*Picea excelsa* Link). Phytopathol. Z. 8: 1-110.

ROTH, L. F., and A. J. RIKER. 1943a. Life history and distribution of *Pythium* and *Rhizoctonia* in relation to damping-off of red pine seedlings. J. Agr. Res. 67: 129-148.

———, and ———. 1943b. Influence of temperature, moisture, and soil reaction on the damping-off of red pine seedlings by *Pythium* and *Rhizoctonia*. J. Agr. Res. 67: 273-293.

———, and ———. 1943c. Seasonal development in the nursery of damping-off of red pine seedlings caused by *Pythium* and *Rhizoctonia*. J. Agr. Res. 67: 417-431.

ROWELL, J. B. 1951. Observations on the pathogenicity of *Rhizoctonia solani* on bent grasses. Plant Disease Reptr. 35: 240-242.

RUHLAND, W. 1908. Beitrag zur Kenntnis des sog. "Vermehrungspilzes." Arb. Kaiserl. biol. Anstalt Land-u Forstwirtsch. 6: 71-76.

RUPPEL, E. G., D. K. BARNES, G. H. FREYRE, and A. SANTIAGO. 1964. Effect of seed protectant and planting depth on Pythium and Rhizoctonia damping-off of *Tephrosia vogelii* in Puerto Rico. Plant Disease Reptr. 48: 714-717.

———, ———, and A. SANTIAGO. 1965. Potential of three *Rhizoctonia* species to incite foliar blight and damping-off on *Tephrosia vogelii* in Puerto Rico. Phytopathology 55: 612-614.

RUSHDI, M., and W. F. JEFFERS. 1956. Effect of some soil factors on efficiency of fungicides in controlling *Rhizoctonia solani*. Phytopathology 46: 88-90.

———, and A. R. SIRRY. 1959. Variation among some isolates of *Rhizoctonia solani* Kühn. Ann. Agr. Sci. 4: 79-87.

RYKER, T. C. 1939. The Rhizoctonia disease of Bermuda grass, sugar cane, rice and other grasses in Louisiana. Congr. Intern. Soc. Sugar Cane Technol. Proc. 6: 198-201.

———, and B. EXNER. 1942. A comparative study of four specias of *Rhizoctonia*. (Abstr.) Phytopathology 32: 24.

———, and F. S. GOOCH. 1938. Rhizoctonia sheath spot

of rice. Phytopathology 28: 233-246.

SACCARDO, P. A. 1899. Sylloge fungorum omnium hucusque cognitorum. Vol. 14. Paria. 1316 p.

SADASIVAN, T. S. 1961. Physiology of wilt disease. Ann. Rev. Plant Physiol. 12: 449-468.

SAKSENA, H. K. 1960. The damping-off organism, *Pellicularia praticola* (Pat.) Flentje, in India. Indian Phytopathol. 13: 165-167.

———. 1961a. Nuclear phenomena in the basidium of *Ceratobasidium praticolum* (Kotila) Olive. Can. J. Botany 39: 717-725.

———. 1961b. Nuclear structure and division in the mycelium and basidiospores of *Ceratobasidium praticolum*. Can. J. Botany 39: 749-756.

———, and O. VAARTAJA. 1960. Descriptions of new species of *Rhizoctonia*. Can. J. Botany 38: 931-943.

———, and ———. 1961. Taxonomy, morphology, and pathogenicity of *Rhizoctonia* species from forest nurseries. Can. J. Botany 39: 627-647.

SAMPSON, K., and J. H. WESTERN. 1954. Diseases of British grasses and herbage legumes. 2d ed. Cambridge Univ. Press, London. 118 p.

SAMUEL, G., and S. D. GARRETT. 1932. *Rhizoctonia solani* on cereals in South Australia. Phytopathology 22: 827-836.

SANFORD, G. B. 1937. Studies on *Rhizoctonia solani* Kühn. II. Effect on yield and disease of planting potato sets infested with sclerotia. Sci. Agri. 17: 601-611.

———. 1938a. Studies on *Rhizoctonia solani* Kühn. III. Racial differences in pathogenicity. Can J. Res. (C) 16: 53-64.

———. 1938b. Studies on *Rhizoctonia solani* Kühn. IV. Effect of soil temperature and moisture on virulence. Can. J. Res. (C) 16: 203-213.

———. 1941a. Studies on *Rhizoctonia solani* Kühn. V. Virulence in steam sterilized and natural soil. Can. J. Res. (C) 19: 1-8.

———. 1941b. Pathogenicity tests on sugar beets of random isolates of *Rhizoctonia solani* Kühn from potato. Sci. Agr. 21: 746-749.

———. 1946. Soil-borne diseases in relation to the microflora associated with various crops and soil amendments. Soil Sci. 61: 9-21.

———. 1947. Effect of various soil supplements on the virulence and persistence of *Rhizoctonia solani*. Sci. Agr. 27: 533-544.

———. 1952. Persistence of *Rhizoctonia solani* Kühn in soil. Can. J. Botany 30: 652-664.

———. 1956. Factors influencing formation of sclerotia by *Rhizoctonia solani*. Phytopathology 46: 281-284.

———. 1959. Root-disease fungi as affected by other soil organisms. P. 367-376. *In* C. S. Holton [ed.], Plant pathology, problems and progress 1908-1958. Univ. Wisconsin Press, Madison.

———, and J. W. MARRITT. 1933. The toxicity of formaldehyde and mercuric chloride solutions on various sizes of sclerotial of *Rhizoctonia solani*. Phytopathology 23: 271-280.

———, and W. P. SKOROPAD. 1955. Distribution of nuclei in hyphal cells of *Rhizoctonia solani*. Can. J. Microbiol. 1: 412-415.

SATO, K., and T. S. SHOJI. 1957. Pathogenicity of *Rhizoctonia solani* Kühn isolated from graminaceous weeds in forestry nurseries. Bull. Forestry Expt. Sta. Meguro 96: 89-104.

———, ———, and N. OTA. 1959. [Studies on the snow molding of coniferous seedlings. I. Gray mold and sclerotial diseases.] [Japanese with English summary.] Bull. Forestry Expt. Sta. Meguro 110: 1-153.

SAYED, M. Q. 1961. The effect of nutrition, pH and

nematodes on damping-off disease of pea, tomato and cucumber. Diss. Abstr. 21: 1701-1702.

SCHAAL, L. A. 1935. Rhizoctoniosis of potatoes grown under irrigation. Phytopathology 25: 748-762.

———. 1939. Penetration of potato-tuber tissue by *Rhizoctonia solani* in relation to the effectiveness of seed treatment. Phytopathology 29: 759-760.

SCHAFFNIT, E., and K. MEYER-HERMANN. 1930. Über den Einfluss der Bodenreaktion auf die Lebenweise von Pilzparasiten und das Verhalten ihrer Wirtspflanzen. Phytopathol. Z. 2: 99-166.

SCHARFETTER, R. 1922. Klimarhytmik, Vegetationsrhytmik, und Formationsrhytmik. Studien zur Bestimmung der Heimat der Pflanzen. Osterreich. Botan. Z. 71: 153-171.

SCHEFFER, R. P., and W. J. HANEY. 1956. Causes and control of root rot in Michigan greenhouses. Plant Disease Reptr. 40: 570-579.

SCHENCK, E. 1924. Über das Auftreten einer Hypochnusart auf Zuckerrübe. Zentr. Bakteriol. (2) 61: 317-322.

SCHLUMBERGER, O. 1927. Die Ubertragung von Kartoffelkrankheiten durch die Pflanzknollen. Mitt. Deut. Landw.-Gessellsch 42: 637-641.

SCHMIEDERKNECHT, M. 1960. Feuchtigkeit als Standortfaktor fur mikroskopische Pilze. Z. Pilzk. (NF) 25: 69-77.

SCHMITTHENNER, A. F. 1964. Prevalence and virulence of *Phytophthora, Aphanomyces, Pythium, Rhizoctonia,* and *Fusarium* isolated from diseased alfalfa seedlings. Phytopathology 54: 1012-1018.

SCHNEIDER, C. L. 1957. Viability of Rhizoctonia and Aphanomyces cultures kept under mineral oil and sealed with parafilm. (Abstr.) Phytopathology 47: 453-454.

SCHNEIDER, R. 1953. Untersuchungen über Feuchtigkeitsanspruche parasitischer Pilze. Phytopathol. Z. 21: 63-78.

SCHROEDER, W. T., and R. PROVVIDENTI. 1961. Rhizoctonia fruit rot of processing tomatoes. Plant Disease Reptr. 45: 160-163.

SCHROTH, M. N., and R. J. COOK. 1964. Seed exudation and its influence on pre-emergence damping-off of bean. Phytopathology 54: 670-673.

———, and W. C. SNYDER. 1961. Effect of host exudates on chlamydospore germination of the bean root rot fungus, *Fusarium solani* f. *phaseoli.* Phytopathology 51: 389-93.

SCHULTZ, H. 1936. Vergleichende Untersuchungen zur Ökologie, Morphologie und Systematik des 'Vermehrungspilzes.' Arb. Biol. Reichsanst. Land- u Forstwirts. 22: 1-141.

SCHUSTER, M. L., S. G. JENSEN, and R. M. SAYRE. 1958. Toothpick method of inoculating sugar beets for determining pathogenicity of *Rhizoctonia solani.* J. Am. Soc. Sugar Beet Tech. 10: 142-149.

SCHUTTE, K. H. 1956. Translocation in the fungi. New Phytologist 55: 164-182.

SCHWINN, F. J. 1961. Uber die "dry-core" Krankheit der Kartoffelknolle. Z. Pflantzenkrankh. 68: 395-406.

SELBY, A. D. 1900. Onion smut — preliminary experiments. Ohio Agr. Expt. Sta. Bull. 122: 71-84.

———. 1902. The prevention of onion smut. Ohio Agr. Expt. Sta. Bull. 131: 47-51.

SELBY, K., C. C. MAITLAND, and K. V. H. THOMPSON. 1963. The degradation of fibrous cotton by extracellular cellulase of *Myrothecium verrucaria.* Biochem. J. 86: 9.

SHARMA, B. B. 1960. On the root rot of *Kochia indica* Wight. Proc. Indian Acad. Sci. (B) 51: 150-156.

SHATLA, M. M. 1959. Tolerance to pentachloronitrobenzene in naturally occurring isolates of *Rhizoctonia solani* Kuehn from cotton. M.S. Thesis, Louisiana State Univ., Baton Rouge. 71 p.

———. 1965. Physiologic specialization and cytology of *Rhizoctonia solani* from cotton. Ph.D. Thesis, Louisiana State Univ. 134 p.

———, and J. B. SINCLAIR. 1963. Tolerance to pentachloronitrobenzene among cotton isolates of *Rhizoctonia solani.* Phytopathology 53: 1407-1411.

———, and ———. 1964. Nuclear condition and division in vegetative hyphae of *Rhizoctonia solani.* (Abstr.) Phytopathology 54: 907.

———, and ———. 1965. Effect of pentachloronitrobenzene on *Rhizoctonia solani* under field conditions. Plant Disease Reptr. 49: 21-23.

———, and ———. 1966. *Rhizoctonia solani*: Mitotic division in vegetative hyphae. Am. J. Botany 53: 119-123.

SHAW, F. J. F. 1912. The morphology and parasitism of *Rhizoctonia.* Mem. Dept. Agr. India (Botan. Ser.) 4: 115-153.

SHAW, M., S. A. BROWN, and D. R. JONES. 1954. Uptake of radio-active carbon and phosphorus by parasitized leaves. Nature (London) 173: 768-769.

———, and D. J. SAMBORSKI. 1956. The physiology of host-parasite relations. I. The accumulation of radioactive substances at infections of facultative and obligate parasites including tobacco mosaic virus. Can. J. Botany 34: 389-405.

SHEPHARD, M. C., and R. K. S. WOOD. 1963. The effect of environment and nutrition of pathogen and host, in the damping off of seedlings by *Rhizoctonia solani.* Ann. Appl. Biol. 51: 389-402.

SHERWOOD, R. T. 1965. Method of producing a phytotoxin. U. S. Pat. Off. Pat. 3,179,653.

———. 1966. Pectin lyase and polygalacturonase production by *Rhizoctonia solani* and other fungi. Phytopathology 56: 279-286.

———, and C. G. LINDBERG. 1962. Production of a phytotoxin by *Rhizoctonia solani.* Phytopathology 52: 586-587.

SHIRAI, M. 1906. On some fungi which cause the so-called white silk disease upon the sprout of the camphor tree. Botan. Mag. Tokyo 20: 319.

SHURTLEFF, M. C. 1953a. Brown patch of turf caused by *Rhizoctonia solani.* Ph.D. Thesis, Univ. Michigan, Ann Arbor. 183 p.

———. 1953b. Factors that influence *Rhizoctonia solani* to incite turf brown patch. (Abstr.) Phytopathology 43: 484.

SIMS, A. C., JR. 1956. Factors affecting basidiospore development of *Pellicularia filamentosa.* Phytopathology 46: 471-472.

———. 1960. Effect of culture substrate on the virulence of single basidiospore isolates of *Pellicularia filamentosa.* Phytopathology 50: 282-286.

SINCLAIR, J. B. 1958. Greenhouse screening of certain fungicides for control of Rhizoctonia damping-off of cotton seedlings. Plant Disease Reptr. 42: 1084-1088.

———. 1960. Reaction of *Rhizoctonia solani* isolates to certain chemicals. Plant Disease Reptr. 44: 474-477.

———. 1965. Cotton seedling diseases and their control. La. Agr. Expt. Sta. Bull. 590: 1-35.

SINGH, B., and R. S. SINGH. 1956. Temperature and moisture relations of the fungi causing seedling-rot, root-rot and wilt of *Cyamopis psoraliodes* D. C. 1. Effect of temperature on the growth of fungi in artificial media. 2. Effect of soil moisture on mortality under controlled conditions. Agra Univ. India J. Res. (Sci.) 5: 135-141.

SINGH, R. S. 1955. Effect of shallow and deep sowing on the incidence of root-rot and wilt of *Cyamopsis psoralioides* DC. Agra Univ. India J. Res. (Sci.) 4: 373-378.

———, and B. SINGH. 1955. Root rot and wilt of *Cyamopsis psoralioides* in relation to thick and thin sowing of the crop. Agra Univ. India J. Res. (Sci.) 4: 379-385.

———, and ———. 1957. Incidence of seedling-rot, root-rot, and wilt of *Cyamopsis psoralioides* under manured and unmanured conditions. Agra Univ. India J. Res. (Sci.) 6: 7-14.

SKILES, R. L. 1953. The strawberry viruses in Minnesota. Diss. Abstr. 13: 149.

SLOOFF, W. C., T. H. THUNG, and J. REITSMA. 1947. Leaf diseases of Sereh (*Andropogon nardus* L.). 1. Banded sclerotial disease, caused by *Rhizoctonia grisea* (Stevens) Matz. Chron. Naturae 103: 6-9.

SMALL, T. 1927. Rhizoctonia "foot-rot" of the tomato. Ann. Appl. Biol. 14: 290-295.

———. 1943. Black scurf and stem canker of potato (*Corticium solani* Bourd. & Galz.). Field studies on the use of clean and contaminated seed potatoes and on the contamination of crop tubers. Ann. Appl. Biol. 30: 221-226.

———. 1945. Black scurf and stem canker of potato (*Corticium solani* Bourd. & Galz.). Further field studies on the use of clean and contaminated seed potatoes and on the contamination of crop tubers. Ann. Appl. Biol. 32: 206-209.

SMITH, A. M., and E. G. PRENTICE. 1929. Investigation on *Heterodera schachtii*. Part I. The infestation in certain areas as revealed by cyst counts. Ann. Appl. Biol. 16: 324-346.

SMITH, D. M. 1951. The influence of seedbed conditions on the regeneration of eastern white pine. Connecticut Agr. Expt. Sta. Bull. 545: 1-61.

SMITH, H. C. 1962. Is barley yellow dwarf virus predisposing factor in the common root rot disease of wheat in Canada. Can. Plant Disease Survey 42: 143-148.

SMITH, J. D. 1959. Fungal diseases of turf grasses. Sports Turf Res. Inst., Bingley, Eng. 90 p.

SMITH, L. F. 1940. Factors controlling the early development and survival of eastern white pine (*Pinus strobus* L.) in central New England. Ecol. Monograph 10: 373-420.

SMITH, L. R., and L. J. ASHWORTH, JR. 1965. A comparison of the modes of action of soil amendments and pentochloronitrobenzene against *Rhizoctonia solani*. Phytopathology 55: 1144-1146.

SMITH, O. F. 1943. Rhizoctonia root canker of alfalfa (*Medicago sativa*). Phytopathology 33: 1081-1085.

———. 1945. Parasitism of *Rhizoctonia solani* from alfalfa. Phytopathology 35: 832-837.

———. 1946. Effect of soil temperature on the development of Rhizoctonia root canker of alfalfa. Phytopathology 36: 638-642.

SMITH, S. E. 1966. Physiology and ecology of orchid mycorrhizal fungi with reference to seedling nutrition. New Phytologist 65: 488-499.

SNEH, B., J. KATAN, Y. HENIS, and I. WAHL. 1966. Methods for evaluating inoculum density of *Rhizoctonia* in naturally infested soil. Phytopathology 56: 74-78.

SNIDER, P. J. 1963. Genetic evidence for nuclear migration in Basidiomycetes. Genetics 48: 47-55.

SNYDER, W. C., and H. N. HANSEN. 1939. The importance of variation in the taxonomy of fungi. Proc. 6th Pacific Sci. Congr. 4: 749-752.

———, and ———. 1940. The species concept in *Fusarium*. Am. J. Botany 27: 64-67.

———, M. N. SCHROTH, and T. CHRISTOU. 1959. Effect of plant residues on root rot of beans. Phytopathology 49: 755-756.

SPALDING, D. H. 1963. Production of pectinolytic and cellulolytic enzymes by *Rhizopus stolonifer*. Phytopathology 53: 929-931.

SPOKAUSKIENE, O. 1961. [Lodging of vegetables-*Rhizoctonia aderholdii* (Ruhl). Kolosh., some biological characters.] [Lithuanian, with Russian summary.] Liet. TSR moks. Akad. Darbai, (C) 1961, 2: 27-35.

SPOONER, A. E., J. E. FIZZEL, and B. A. WADDLE. 1959. Gibberelic acid and cotton seedling growth. *In* Using gibrel and other gibberelins on cotton. Merk. and Company, Rahway, N.J. 26 p.

STAFFELDT, E. E., and C. R. MAIER. 1959. Cultural and pathogenic variability of *Rhizoctonia solani* isolates from cotton-growing areas of New Mexico. Plant Disease Reptr. 43: 1063-1066.

STAHEL, G. 1940. *Corticium areolatum,* the cause of the areolate leaf spot of citrus. Phytopathology 30: 119-130.

STATEN, G., and J. F. COLE, JR. 1948. The effect of pre-planting investigations on pathogenicity of *Rhizoctonia solani* in seedling cotton. Phytopathology 38: 661-664.

STEINBERG, R. A. 1948. Essentiality of calcium in the nutrition of fungi. Science 107: 423.

———. 1950. Growth on synthetic nutrient solutions of some fungi pathogenic to tobacco. Am. J. Botany 37: 711-714.

STEWART, R. B., and M. D. WHITEHEAD. 1955. Nub-root—the expression of seedling disease in mature cotton and flax plant. Phytopathology 45: 413-416.

STOREY, I. F. 1941. A comparative study of strains of *Rhizoctonia solani* (Kühn) with special reference to their parasitism. Ann. Appl. Biol. 28: 219-228.

STRETTON, H. M., N. T. FLENTJE, and A. R. McKENZIE. 1967. Homothallism in *Thanatephorus cucumeris*. Australian J. Biol. Sci. 20: 113-120.

———, A. R. McKENZIE, K. F. BAKER, and N. T. FLENTJE. 1964. Formation of the basidial stage of some isolates of *Rhizoctonia*. Phytopathology 54: 1093-1095.

STRONG, M. C. 1961. Rhizoctonia stem canker of tomatoes. Plant Disease Reptr. 45: 392.

STROUBE, W. H. 1954a. Rhizoctonia aerial blight, a threat to Lespedeza production in Louisiana. (Abstr.) Phytopathology 44: 507.

———. 1954b. Host range of the Rhizoctonia aerial blight fungus in Louisiana. Plant Disease Reptr. 38: 789-790.

SULBA RAO, N. S. 1964. Maintenance of fungi on synthetic sponge. Indian Phytopathol. 17: 183.

SUNDARAM, N. V. 1953. Thread blight of ginger. Indian Phytopathol. 6: 80-85.

TAKAHASHI, M., and Y. KAWASE. 1964. Ecologic and taxonomic studies on *Pythium* as pathogenic fungi. I. Vertical distributions of several pathogenic fungi in soil. [Japanese, with English summary.] Ann. Phytopathol. Soc. Japan 29: 155-161.

TAKAHASHI, T. 1961. *Rhizoctonia solani*. J. Agr. Chem. Soc. Japan. 35: 1397.

———. 1963. Studies on the microbiological transformations of steroids. Part VIII. 19-Hydroxylation of $17\alpha$, $20\alpha$, 21-tri-hydroxypregn-4-en-3-one by *Pellicularia filamentosa* f.s. *microsclerotia*. Agr. Biol. Chem. Japan 27: 639-646.

———. 1964. Studies on the microbiological transformation of steroids. Part IX. Transformation of $C_{19}$-Steroids by

*Pellicularia filamentosa* f. sp. *microsclerotia* IFO 6298. Agr. Biol. Chem. Japan 28: 38-47.

TAKETA, K., and B. M. POGELL. 1966. The effect of palmityl coenzyme A on glucose 6-phosphate dehydrogenase and other enzymes. J. Biol. Chem. 241: 720-726.

TALBOT, P. H. B. 1954. Micromorphology of the lower Hymenomycetes. Bothalia 6: 249-299.

———. 1958. Studies of some South African resupinate Hymenomycetes. Part II. Bothalia 7: 131-187.

———. 1965. Studies of *"Pellicularia"* and associated genera of Hymenomycetes. Persoonia 3: 371-406.

TAMBLYN, N., and E. W. B. DaCOSTA. 1958. A simple technique for producing fruit bodies of wood-destroying basidiomycetes. Nature (London) 181: 578-579.

TANDON, I. N. 1961. A new seedling blight of guava and its control. Indian Phytopathol. 14: 102-103.

TAYLOR, D. P., and T. D. WYLLIE. 1959. Interrelationship of root knot nematodes and *Rhizoctonia solani* on soybean emergence. (Abstr.) Phytopathology 49: 552.

TAYLOR, S. A. 1958. The activity of water in soils. Soil Sci. 86: 83-90.

———, D. D. EVANS, and W. D. KEMPER. 1961. Evaluating soil water. Utah Agr. Expt. Sta. Bull. 426. 67 p.

TERVET, I. W. 1937. An experimental study of some fungi injurious to seedling flax. Phytopathology 27: 531-546.

THATCHER, F. S. 1942. A stem-end rot of potato tubers caused by *Rhizoctonia solani*. Phytopathology 32: 727-730.

THAXTER, R. 1890. The smut of onions (*Urocystis cepulae* Frost.). Ann. Rept. Connecticut Agr. Expt. Sta. 1889: 129-153.

THIRUMALACHAR, M. J. 1953. *Rhizoctonia solani* infections of potato tubers in India. Phytopathology 43: 645-647.

THOMAS, C. A., and R. G. ORELLANA. 1964. Phenols and pectin in relation to browning and maceration of castor bean capsules by *Botrytis*. Phytopathol. Z. 50: 359-366.

THOMAS, K. S. 1925. Onderzoekingen over *Rhizoctonia*. Electr. Drukkerij "De Industrie" J. Van Druten, Utrecht. 98 p.

THORNTON, R. H. 1953. Features of growth of *Actinomyces* in soil. Research 6: A1-A3.

———. 1956. *Rhizoctonia* in natural grassland soils. Nature (London) 177: 230-231.

———. 1958. A soil fungus trap. Nature (London) 182: 1690.

THRUPP, A. C. 1927. Notes on western yellow pine distribution. Forestry Chron. 3(4): 21-22.

TIMS, E. C., and F. BONNER. 1942. Studies of fig leaf blights. Proc. Louisiana Acad. Sci. 6: 13-34.

———, and P. J. MILLS. 1938. Leaf blights of fig in Louisiana. (Abstr.) Phytopathology 28: 663.

———, and ———. 1943. Corticium leaf blights of fig and their control. Louisiana Agr. Expt. Sta. Bull. 367: 1-19.

TISDALE, S. L., and W. L. NELSON. 1960. Soil fertility and fertilizers. The Macmillan Company, New York. 430 p.

TOGASHI, K. 1949. Biological characters of plant pathogens, temperature relations. Meibund Company, Tokyo. 478 p.

TOLBA, M. K., and A. H. MOUBASHER. 1955. Influence of the origin of the isolate of *Rhizoctonia solani* on its pathogenicity. Nature (London) 176: 211.

———, and A. M. SALAMA. 1960. On the mechanism of

sucrose utilization by mycelial felts of *Rhizoctonia solani*. I. Utilization of sucrose, maltose and raffinose. Arch. Mikrobiol. 36: 23-30.

———, and ———. 1961a. Effect of dihydro-streptomycin on the growth and carbohydrate metabolism of mycelial mats of *Rhizoctonia solani*. Arch. Mikrobiol. 38: 283-288.

———, and ———. 1961b. On the mechanism of sucrose utilization by mycelial felts of *Rhizoctonia solani*. II. Effects of glucose, fructose, silver nitrate, sodium fluoride and preheating of mycelial mats on sucrose utilization. Arch. Mikrobiol. 38: 289-298.

———, and ———. 1962a. Studies on the mechanisms of action of sulfanilamide on mycelial felts of *Rhizoctonia solani*. Arch. Mikrobiol. 43: 336-348.

———, and ———. 1962b. Studies on the mechanism of fungicidal action of mercuric chloride on mycelial felts of *Rhizoctonia solani*. Arch. Mikrobiol. 43: 349-364.

———, and ———. 1964. Further studies on the influence of the origin of isolates of *Rhizoctonia* on its pathogenicity. J. Botany United Arab Republic 7: 1-20.

———, and ———. 1965. Effect of age on growth and metabolism of mycelial felts of *Rhizoctonia solani*. J. Exptl. Botany 16: 163-168.

TOLMSOFF, W. J. 1962a. Biochemical basis for biological specificity of Dexon (p-dimethylamino-benzenediazo sodium sulfonate) as a fungistat. (Abstr.) Phytopathology 52: 755.

———. 1962b. Respiratory activities of cell-free particles from *Pythium ultimum*, *Rhizoctonia solani*, and sugar beet seedlings. (Abstr.) Phytopathology 52: 775.

———. 1965. Biochemical basis for biological specificity of Dexon (p-dimethylaminobenzenediazo sodium sulfonate), a respiratory inhibitor. Ph.D. Thesis, Univ. California, Davis. 207 p.

TOMPKINS, C. M. 1959. Leaf rot of poinsettia cuttings caused by *Rhizoctonia solani* and its control. Plant Disease Reptr. 43: 1936-1037.

———, and P. A. ARK. 1946. Seedling disease of yellow calla, caused by *Corticium solani*, and its control. Phytopathology 36: 699-702.

TOUSSOUN, T. A., and Z. A. PATRICK. 1963. Effect of phytotoxic substances from decomposing plant residues on root rot of bean. Phytopathology 53: 265-270.

TOWNSEND, B. B. 1957. Nutritional factors influencing the production of sclerotia by certain fungi. Ann. Botany (NS) 21: 153-166.

———, and H. J. WILLETTS. 1954. The development of sclerotia of certain fungi. Trans. Brit. Mycol. Soc. 37: 213-221.

TOWNSEND, G. R. 1934. Bottom rot of lettuce. New York (Cornell) Agr. Expt. Sta. Mem. 158. 46 p.

TRIBE, H. T. 1960a. Aspects of decomposition of cellulose in Canadian soils. I. Observations with the microscope. Can. J. Microbiol. 6: 309-316.

———. 1960b. Decomposition of buried cellulose film, with special reference to the ecology of certain soil fungi. P. 246-256. *In* D. Parkinson and J. S. Waid [eds.], The ecology of soil fungi. Liverpool Univ. Press.

———. 1966. Interactions of soil fungi on cellulose film. Trans. Brit. Mycol. Soc. 49: 457-466.

TSIANG, CHEN-TONG. 1947. Studies on root rot of flax caused by *Rhizoctonia solani* Kühn. Ph.D. Thesis, Univ. of Minnesota, Minneapolis. 94 p.

TUPENEVICH, S. M. 1958. [The increase in resistance of winter wheat and table beet to diseases by working the soil according to T. S. Mal'tsev.] Sborn. Rab. Inst. prikl. Zool. Fitopatol. 5: 175-194.

TYNER, L. E., and G. B. SANFORD. 1935. On the production of sclerotia by *Rhizoctonia solani* Kühn in pure culture. Sci. Agr. 16: 197-207.

UI, T. 1966. Formation of sclerotia and mycelial strands in *Rhizoctonia solani* Kühn. [Japanese, with English summary.] Ann. Phytopathol. Soc. Japan 32: 203-209.

———, Y. MITSUI, and Y. HARADA. 1963. Studies on the viscissitudes of *Pellicularia filamentosa* in soil. Part II. The alternation of strains of *Rhizoctonia solani* in the soil of a particular flax field. [Japanese, with English summary.] Ann. Phytopathol. Soc. Japan 28: 270-279.

———, and A. OGOSHI. 1964. A comparison of techniques for isolating *Rhizoctonia* from soil. [Japanese, with English summary.] Mem. Fac. Agr. Hokkaido Univ. 5: 5-16.

———, and Y. TOCHINAI. 1955. The relation between the occurrence of root-rot of sugar beets and the vicissitude of population of the causal fungus, *Pellicularia filamentosa*, in soil. [Japanese, with English summary.] Ann. Phytopathol. Soc. Japan 19: 109-113.

ULLSTRUP, A. J. 1936. Leaf blight of China aster caused by *Rhizoctonia solani*. Phytopathology 26: 981-990.

———. 1939. The occurrence of the perfect stage of *Rhizoctonia solani* in plantings of diseased cotton seedlings. Phytopathology 29: 373-374.

VAARTAJA, O. 1952. Forest humus quality and light conditions as factors influencing damping-off. Phytopathology 42: 501-506.

———. 1964a. Chemical treatment of seed beds to control nursery diseases. Botan. Rev. 30: 1-91.

———. 1964b. Survival of *Fusarium, Pythium*, and *Rhizoctonia* in very dry soil. Bi-monthly Prog. Rept. Dept. Forestry Can. 20: 3.

———, and W. H. CRAM. 1956. Damping-off pathogens of conifers and of *Caragana* in Saskatchewan. Phytopathology 46: 391-397.

———, ———, and G. A. MORGAN. 1961. Damping-off etiology especially in forest nurseries. Phytopathology 51: 35-42.

VALDEZ, R. B. 1955. Sheath spot of rice. Philippine Agr. 39: 317-336.

VAN BEEKOM, C. W. C. 1945. Rhizoctonia-ziekte in Aardappelenen Bemesting. Tijdschr. Plantenziekten 51: 82-84.

VAN DER PLANK, J. E. 1963. Plant diseases: epidemics and control. Academic Press Inc., New York, London. 349 p.

VAN EMDEN, J. H. 1958. Control of *Rhizoctonia solani* Kühn in potatoes by disinfection of seed tubers and by chemical treatment of the soil. European Potato J. 1: 52-64.

VAN ETTEN, H., and D. F. BATEMAN. 1965. Proteolytic activity in extracts of Rhizoctonia-infected hypocotyls of bean. (Abstr.) Phytopathology 55: 1285.

———, D. P. MAXWELL, and D. F. BATEMAN. 1967. Lesion maturation, fungal development, and distribution of endopolygalacturonase and cellulase in Rhizoctonia-infected bean hypocotyl tissues. Phytopathology 57: 121-126.

VAN ETTEN, J., and H. P. MOLITORIS. 1965. Changes in chemical composition during aging of *Rhizoctonia solani* and *Sclerotium bataticola*. Phytopathology 55: 1080.

VANTERPOOL, T. C. 1953. *Corticium praticola* and the *Rhizoctonia solani* problem. (Abstr.) Phytopathology 43: 488.

VARNEY, E. H. 1961. Vermiculite media for growing fungi. Plant Disease Reptr. 45: 393.

VASUDEVA, R. S. 1936. Studies on the root-rot disease of cotton in the Punjab. II. Some studies in the physiology of the causal fungi. Indian J. Agr. Sci. 6: 904-916.

———. 1944. Studies on the root-rot diseases of cotton in the Punjab. XIII. Leaf temperatures of healthy and root-rot affected plants. Indian J. Agr. Sci. 14: 385-388.

———, and M. ASHRAF. 1939. Studies on the root-rot disease of cotton in the Punjab. VII. Further investigation of factors influencing incidence of the disease. Indian J. Agr. Sci. 9: 595-608.

———, and M. R. SIKKA. 1941. Studies on the root-rot disease of cotton in the Punjab. X. Effect of certain fungi on the growth of root-rot fungi. Indian J. Agr. Sci. 11: 422-431.

VENKATARAMANI, K. S., and C. S. VENKATA RAM. 1959. *Rhizoctonia solani* inciting a collar rot of tea. Phytopathology 49: 527-528.

VENKATARAYAN, S. V. 1949. The validity of the name *Pellicularia koleroga* Cooke. Indian Phytopathol. 2: 186-189.

VERDEREVSKI, D. D. 1959. [Immunity of plants to parasitic diseases.] Moscow, 372 p.

VERHOEFF, K. 1963. Voetrot bij tomaat, veroorzaakt door *Rhizoctonia solani*. Neth. J. Plant Pathol. 69: 265-278.

VERONA, O. 1952. Azione della vitamina $K_5$ su alcuni funghi, incluso funghi fitopatogeni. Nuovo Giorn. Botan. Ital. (NS) 59: 522-524.

VIENNOT-BOURGIN, G. 1949. *Corticium solani* (Prill. et Del.) Bourd et Galz. *In* Les Champignons parasites des plantes cultivées 2: 1191-1199. Masson and Cie., Paris.

WAGNER, F. 1955. Untersuchungen uber die Einwirkung von 2, 4-D- und MCPA- Preparaten auf Wachstrum und Conidienbildung phytopathogener Pilze. Arch. Mikrobiol. 22: 313-323.

WAKEFIELD, E. M. 1939. Nomina generica conservanda. Contributions from the Nomenclature Committee of the British Mycological Society. I. Trans. Brit. Mycol. Soc. 23: 215-232.

WAKSMAN, S. A. 1952. Soil Microbiology. John Wiley & Sons, Inc., New York. 356 p.

WALKER, A. C., L. R. HAC, A. ULRICH, and F. J. HILLS. 1950. Nitrogen fertilization of sugar beets in the Woodland area of California. Proc. Am. Soc. Sugar Beet Tech. 1950: 362-371.

WALKER, J. C. 1957. Plant pathology. P. 437-447. 2d ed. McGraw-Hill Book Company, New York.

———. 1963. The physiology of disease resistance. West Virginia Agr. Expt. Sta. Bull. 488 T: 1-25.

WALKER, M. N. 1928. Soil temperature studies with cotton. III. Relation of soil temperature and soil moisture to soreshin disease of cotton. Florida Agr. Expt. Sta. Tech. Bull. 197: 345-371.

WALLACE, J., J. KUC, and H. N. DRANDT. 1962. Biochemical changes in the water-insoluble material of maturing apple fruit and their possible relationship to disease resistance. Phytopathology 52: 1023-1027.

WARCUP, J. H. 1950. The soil-plate method for isolation of fungi from soil. Nature (London) 166: 117-118.

———. 1951. The ecology of soil fungi. Trans. Brit. Mycol. Soc. 34: 376-399.

———. 1955. Isolation of fungi from hyphae present in soil. Nature (London) 175: 953-954.

———. 1957. Studies on the occurrence and activity of fungi in a wheat-field soil. Trans. Brit. Mycol. Soc. 40: 237-259.

———. 1959. Studies on Basidiomycetes in soil. Trans.

Brit. Mycol. Soc. 42: 45-52.

———. 1960. Methods for isolation and estimation of activity of fungi in soil. P. 3-21. *In* D. Parkinson and J. S. Waid [eds.], The ecology of soil fungi. Liverpool Univ. Press, Liverpool, Eng.

———, A KERR, and J. F. DE BEER. 1963. Studies on the biology of *Rhizoctonia solani* in soil. Waite Agr. Res. Inst. Rept. 1962-1963: 54-55.

———, and P. H. B. TALBOT. 1962. Ecology and identity of mycelia isolated from soil. Trans. Brit. Mycol. Soc. 45: 495-518.

———, and ———. 1966. Perfect states of some Rhizoctonias. Trans. Brit. Mycol. Soc. 49: 427-435.

WARREN, J. R. 1948. A described species of *Papulospora* parasitic on *Rhizoctonia solani* Kühn. Mycologia 40: 391-401.

WASTIE, R. L. 1961. Factors affecting competitive saprophytic colonization of the agar plate by various root-infesting fungi. Trans. Brit. Mycol. Soc. 44: 145-159.

WEBER, G. F. 1923. Potato diseases and insects. Florida Agr. Expt. Sta. Bull. 169: 101-164.

———. 1931. Bottom rot and related diseases of cabbage caused by *Corticium vagum* B. & C. Florida Agr. Expt. Sta. Bull. 242: 1-31.

———. 1932a. Diseases of pepper in Florida. Florida Agr. Expt. Sta. Bull. 244: 1-46.

———. 1932b. Somes diseases of cabbage and other crucifers in Florida. Florida Agr. Expt. Sta. Bull. 256: 1-62.

———. 1939. Web-blight, a disease of beans caused by *Corticium microsclerotia*. Phytopathology 29: 559-575.

———. 1951. *Corticium microsclerotia* nom. nov. Mycologia 43: 727-728.

———, and L. ABREGO. 1958. Damping-off and thread blights of coffee in El Salvador. Plant Disease Reptr. 43: 1378-1381.

———, and A. C. FOSTER. 1928. Diseases of lettuce, romaine, escarole, and endive. Florida Agr. Expt. Sta. Bull. 195: 301-333.

———, and G. B. RAMSEY. 1926. Tomato diseases in Florida. Florida Agr. Expt. Sta. Bull. 185: 59-138.

WEBSTER, R. E., and R. W. ROTH. 1965. Pathogenicity of *Rhizoctonia solani* on lima bean seedlings. (Abstr.) Phytopathology 55: 506.

WEDDING, R. T., and J. B. KENDRICK, JR. 1959. Toxicity of N-methyl dithiocarbamate and methyl isothiocyanate to *Rhizoctonia solani*. Phytopathology 49: 557-561.

WEI, C. T. 1934. Rhizoctonia sheath blight of rice. Nanking Univ. Coll. Agr. Forestry Bull. (NS) 15. 21 p.

WEINDLING, R. 1932. *Trichoderma lignorum* as a parasite of other soil fungi. Phytopathology 22: 837-845.

———, and H. S. FAWCETT. 1936. Experiments in the control of Rhizoctonia damping-off of citrus seedlings. Hilgardia 10: 1-16.

WEINHOLD, A R., and T. BOWMAN. 1967. Virulence of *Rhizoctonia solani* as influenced by nutritional status of inoculum. (Abstr.) Phytopathology 57: 835-836.

———, and F. F. HENDRIX, JR. 1962. Inhibition of *Rhizoctonia solani* by potato-dextrose agar previously exposed to light. (Abstr.) Phytopathology 52: 32.

———, and ———. 1963. Inhibition of fungi by culture media previously exposed to light. Phytopathology 53: 1280-1284.

WEISS, F., and M. J. O'BRIEN. 1953. Index of plant diseases in the United States. U. S. Dept. Agr. Plant Disease Surv. Spec. Publ. I: 1195-1263.

WELLMAN, F. L. 1932. Rhizoctonia bottom rot and head rot of cabbage. J. Agr. Res. 45: 461-469.

WELLS, K. 1964. The basidia of *Exidia nucleata*. I. Ultrastructure. Mycologia 56: 327-341.

WESTENDORP, G. D. 1852. Notice sur quelques cryptogames inedifes ou nouvelles pour la flore belge. Bull. Acad. Roy. Belg. 18: 384-417.

———, and A. C. F. WALLAYS. 1846. Herbier Cryptogamique. Fasc. 5, No. 225.

WESTER, R. E., and R. W. GOTH. 1965. Pathogenicity of *Rhizoctonia solani* on lima bean seedlings. Phytopathology 55: 506.

WHITE, L. V. 1962. Root knot and the seedling disease complex of cotton. Plant Disease Reptr. 46: 501-504.

WHITNEY, H. S. 1959. An effect of soil-packing in densely crowded coniferous seedlings. Bi-monthly Progr. Rept. Can. Dept. Agr. 15: 5.

———. 1963. Heterokaryosis and variation in *Rhizoctonia*. Ph.D. Thesis, Univ. of California, Berkeley. 202 p.

———. 1964a. Sporulation of *Thanatephorus cucumeris* (*Rhizoctonia solani*) in the light and in the dark. Phytopathology 54: 874-875.

———. 1964b. Physiological and cytological studies of basidiospore repetition in *Rhizoctonia solani* Kühn. Can. J. Botany 42: 1397-1404.

———, and J. R. PARMETER, JR. 1963. Synthesis of heterokaryons in *Rhizoctonia solani* Kühn. Can. J. Botany 41: 879-886.

———, and ———. 1964. The perfect stage of *Rhizoctonia hiemalis*. Mycologia 56: 114-118.

WILDE, S. A., and D. P. WHITE. 1939. Damping-off as a factor in the natural distribution of pine species. Phytopathology 29: 367-369.

WILHELM, S. 1956. A sand-culture technique for the isolation of fungi associated with roots. Phytopathology 46: 293-295.

———. 1957. Rhizoctonia bud rot of strawberry. Plant Disease Reptr. 41: 941-944.

———. 1959. Parasitism and pathogenesis of root-disease fungi. P. 356-366. *In* C. S. Holton [ed.], Plant pathology problems and progress, 1908-1958. Univ. Wisconsin Press, Madison.

———, W. J. KAISER, S. G. GEORGOPOULOS, and K. W. OPITZ. 1962. Verticillium wilt of olives in California. (Abstr.) Phytopathology 52: 32.

———, R. C. STORKAN, and J. E. SAGEN. 1961. Verticillium wilt of strawberries controlled by fumigation of soil with chloropicrin and chloropicrin-methyl bromide mixtures. Phytopathology 51: 744-748.

WILKEN, D. R., T. KAGAWA, and H. A. LARDY. 1965. The role of adenine nucleotides in control of choline oxidation by mitochondria. J. Biol. Chem. 240: 1843-1846.

WILLIAMS, P. H., and J. C. WALKER. 1966. Inheritance of Rhizoctonia bottom rot resistance in cabbage. Phytopathology 56: 367-368.

WILLIAMS, W. A., and D. RIRIE. 1957. Production of sugar beets following winter green manure cropping in California: I. Nitrogen nutrition, yield, disease, and pest status of sugar beets. Proc. Soil Sci. Soc. Am. 21: 88-94.

WILSENACH, R., and M. KESSEL. 1965. On the function and structure of the septal pore of *Polyporus rugulosus*. J. Gen. Microbiol. 40: 397-400.

WINTER, A. G. 1950. Untersuchungen über die Ökologie von *Rhizoctonia solani* im natürlichen Boden. Nachrl. Deut. Pflanzenschutzdienst 2: 8-9.

———. 1951. Untersuchugen über die Ökologie und den Massenwechsel bodenbewohnender mikroskopischer Pilze. II. Mitteilung. Der Einfluss organischer Nährstoffe auf die Entwicklung einiger Pilze (*Cladosporium* sp., *Chaetomium* sp. *Fusarium* sp., *Verticillium glaucum, Penicillium* sp., *Rhizoctonia solani* u.a.) in natürlichen und partiell sterilisierten Böden. Arch. Microbiol. 16: 136-162.

WOLF, F. A. 1914. Fruit rots of eggplant. Phytopathology 4: 38.

WOLLENWEBER, H. W. 1913. Pilzparasitäre Welkekrankheiten der Kulturpflanzen. Berlin Deut. Botan Ges. 31: 17-34.

WOOD, F. A., and R. D. WILCOXSON. 1960. Another screened immersion plate for isolating fungi from soil. Plant Disease Reptr. 44: 594.

WOOD, R. K. S. 1951. The control of diseases of lettuce by the use of antagonistic organisms. II. The control of *Rhizoctonia solani* Kühn. Ann. Appl. Biol. 38: 217-230.

———, and M. TVEIT. 1955. Control of plant diseases by use of antagonistic organisms. Botan. Rev. 21: 441-492.

WOODCOCK, W. P. 1962. Influence of aeration on exudation from seeds and on growth of root-rot fungi. (Abstr.) Phytopathology 52: 927.

WRIGHT, E. 1941. Control of damping off of broadleaf seedlings. Phytopathology 31: 857-858.

———. 1944. Damping-off in broadleaf nurseries of the Great Plains region. J. Agr. Res. 69: 77-94.

———. 1945. Relation of macrofungi and microorganisms of soils to damping-off of broadleaf seedlings. J. Agr. Res. 70: 133-141.

———. 1957. Influence of temperature and moisture on damping-off of American and Siberian elm, black locust, and desert willow. Phytopathology 47: 658-662.

WYLLIE, T. D. 1959. Infection of soybean roots by *Rhizoctonia solani*. Phytopathology 49: 555.

———. 1961. Host-parasite relationships between soybean and *Rhizoctonia solani*. Diss. Abstr. 21: 2854.

———. 1962. Effect of metabolic by-products of *Rhizoctonia solani* on the roots of Chippewa soybean seedlings. Phytopathology 52: 202-206.

YARWOOD, C. E., and L. JACOBSON. 1955. Accumulation of chemicals in diseased areas of leaves. Phytopathology 45: 43-48.

YOUNG, H. C., and C. W. BENNETT. 1922. Growth of some parasitic fungi in synthetic culture media. Am. J. Botany 9: 459-469.

YU, T. F. 1940. The relation of soil temperature to pathogenicity of *Rhizoctonia solani* Kühn on broad bean seedlings. Nanking J. 9: 269-280.

ZALESKI, K., and W. BLASZCZAK. 1954. Wstepne doswiadczenia nad patologia, szkodliwoscia i niektorymi sposobami zwalczania rizoktoniozy ziemniakow (*Rhizoctonia solani* Kuehn). [Polish, with English summary.] Roczn. Nauk Rol. 69: 529-556.

ZYNGAS, J. P. 1963. The effect of plant nutrients and antagonistic micro-organisms on the damping-off of cotton seedlings caused by *Rhizoctonia solani* Kühn. Diss. Abstr. 23: 3587.

# Index